Innovations in Urban and Regional Systems

Jean-Claude Thill
Editor

Innovations in Urban and Regional Systems

Contributions from GIS&T, Spatial Analysis and Location Modeling

Springer

Editor
Jean-Claude Thill
University of North Carolina at Charlotte
Charlotte, NC, USA

ISBN 978-3-030-43692-6 ISBN 978-3-030-43694-0 (eBook)
https://doi.org/10.1007/978-3-030-43694-0

This Springer imprint is published by the registered company Springer Nature Switzerland AG
The registered company address is: Gewerbestrasse 11, 6330 Cham, Switzerland

Contents

Introduction

Research on Urban and Regional Systems: Contributions from GIS&T, Spatial Analysis, and Location Modeling

Jean-Claude Thill

1 Generalization and Contextualization

The changes that permeated knowledge creation during the age of enlightenment and the concomitant scientific revolution have shaped the emergence of contemporary disciplinary perspectives. In particular, one of the guiding principles of the scientific method has been the generalization beyond the particular facts and the construct of a theoretical body of knowledge on the basis of accumulated empirical evidence. Under this line of reasoning, the particularities of a specific slice in time, of a specific location, and of a particular agent or element have become subsumed by the imperative of drawing rules and laws.

As positivism and neo-positivism affirmed their preeminence in intellectual discourse, some of the mainstream scientific disciplines—both natural and social—assumed away the role of context, space, location, and place in the interest of expediency and in search for rules that would abide by strict cannons of generalizability. A case in point is economics which, armed with strong deductive and axiomatic principles, articulated economic behaviors and economic structures "in a vacuum" devoid of geographic context, of territorial extent, and of any apprehension of spatial proximity or remoteness with respect to other relevant activities or amenities. Against this backdrop, geography and other associated disciplines credibly established a counterpoint narrative on socioeconomic processes as well as on natural environmental processes. It is by anchoring the situational and contextual constructs of region, space, place, distance, and a number of derived concepts in a neo-positivist frame of reference that success was garnered.

J.-C. Thill (✉)
Department of Geography and Earth Sciences, University of North Carolina at Charlotte, Charlotte, NC, USA
e-mail: jean-claude.thill@uncc.edu

School of Applied Economics, Renmin University of China, Beijing, China

J. Thill (ed.), *Innovations in Urban and Regional Systems*,
https://doi.org/10.1007/978-3-030-43694-0_1

Our contention here is that the active practice of internal and external contextualizations (Elger 2010a) enhances the scientific understanding of a specific case by placing it in the relevant context of other influential processes that may be at play in this case, as well as in the wider setting that lays outside of the scope of the case. This does not call into question the neo-positivist approach to research nor does it advocate the violation of the principle of generalization by any means. However, it certainly underscores the need for research designs and approaches to data analytics that are fully capable of accounting for internal and external contexts and of analyzing them in ways that support replication and generalization. We now further elaborate on the distinctive nature of internal contextualization and external contextualization, as this is a critical premise of the spatially explicit and place-based perspective that is advocated in this work. In this chapter, we aim to restate the scientific merit of such endeavor and reassert its unique value in translating academic research into applied outcomes relevant to specific communities and decision makers.

2 Internal Contextualization and Systems Analysis

With internal contextualization, the focus is on "addressing the interrelationships between different aspects of the case and analyzing the significance of specific events, narratives, and processes in relation to this wider configuration" (Elger 2010a, p. 231). This concept has usually been leveraged in the conduct of qualitative research of discourse analysis following ethnographic paradigms and practices of engagement and observation (Strathern 1987; Dooley and Levinsohn 2001). However, as a construct, its contribution to knowledge discovery is equally valuable in the tradition of more quantitative scientific inquiry. In particular, we can discern a close alignment between internal contextualization and the systemic view on phenomena that have been espoused as a powerful analytical framework across numerous disciplines, spanning the social and economic realms, the natural and physical sciences, and the more applied disciplines of engineering.

The systems approach, or systems theory (Luhmann 2013), posits that facts and events do not happen in isolation of each other nor of their setting. They are part of an ad hoc grouping of interrelated entities that form a unified ensemble, the system. The relational properties of the system encapsulate how the entities interact with each other, laterally as members of the same classes of entities, as well as vertically across classes and hierarchical layers of entities. Because of these interactions, the system is not merely the aggregation of its parts and the system as a whole can exhibit behaviors and dynamical trends that differ from that of its elements taken individually. Complex structures can emerge from the bottom up—as in multi-agents systems (Shoham and Leyton-Brown 2008; Wooldridge 2002; Benenson and Torrens 2004), along causal pathways that follow nonlinear, recursive, and adaptive (i.e., changing with the context and under the mutual influence of the system's elements) trajectories and patterns. They have become the hallmark of the system dynamics approach (Forrester 1961, 1969, 1971; Meadows 1972) in line with the precepts of complex systems theory (Miller and Page 2007).

When events and phenomena are complex, strongly interconnected and adaptive, and incompletely specified, systems analysis has been advantageously leveraged to identify possible futures, the range of their variabilities, and the critical points that condition evolutionary trajectories leading to specific outcomes. Systems elements do not only encompass the natural (e.g., lakes, mountains) and built components (roads, power plants, etc.) but also the human, societal, and institutional elements (Ottens et al. 2006); hence, the term of socio-technical systems (Kaghan and Bowker, 2001) is often used today to reflect that the human and social perspectives are explicitly incorporated into physical and technical systems (e.g., ONeill and Gibbs 2013; Marletto 2014; Li et al. 2015). These "soft" elements may entail user acceptance and attitudes, value systems, cultural norms, human and social capital, and others.

3 External Contextualization and Spatial Differentiation

Evidence-based research proceeds through the conduct of case studies as a fundamental tenet of its design principles (Yin 1994). Because of the unavoidable spatial and/or temporal bounding of the case (Elger 2010b) and because interesting phenomena happen across a plurality of locations littering the geographic space, placing any case study in its wider context—be it social or natural—and characterizing this context in terms that may in fact condition or affect the causal processes embedded in the case on hand is an essential component of the research activity (Elger 2010a). The wider context of one or multiple case studies that form a research corpus is what makes external contextualization a research imperative.

Social phenomena and natural circumstances are known to not display the same properties across large geographic territories. This differentiation or variability of traits, events, or relationships is the rule, not the exception. In other words, the geographic space is inherently differentiated. The study of the modalities of geographic differentiation informs on the external contextualization of research, on generalization that can be drawn from case studies, as well as on potential biases in inference.

Sometimes, the observed heterogeneity of the social or natural landscape is misconstrued to a certain extent and it is then said that "geography explains." For instance, a contextual factor of a certain phenomenon may exhibit some geographic variability, which would translate into a mapping of the phenomenon in question that exhibits similar spatial heterogeneity. In fact, while the contextual factor may well have a causal relationship to the phenomenon, the heterogeneous map of the phenomenon is merely the spurious expression of the spatial heterogeneity of the causal factor.

In spatial analysis (Fotheringham and Rogerson 2009), the notion of spatial differentiation takes on a narrower meaning tied to the causal social or physical processes that operate in the production of certain outcomes. These so-called spatial effects fall into two sets of related and complementary considerations, namely spatial dependence and spatial heterogeneity. Spatial dependence (also known as spatial autocorrelation) refers to the statistical dependence of the value of a trait at a certain geospatial location on the values observed at other nearby locations for the same trait (Anselin 1988; LeSage and Pace 2009). In a sense, this is the geospatial equivalent to the proverb "Birds of a feather flock together." Statistically, it is a second-order property of the spatial process generating the attribute. Spatial autocorrelation statistics (e.g., Cliff and Ord 1973; Getis and Ord 1992; Anselin 1995) are effective tools of spatial analysis to assess the degree of spatial differentiation that may exist in various phenomena, and therefore, the extent of external contextualization is required to comprehend the leading factors of the phenomenon of interest. The mutual interaction between entities that are neighbors in the geographic space produces the so-called spatial spillover effect (Anselin 2003), which encapsulates the notion of spatial dependence and is a well-known form of spatial externality. Examples would include the process of knowledge and innovation diffusion through learning by one socioeconomic agent from another, or real property values or gentrification across a metropolitan area, and many others. Spatial lag models (Anselin 1988; LeSage and Pace 2009) are spatially explicit in the sense that they recognize the proximity relationships that exist between data points and statistically separate the spillover effects of spatial dependence from the other causal pathways (Fischer 2005).

While the empirical assessment of the degree of spatial dependence is usually done via econometric means, a well-grounded conceptualization of the geographic space and of geospatial relationships that it supports is a prerequisite of this endeavor. We will come to this point later in this chapter. It is pertinent here, however, to point out that whether distance or geospatial proximity between place-based entities matters has been the focal point of considerable debate over the past two decades. This debate was largely triggered by the provocative title of Cairncross's book (1997) "The Death of Distance." Cairncross argued that modern communication technologies were about to make distance "irrelevant to the way we conduct our business and personal lives." There is no doubt today that this prophecy did not come to pass and that it was grounded in a sheer misunderstanding of the "soft side" of interpersonal communication in human societies and of the social construction of space and time (Harvey 1994).[1] As early as the mid-1960s, Janelle (1969) noted that technological innovation was altering the relationship between space and time (the speed at which interaction between places following an historic upward trend) and in turn prompted an economic and social reorganization of the geographic space. This so-called time–space convergence (also known as time–space compression) and its implications were more theoretically articulated by Harvey (1990) on the ground of a broader set of societal and technological transformations. In any case, there is a plethora of evidence that spatial dependence matters and stands at the heart of increasing disparities across territorial expanses. This is particularly manifested between rural and metropolitan areas owing to strikingly contrasted opportunities for citizens in

[1]Incidentally, this is also what prompted the emergence of socio-technical systems as a framework for the analysis and design of complex multi-level and multi-scalar systems.

the areas of employment, education, or health care and for business to tap into urban agglomeration economics (Glaeser 2010). While the tyranny exerted by distance (Bunge 1961; Blainey 1966) may have faded considerably over the ages, globally and the finer granularity of individual cities, distance continues to matter. Olson and Olson's (2000) extensive research on the comparative efficiency of various models of work organization within corporate entities reached the same conclusion.

As the second core pillar of spatial differentiation, spatial heterogeneity sometimes describes the unevenness of the distribution of events of interest across the geographic space. In a strict sense, however, it refers to the distribution of a relationship across a spatial (social or natural) landscape. Applied statistics allows for variability in the strength of the relationship between a dependent variable and independent predictors, but a fundamental assumption of the prescribed guidelines is that this variability is not systematic and follows a Gaussian distribution. Relaxation of this constraint may be particularly well suited to align with the notion of external contextualization. Various approaches to the spatial variability of regression parameters have been advanced over the past few decades (Casetti 1970; Fotheringham et al. 2002). They can be particularly effective at detecting spatial non-stationarity in multivariate relationships, which in turn usually indicates that the model is underspecified and that an influential factor exhibiting strong spatially variability would be missing from the model.

4 Space and Distance

We have made a case for spatially explicit research to address the need for internal and external contextualizations. With this in mind, a well-thought-out conceptualization of the geographic space is a prerequisite to any thoughtful use of contextualized social science. As recently restated by Thill (2011), the concept of space has a history that spans over 2000 years. Greek philosophers and mathematicians debated on the formalism of space, its nature, and properties, but also its semantics in relation to other core notions like time. Greek understanding of the concept of space was that of a primal, unchanging, and empty expanse occupied by entities referenced to a simple coordinate system. Euclidean geometry provided the standard framework for abstract reasoning and analysis in and on this absolute space. At the core, space was held as objective, infinite, homogeneous, and isotropic—a container of events, phenomena, and processes. Kant (1929) forcefully reasserted this position, and, in the wake of the quantitative revolution in social sciences, spatial analysis embraced the same view that "space is consistently measurable, immutable, uniform, and independent of the social or environmental qualities" (Thill 2011, p. 5). The vast majority of spatial econometrics uses this view as a bedrock foundation (Anselin 1988; LeSage and Pace 2009). It is commonplace for spatial effects to be apprehended in rather simplistic ways through a set of spatial relationships between discrete entities that form a complete tessellation laid over the study territory. The so-called spatial weight W matrix is conventionally constructed as a time-independent table of pairwise polygonal contiguities. This and related conceptualizations of space and spatial relationships were found by Gatrell (1983) to be "naïve." Yet the notion of absolute space became the cornerstone of geographic information systems (GIS), and it still is largely so today. The fact that GIS thinking still very much revolves around a separation between the spatial information (or data) (i.e., coordinates) that provides the frame of reference tied to a datum that is "set in stone," along with a georeferencing system and attributes that are collected, assembled, and linked back to the datum is the most obvious aspect of this legacy. A more detailed discussion of this state of affairs can be found in Thill (2011).

Alternatively, some scholars have advocated for a relativist notion of space. These views were very fundamental to philosopher and physicist Gottfried Leibniz's conception, socio-political theories of sociologists Emile Durkheim and Henri Lefebvre, but also Albert Einstein's theory of relativity. The novel thinking centered on a relative space was promoted by a broad coalition of humanists, structuralists, and behavioralists on the ground that spatial patterns that can be discerned in semantic structures are not dictated by the exogenous properties of the physical space (a sort of tyranny of the physical space that would be labeled as laws!). Space can be constructed by perceptions, experiences, interactions, and relationships (Ullman 1954; Gould 1991; Harvey 1994).

In the context of Geography and Spatial Sciences, the early and insightful contribution made by Ullman (1954) stands to be noted as it is often under-appreciated. In his seminal 1954 essay on "Geography as Spatial Interaction," Ullman posits that the concept of situation is one of two foundational notions of geographic theory. He elaborates further by saying that:

> Situation refers to the effects of one area, or rather phenomena in one area, on another area. It should logically focus on the connections between areas and leads to such terms as "circulation" and "regional interdependence," or to specific aspect such as "diffusion" or "centralization."
>
> This situation concept is defined here as "spatial interaction," and is intended as more positive and dynamic concept than either situation, relations, or even circulation. In a sense it provides a "motor" for situation concepts. (p. 1)

Ullman blurs the line between the semantic information and the geospatial frame of reference. He goes on to acknowledge that credit for this vision of situation or spatial interaction as a structuring concept of geography is owed to others, including P. R. Crowe, R. S. Platt, J. R. Whitaker, and even R. Hartshorne. What sets Ullman apart from these precursors, however, is that he did not merely verbalize that regional interdependence or spatial interaction is an element to be reckoned with in comprehending geographies; he proceeded to articulate an empirically validated modeling system based on the principles of complementarity, intervening opportunity, and distance to operationalize this new theory. This conceptualization provided the basis for the establishment of one of the most extensively tested theoretical frameworks in geography, regional, and spatial sciences (Fotheringham and O'Kelly 1989; Roy and Thill 2004). Spatial interaction modeling and its disaggregate variants have come to stand for the very fundamental principles that govern the dynamics of market-driven economic systems. This simplicity is precisely what suits them well at all scales, ranging from the local (retail trade and shopping behavior, commuting patterns) to the global (continental migrations, international trade).

Flows of spatial interaction between places reveal relationships between these places. Two cities that entertain close trade relationships are bound to exhibit one or more of the requisite properties of spatial interaction: They are geographically close, their economic structures are close substitutes, or there are no intervening trading partners. The collective pattern of spatial interactions of various sorts (freight, tourists, business travelers, ideas, capital, phone calls, migrants, etc.) may well appear chaotic at first sight. It is rather tempting for instance to deem the World Wide Web ruled by confusion and disorder, given the absence of any central coordinating agent. In fact, the World Wide Web exhibits a well-defined organization that emerges spontaneously from a myriad of decisions made by decentralized agents. It is a self-organizing system. This is not unlike fields of free economic agents that operate markets and generate global trade patterns. Human spatial organization (Abler et al. 1971) of one form or another is the end result of spatial interactions. Space is a socioeconomic construct.

One of the more notable contributions to this discourse is Castells (1996). In his influential book, Castells draws on social processes, including conflicts and strategies of diverse social actors, whose dynamics play out in ways that create a new and fluid spatial logic dubbed space of flows. Through this theorizing, space is the expression of society, of its social practices that operate through flows. As in Ullman's spatial interaction system, flows can be understood to include any "purposeful, repetitive, programmable sequences of exchange and interaction between physically disjointed positions held by social actors in the economic, political, and symbolic structures of society" (Castells 1996, p. 412). At the global scale, the logic of social dominance creates a space of flows comprised as a three-tier ensemble of operational to strategic networks. Space is increasingly understood to be effectively conceptualized as a network framed by relationships of various sorts, such as trade flows (e.g., Jung et al. 2018) or shared imaginaries in the collective discourse of a population group, such as web pages in contemporary communication (e.g., Zhang and Thill 2019). Space is about relations, which can be physical, economic, social, political, or virtual in nature.

Following a resolutely neo-positivist trajectory, Gatrell (1983) came to adopt a rather concordant perspective. Here, space is viewed as "a relation defined on a set of objects" (p. 3). As Gatrell sees it, this perspective presents two advantages over the absolute space that broaden the relevance of the approach. First, the relations that frame such space are not restricted to any particular semantic type, so that many geographies can be explored. Second, relations are not bound to exhibit strict metric properties.

Thus, generally speaking, multiple concepts of relative space can be conceived of and no concept should be favored over another; they are complementary and collectively celebrate the richness of processes creating spatial structures resulting from a variety of socioeconomic and natural processes affecting entities and actors located variously across the territory under study. Forer (1978), Gatrell (1983), Couclelis

(1992), and others effectively argued that there is no single best perspective for discussing and comprehending space. Selected schemes are examined later in this section. Since the 1960s, the idea of relative space has transpired through numerous mathematical models of socioeconomic processes (Couclelis 1992), but much less so for environmental processes (Meentemeyer 1989).

If the geographic space is indeed relative and if it is a creation of perceptions, experiences, social and physical interactions and relationships (Gould 1991), there is no universal spatial "ground truth" anymore. Instead, space is eminently plastic (Forer 1978), fungible, and malleable. The prospect of working with a relative space of one form or another (hence, one of many possible *spaces*) may appear rather daunting given the contrasting differences that exist between absolute and relative spaces. Table 1 summarizes some of the salient characteristics of the two theorizations of space. Challenges presented by non-physical spaces abound at many levels. They range from capturing the relevant semantic information that frames such spaces, to measuring their latent semantic dimensions, developing compatible data models and visualization forms, developing semantically truthful spatial reasoning and analytical tools, and processes of multi-spectral information fusion that are not restricted to Euclidean geometries.

Table 1 Summary of properties of absolute and relative spaces

Absolute space	Relative space
Exists independent of any objects	In reference to relational processes only
A "container"	Contents
Geographical space	Attribute space, social space, behavioral space, etc.
Mathematical (Euclidean) space	Non-Euclidean space
Geographical distance	Relative (non-geographical) distance
Constant friction of distance, linear time	Time–space convergence/divergence, nonlinear time
Independent of observer, non-judgmental observation	Perceptions and experiences shape one's view of the world
Time is independent, constant, and universal	Coexistence and connection of space and time: space–time
Space is fixed, a background for all features	Space and time as attributes of features
Objective	Objective or subjective
Compatible with field-based and object-based data models	Not directly compatible
Associated primarily with study of earth, conventional mathematical models	Associated primarily with studies of functions and interactions
Rigid geometric structure within which events and features are referenced	Locations referenced in terms of relationships between features, topological, and temporal

Source Thill 2011

Spatial scientists have worked toward formalizing spaces in ways that encapsulate a variety of forms of relations and experiences, from cognitive to socioeconomic, political, virtual, and physical. Constructing a relative space involves a number of tasks on the part of the analyst, namely the observation, interpretation, and synthesis of processes and patterns of fluxes within a certain knowledge domain (Wachowicz 1999). Consequently, as a visual device, a map can be regarded as "a visual or graphical representation of a space" (Gatrell 1983, p. 4) and may exhibit only remote resemblance to the corresponding absolute (Euclidean most of the time) map because of distortions imparted by semantic relations between features. Map transforms and cartograms are common mapping techniques called upon to express cartographically a relative view of space. Various methods exist to create synthetic attribute maps. Some techniques process semantic attributes of features by collapsing measured dimensions to a manageable number of new latent dimensions. This is a process known as "spatialization" (Skupin and Fabrikant 2007). Through this exercise of unsupervised data reduction, an attribute space is elicited and observations are projected onto it. For instance, self-organizing maps (Agarwal and Skupin 2008) serve to reveal the complex, unknown, and sometimes concealed structures of competition between airlines serving the US airports in Yan and Thill (2009). The dynamics of urban neighborhoods can also synthetically be revealed and analyzed through this analytical process (Delmelle et al. 2013). Novel techniques of network science can also be very effective at reducing the dimensional of large datasets including data of different granularity and thematic heterogeneity. A good example of such effort would be the recent work by Zhang and Thill (2017). A more classical alternative would be the method of multidimensional scaling (MDS), which works on a matrix of dissimilarities among observations (Gatrell 1983). It provides a spatial configuration of observations such that pairwise distances between observations in this configuration match input dissimilarities as well as possible. Thill (2011) reports on the use of MDS to infer the properties of a population migration space for the United States, where a three-dimensional rendering outperforms more conventional two-dimensional forms, given the rather unusual nature of the configuration of migration flows in the United States.

Gatrell (1983) reports on countless studies that transcend and enrich the "flat" and static view of absolute space. These include mappings of time spaces exhibiting space-stretching of urban regional environments along high-speed and high-capacity corridors and space-shrinking in zones where mobility is low and congestion is high, mappings of functional economic spaces shaped by competitive processes and institutional controls, mappings of cognitive spaces, and many others. Cognitive maps have for instance been instrumental in establishing that physical distances tend to be perceived logarithmically rather than linearly (remote locations appear proportionately closer than nearby ones) and that their perception is influenced by activity patterns, as well as one's socio-demographic profile. Maps that people carry in their minds are decidedly not Euclidean, as argued earlier on by Downs and Stea (1973)! Like time spaces and experiential spaces, they are subjective, whereas the economic space framed by transactional costs of commerce between places has an objective quality.

Thus, the relative view of space is a radical departure from the absolutism of the Kantian space. While choosing between the two views may depend on the purpose (Galton 2001), the relativist concept opens up the possibility to get to the core of processes that recursively create semantic realities with a geographic expression. For a broad class of spatio-temporal analyses and in relation to Ullman's systems of spatial interaction, the concept of relative space strikes as more general and empirically more useful. In fine, both the absolute space and the relative spaces are tremendous value as fundamental underpinnings of innovative research in social sciences.

Whereas settling on an absolute representation of space or a relative representation is a matter that is determined by the purpose of the analysis, in either instance there is a shared property that elements and entities that are within the scope of the analysis are differentiated. Space matters! The degree of this differentiation can be measured by their similarity, which boils down to measuring how close there are from one another in the assumed space. Distance (or its reverse, proximity) and related notions of adjacency, contiguity, and connectivity (depending on the data model assumed as well as geometric and topological property of space) are essential properties that enable the analysis of spaces (Gatrell 1983) and of the entities that they encompass. In particular, they constitute the operational concepts permitting the assessment of spatial dependence and spatial heterogeneity, and thus contextualization.

5 Region and Regional Systems

The region is a concept that is both convenient and attractive for scholarship that is contextually aware. In a generic sense, it encapsulates the variability of conditions and circumstances of a particular phenomenon across the geographic space (external context), while also recognizing that these conditions and circumstances may be quite similar internally and may exhibit a high degree of mutual interdependence. Thus, it has been defined as "a land area with certain attributes and characteristics that differentiate it from others that are spatially separate" (Thill 2017). The concept of region has a very long tradition in the discipline of geography, where it is often seen as the essence of the discipline itself. Yet, its meaning has changed over time. This is not the place to retrace this history, as this has been done quite competently by others already (see, for instance, Tomaney 2009).

The region is a geographic territory. It is generally understood that this territory is in one piece, although some outlying parts of it may not be contiguous. The region is not an arbitrary territory; to borrow the words from Isard, it "is not merely an arbitrarily demarcated area; rather it is an area that is meaningful because of one or more problems associated with it which we as regional scientists want to examine and help solve" (1975, p. 1). The region must be purposefully delineated with the express objective of generating areas that share similar properties, in contrast with other areas. Thus, the region is bound to change with the properties used to define it; it is also bound to change over time as the spatial variation of these properties evolves.

The delimitation of regions is inherently a problem in classification (Abler et al. 1971), whereby small and elemental units on the surface of the earth are grouped together based on shared traits. The traits used in each regionalization exercise include one or more characteristics that are semantically related to the purpose of the regionalization as well as location in the geographic space (Chorley and Haggett 1967; Haggett et al. 1977). Methods of regionalization abound (Johnston 1970; Duque et al. 2007) and, because of their combinatorial properties, their computational complexity may force to resort to heuristic algorithms and machine learning processes in many real-world applications.

It is fitting to recognize that, however complicated and important the analytics of regionalization can be, the region is more than a generalized entity of the data. Measurement considerations need to be fully aligned with the concept that underpins the construct. In substance, as pointed by Thill (2017), "each region is supported, and in a sense defined, by a distinctive socioeconomic organization that regulates its workings and its dynamics." While the areal classification of static phenomena producing "formal regions" whose properties are relatively uniform throughout may be of interest, this static view may be quite limiting (Abler et al. 1971).

Predictably, Ullman's paradigm of spatial interaction shifted attention away from the classical geographic concept of formal region. When place properties are no longer confined to site-specific attributes and relational properties come into play, functional or nodal regions take shape. "The essence of the functional region is … a network of social and economic tendons that bind" (Noronha and Goodchild 1992, p. 88) places together, while the special case of nodal regions also encapsulates dominance relationship between places. Along the same line of thought, Gould (1991) stated that "…a place, or a region, only takes on human and geographic meaning in relation to other places and regions, and relations mean, once again, connections over geographic and all sorts of other spaces. No place or region exists meaningfully in and of itself, disconnected and floating in a void, but always in relation to, and connected with, others" (p. 4). Nystuen and Dacey (1961) proposed an algorithm to delineate nodal regions in networks traversed by flows. A plurality of approaches exists today to define functional regions, all of which aim at maximizing within-region interactions and/or minimizing between-region interactions. Noronha and Goodchild (1992) proposed an approach that exploits the properties of an extension of the well-known spatial interaction model. Hierarchical algorithms of data mining have also been advanced to extract the community structure of the network transform of the system of interacting nodes. Chi et al. (2016) present the results of one such algorithm (Clauset et al. 2004) for the optimal regionalization of communication flows in Heilongjiang Province, China. This analysis is applied to more than ten million cell-to-cell wireless phone calls made during a three-month period of 2007. By comparing the actual network of cellular communication to a randomized network, a highly significant community structure is revealed with 20 regions. Although no physical movement is directly involved in the articulation of these flows, the spatial structure of these communities is strikingly ordered. Overwhelmingly, communities are contiguous territorial sets. Also, these regions exhibit a surprisingly strong concordance with mid-level prefectural administrative divisions.

As indicated earlier, each region is the spatial context and conduit where a number of intertwined social, economic, political, institutional, and natural processes operate. In his seminal book, Isard (1975) insisted that regions should not be viewed in isolation from the rest of the world. Regions entertain relationships with others to form larger ensembles, or systems, that intersect in different and complex ways, including across geographic scales. A foundational principle of regional science is that the region and its socioeconomic complexion result from the cumulative interaction between its own human, social, and natural resources, and initiatives (local and intraregional processes), how these processes operate over greater distances across the boundaries of regions (interregional processes), and how they relate with larger regional ensembles or the rest of the world (the global) (multi-scalar processes) (Thill 2017). Some of these systemic processes may be institutionally driven or supported (top down) through a variety of public policies enacted by government entities, while others emerge from the bottom up, under the initiative of individual citizens (consumers, travelers, workers) or individual business entities. The net outcome of the interactions and functional relationships among and between regions and individual agents therein is some form of order expressed through "spatial organization."

Research that focuses on the region or regions is contributed to by scholars from various disciplinary backgrounds. Given the multifaceted nature of the regional construct, it should not strike as a surprise that the scholarship of regional science (Thill 2017) is also very diverse. Five main strands can be identified in this respect, namely.

- Environmental and resource analysis
- Location analysis
- Regional analysis
- Urban analysis
- Transportation and mobility analysis.

Several of these foci intersect quite explicitly with the urban phenomenon. Further elaboration on the research traditions of regional science can be found in Florax and Plane (2004), and Mulligan (2014) more recently. Emerging trends and future developments in regional science research are discussed in Jackson and Schaeffer (2017a, b).

6 City and Urban Systems

It is beyond our purpose to articulate a fleshed-out operational definition of a city or urban area. Many worthy considerations have been advanced to that effect before (Caves 2005). Suffices it to say here that a city or urban area can be distinguished from other human settlements by its relatively larger population, higher population density, but also by its social, economic, and political functions, and by the form of its built environment. In his recent book, Batty (2013) advocated for a science of cities (or urban science) that breaks away from a view steep into static locational considerations in favor of theories and models that focus on concepts of why people come together in certain locations that we call cities to trade, exchange ideas, learn, realize social, and innovative contacts. "Locations are built on interactions. What happens in locations is a synthesis of what happens through networks and of how activities interact with one another" (p. 15). This vision strikes as rather similar to Ullman's "Geography as Spatial Interaction" thesis. One key difference, however, is that the complexity of interactions that Batty talks about is instrumental in the creation of structures and spaces that may not be regions in the traditional sense of the term. They may create networks that link physically disconnected locations, as in the case of world city networks (Beaverstock et al. 2000).

Undoubtedly, conceiving urban areas as systems predates Batty's (2013) call. Certainly, some elements of it can be found in Christaller's (1933) central place theory, and even earlier. Starting in the 1960s and 1970s, subsequent studies started to explicitly acknowledge the lineage with systems theory and complexity theory, and also to articulate more polished theoretical arguments rooted in industrial organization, the new economic geography, and other new theoretical developments (e.g., Pred 1977; Bourne and Simmons 1978; Batty 1971; Fujita and Mori 1997; Capello 2000; Peris et al. 2018). The study by Berry (1964) has been particularly path-breaking and influential in setting the agenda for the theme that permeates through this particular article. Berry indeed articulates very cogently the relationship that exists between the city as a system formed of multiple parts or elements (neighborhoods, business districts, etc.) and the broader context that brings cities to interact with each other and form a higher-level network. In the latter, cities are nodes scattered over continental or global territories, that interact in different fashions leading to the emergence of meso- and macro-scale structures. Berry (1964) is also path-breaking research in urban science in another respect. It blurs the boundary between the urban space (city) and the space that surrounds it. The latter, or exurbia, under conditions of population growth, may see its minor points of settlement absorbed within the urban fabric over time. Thus, a larger urban region (Davoudi 2009) may replace a traditional monocentric city, with the rise of edge cities complementing existing urban cores. These new forms or urbanization have been studied by various authors under somewhat related terminologies, such as "urban agglomerations," "polycentric urban regions" (PURs), "conurbations," "megalopolis," or "megacities" (e.g., Kloosterman and Musterd 2001; Hall and Pain 2006; Fang and Yu 2017). They support the fast urbanization of world populations, especially in developing countries.

7 The Book

This book contributes to furthering contextualized knowledge on cities and regions, thus expanding scholarship in urban science and regional science, through a collection of nineteen distinctive research contributions by noted scholars from all world regions. All the chapters of this book are original pieces of research presenting innovative research developments in the context of urban and regional systems. They were selected for their complementary focus on the urban or regional context. Each chapter was subjected to a rigorous peer-review process and was also reviewed by the editor of the volume. All the contributions follow a research perspective centered on spatially explicit analysis, thus featuring many of the analytical enhancements mentioned earlier in this article that permit appropriate handling of internal and external contextualizations in social science research. Specifically, the research designs and methodologies that are pursued in this book follow the three following traditions, Geographic Information Science and Technologies (GIS&T), location modeling, and spatial analysis. These traditions constitute the most compelling and effective approaches to scientific spatial thinking. Their unique contribution to urban and regional research was previously discussed in Thill (2018). This volume makes a scientific contribution owing to its topical and methodological focus, as well as the geographic diversity of authors and case studies presented.

The rest of the book is organized in two parts. The first part contains nine chapters on critical issues of urban growth and urban management. Therefore, the research setting is either a single city used as case study or multiple cities that form an entire system. They take advantage of the considerable data infrastructure that has accumulated over the past decades that is geospatial in nature, across many countries. Cities constitute particularly data-rich environments. Contributing to this state of affairs are data sourced from three primary channels: "designed" collection efforts undertaken by national census bureaus whose products are increasingly easily accessible at fine spatial and temporal granularity, data collected by more local towns and cities and share with the public as part of e-government initiatives aimed at greater public participation and transparency, and finally "organic" data from information and communication technologies that have fueled the growth of the big data revolution. All these sources are triggered a renewed interest for urban science, particularly when it aligns with the smart city movement. Contribution includes here look at macro-scale properties of a city system and meso-scale structures that can be identified in cities when the city is conceived as the product of interactions among actors. They also study the spatial distribution of public services and other facility location issues in an urban environment, localized growth factors, infrastructure and urban growth, rent dynamics and zoning regulations, and finally wealth inequality in urban areas.

The second part contains ten chapters that all broaden the scope of the analysis to a region or multiple regions within a spatially defined study area. Collectively, they treat a variety of topics of contemporary significance in this context, namely matters of economic restructuring, convergence, growth, and environment assessment and impact. In more specific terms, research reported in these chapters concerns spatial inequality and institutional interventions through policies and subsidies, taxation, access and use of internet-based information services, population migration, carbon emission and trade, climate change, land-use patterns, and water pollution.

The chapters assembled in this volume form a selection of some of the most compelling developments in spatially explicit research on urban and regional systems today. They are designed to provide a critical look in their subject matter through the use of advanced methods of spatial analysis, location modeling, and GIS&T. This scholarship adds to our knowledge-based accumulated over the past sixty years or so. The reliance on well-thought-out theories and the use of rigorous data analytic methodologies converge to produce pertinent results that are informative in regional and urban contexts and inform policy decision makers. They demonstrate the relevance of urban and regional research to the well-being of contemporary societies, and the value of research contributed by urban and regional scientists to address issues of pressing need to human societies. We contend that cutting-edge scholarship such as this will continue to enhance and bridge various social and natural sciences in specific spatial contexts in the future.

References

Abler, R., Adams, J. S., & Gould, P. (1971). *Spatial organization*. Prentice-Hall, Englewood Cliffs, NJ: The Geographer's View of the World.

Agarwal, P., & Skupin, A. (Eds.). (2008). *Self-organising maps: Applications in geographic information science*. Chichester: Wiley.

Anselin, L. (1988). *Spatial econometrics: Methods and models*. Dordrecht, The Netherlands: Kluwer Academic Publishers.

Anselin, L. (1995). Local indicators of spatial association—LISA. *Geographical Analysis, 27*, 93–115.

Anselin, L. (2003). Spatial externalities, spatial multipliers, and spatial econometrics. *International Regional Science Review, 26*(2), 153–166.

Batty, M. (1971). Modelling cities as dynamic systems. *Nature, 231*, 425–428.

Batty, M. (2013). *The new science of cities*. Cambridge, MA: MIT Press.

Beaverstock, J. V., Smith, R. G., & Taylor, P. J. (2000). World-city network: A new metageography? *Annals of the Association of American Geographers, 90*, 123–134.

Benenson, I., & Torrens, P. (2004). *Geosimulation: Automata-based modeling of urban phenomena*. Hoboken, NJ: Wiley.

Berry, B. J. L. (1964). Cities as systems within systems of cities. *Papers and Proceedings of the Regional Science Association, 13*, 147–163.

Blainey, G. (1966). *The tyranny of distance: How distance shaped Australia's history*. Melbourne, Australia: Macmillan.

Bourne, L. S., & Simmons, J. W. (1978). *Systems of cities: Readings on structure, growth, and policy*. Oxford, UK: Oxford University Press.

Bunge, W. (1961). The structure of contemporary American geographic research. *The Professional Geographer, 13*(3), 19–23.

Cairncross, F. (1997). *The death of distance: How the communications revolution will change our lives*. Boston, MA: Harvard Business School Press.

Capello, R. (2000). The city network paradigm: Measuring urban network externalities. *Urban Studies, 37*, 1925–1945.

Casetti, E. (1970). Generating models by the expansion method: Applications to geographical research. *Geographical Analysis, 4*(1), 81–91.

Castells, M. (1996). *The rise of the network society*. Oxford: Blackwell Publishers.

Caves, R. W. (Ed.). (2005). *Encyclopedia of the city*. London, UK: Routledge.

Chi, G., Thill, J. -C., Tong, D., Shi, L., & Liu, Y. (2016). Uncovering regional characteristics from mobile phone data: A network science approach. *Papers in Regional Science, 95*(3), 613–632.

Chorley, R. J., & Haggett, P. (1967). *Models in geography*. London: Methuen.

Christaller, W. (1933). *Central places in Southern Germany* (C. W. Baskin, Trans.). Englewood Cliffs, NJ: Prentice-Hall

Clauset, A., Newman, M. E. J., & Moore, C. (2004). Finding Community structure in very large networks. *Physical Review E, 70*(6), 1–6.

Cliff, A. D., & Ord, J. K. (1973). *Spatial autocorrelation*. London: Pion.

Couclelis, H. (1992). Location, place, region and space. In R. F. Abler, M. G. Marcus, & J. M. Olson (Eds.), *Geography's inner worlds* (pp. 215–233). New Brunswick, NJ: Rutgers University Press.

Davoudi, S. (2009). City-region. In R. Kitchin & N. Thrift (Eds.), *International encyclopedia of human geography* (pp. 255–265). Amsterdam: Elsevier.

Delmelle, E. C., Thill, J. -C., Furuseth, O., & Ludden, T. (2013). Trajectories of multidimensional neighborhood quality of life change. *Urban Studies, 50*, 923–941.

Dooley, R. A., & Levinsohn, S. H. (2001). *Analyzing discourse: A manual of basic concepts*. Clayfield, Australia: SIL International Digital Resources.

Downs, R. M., & Stea, D. (1973). *Image and environment: Cognitive mapping and spatial behavior*. Chicago: Aldine Publishing Co.

Duque, J. C., Ramos, R., & Surinach, J. (2007). Supervised Regionalization methods: A survey. *International Regional Science Review, 30*(3), 195–220.

Elger, T. (2010a). Conceptualization. In A. J. Mills, G. Durepos, & E. Wiebe (Eds.), *Encyclopedia of case study research* (pp. 231–233). Thousand Oaks, CA: Sage.

Elger, T. (2010b). Bounding the case. In A. J. Mills, G. Durepos, & E. Wiebe (Eds.), *Encyclopedia of case study research* (pp. 55–59). Thousand Oaks, CA: Sage.

Fang, C., & Yu, D. (2017). Urban agglomeration: An evolving concept of an emerging phenomenon. *Landscape and Urban Planning, 162*, 126–136.

Fischer, M. M. (2005). Spatial analysis in geography. In N. J. Smelser & P. B. Baltes (Eds.), *International encyclopedia of the social & behavioural sciences* (2nd ed., Vol. 23, pp. 94–99). Amsterdam: Elsevier.

Florax, R. J. G. M., & Plane, D. A. (Eds.). (2004). *Fifty years of regional science*. Heidelberg, Germany: Springer.

Forer, P. (1978). A place for plastic space? *Progress in Human Geography, 3*, 230–267.

Fotheringham, A., Stewart, C. B., & Charlton, M. E. (2002). *Geographically weighted regression: The analysis of spatially varying relationships*. Chichester, UK: Wiley.

Fotheringham, A. S., & O'Kelly, M. E. (1989). *Spatial interaction models: Formulations and applications*. Dordrecht: Kluwer.

Fotheringham, A. S., & Rogerson, P. A. (Eds.). (2009). *The SAGE handbook of spatial analysis*. Los Angeles, CA: Sage.

Forrester, J. W. (1961). *Industrial dynamics*. Waltham, MA: Pegasus Communications.

Forrester, J. W. (1969). *Urban dynamics*. Waltham, MA: Pegasus Communications.

Forrester, J. W. (1971). *World dynamics*. Cambridge, MA: Wright-Allen Press.

Fujita, M., & Mori, T. (1997). Structural stability and evolution of urban systems. *Regional Science and Urban Economics, 27*, 399–442.

Galton, A. (2001). Space, time, and the representation of geographical reality. *Topoi, 20*, 173–187.

Gatrell, A. C. (1983). *Distance and space: A geographical perspective*. Oxford, UK: Clarendon Press.

Getis, A., & Ord, K. J. (1992). The analysis of spatial association by use of distance statistics. *Geographical Analysis, 24*(3), 189–206.

Gould, P. (1991). Dynamic structures of geographic space. In S. D. Brunn, T. R. Leinbach (Eds.), *Collapsing space and time: Geographic aspects of communications and information*. London, UK: Harper-Collins.

Glaeser, E. L. (Ed.). (2010). *Agglomeration economics*. Chicago, IL: The University of Chicago Press.

Haggett, P. A., Cliff, D., & Frey, A. B. (1977). *Locational analysis in human geography* (2nd ed.). London: Arnold.

Hall, P., & Pain, K. (2006). *The polycentric metropolis: Learning from mega-city-regions in Europe*. London: Earthscan.

Harvey, D. A. (1990). *The condition of postmodernity: An enquiry into the origins of cultural change*. Cambridge, MA: Blackwell.

Harvey, D. A. (1994). The social construction of space and time: A relational theory. *Geographical Review of Japan Series B, 67*(2), 126–135.
Isard, W. (1975). *Introduction to regional science*. Englewood Cliffs, NJ: Prentice-Hall.
Jackson, R., & Schaeffer, P. (Eds.). (2017a). *Regional research frontiers: Vol. 1—Innovations, regional growth and migration.* Heidelberg, Germany: Springer.
Jackson, R., & Schaeffer, P. (Eds.). (2017b). *Regional research frontiers: Vol. 2—Methodological advances, regional systems modeling and open sciences.* Heidelberg, Germany: Springer.
Janelle, D. G. (1969, June). Spatial reorganization: A model and concept. *Annals of the Association of American Geographers, 59*, 348–364.
Johnston, R. J. (1970). Grouping and regionalizing: Some methodological and technical observations. *Economic Geography, 46*, 293–305.
Jung, P. H., Kashiha, M., & Thill, J. C. (2018). Community structures in networks of disaggregated cargo flows to maritime ports. In V. Popovich, M. Schrenk, J. C. Thill, C. Claramunt, & T. Wang (Eds.), *New frontiers in information fusion and intelligent GIS: From maritime to land-based research, Lecture Notes in Geoinformation and Cartography* (pp. 167–186). Cham: Springer.
Kaghan, W., & Bowker, G. (2001). Out of machine age? Complexity, sociotechnical systems and actor network theory. *Journal of Engineering and Technology Management, 18*, 253–269.
Kant, I. (1929). *Critique of pure reason* (N. K. Smith, Trans.). Basingstoke and London, UK: Macmillan.
Kloosterman, R. C., & Musterd, S. (2001). The polycentric urban region: Towards a research agenda. *Urban Studies, 38*, 623–633.
LeSage, J., & Pace, K. (2009). *Introduction to spatial econometrics*. Boca Raton, FL: CRC Press.
Li, FGN., Trutnevyte, E., Strachan, N. (2015). A review of socio-technical energy transition (STET) models. *Technological Forecasting and Social Change, 100*, 290–305.
Luhmann, N. (2013). *Introduction to systems theory*. Cambridge, UK: Polity Press.
Marletto, G. (2014). Car and the city: Socio-technical transition pathways to 2030. *Technological Forecasting and Social Change, 87*, 164–178.
Meadows, D. H. (1972). *Limits to growth*. New York: Universe Books.
Meentemeyer, V. (1989). Geographical perspectives of space, time, and scale. *Landscape Ecology, 3*, 163–173.
Mulligan, G. (2014). Regional science at sixty: Traditional topic and new directions. *Australasian Journal of Regional Science, 20*(1), 4–67.
Miller, J. H., & Page, S. E. (2007). *Complex adaptive systems: An introduction to computational models of social life*. Princeton, NJ: Princeton University Press.
Noronha, V. T., & Goodchild, M. F. (1992). Modeling interregional interaction: Implications for defining functional regions. *Annals of the Association of American Geographers, 82*, 86–102.
Nystuen, J. D., & Dacey, M. F. (1961). A graph theory interpretation of nodal regions. *Papers and Proceedings of the Regional Science Association, 7*, 29–42.
Olson, G. M., & Olson, J. S. (2000). Distance matters. *Human-computer Interaction, 15*, 139–178.
ONeill, K. J., & Gibbs, D. C. (2013). Towards a sustainable economy? Socio-technical transitions in the green building sector. *Local Environment, 19*, 572–590.
Ottens, M., Franssen, M., Kroes, P., & Van De Poel, I. (2006). Modelling infrastructures as socio-technical systems. *International Journal of Critical Infrastructures, 2*(2), 133–145.
Peris, A., Meijers, E., & van Ham, M. (2018). The evolution of the systems of cities literature since 1995: Schools of thought and their interaction. *Networks and Spatial Economics, 18*, 533–554.
Pred, A. (1977). *City systems in advanced economies: Past growth, present processes, and future development options*. Chichester: Wiley.
Roy, J. R., & Thill, J. -C. (2004). Spatial interaction modelling. *Papers in Regional Science, 83*(1), 339–361.
Shoham, Y., & Leyton-Brown, K. (2008). *Multiagent systems: Algorithmic, game-theoretic, and logical foundations*. Cambridge, UK: Cambridge University Press.
Skupin, A., & Fabrikant, S. (2007). Spatialization. In J. P. Wilson, & A. S. Fotheringham (Eds.), *Handbook of geographic information science* (pp. 61–79). Wiley.

Strathern, M. (1987). Out of context: Persuasive fictions in anthropology. *Current Anthropology, 28*(3), 251–270.

Thill, J. -C. (2011). Is spatial really that special? A tale of spaces. In V. Popovich, C. Claramunt, T. Devogele, M. Schrenk, & K. Korolenko (Eds.), *Information fusion and geographic information systems: Towards the digital ocean* (pp. 3–12). Heidelberg: Springer.

Thill, J. -C. (2017). Regional science. In N. Castree, M. Goodchild, W. Liu, A. Kobayashi, R. Marston, & D. Richardson (Eds.), *International encyclopedia of geography*. Wiley-Blackwell. https://doi.org/10.1002/9781118786352.wbieg0979.

Thill, J. -C. (2018). Innovations in GIS&T, spatial analysis, and location modeling. In J. -C. Thill (Ed.), *Spatial analysis and location modeling in urban and regional systems* (pp. 1–6). Cham: Springer.

Tomaney, J. (2009). Region. In R. Kitchin & N. Thrift (Eds.), *International encyclopedia of human geography* (pp. 136–150). Amsterdam: Elsevier.

Ullman, E. L. (1954). Geography as spatial interaction. In D. Revzan & E. S. Englebert (Eds.), *Interregional linkages* (pp. 1–12). Berkeley, CA: University of California.

Wachowicz, M. (1999). *Object-oriented design for temporal GIS*. London: Taylor and Francis.

Wooldridge, M. (2002). *An introduction to multiagent systems*. Chichester, UK: Wiley.

Yan, J., & Thill, J. -C. (2009). Visual data mining in spatial interaction analysis with self-organizing maps. *Environment and Planning B, 36*, 466–486.

Yin, R. K. (1994). *Case study research: Design and methods* (2nd ed.). Thousand Oaks, CA: Sage.

Zhang, W., & Thill, J. -C. (2017). Detecting and visualizing cohesive activity-travel patterns: A network analysis approach. *Computers, Environment and Urban Systems, 66*, 117–129.

Zhang, W., & Thill, J. -C. (2019). Comparative mesoscale analysis of world city networks: An analytical paradigm shift to detect stochastic hierarchies in networks. *Annals of the American Association of Geographers, 109*(3), 887–908.

Urban Growth and Management

On Locational Commonalities of Largest Cities Worldwide

Boris A. Portnov

1 Introduction

Each country has its largest city. In some countries, the largest city is just fractionally larger than its strongest competitor in the city population size hierarchy, whereas more often, it far exceeds the size of any other city in the nation.[1] Many leading cities are national capitals, such as Moscow, Tokyo, London, and Paris. However, other largest cities, such as New York City (NYC), Shanghai, Mumbai, Karachi, and Toronto, do not have capital status. Some largest cities are centers of international banking, commerce, and culture (London, Tokyo, Sydney, New York and Paris), while others (e.g., Kathmandu, Ashgabat and Thimphu, the capital of Bhutan) are economic peripheries. Some largest cities have deep roots in history (such as Rome, Istanbul, and Athens), while others are relatively new (e.g., New York, Toronto, and Sydney). Some largest cities are located on the shore (São Paulo, Sydney and Tokyo), while others (e.g., Paris, Moscow, and Berlin) are deep inland.

The largest cities have elevated importance as major national *loci* of population, political power, employment, culture, and socio-economic development (Portnov 2011). As development differentials between densely populated metropolitan areas and peripheral regions are increasing, overcoming inequalities in socio-economic

[1]The dominant status of largest cities may change over time. Thus, some 70 years after the City of Philadelphia was founded in 1682, it passed Boston and became the largest city in British America. However, by 1790, New York City (NYC) had held a slight edge over Philadelphia becoming the most populous city in the country. Yet, as an urban system matures, changes in the top level of urban ranking become rather uncommon, representing an exception to the rule. Supporting examples include NYC, which has retained its leading position in the national urban hierarchy for the past 200+ years, and other major cities across the globe (e.g., London, Paris, Tokyo, Vienna, Moscow, and Budapest).

B. A. Portnov (✉)
Department of Natural Resources and Environmental Management, University of Haifa, Mt. Carmel, 3498838 Haifa, Israel
e-mail: portnov@research.haifa.ac.il

J. Thill (ed.), *Innovations in Urban and Regional Systems*,
https://doi.org/10.1007/978-3-030-43694-0_2

development has become a key issue for urban and regional planning worldwide (Puga 1999; Felsenstein and Portnov 2005). In many sparsely populated peripheral areas, the inhabitants are denied access to social amenities, which are available in more populated areas. As the population of a community increases, it crosses the threshold for higher-level services and starts offering richer opportunities for employment, education, and leisure. In this respect, knowledge about factors that affect the population growth of cities may have important policy implications. For instance, this knowledge may guide regional development policies aimed at enhancing urban growth in regions where this objective is desirable (Portnov and Schwartz 2009b).

The questions we attempt to answer in this paper are as follows: *Are there any particular locational attributes that are significantly associated with the largest cities of nations? If so, what common locational denominators may help to distinguish main cities of countries from rest of the national urban systems?*

One clarification is important. When using the term 'main city,' urban and regional scientists often refer to any city above a given size, e.g., 500,000 or 1,000,000 residents. In this sense, every country (especially populous ones) may have *several* major cities. Our definition of the 'main city' is different: By using this term we refer to a place that is *the absolute largest in the national urban hierarchy*. Thus, every country can only have *one* such city. This definition has a clear advantage: Keeping the city's status fixed (as either main city or not), while removing all other selection criteria, can help us investigate how various factors affect the probability that a city may gain such leading status.

For our analysis, we use newly assembled data on some 82,000 populated places worldwide, which, to the best of our knowledge, have been analyzed here for the first time.

As the largest cities of countries are the absolute winners in the national races for potential investors and migrants, a better understanding of factors underlying their success may help improve development policies, thus adding a practical dimension to our inquiry.

There are two basic approaches to the analysis of urban geographies. The *first approach* emphasizes the importance of *individual attributes, unique features, and histories* that help to 'propel' a city to the dominant position in the national urban hierarchy i.e., the location of the royal court and governmental institutions in the case of Bangkok (Fry 1983); the role of colonial policies directing urban growth to coastal cities and aimed at facilitating the extraction and export of natural resources (Timberlake and Tinker 1985; Njoh 2007); or advantages of a particular port compared to ports in other coastal cities (Glaeser 2005b). The *second approach* emphasizes the *commonalities in urban growth patterns*, rather than unique features an individual urban place may possess. Although both approaches complement each other, the present paper adopts the latter, *commonalities-based approach*, emphasizing the importance of locational attributes as factors of urban growth. In fact, the appearance that 'new-age' forces 'dwarfed' traditional location attributes (such as distances and transport costs) may be misleading, due to hardly diminishing needs in physical commuting and increasing importance of environmental amenities for migration and residential choices (Clark et al. 2002; Portnov and Schwartz 2008).

Our hypothesis posited for empirical testing in this study was as follows: *If global locational attributes are important determinants of urban success, as expected from the outset of the analysis, then we should see significant commonalities in the locational attributes of largest cities (H_1). If, however, our assumption is incorrect, locational attributes will not help to differentiate between geographic attributes of largest cities of countries and those of smaller localities (H_0).*

The rest of the paper is organized as follows. It begins with a brief discussion of locational attributes associated with main city status in the national urban hierarchies. In the following sections, empirical data are used to identify the effect of different locational and non-locational attributes on the locational patterns of major cities worldwide. We conclude our analysis by discussing the implications of our findings for regional and urban development policies.

2 Locational Attributes as Factors of Urban Growth

In an early paper, the US geographer Jefferson (1939) termed the largest cities of nations 'primate cities,' attributing their success to sheer population size, which becomes a strong pull factor. This simple explanation is supported by recent urban growth theories. Thus, according to the 'neo-cultural' growth approach, which emphasizes the role of cultural diversity, human capital, innovation, and creativity (Glaeser et al. 1992; Florida 2002; Glaeser and Saiz 2003; Markusen 2006; Florida et al. 2008; Storper and Scott 2009; Mellander and Florida 2009), the most successful cities capitalize on such factors, irrespective of locational, historical or physical attributes. A similar explanation is advocated by the *endogenous growth theory*, which postulates that cities develop because of the diversity of the production factors they host, and that urban success (or lack thereof) has little to do with circumstances outside of the cities themselves (Jacobs 1969; Henderson 1974; Henderson et al. 1995; Glaeser et al. 1992; Laincz and Peretto 2006; Ha and Howitt 2007; Madsen et al. 2010).

In contrast, locational factors underlie most early studies in urban geography, as these early studies emphasized transport costs, distance 'friction,' commuting limits, and geographic barriers to trade (von Thünen 1826; Christaller 1933; Lösch 1938; Isard 1956; Beckmann 1968; Duranton 1999; Portnov and Schwartz 2009a, b). According to these studies, cities succeed because of location advantages, such as access to infrastructures, local natural resources, climatic advantages, and functional connections with other cities (Portnov et al. 2007; Portnov and Schwartz 2008, 2009a, b).

However, in recent urban development discourse, the effects of location on urban performance appear neither obvious nor straightforward. Thus, the proponents of a 'non-spatial' approach to urban development deny that urban location presents any 'natural growth advantages.' According to them, as people of similar backgrounds, incomes, and environmental preferences 'flock together,' location differences emerge (Gotlieb 1996; Hess 2004). The view behind the approach is that societal processes of repulsion and attraction, and individual drives to utility maximization are the drivers of urban success, rather than physical location per se. Furthermore, major forces shaping economies in the second half of the twentieth century—globalization, the dominance of service industries, information technologies, etc.—are often deemed unconstrained by location (Sassen 1997, 2000). That view is shared by large by the 'new economic geography' (NEG), which assumes that forces of concentration and de-concentration generate multiple *equilibria*, allowing development to occur in different *foci* (Krugman 1993, 1995, 1999; Fujita et al. 2001; Fujita and Mori 1996, 2005; Behrens and Robert-Nicoud 2008).

Although the explanations for urban primacy, suggested by different schools of thought, may differ substantially, largest cities are nevertheless expected *to be unique in their attributes and therefore have little in common.* In the following sections, we shall attempt to verify whether this assumption is correct.

3 Research Methodology

3.1 Data Sources

Data for the present analysis were drawn from the following two main sources:

(1) *Geonames Database* (2007), which covers some 82,000 populated places worldwide with at least 1000 residents (see Fig. 1), and includes geographic location, elevation above sea level and population size (as of the year 2000),[2] and
(2) *Geo-referenced database* maintained by ESRI (2007) for individual countries and including development indicators, such as a country's land area, population size, per capita income levels.

The ESRI database was also used to calculate the location attributes of places, such as latitude, distance from the seashore (exclusive of lakes and inland waterways), and proximity to the nearest international border (ESRI 2000–2017).[3] The calculation was performed in the ArcGIS 10.x software, using the 'spatial join' tool which

[2]The selection of observation units (municipalities, built contiguities, or metropolitan areas) is critical for analysis. As study by Portnov et al. (2012) indicated, the choice of spatial units is important because different territorial units (e.g., urban areas or municipalities) have different mechanisms behind their population growth patterns. Thus, selecting urban areas as units of the analysis reflects their functioning as economically integrated units. However, development disparities between local administrative units (i.e., municipalities) may also have a profound effect upon population growth, as favorably located and attractive municipalities may provide better services and facilities, thus appealing to more migrants and businesses. In this study, we used municipalities as primary spatial units—as, for most countries, they are the smallest territorial units for which comparable population data are available. However, subsequent studies may be needed in order to determine whether the use of metropolitan zones as spatial units may change the results.

[3]For island countries values of the international border variable were set equal to those of the seashore distance.

Fig. 1 Main cities of countries and the rest of localities under study. *Source* Mapped using data from Geonames (2007) and ESRI (2007) databases

helps to link attributes of different geographic layers (maps) based on their spatial association (ESRI 2000–2017).

3.2 Explanatory Variables

The analysis of the factors affecting the leading status of a locality in the country (i.e., either largest city or not) was run in two stages: First, without country attributes, and, next, with country attributes added. The following explanatory variables were used as inputs in the first set of models: latitude (in decimal degrees—dd) (which in the absence of specific climatic data served as a proxy for climatic harshness); elevation above sea level (in meters); distance to the seashore (dd); distance to the nearest international border (dd); and capital city status (yes/no).[4]

In the second set of models, the following variables served as country-specific attributes being assigned to all places in a given country: land area in square kilometers (ln); country's total population (ln); per capita GDP in USD (ln); and percent urban (%). Although these variables cannot differentiate between individual localities in a given country, they help, nevertheless, to reflect *cross-country* differences in terms of population, income levels, and urban development.

Lastly, in the third set of models, the continent (or world region) in which a country is located was further indexed as binary variables—North America and Western Europe, Middle East, Africa, Eastern Europe, and Asia—taking on the value 1 if a country is in that continent or region and 0 otherwise.

While these indicators do not include *all* possible predictors of urban development—inasmuch as economic characteristics such as industrial structure or measures of productivity are not included, due to restriction on data availability and comparability—geographical aspects, such as population size, location, and environmental conditions, are used here to determine the ability of such measures to explain largest city status, as the following discussion demonstrates.

Seashore proximity may facilitate regional and international trade, enabling urban growth to be sustained and thus improve the overall economic performance of a locality (Fujita and Mori 1996). Seashore proximity may be especially crucial in countries lacking a developed inland transportation network (Gallup et al. 1999), thus increasing the probability of a locality achieving leading status in the national urban system.

The effect of rugged terrain on economic development is also well established. Such terrain increases construction costs and makes cultivation more difficult (Nunn and Puga 2009). Urban places located at *lower elevations*, which are also normally

[4]The use of decimal degrees (dd) for calculating distances resulted from the use of maps with geographic coordinate systems based on three-dimensional spherical projections (Kennedy and Kopp 2000). Although the actual (metric) distances for each *dd* differ across locations, this was considered as an advantage, helping to account for (potentially) greater 'friction of distance' in more extreme latitudes.

associated with *seashore proximity*, may thus enjoy easier access to other developed *loci* while experiencing less problems with agricultural and water supply.

The *harsh climate* of some geographic areas places limitations on physical interurban exchanges, as well as on access to urban amenities and on additional aspects of human comfort. Moreover, towns located at high elevations in northern latitudes may be hindered (at least seasonally) in their access to national *loci* of employment and cultural life (Glaeser 2005a; Cheshire and Magrini 2006).

The *latitude* attribute was expected to encompass such influences. There may also be an interaction effect between high elevations and climate, with elevated places mitigating ambient heat, thus driving a place closer to the 'global average' of climatic comfort. Furthermore, other conditions being equal, a city located near an *international border* is likely to be better-off and thus may gain leading city status more easily than its off-border counterparts due to economic returns associated with cross-border exchanges. However, the influence of international border proximity on the status of cities may be different than hypothesized above, as many main cities have historically been far from international borders to minimize security risks (Salehyan and Gleditsch 2006; Nunn and Puga 2009).

In addition to the above explanatory factors, country-specific attributes (*per capita in come levels, percent urban population*, and *regional location*) may help to account for differences between developing and developed countries, as well as between countries of *Old Europe* versus *New Europe*, that is, Western Europe versus the former Soviet bloc—reflecting differences in initial economic systems and subsequent specialization. It is thus hypothesized that these variables comprise an adequate set of predictors, although they are restricted to a relatively small number of explanatory variables. The descriptive statistics for this dataset are reported in Appendix 1.

3.3 Statistical Analysis

The analysis was performed in *three* phases. The first phase consisted of a comparison of the spatial distribution patterns of main cities (209) with those of the other populated places in the sample (82,131 localities).[5] Binary logistic regression (BLR) was next employed in order to determine which locational and non-location related factors help to distinguish such cities from the rest of the national urban systems.

BLR is a statistical tool, often used to predict a discrete outcome from common and rare events (such as group membership), from a set of variables which may be continuous, discrete, dichotomous, etc. In the logistic analysis, the dependent (or response) variable is dichotomous, such as presence/absence or success/failure (Agresti 1996), with the logistic regression model taking the following general form (Eq. 1):

[5]The definitions of metropolitan areas or commuting contiguities, which may be considered economically integrated units, are likely to differ from one country to the next. As country-specific definitions were unavailable for the present study, we used municipalities as main units of analysis. The advantages and limitations of this approach will be discussed in some detail in the concluding section of the paper.

$$\log\left[\frac{p_i}{1-p_i}\right] = \beta_0 + \beta_1 x_{1i} + \cdots + \beta_n x_{in}, \tag{1}$$

where p_i is the probability that the ith case has largest city status; x_{ij} is the jth predictor for the ith case, β_j is the jth coefficient, and n is the number of predictors.

For the next step, we considered the ratio between the size of the largest city and the second-largest city in each country in order to measure the degree to which the main city is extraordinarily large in terms of population size within its national context. This analysis was performed for 163 countries, for which complete and comparable development data were available. The analysis excluded mostly small developing countries with incomplete data and city-states and was performed using discriminant analysis (Cacoullos 1973; Huberty and Olejnik 2006). The primary goal of the analysis was to determine factors and development attributes which help to distinguish between groups of countries characterized by different gaps in the population size between first and second-largest cities, i.e., the *degree of extraordinariness* of the largest city in the national context.

4 Results

4.1 General Trends

Figure 2 compares the locational attributes of largest cities and other localities, according to the following three attributes: latitude (Fig. 2a); elevation above sea level (Fig. 2b), and distance to the seashore (Fig. 2c).

As Fig. 2a shows, largest cities are more often found in tropical and subtropical latitudes, as compared to other localities. In particular, more than one-third of all largest cities of countries are located within the 0–20 dd latitude belt, north of the equator, while smaller localities are more often found further north, in colder areas, between 30 and 60 dd (see Fig. 2a).

Moreover, as Fig. 2 demonstrates, approximately 71% of largest cities are found at low elevations (i.e., less than 200 m above sea level), and much fewer—at higher elevations (see Fig. 2b). Also, compared to localities of smaller size, largest cities tend to be closer to the seashore (<0.1 dd from the coast/ocean shoreline), while fewer are located inland (Fig. 2c).

4.2 Ordinariness Index

For a more formal test of the above observation that—in terms of their basic locational attributes—main cities of countries are *farther from geographic extremes than are their smaller counterparts*, a simple quantitative measure was devised. The proposed

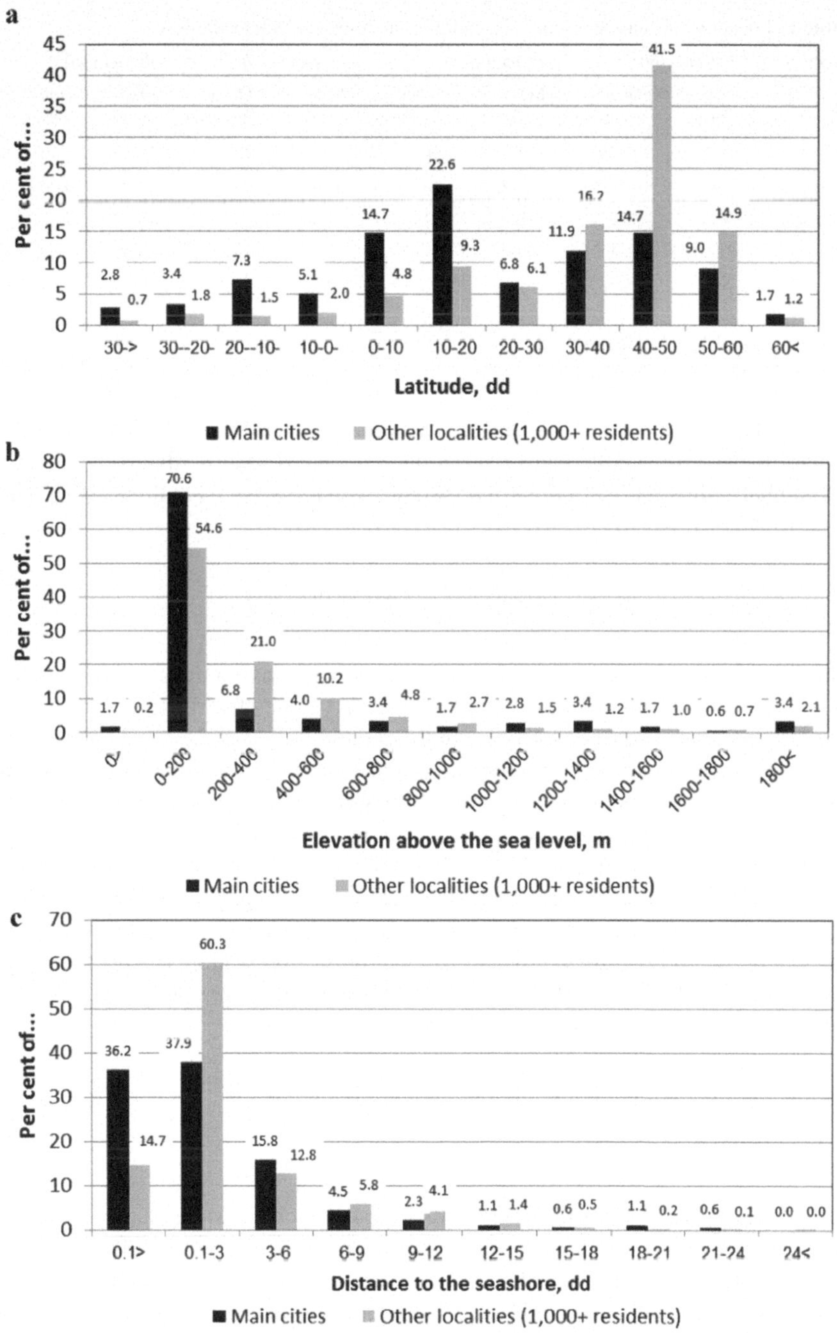

Fig. 2 Locational attributes of main cities of counties and of other localities: **a** latitude, **b** elevation above sea level, and **c** distance to the seashore

Table 1 Location Ordinariness Index (ILO) for main cities and other localities

Statistic	Latitude (dd)		Elevation (m)		D_shore (dd)		Border (dd)	
	Main cities	Other localities	Main cities	Other localities	Main cities	Other localities	Main cities	Other localities
Mean	18.02	35.59	285.27	313.27	1.69	2.42	1.62	3.60
Minimum	−36.87	−54.80	−2.00	−279.00	0.00	0.00	0.00	0.00
Maximum	64.15	78.22	2,620.00	4,824.00	12.25	37.58	10.76	79.12
ILO	3.60	1.31	16.00	14.45	21.24	14.59	48.00	21.03
No. of cases[a]	209	81,923	209	81,923	209	81,923	209	81,923

[a]Total number of localities—82,132

index which we defined as the Index of Locational Ordinariness (or *ILO*) follows the general logic of the coefficient of variation (standard deviation divided by mean), but compares the observed values of a variable with its global maximum, rather than with its mean:

$$\mathrm{ILO}_j = \mathrm{mean}_j^{-1}\sqrt{\sum\left(X_{ij} - \max_j\right)^2 / N} \tag{2}$$

where mean and max are, respectively, the mean and maximum values of variable j (e.g., latitude, elevation, etc.); X_{ij} is the observed value of variable j in locality i; and N is the total number of localities in the sample.

The values of the proposed index are easy to interpret: Large positive values of this index indicate considerable departure from its extreme (maximum) values, while values close to *zero* indicate that the observations cluster around the maximum.

The results of the analysis for two subsamples of localities—main cities of countries and other localities with over 1000 residents—are reported in Table 1, which shows that largest cities consistently exhibit larger ILO values for all location variables analyzed. This is fully consistent with our initial observation, namely that *main cities appear to have less extreme values for their location attributes than do their smaller (less populous) counterparts*.

4.3 Logistic Analysis

Table 2 reports the results of logistic regression, in which *city status* (i.e., a binomial measure indicating whether a city is the largest city in a country or not) is the dependent variable, while location and country-specific attributes are predictors. Model 1 is the only model calculated only for locality-specific attributes; other models also include country-specific attributes (Model 2), or both country-specific and regional attributes (Model 3). All models provide reasonably good fit (with the Nagelkerke

Table 2 Factors influencing main city status—results of binomial logistic regression analysis (dependent variable: main city versus other locality status)

Variable	Model 1		Model 2		Model 3	
	OR[a]	(95%CI)[b]	OR[a]	(95%CI)[b]	OR[a]	(95%CI)[b]
Constant	0.003***		137.048*		11,206	
Latitude, dd	0.975***	(0.965–0.985)	0.981**	(0.967–0.995)	0.978*	(0.959–0.997)
Capital status (yes)	10 386***	(5973–18,058)	11,059***	(5334–22,928)	11,095***	(5268–23,370)
Elevation, m	0.999***	(0.999–1.000)	0.999***	(0.998–1.000)	0.999**	(0.998–1.000)
Seashore, dd	0.841***	0 (0.756–0.936)	0.792***	(0.692–0.907)	0.789***	(0.685–0.907)
Border, dd	0.847**	(0.761–0.941)	0.887	(0.785–1.002)	0.892	(0.792–1.004)
Land area, km^2 (ln)	–	–	1.014	(0.781–1.317)	1.034	(0.785–1.362)
Total population, 1000 (ln)	–	–	0.640**	(0.475–0.861)	0.605**	(0.437–0.836)
Per capita GDP, USD (ln)	–	–	0.655*	(0.434–0.989)	0.577	(0.325–1.025)
Percent urban	–	–	1.000	(0.979–1.022)	0.996	(0.974–1.018)
Western Europe and North America	–	–	–	–	0.662	(0.161–2.727)
Middle East	–	–	–	–	0.364	(0.084–1.579)
Africa	–	–	–	–	1.818	(0.542–6.091)
Eastern Europe	–	–	–	–	1.019	(0.268–3.872)
Asia	–	–	–	–	0.657	(0.2122–0.039)
No. of cases	82,132		82,132		82,132	
−2 Log likelihood	832.282		518.698		514.680	
Nagelkerke R^2	0.718		0.748		0.750	
Hosmer–Lemeshow goodness-of-fit test						
χ^2	8.355		5.398		3.877	

(continued)

Table 2 (continued)

Variable	Model 1		Model 2		Model 3	
	OR[a]	(95%CI)[b]	OR[a]	(95%CI)[b]	OR[a]	(95%CI)[b]
Percentage correct						
Main cities	77.990		76.978		76.978	
Other localities	99.967		99.975		99.975	
Overall percentage	99.911		99.934		99.934	

Model 1: Country's attributes and population size variable excluded; Model 2: Country's attributes added; population size variable excluded; Model 3: Country's and regional attributes added; population size variable excluded

[a]Odds ratio (Exp(B))

[b]95% confidence interval (CI)

Significance levels: *(0.05), **(0.01), ***(0.001)

R^2 ranging between 0.71 and 0.75 and the Hosmer–Lemeshow goodness-of-fit test resulting in $\chi^2 > 3.0$ and $P > 0.3$).

As Model 1 demonstrates, the probability of a locality having a status of its country's main city *declines* with latitude (OR (Odds Ratio) = 0.975; $P < 0.001$), elevation (OR = 0.999; $P < 0.01$), and distances to the seashore and international borders (OR = 0.841 and 0.847, respectively; $P < 0.01$). The high statistical significance of these variables indicates that, in line with our initial research assumption, *locational attributes, covered by the study, do help, in fact, to distinguish between geographic patterns of largest cities of countries and those of smaller localities.* Furthermore, quite expectedly, the capital city status substantially also raises the probability that a locality is the largest city of its respective country (OR = 10,386; $P < 0.001$).

The addition of country-specific (Model 2) and regional attributes (Model 3, Table 2) does not substantially change the outcome; capital city status, elevation, distance to the *seashore,* and *latitude* maintain their statistical significance (OR = 0.792 and 0.981; $P < 0.05$). In the 'country-adjusted' model (Model 2, Table 2), the country's total population and *per capita* GDP emerge as statistically significant ($P < 0.01$), indicating that for any locality, the probability of achieving largest city status *decreases* in more developed and populous nations.

4.4 *Differences Between Countries' Subgroups*

The influence of individual factors on city status may depend on local economic setting, with some factors (e.g., access and proximity) being more important in less developed countries than in the developed world (Portnov and Schwartz 2008). To test this hypothesis, we split our analysis for localities into low income (GDP_{PC} (per capita) ≤ USD$3500); medium income (USD$3500 < GDP_{pc} ≤ USD$14,000); and high-income countries (GDP_{pc} > USD$14,000).[6] The results are reported in Table 3.

As Table 3 shows, the models provide relatively good fits; Nagelkerke R^2 = 0.745–0.842 and the Hosmer–Lemeshow goodness-of-fit test resulted in $\chi^2 = 5.7$–9.5 and $P > 0.2$. The models appear to predict correctly the occurrence of most largest cities of nations resulting in approximately 74% correct predictions in the low-income country group (Model 4), approximately 83% correct predictions in the medium-income group (Model 5), and 62.5% correct predictions in the high-income group (Model 6). These relatively high prediction levels (74–83%) indicate that the models reported do appear to cover most essential predictors helping to differentiate between largest cities of countries and smaller localities.

Although the results for Models 4–6 are essentially similar, they do differ with respect to three variables: *elevation*, *distance to the seashore*, and *distance to the nearest international border*. Notably, the statistical significance of *elevation* drops, in line with rising incomes ($P < 0.001$ in the low-income group versus $P > 0.5$ in the

[6]The per GDP break points were determined as fractions of USD$7000, the average per capita GDP in our sample.

Table 3 Factors influencing main city status across different income groups of countries—results of binomial logistic regression analysis (dependent variable: main city versus other locality status)

Variable	Model 4		Model 5		Model 6	
	OR[a]	(95%CI)[b]	OR[a]	(95%CI)[b]	OR[a]	(95%CI)[b]
Constant	1.568		0.078		0.000*	
Latitude, dd	0.984	(0.967–1.002)	0.978*	(0.955–1.000)	0.985	(0.959–1.012)
Capital status (yes)	0.001***	(0.000–0.002)	0.000***	(0.000–0.002)	0.002***	(0.000–0.012)
Elevation, m	0.999***	(0.998–0.999)	0.998**	(0.997–0.999)	0.999	(0.995–1.003)
Seashore, dd	0.818**	(0.690–0.969)	0.737	(0.522–1.041)	0.960	(0.686–1.345)
Border, dd	0.912	(0.786–1.058)	0.809*	(0.658–0.994)	0.702**	(0.525–0.938)
No. of cases	17,650		20,760		43,721	
−2 Log likelihood	425.230		106.791		100.200	
Nagelkerke R^2	0.745		0.842		0.755	
Hosmer–Lemeshow goodness-of-fit test						
χ^2	9.542		5.780		6.402	
Percentage correct						
Main cities	73.913		82.979		62.500	
Other localities	99.909		99.986		100.000	
Overall percentage	99.705		99.947		100.000	

Note See comments to Table 2; Model 4: Low-income countries (GDP_{pc} < \$3500); Model 5: Medium-income countries (\$3500 ≤ GDP_{pc} < \$14,000); Model 6: High-income countries (GDP_{pc} ≥ \$14,000)

high-income group of countries—see Table 3). One possibility is that an increase in national wealth translates to lowered constraints associated with additional costs of both accessibility and the use of heating/air conditioning which are often imposed by elevation. The *proximity to the seashore* variable shifts similarly, as its statistical significance declines with national wealth, indicating that localities closer to the coast are more likely to become largest cities in poorer countries. Again, more study is needed to determine additional associations and causality, but one possible reason for this is that maritime traffic may compensate for poor inland infrastructures.

Proximity to international borders appears more essential for the formation of main cities in medium- and high-income countries; for Model 6 (high-income countries), OR = 0.702 and $P < 0.05$, and the results show a higher odds ratio and the same significance level for medium-income countries (Model 5), with OR = 0.809 and $P < 0.05$. However, for Model 4 (low-income countries), this variable is not significant. Further research could assist in determining the reasons for this phenomenon.

However, it is also possible that this results reflect differences between closed and open economies. In addition, for some low-income countries, this result may reflect the realities of security concerns related to neighboring states, in cases where political instability, war, or border disputes exist, or have historically existed.

5 Additional Factors

When a main city is extreme in one attribute (e.g., located at a higher elevation), such a location may mitigate some other extreme characteristic, e.g., hot climate. Many other examples of such potential interactions could be made. To explore this hypothesis, we included in the analysis an additional explanatory variable: *elevation*tropics*. This variable was calculated as the interaction term between elevations above sea level and indicator variable for tropical locations, which takes a value of 1 for cities and towns located between latitudes $-20°$ and $20°$, and 0 otherwise. Our working hypothesis was that due to effect mitigation, this variable would be positively correlated with main city status. Concurrently, the relationship between main city status and the elevation-only variable was expected to be negative, reflecting our hypothesis that main cities are *less likely to be found at extreme elevations.*

Several studies have provided extensive empirical evidence regarding the effects of *proximity to population centers*, often referred to as *market potential or socio-economic potential*, (see, *inter alia,* Babarović 1979; Redding and Venables 2004; Head and Mayer 2010). Following the logic of the *socio-economic potential* index, proposed by Babarović (1979), we calculated the size-proximity index (SP) for each populated place in our sample (82,132), using the following formula (Eq. 3):

$$\mathrm{SP}_i = \sum_{j=1}^{n} \left(\frac{P_j}{d_{ij}^2} \right), \tag{3}$$

where SP_i is size-proximity index for place i, P_j is population size of place j located within a commuting range from place i (assumed to be 1 dd),[7] n is the total number of populated places located within the commuting range from i, and d_{ij} is the great circle distance between i and j. (In the calculation, each place i was assumed to be its own neighbor, with proximity distance set small ($d_{ii} = 0.01$), to avoid division by zero.) Since other populated places can serve as markets for goods and services produced in the locality and as places of employment for its residents, the size-proximity indicator should expectedly correlate positively, *ceteris paribus*, with the main city status.

Table 4 presents the results of the adjusted analysis. In particular, both in the model with regional dummies (Model 8), and in the model without regional dummies (Model 7), *extreme latitudes, elevations and remoteness from seashores and international borders are all negatively associated with main city status* (OR < 1; $p < 0.01$), while capital city status and size-proximity (SP) index are positively associated with urban primacy ($P < 0.001$). The *elevation*tropics* interaction variable also emerged as positively associated with the main city status, as expected, though this variable did not emerge as statistically significant ($P > 0.4$).

[7] See footnote 4.

Table 4 Factors influencing main city status, including size-proximity and elevation with interaction term—results of binomial logistic regression analysis (dependent variable: main city versus other locality status)

Variables	Model 7		Model 8	
	OR[a]	(95% CI)[b]	OR[a]	(95% CI)[b]
Constant	0.000***		0.000***	
Latitude, dd	0.970***	(0.959–0.980)	0.984*	(0.970–0.998)
Capital status (yes)	7233***	(4115–12,714)	5,187***	(2929–9187)
Elevation, m	0.999**	(0.998–1.000)	0.999**	(0.998–0.999)
Elevation*tropics	1.000	(0.999–1.001)	1.000	(0.999–1.001)
Seashore, dd	0.822***	(0.728–0.927)	0.795***	(0.700–0.904)
Border, dd	0.833***	(0.745–0.932)	0.862**	(0.778–0.955)
Size-proximity (SP) index (ln)	1.613***	(1.385–1.878)	1.743***	(1.499–2.028)
Western Europe and North America			0.168***	(0.075–0.377)
Middle East			1.051	(0.298–3.700)
Africa			1.259	(0.494–3.208)
Eastern Europe			0.566	(0.197–1.622)
Asia			0.244**	(0.096–0.617)
No. of cases	82,132		82,132	
−2 Log likelihood	795.510		768.835	
Nagelkerke R^2	0.731		0.740	
Hosmer–Lemeshow test				
χ^2	5.573		3.692	
Sig	0.695		0.884	
Percentage correct				
Main city	77.512		76.555	
Other localities	99.967		99.967	
Overall percentage	99.910		99.907	

Note See comments to Table 2; Model 7: Regional dummies excluded; Model 8: Regional dummies added

5.1 Discriminant Analysis

In a divergence from considering largest cities versus all other localities (with over 1000 residents), as we did in the previous sections, in this section we focus upon the factors affecting the leading status of largest cities in their respective countries. Through discriminant analysis, we assess leading city status by considering the *gap in population size* between each largest city and its nearest neighbor in the city-size hierarchy.

Discriminant analysis is a popular multivariate statistical technique (Cacoullos 1973; Huberty and Olejnik 2006), used to determine the best linear combination of explanatory variables, differentiating groups of cases into which the dependent variable is subdivided. An advantage of this technique as compared to techniques such as regression analysis, is that it reveals *common characteristics* of the cases forming subgroups of the dependent variable (Appendix 2).

In this analysis, we split the 163 countries in our sample (see research methodology section) into three subgroups, defined by the Jenks natural break method (North 2009). The method helps to transform a single continuous variable into relatively homogeneous categories, using *natural breaks* in the distribution of that value. The following three categories were defined for the ratio between the population size of the *second and first largest cities* of a country (termed the *B-ratio*):

- Group 1: *B-ratio* < 0.3 (first and second cities are substantially different in size);
- Group 2: $0.6 >$ *B-ratio* ≥ 0.3(moderate difference in size between first and second cities); and
- Group 3: B-ratio ≥ 0.6 (First and second cities of a country are not substantially different in size).

The canonical discriminant functions (CDFs), denote the importance of each predictor in the discriminant analysis, are reported in Table 5, where standardized coefficients with large absolute values (in two leftmost columns) correspond to variables with greater discriminating ability. That is, a variable with large standardized coefficients performs better in explaining why a particular country belongs to, for example, Group 1 (in which first and second cities are substantially different in size) or to Group 3 (in which first and second cities are not substantially different in size). The structure matrix coefficients (the two rightmost columns in Table 5) indicate simple correlations between variables and discriminant functions.

As Table 5 shows, the standardized coefficients indicate that the variables with the greatest ability to discriminate are *latitude, population, land area, GDP per capita*, and *capital city status*; in addition, according to the χ^2 goodness-of-fit test discriminant function CDF2 is non-significant. Therefore, CDF2 is demonstrated not to discriminate well between the groups of cases. CDF1 explains most of the data variance (approx. 82%) and has strong discriminant power, with $\chi^2 = 77.4$ and $P < 0.001$. Aside from regional dummies, three variables load on CDF1, two of which are correlated positively with CDF1—*per capita* GDP and *percent urban*; the variable negatively correlated with CDF1 is *capital city*. The functions were thus labeled *development status* and *local geography*, respectively.

What groups of countries are discriminated by these discriminant functions? The graph, known as a *territorial map* (Appendix 3), helps to answer this question. Numbers 1, 2, and 3 on this map represent individual countries falling into Groups 1, 2, or 3, respectively. The graph's *X*-axis and *Y*-axis feature values which the CDF1 and CDF2 can take—from negative (lower left corner of the diagram) to positive (upper right corner). In this diagram, each country is mapped in two-dimensional space, according to the values of the discriminant functions, as presented in Table 5.

Table 5 Discriminant analysis for countries according to population size ratio (second largest city/first largest city)—Canonical Discriminant Functions (CDFs)

Variable	Standardized coefficients		Structure matrix coefficients	
	CDF 1	CDF 2	CDF 1	CDF 2
Latitude (dd)	−0.464	0.359	−0.112	0.436[a]
Population, 1000 (ln)	0.618	0.895	0.084	0.741[a]
Land area, km^2 (ln)	−0.703	−0.130	−0.053	0.556[a]
GDP per capita, USD (ln)	0.523	0.491	0.435[a]	0.339
Percent urban	0.014	−0.207	0.272[a]	0.116
Capital city	−0.706	0.297	−0.724[a]	0.205
Asia	0.028	−0.171	−0.070	−0.084[a]
Africa	0.135	0.273	−0.070[a]	0.061
Eastern Europe	0.096	−0.137	−0.186[a]	0.087
Middle East	0.350	−0.095	0.212[a]	−0.047
Western Europe and North America	0.226	0.058	0.312	0.457[a]
% of Variance	82.398	17.602		
χ^2	77.354	15.455		
Sig	<0.001	0.116		

[a]Indicates a variable's largest absolute correlation with one of the CDFs. See text for CDF explanations and countries' grouping

As Appendix 3 shows, the clustering of countries belonging to Groups 2 and 3 (moderate-to-small population size differences) occurs where CDF1 (featured on the *X*-axis) takes on *positive* values, while the clustering of countries belonging to Groups 1 and 2 (moderate-to-large population size differences) occurs if the values of CDF1 are *negative*. As mentioned, this discriminant function is positively correlated with per capita GDP and percent urban, and negatively correlated with the capital city status of its largest city (see CDF1 structure matrix coefficients in Table 5). Therefore, less developed countries in which the largest city functions as national capital are likely to exhibit a large population size difference between the second-largest and largest cities. In contrast, in countries which are both high-income and urbanized, and in which the largest city is not the capital city, second-largest and largest cities tend to be more similar in size. The analysis, therefore, singles out the variables, which explain to what degree the largest city is outstanding in terms of its population size: whether it is the capital of the country, coupled with the national urbanization level and GDP per capita.

6 Discussion and Conclusions

As the present analysis indicates, over one-third of main cities of the world are located within the 0°–20° latitude belt, north of the equator, while others (i.e., smaller localities) are often found in colder regions, between 30° and 60° latitudes. Moreover, nearly 71% of all main cities are at low elevations (i.e., less than 200 m above sea level), and fewer (less than 29%) are located at higher elevations. There are also more main cities (36.2%, as opposed to only 14.7% among other localities) located close to the seashores, and fewer located inland.

A multivariate logistic analysis, using geo-referenced data on approximately 82,000 populated places worldwide, further reveals that a locality's likelihood of being the largest city decreases with latitude and higher elevations (for both, $P < 0.001$), while capital city status substantially increases that likelihood ($P < 0.001$). The likelihood of gaining largest city status *declines*, *ceteris paribus*, as distance from the shore increases ($P < 0.05$).

The largest cities of countries thus appear to be *quite ordinary in their locational attributes:* They are seldom found at high elevations, far from the seashore or in extremely cold or extremely hot areas. This may be, in fact, essential to their primacy, by making them attractive to many. However, one clarification is required: In terms of location attributes (such as climate, elevation, proximity to the coast), by *locational ordinariness* we refer to *distancing from extremes*—not a simple arithmetic averaging of geographic attributes.

The question regarding what constitutes an *ordinary location* should also be clarified. Ordinariness in terms of such considerations as climate may result from interacting variables, the relationships between which are not necessarily similar or symmetric among largest cities, but the various combinations of which produce generally suitable results for essential aspects of urban life, which may appeal to a wide swath of the population. Thus, climate-related locational ordinariness means a tendency to temperate climate defined on the global scale, rather than at some local scale. According to this notion, main cities are likely to be located at higher elevations in hot countries (e.g., Addis Ababa and Mexico City) and in the southernmost parts of cold countries (Oslo, Helsinki, and Stockholm).

In terms of other location variables (seashore proximities, etc.), our premise is that *locational ordinariness* can be translated as distancing from remote and/or *risky* locations associated with high probabilities of extreme events, such as flooding or water shortages. In line with this understanding, location close to the shoreline, for example, may be mitigated by an elevation slightly above sea level (which prevents flooding); the combination of such attributes provides an *ordinary location*, itself not necessarily at the center of a continuum between locational extremes, as simple attribute averaging might imply.

Extreme environmental conditions usually have a negative impact upon population growth, making settlements located in climatically extreme areas less attractive and desirable for migrants (Portnov and Erell 2001). Should we adopt the *firms follow workers* approach substantiated by recent studies (Glaeser and Saiz 2003; Kolko

2010), we would thus expect ordinary places to have better chances of economic survival and growth than extreme ones.

Furthermore, even if 'first nature' (i.e., factors attributed to location and natural resources in general) is important, a dynamic explanation of why largest cities emerge in particular places is still required. Otherwise, reservations may, of course, be raised against oversimplifications. For instance, development *lock-in effects* and *path dependence* may be mentioned (Martin and Sunley 2006); in addition, uneven access to natural resources (in both the past and the present) and accessibility may further complicate the issue. The role of administrative functions (such as national capital city status) should also be taken into account.

Several questions may thus arise: *Is it possible that secondary or tertiary cities can become largest cities without first nature geography changing? Why does a particular city became the largest city of a country, if several cities within the same country may be 'ordinary' in terms of their geographic location?*

Our answers to these questions are relatively simple. While location ordinariness (in terms of first nature geography) may be a precondition for success, the *forces of second nature are capable of ensuring it.* Assigning capital city status and higher values of the size-proximity index, as the present analysis indicates, does appear to contribute significantly to *urban primacy.*[8] Future research should thus focus upon introducing more second nature factors (such as infrastructures and unique facilities—universities, hospitals, industrial clusters, etc.), in addition to first nature factors, into empirical analysis.

The main contribution of the present study is that seems to further urban theory by introducing a new idea that *largest cities may combine an ordinary first nature with a unique second nature,* and providing some empirical evidence to support this claim.

The influence of individual factors on city status may depend upon the local economic setting, with some factors (e.g., access and proximities) being more important in less developed countries, where road infrastructures are less advanced than in the developed world. To verify this hypothesis, we ran our analysis for localities in three distinct groups of countries, defined in terms of income levels. As the analysis indicated, coastal settlements have more chances, all other factors being equal, of becoming the largest cities in low-income countries, a result that may indicate high dependency upon access to maritime transport. In contrast, proximity to international borders appears to be more important to largest city status in high-income countries than elsewhere. One possibility is that this reflects differences between closed and open economies.

Using data for 163 countries, we also investigated whether main cities are unique in terms of population size in their respective national contexts. We found that high development levels are likely to reduce the population gap between the first and

[8]In the present study, the size-proximity index was *not* included in the discriminant analysis, which was carried out at the country level. Perhaps, in future studies, this factor needs to be analyzed again at a different scale. Furthermore, data availability or lack thereof should not close the book on other second nature factors—the literature needs to be investigated more strenuously, and solutions for analyzing this need to be developed, to move forward with future research.

second city in a country, whereas when the first city is the national capital this gap tends to widen. The explanation for these trends may be fairly straightforward. The importance of central power to the formation of cities is well known to urban studies. In his *Al-Muqaddimah* (Introduction to History), the fourteenth-century scholar, Ibn Khaldūn (Rosenthal (tr) 1967), noted that major cities tended to be the *capitals* of their respective nations; the concentration of power and wealth, attributed to taxes collected by the prince's court, raised demand for goods and services. Craftsmen, artists, and intellectuals settled near these *foci* of wealth and power, to learn from masters or to receive economic support. This *urban causation* process, discussed by Weber (1921), was reiterated by recent urban and regional development theories ranging from *cumulative causation* to *endogenous growth theory*, *learning cities,* and *new economic geography*. Most theories emphasize the accumulation of knowledge and skills, as well as the role of interpersonal contacts and *information sharing*, in the growth of cities.

In contrast, economic wealth and high levels of urbanization may operate in the opposite direction, effectively minimizing the gap in population sizes between largest and next largest cities. As a country becomes wealthier and its settlement system matures, more cities are likely to succeed along with the largest one. A similar process may occur on the regional scale as well; even in developed nations, there are sparsely developed regions in which major cities are few and far surpass in size their local counterparts.

Acknowledgement The author is grateful to two anonymous reviewers for their numerous helpful comments and suggestions. The author's gratitude is also due to Prof. John Parr and Prof. Moshe Schwarts for their comments on the early draft of the paper.

Appendix 1

Descriptive statistics for the research variables

Variable	No. of cases	Minimum	Maximum	Mean	SD
A. Main cities					
Latitude (dd)	209	−36.87	64.15	18.02	24.36
Population, residents	209	1,232	14,608,512	1,926,252	2,965,911
Elevation above the sea level (m)	209	−2.00	2620.00	283.95	517.57
Seashore (dd)	209	0.00	12.25	1.69	2.04
Border (dd)	209	0.00	10.76	1.62	1.82
Capital city status (N/(%))	168/(80.38)	–	–	–	–
B. The rest of localities					
Latitude (dd)	82,131	−54.80	77.30	35.59	18.86

(continued)

(continued)

Variable	No. of cases	Minimum	Maximum	Mean	SD
Population, residents	82,131	1001	10,927,986	25,771	129,714
Elevation above the sea level (m)	82,131	−279.00	4824.00	313.28	441.40
Seashore (dd)	82,131	0.00	37.58	2.42	3.24
Border (dd)	82,131	0.00	79.12	3.60	4.69
Capital city status (N/(%))	42/(0.05)	–	–	–	–
C. Countries' attributes[a]					
Land area(ln)	163	5.35	16.64	13.67	1.84
Total population, 1000 (ln)	163	11.93	20.94	18.00	1.40
Per capita GDP (ln)	163	6.27	10.40	9.30	1.02
Percent urban	163	6.16	100.00	67.80	24.50
Regional breakup of countries (N/(%)):	163/(100.0)	–	–	–	–
West	25/(15.3)	–	–	–	–
Middle East	18/(11.0)	–	–	–	–
Africa	44/(27.0)	–	–	–	–
Eastern Europe	17/(10.4)	–	–	–	–
Asia	17/(10.4)	–	–	–	–
Other	42/(25.8)	–	–	–	–

[a]With available development attributes only

Appendix 2

Discriminant Analysis

In discriminant analysis, the linear combinations of predictors separating individual values of the dependent variable are called 'discriminant functions' and are defined as follows:

$$d_{ik} = b_{0k} + b_{1k}x_{i1} + \cdots + b_{pk}x_{ip},$$

where d_{ik} is the value of the kth discriminant function for the ith case, p is the number of predictors, b_{jk} is the value of the jth coefficient of the kth function, and x_{ij} is the value of the ith case of the jth predictor (IBM 2019).

Appendix 3

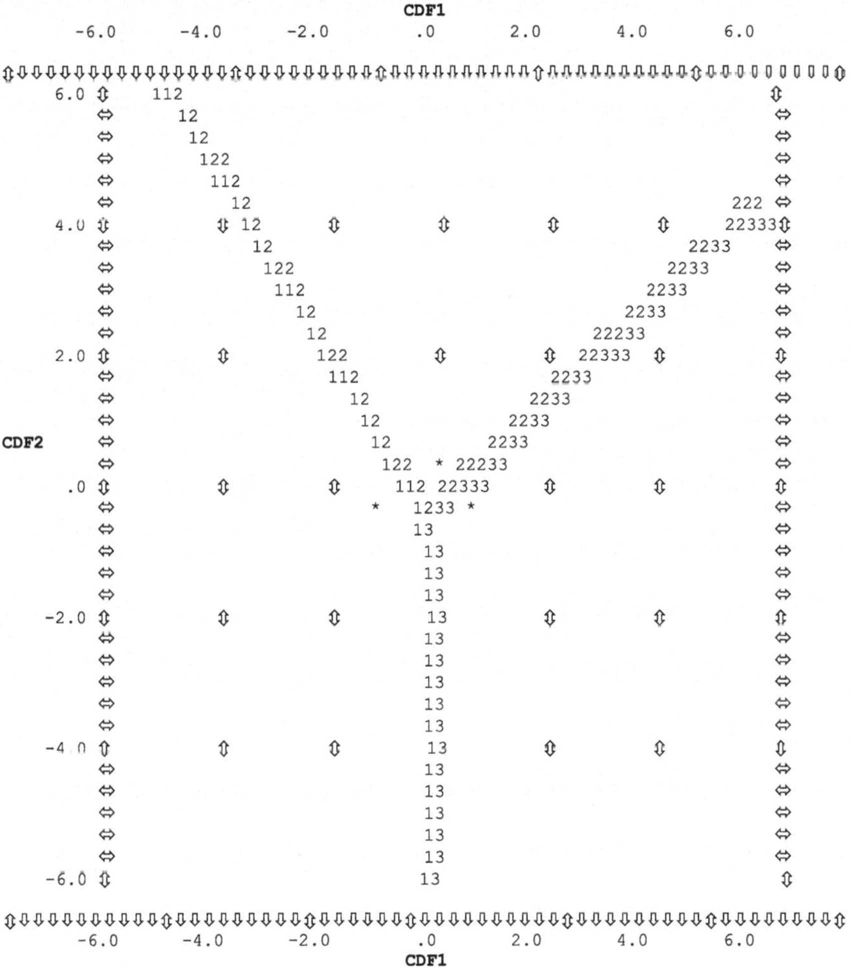

Territorial map of Canonical Discriminant Functions (CDFs) for world countries according to their *B*-ratio classifications (Groups 1, 2, or 3)

References

Agresti, A. (1996). *An Introduction to categorical data analysis*. Wiley Inc.

Babarović, I. (1979). Rural marginality and regional development policies in Brazil. In A. Kuklinski (Ed.), *Regional policies in Nigeria, India, and Brazil* (pp. 186–319). The Hague: Mouton Publishers.

Beckmann, M. (1968). *Location theory*. NY: Random House.

Behrens, K., & Robert-Nicoud, F. (2008). *Survival of the fittest in Cities: Agglomeration, selection, and polarisation*. CEPR Discussion Papers 7018.

Cacoullos, T. (Ed.). (1973). *Discriminant analysis and applications*. NY: Academic.

Cheshire, P. C., & Magrini, S. (2006). Population growth in European cities: Weather matters—But only nationally. *Regional Studies, 40*(1), 23–37.

Christaller, W. (1933; 1966 English edition). *Central places in Southern Germany*. Englewood Cliffs, NJ: Prentice Hall Inc.

Clark, T. N., et al. (2002). Amenities drive urban growth. *Journal of Urban Affairs, 24*(5), 493–515.

Cronon, W. (1991). *Nature's metropolis: Chicago and the Great West*. New York: WW Norton and Company.

Duranton, G. (1999). *Distance, land, and proximity, economic analysis and the evolution of cities*. Research Papers in Environmental and Spatial Analysis No. 53 (Department of Geography and Environment, London School of Economics).

ESRI. (2000–2017). *ESRI data and maps*. Redlands, CA: Environmental Science Research Institute.

Felsenstein, D., & Portnov, B. A. (Eds.). (2005). *Regional disparities in small countries*. Heidelberg: Springer.

Florida, R., Mellander, C., & Stolarick, K. (2008). Inside the black box of regional development—Human capital, the creative class and tolerance. *Journal of Economic Geography, 8*(5), 615–649.

Florida, R. L. (2002). *The rise of the creative class and how it's transforming work, leisure, community, and everyday life*. NY: Basic Books.

Fry, G. W. (1983). Bangkok: The political economy of a hyperurbanized primate city. *The Asian Journal Of Public Administration, 5*(1), 14–32.

Fujita, M., & Mori, T. (1996). The role of ports in the making of major cities: Self-agglomeration and hub-effect. *Journal of Development Economics, 49*, 93–120.

Fujita, M., & Mori, T. (2005). *Frontiers of the new economic geography*. KIER Working Papers 604, Kyoto University, Institute of Economic Research.

Fujita, M., Krugman, P., & Venables, A. J. (2001). *The spatial economy: Cities, regions, and international trade*. Cambridge, MA: The MIT Press.

Gallup, J. L., Sachs, J., & Mellinger, A. (1999). Geography and economic development. *International Regional Science Review, 22*, 179–232.

Geonames. (2007). *Geonames project database*. https://www.geonames.org.

Glaeser, E. L. (2005a). *Smart growth: Education, skilled work and the future of cold weather cities*. Policy Brief. John F. Kennedy School of Government, Harvard University.

Glaeser, E. L. (2005b). Urban *colossus: Why is New York America's largest city*. FRBNY Economic Policy Review.

Glaeser, E. L., & Saiz, A. (2003). *The rise of the skilled city*. Harvard Institute of Economic Research. Discussion Paper #2025/2003.

Glaeser, E. L., Kallal, H. D., Scheinkman, J. A., & Shleifer, A. (1992). Growth in Cities. *The Journal of Political Economy, 100*(6), 1126–1152.

Gotlieb, Y. (1996). *Development, environment, and global dysfunction: Toward sustainable recovery*. Delray Beach, Fl: St. Lucie Press.

Ha, J., & Howitt, P. (2007). Accounting for trends in productivity and RandD: A Schumpeterian critique of semi-endogenous growth theory. *Journal of Money Credit and Banking, 39*, 733–774.

Head, K., & Mayer, T. (2010). Gravity, market potential and economic development. *Journal of Economic Geography, 11*(2), 281–294.

Henderson, J. V. (1974). The sizes and types of cities. *The American Economic Review, 64*(4), 640–656.
Henderson, V., Kuncoro, A., & Turner, M. (1995). Industrial development in cities. *Journal of Political Economy, 103*, 1067–1090.
Hess, M. (2004). Spatial relationships? Towards a reconceptualization of embeddedness. *Progress in Human Geography, 28*(2), 165–186.
Huberty, C. J., & Olejnik, S. (2006). *Applied MANOVA and discriminant analysis* (2nd ed.). NY: Wiley-Interscience.
IBM. (2019). IBM SPSS statistics v.25Core system user's guide. Armonk, New York: IBM Corp.
Isard, W. (1956). *Location and space-economy; a general theory relating to industrial location, market areas, land use, trade, and urban structure* (1967; 5th ed.). Cambridge, MS: The MIT Press.
Jacobs, J. (1969). *The economy of cities*. NY: Random House.
Jefferson, M. (1939). The law of the largest city. *Geographical Review, 29*(2), 226–232.
Kennedy, S., & Kopp, M. (2000). *Understanding map projections* (pp. 110). ESRI Press.
Kolko, J. (2010). Urbanization, agglomeration, and co-agglomeration of service industries. In E. L. Glaeser (Ed.), *Agglomeration economics* (pp. 151–180). London: University of Chicago Press.
Krugman, P. (1993). First nature, second nature, and metropolitan location. *Journal of Regional Science, 33*(2), 129–144.
Krugman, P. (1995). *Development, geography and economic theory*. Cambridge, MS: The MIT Press.
Krugman, P. (1999). The role of geography in development. *International Regional Science Review, 22*(2), 142–161.
Laincz, C. A., & Peretto, P. F. (2006). Scale effects in endogenous growth theory: An error of aggregation not specification. *Journal of Economic Growth, 11*(3), 263–288.
Lösch, A. (1938; 1971 English Edition). *The economics of location*. New Haven and London: Yale University Press.
Madsen, J. B., Saxenaa, S., & Anga, J. B. (2010). The Indian growth miracle and endogenous growth. *Journal of Development Economics, 93*(1), 37–48.
Markusen, A. (2006). Urban development and the politics of a creative class: Evidence from the study of artists. *Environment and Planning A, 38*(10), 1921–1940.
Martin, R., & Sunley, P. (2006). Path dependence and regional economic evolution. *Journal of Economic Geography, 6*(4), 395–438.
Mellander, C., & Florida, R. (2009). Creativity, talent, and regional wages in Sweden. *Annals of Regional Science, 46*(3), 637–660.
Njoh, A. J. (2007). *Planning power: Town planning and social control in colonial Africa* (Chap. 4). London and New York: UCL Press.
North, M. A. (2009). A method for implementing a statistically significant number of data classes in the Jenks algorithm. In *Proceedings of the 2009 Sixth International Conference on Fuzzy Systems and Knowledge Discovery* (vol. 1, pp. 35–38).
Nunn, N., & Puga, D. (2009). *Ruggedness: The blessing of bad geography in Africa*. NBER Working Paper Series, Working Paper 14918. Cambridge, MA: National Bureau of Economic Research.
Poot, J. (2008). Demographic change and regional competitiveness: The effects of immigration and ageing. *International Journal of Foresight and Innovation Policy, 4*(1–2), 129–145.
Portnov, B. A., & Erell, E. (2001). *Urban clustering: The benefits and drawbacks of location* (p. 317). Aldershot: Ashgate.
Portnov, B. A. (2011). Does Zipf's law hold for primate cities? Some evidence from a discriminant analysis of world countries. *JahrbRegwiss (Review of Regional Studies), 31*, 113–129.
Portnov, B. A., & Schwartz, M. (2008). On the relativity of urban location. *Regional Studies, 42*(4), 605–615.
Portnov, B. A., & Schwartz, M. (2009a). On the Importance of 'location package' for urban growth. *Urban Studies, 46*(8), 1665–1679.

Portnov, B. A., & Schwartz, M. (2009b). Urban clusters as growth *foci*. *Journal of Regional Science, 49*(2), 287–310.
Portnov, B. A., Adhikari, M., & Schwartz, M. (2007). Urban growth in Nepal: Does location matter? *Urban Studies, 44*(5/6), 915–937.
Portnov, B. A., Reiser, B., & Schwartz, M. (2012). Does Gibrat's law for cities hold when location counts? *Annals of Regional Sciences, 48*, 151–178.
Puga, D. (1999). The Rise and Fall of Regional Inequalities. *European Economic Review, 43*(2), 303–334.
Redding, S., & Venables, A. J. (2004). Economic geography and international inequality. *Journal of International Economics, 62*(1), 53–82.
Rosenthal, F. (tr). (1967). *Ibn Khaldun. The Muqaddimah: An introduction to history*. NJ: Princeton University Press.
Salehyan, I., & Gleditsch, K. S. (2006). Refugees and the spread of civil war. *International Organization, 60*(2), 335–366.
Sassen, S. (1997). Electronic space and power. *Journal of Urban Technology, 4*(1), 1–17.
Sassen, S. (2000). Cities in the global economy. In R. Simmonds & G. Hack (Eds.), *Global city regions: Their emerging forms* (pp. 269–276). London and NY: Spon Press.
Storper, M., & Scott, A. J. (2009). Rethinking human capital, creativity and urban growth. *Journal of Economic Geography, 9*(2), 147–167.
Timberlake, L., & Tinker, J. (Eds.). (1985). *Africa in crisis; the causes, the cures of environmental bankruptcy*. London: Earthscan.
Von Thünen, J. H. (1826; 1966 English edition). *The isolated state*. Oxford: Pergamon Press.
Weber, M. (1921; 1986 reprint). *The city*. Glencoe, IL: Free Press.

Exploring Regional Variations in "Socio-Spatial" Interaction and Geographic Homophily Using Location-Sharing Services Data

Laurie A. Schintler, Rajendra Kulkarni, Kingsley Haynes, and Roger Stough

1 Introduction

Web 2.0 technologies and platforms, along with "smart" communications devices, enable us to share content, collaborate, and connect in unprecedented ways. Accordingly, we are now linked through a vast, complex, and dynamic system of digital networks that span the globe. These trends and developments raise new questions about the geography of digital socialization, and to what extent and under what circumstances distance is still pertinent in the Web 2.0 era. Research already done in this subject suggests that space is still very much relevant in many regards. In particular, geographic proximity affects the structure and dynamics of digital networks; social ties influence location preferences and activity choices; digital friends tend to co-locate in space, and there are distinct socio-spatial communities in digital networks. While we are gaining a better understanding of how digital interaction is intertwined with spatial interaction, much of the attention has been focused at national and international levels. Less is known about socio-spatial interaction within cities and regions, and specifically, whether or not there are sub-regional variations in these patterns of association.

In this chapter, we attempt to shed more insight into this topic. Specifically, using the Atlanta metropolitan statistical area as a case study, we develop a novel methodology to address the following research questions:

- To what extent is geographic homophily present at the sub-regional level?
- Is there a distance-decay effect present at this level? Specifically, is there an inverse relationship between distance and the similarity of locations visited by individuals and their social ties?

R. Stough: Deceased September 6, 2019.

L. A. Schintler (✉) · R. Kulkarni · K. Haynes · R. Stough
The Schar School of Policy and Government, George Mason University, Fairfax, VA, USA
e-mail: lschintl@gmu.edu

J. Thill (ed.), *Innovations in Urban and Regional Systems*,
https://doi.org/10.1007/978-3-030-43694-0_3

- Are there variations in the degree and nature of geographic homophily at the sub-regional level?

To examine these issues, we develop and use a methodology based on bipartite network modeling. More specifically, we formulate a bipartite graph that has locations as one set of vertices and individuals who visit those locations as the other set of vertices. From this network, we extract a one-mode network with locations as vertices and individuals common to those locations as edges. This network allows us to quantify the similarity (or dissimilarity) of locations based on those individuals who visit them. To examine whether or not a distance-decay effect is present, we formulate an additional set of networks, each of which contains only edges between adjacent locations based on different distance thresholds. With these networks, we use local indicators of spatial association, geographically weighted regression, and community detection techniques to assess regional variations in geographic homophily and socio-spatial interaction.

For this study, we draw upon location-sharing services (LSS) data, which is very rich in details about how people interact in physical and digital spaces. LSS platforms allow individuals to "check-in" to locations and then convey this information to their friends in real time. When an individual "checks-in" to a location, the latitude and longitude coordinates of the location visited, time and day of the visit, user-id, and comments provided by the user are recorded. Specifically, we use a sample of digital transactions produced by the LSS platform, Gowalla, from the Stanford Large Network Database (SNAP) (Cho et al. 2011). The full data set contains the locations and times of 6,442,890 "check-ins" for 196,591 Gowalla users for the period February 2009 to October 2010. The ego network for Gowalla users during this period was also available.

One concern about location-sharing services data, and more generally, volunteered geographic information (VGI), is that the information tends to be biased (Elwood et al. 2011). First, the users of such services often are not representative of the entire population. Second, users vary in the frequency with which they report information. At the same time, some studies have found that that bias in location-sharing services data is not as bad as some claim. For example, Cho et al. (2011) find that mobility patterns revealed by Gowalla and Brightkite "check-ins" are very similar to those based on data from cell phone transactions. While potential bias in the sample we use for this study is undoubtedly a concern, it is beyond the scope of our efforts to conduct a robust assessment of the veracity and quality of the data.

Following this introduction, we provide a brief synopsis of the literature on the geography of digital socialization. Next, we introduce our methodology based on bipartite network modeling. We then apply the methods to the Atlanta metropolitan area as an illustration. The paper concludes with some directions for future research.

2 Geography of Digital Socialization: A Review of the Literature

Geography is often a source of homophily, i.e., the tendency for people to pursue, engage with, or be attracted to others who are similar to themselves. In face-to-face communities, people who are located close to one another or who have similar location choice behavior have a higher propensity to connect and interact than those who are separated by space (McPherson et al. 2001). In digital networks, we see similar phenomena. There is homophily with respect to the location choices of digital friends and the locations visited by individuals reveal that there are strong social and spatial linkages (Pelechrinis and Krishnamurthy 2016). Further, geographic proximity increases the chance of friendship in digital networks (Onnela et al. 2011; Kaltenbrunner et al. 2011; Brown et al. 2012b; Takheteyev et al. 2012). However, there are variations in the strength of this association across different communication vehicles and platforms. For instance, Brown et al. (2012b) found that friendship in social media is less influenced by space than it is in digital location-sharing networks. As the authors point out, friendship in social media platforms often is based more on shared interests and conversational themes rather than location choices and preferences. It is generally the reverse for location-sharing communities. Friends in location-based networks tend to base their location choices on the recommendations and behavior of their social ties in such communities (Cho et al. 2011). The relationship between distance and probability of friendship in digital networks is nonlinear, although the precise nature of this nonlinearity is unknown.

At the sub-regional level, some studies indicate that geographic proximity has little influence on the probability of friendship (Wellman 1997). Others have found that there appear to be distinct distance thresholds beyond which space plays a critical role in dictating social ties. For example, Kaltenbrunner et al. (2011) find this cut-off point to be about 30 km for social media networks. On the other hand, Takheteyev et al. (2012) find that geographic proximity is an excellent predictor of friendship in digital communities, especially for people located within 100 km of one another, and interestingly, beyond 1000 km the number of connections is higher than what would be expected at random (Takheteyev et al. 2012).

The association between geographic proximity and intensity of social interaction has also been examined. Wellman (1997) found that local contacts, i.e., those who live within five kilometers of one another, tend to call each other more often than more distant friends. For online social networks, however, spatial proximity appears to be only weakly associated with the degree of interaction (Onnela et al. 2011). Further, more active online social media users, i.e., those positioned in the core of the network, have social ties that are at a lesser distance than those who are on the periphery (Volkovich et al. 2012). As the authors point out, age is likely the primary reason for this finding. Less active people tend to be older, and in contrast to younger people, they have ties that are geographically dispersed and go back further in time.

Space also plays a role in the formation and behavior of communities in digital networks. Such communities tend to be concentrated in space, although there appear other factors at play, such as those related to language (Expert et al. 2011). There are also temporal features of spatial digital communities. For example, they tend to shift during the day based on mobility patterns within regions (Walsh and Pozdnoukhov 2011). Geography also influences the size of digital communities. Onnela et al. (2011) find that the geographic span of a community is positively correlated with its size. Further, members of larger groups tend to be spatially more dispersed than those in smaller communities. However, the relationship is nonlinear. More specifically, the geographic span of a community gradually increases with membership up to a group size of about 30 people, at which point the association becomes much stronger. In other words, the tendency for individuals within communities to be geographically proximate is stronger for smaller rather than larger groups. Brown et al. (2012a) introduce the idea of a "place-focused" community, which they define as a group of individuals who have shared locations. They find a negative association between the size of such communities and the degree to which members co-locate in space, but this appears to only pertain to relatively small communities, i.e., comprised of less than 20–30 people. Both within and across cities, place-based communities have distinct spatial boundaries; however, the degree and nature of socio-spatial delineation and stratification vary from one region to another (Lawlor et al. 2012). Our work builds on that of Lawlor et al. (2012) in that we examine regional patterns of socio-spatial interaction based on the extent to which digital friends are similar in terms of their location choices and behavior but add to this by explicitly incorporating information on social ties. Moreover, the focus of our study is on the sub-regional level, i.e., within a city.

3 "Socio-Spatial" Bipartite Network Topology

A bipartite network (or two-mode network) contains vertices that are partitioned into two disjoint sets. It includes only connections from vertices in one set to vertices in the other set. In other words, there are no connections within sets (Freeman and Duqueene 1993). One of the earliest applications of bipartite theory was the Davis study, which looked at the participation of 15 women in 14 events in the South (Davis and Gardner 1941). Since then bipartite network modeling has been applied extensively in various contexts, including examining socio-spatial interaction and related dynamics (see, e.g., Pelechrinis and Krishnamurthy 2016).

For this study, specify a bipartite graph $G = (L, S, E)$, where $L = \{l_1,\ldots, l_n\}$ is a set of locations, $S = \{s_1,\ldots, s_n\}$ is the set of social ties (or individuals) associated with those locations, and $E = \{l_{ij}, s_{ks}\}$ is the collection of edges between locations and individuals. In this model, an edge exists between a location and an individual if either the person visited the location or he/she is an immediate social tie of someone who went there. We do not consider the frequency with which an individual visits a single location, but rather only if he/she has been present at that place at any time. Thus, we formulate an un-weighted 2-mode network. One can think of this network as a "spatial" ego network, which captures the immediate social ties of locations rather than those of individuals. We refer to this structure as a "socio-spatial" bipartite network topology.

From any bipartite network, it is possible to extract two structurally equivalent one-mode networks (Wasserman and Faust 1994). For the bipartite network, we specify one such a network has individuals as vertices and locations common to those individuals as edge weights. The other network, which we utilize for the purpose of this study, has locations as vertices and individuals common to locations as edges. The un-weighted version of this network is described by a binary adjacency matrix, A_{ij}, where a value of one for any pair of locations indicates that they have at least one person in common, and zero otherwise. We can also specify a weighted network, which is described by a real adjacency matrix A^w_{ij} in which the cells indicate the number of social ties each pair of locations has in common.

For the weighted network, we normalize the edge weights using the following formula:

$$e_{ij}^{\text{Norm}} = \frac{e^w_{ij}}{\sqrt{l_{ii}} * \sqrt{l_{jj}}} \tag{1}$$

where e^w_{ij} is the number of social ties location i and j have in common, l_{ii} is the number of individuals associated with location i, and l_{jj} is the number tied to location j. Thus, each edge weight represents the degree of association between locations i and j, controlling for the number of social ties associated with each location. This normalization helps to minimize skew in the distribution of social ties and location choices.

It is also possible to extract a subgraph from the full one-mode network that contains only edges between those locations that are contiguous to one another. This constrained network allows us to examine how similar locations are to one another within a certain geographic range (e.g., local vs. regional), and to assess whether or not there is a distance-decay effect where the similarity of locations declines with distance. In this paper, we specify four networks based on 25th, 50th, 75th, and 100th percentile distance thresholds. We use percentile distances given that the distribution of distances in our sample is highly skewed, with many distances being relatively short and just a few being long distance. Of course, the network based on 100th percentile distance threshold is just the full 1-mode network.

4 Case Study: "Socio-Spatial" Interaction in Atlanta

We now apply the bipartite network modeling approach to a sample of Gowalla location-sharing services data for Atlanta for June in 2010 to examine the research questions laid out in the introduction of this paper. This sample comprises 207,441 "check-ins" and the social ties of 10,940 users. We were limited to using only one month of data because of the computational constraints of the software we used for the network analysis.

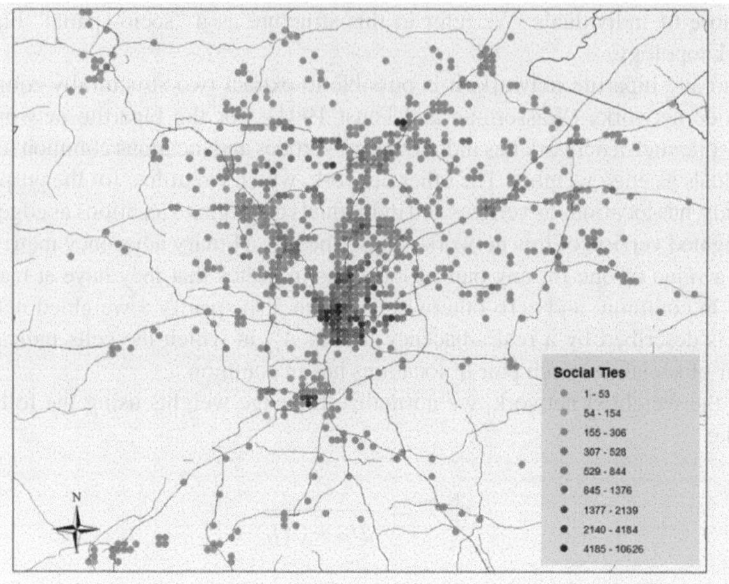

Fig. 1 Spatial ego network for Atlanta based on Gowalla "check-ins" for June 2010

Figure 1 shows the spatial ego network for Atlanta. Eight hundred ninety-four locations had at least one Gowalla "check-in" during June 2010. Most locations had more than one visitor. Not surprisingly, the highest numbers of social ties were at locations in the center of the city and around the Hartsfield-Jackson Atlanta International Airport (located in the southwest quadrant of the map).

4.1 Measuring the Degree of Geographic Homophily and Distance Decay Effects

From the bipartite network constructed for Atlanta, we extracted the one-mode network which has locations as vertices and individuals common to locations as edges. Again, for the weighted network, we normalize the edge weights using Eq. (1). Table 1 reports summary statistics for each network. Un-weighted density is the total number of ties in relation to the maximum number of pairs possible, while weighted density is simply the summation of edge weights. Using these densities, we can calculate the average edge weight $E(e_{ij}^{w})$ for each of the networks.

Interestingly, the network based on the 50th percentile distance has the highest average edge weight. In other words, locations are most similar in terms of visitors and their social ties within this range of distance. Beyond this threshold, average correlation declines suggesting that there may be a distance-decay effect but only

Table 1 Network statistics by percentiles for distance

Metric	25th	50th	75th	100th
Vertices, V	880	886	889	894
Un-weighted density, D	50,046	99,627	150,832	201,420
Weighted density, D_{w}	5080	15,863	19,290	23,628
Average edge weight, $E(e_{ij}^{\mathrm{w}})$	0.102	0.159	0.128	0.117
Un-weighted normalized density, D_{Norm}	0.1294	0.2541	0.3821	0.5046
Density weighted against observed maximum, $D_{\mathrm{w},Norm}$	0.0183	0.0295	0.0398	0.0482
Hypothetical maximum weighted density, $D_{\mathrm{w}}^{\mathrm{Max}}$				399,171
Weighted density in relation to hypothetical maximum, $D_{\mathrm{w},Norm}^{\mathrm{Max}}$	0.0127	0.040	0.048	0.059

beyond a specific geographic range. To be able to assess this more accurately, we would have to consider a much more fine-grained distribution of distance thresholds.

Table 1 also reports the normalized densities for both sets of un-weighted and weighted networks. Normalized density provides a measure of the degree of geographic homophily within a region. Because the measure controls for the number of locations, we can use it to draw comparisons across different metropolitan areas or regions. However, it should be noted that normalized densities are not comparable between networks defined by different distance thresholds, as normalized density can only increase or stay the same as the range of distance considered expands. To address this problem, we formulate an "adjusted" normalized density, which normalizes on the maximum weighted density possible given the total number of locations that have at least one individual associated with them in the study region. The following equation gives the maximum density for any network:

$$D_{\mathrm{w}}^{\mathrm{Max}} = \frac{n(n-1)}{2} \tag{2}$$

where n is the number of nodes. In Atlanta, n is equal to 894. For each distance threshold, we can then compute the actual weighted density in relation to the maximum density possible, given the total number of locations:

$$D_{\mathrm{w,Norm}}^{\mathrm{Max}} = \frac{2D_w}{n(n-1)} \tag{3}$$

where D_{w} is weighted density. Note that we can use weighted density since the edge weights, based on (1), have an upper limit of one.

From the normalized densities calculated using Eq. (1), one can see that the degree of similarity between locations increases with distance. This finding seems to contradict what we found in terms of average correlations. However, in our computation of average correlations, we did not include locations with no "check-ins." The inclusion of this in the calculations would yield different correlations. Further, the rate at which the adjusted normalized density increases is not linear. Between the 25th and 50th percentile, density jumps over threefold but then tapers off beyond this point.

5 "Socio-Spatial" Community Structure in Atlanta

A community detection algorithm attempts to segment a network into groups of densely connected nodes or edges separated by sparse connections. There are many different types of community detection algorithms (Fortunato 2009). For this study, we use Blondel's modularity optimization algorithm, which has several appealing properties. First, it yields information on the hierarchical structure of communities. Second, it is computationally efficient and thus can be applied to relatively large-scale networks. Lastly, it produces higher quality communities than other algorithms (Blondel et al. 2008).

Blondel community detection is based on optimization of a modularity function. For an un-weighted network, this function is:

$$Q = \frac{1}{2m}\sum_{i,j}\left[e_{ij} - \frac{d_i d_j}{2m}\right]\delta_{c_i c_j} \tag{4}$$

where e_{ij} is the edge between node i and j, $d_i = \sum_j e_{ij}$, $d_j = \sum_j e_{ji}$, $m = \frac{1}{2}\sum_{ij} e_{ij}$, c_i is the community node i is assigned to and c_j the community node j belongs to. The Kronecker delta $\delta(u, v)$ is equal to one if $u = v$ and zero otherwise. This function measures the density of edges within communities in relation to those between communities. It has a lower bound of -1 and upper bound of 1, where a value closer to one characterizes a more well-defined, sophisticated community structure (Blondel et al. 2008).

We now apply this community detection algorithm to each of the un-weighted networks. Table 2 reports summary statistics for the community structures detected for each network. The network that uses the 25th percentile distance threshold produced

Table 2 Summary statistics for un-weighted Blondel community detection

Percentile	Number of hierarchies	Number of communities	Maximum size of community (nodes)	Shannon entropy	Normalized Shannon entropy	Modularity
25	4	11	267	0.710	0.682	0.382
50	3	8	328	0.547	0.606	0.206
75	3	7	353	0.499	0.590	0.109
100	3	5	272	0.605	0.866	0.083

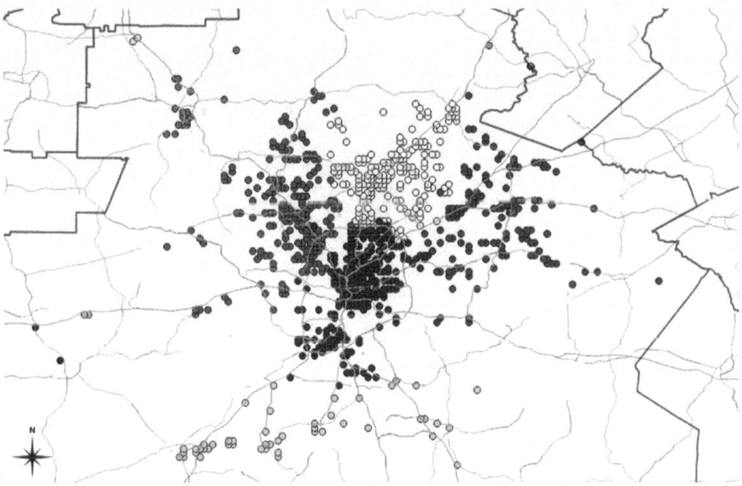

Fig. 2 Blondel communities using network based on 25th percentile distances

the most sophisticated community structure in terms of numbers of hierarchies, the number of communities, and modularity. We also computed both un-normalized and normalized Shannon information entropy for each set of communities, wherein each case pi in the entropy formulation corresponds to the fraction of nodes in the community *i*. The results indicate that the communities detected for the networks based on the 50th and 75th percentile distances are the least diverse. In other words, for these networks, there are a few communities that contain the majority of the nodes.

We mapped the communities to examine whether or not they are concentrated in space, and if their spatial boundaries shift given different distance thresholds. From the maps displayed in Figs. 2, 3, 4 and 5, one can see clearly that the communities are spatially compact, but that the degree of spatial association declines as the distance threshold is relaxed. However, even though the network based on the 100th percentile distance threshold exhibits the most spatially dispersed pattern, there are still some locations where parts of communities are concentrated. It should also be pointed out that the 50th percentile community structure shows the most significant degree of geographic homophily, as evidenced by the fact that there are only a few communities with the majority of nodes, and these communities are spatially concentrated.

6 Regional Variations in Geographic Homophily

One shortcoming of the Blondel community detection algorithm, as well as many others, is that community structure is based only on assortative relationships. That is, it does not consider disassortative connections. The similarity (or dissimilarity) of locations based on visitors and the social ties of the visitors can be assessed through

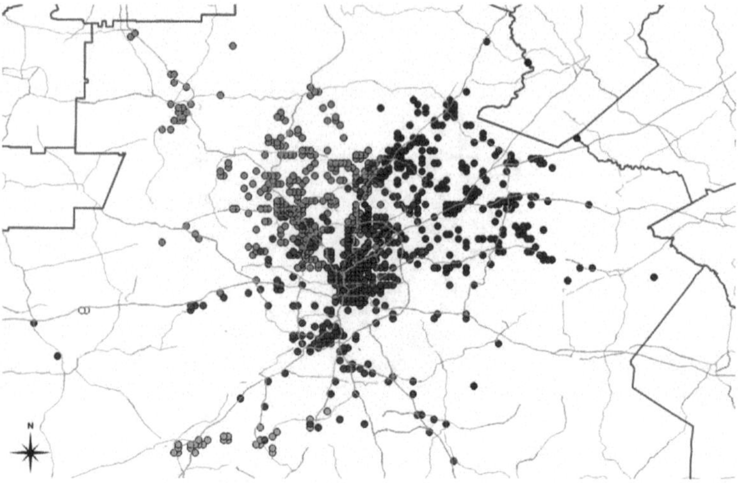

Fig. 3 Blondel communities using network based on 50th percentile distances

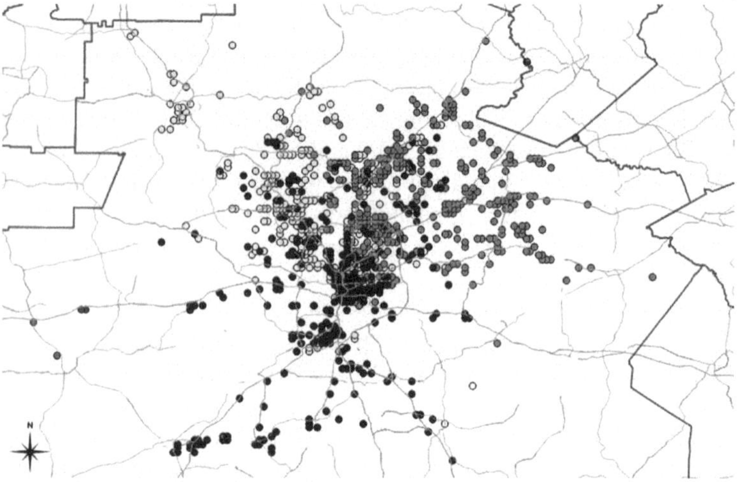

Fig. 4 Blondel communities using network based on 75th percentile distances

the use of local indicators of spatial association (LISA). These statistics also enable us to measure the spatial extent of geographic homophily.

For each location, we first computed the average degree of correlation based on the average edge weights ($E(e_{ij}^{w})$) for each set of locations associated with the different networks and distance thresholds. In each case, we used eight nearest neighbors to determine the contiguity of locations. Table 3 reports the results of this part of the analysis.

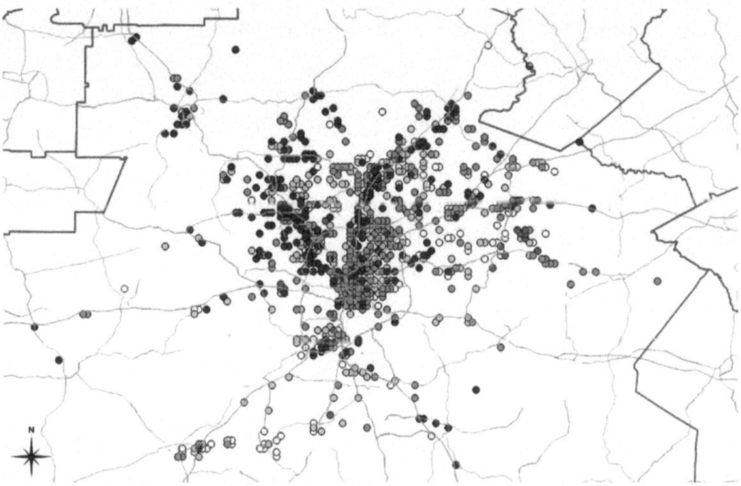

Fig. 5 Blondel communities using network based on 100th percentile distances

Table 3 Moran's I statistic, and counts and percentages by LISA categories for average correlation, $Avcorr_i$

	25th	50th	75th	100th
Moran's I statistic	0.4477***	0.3648***	0.2417***	0.1345*
LISA categories				
High-high	74 (8.52%)	57 (6.56%)	41 (4.72%)	37 (4.26%)
Low-low	102 (11.74%)	112 (12.89%)	99 (11.39%)	94 (10.82%)
High-low	18 (2.07%)	15 (1.73%)	12 (1.38%)	17 (1.96%)
Low–high	12 (1.38%)	12 (1.38%)	16 (1.84%)	17 (1.96%)
Insignificant ($p > 0.1$)	663 (76.29%)	673 (77.45%)	701 (80.67%)	704 (81.01%)

Significant at $^*p < 0.05$, $^{**}p < 0.01$, and $^{***}p < 0.001$

The results of the spatial statistical analysis are consistent with the findings reported in the previous section. More specifically, the LISA statistics provide additional evidence of geographic homophily. That is, within the Gowalla social network, the same people tend to visit the same locations. Moreover, this relationship exhibits a certain degree of positive spatial association. For each distance threshold, there are statistically significant spatial clusters of locations that share common visitors. However, in each case, the majority of locations exhibit either a negative association or no statistically significant association. Further, as the distance threshold is relaxed, the degree of spatial association declines.

Next, we use geographically weighted regression to examine whether or not there are regional variations in the relationship between distance and spatial co-location. Geographically weighted regression (GWR) is different from the basic ordinary least squares (OLS) model in that it incorporates information on each point in space.

Accordingly, it allows for parameter estimates to vary by location, as given by:

$$Y_i = a_{i0} + \sum_{k=1}^{K} X_{ik} + \varepsilon_i \tag{5}$$

where i is a location in space and k is the number of independent variables (Fotheringham and Charlton 1997a, b).

We estimate the following GWR model:

$$\text{Avcorr}_i = a_{i0} + \sum_{j=1}^{J} a_{ij}\text{Avdist}_i + \varepsilon_i \tag{6}$$

where Avcorr_i is the average correlation (or average edge weight $E(e_{ij}^{w})$) at location i, and Avidist_i is the average distance at location i. We use information from the 100th percentile distance 1-mode network to estimate this model. Further, for the estimation, we use an adaptive kernel with the following standard weighing scheme:

$$w_{ij} = \exp\left(-b^{-1}d_{ij}^2\right) \tag{7}$$

where b refers to bandwidth and d_{ij} is the distance of point i from j.

Table 4 displays the results for both the GWR models and the global OLS model. The AIC statistic for the geographically weighted regression (GWR) is lower than that of the ordinary least squares (OLS) model, indicating that of the two GWR provides the best fit. Additionally, the adjusted R2 for the GWR is much higher than that of the OLS model. This may suggest that the relationship between distance and spatial co-location is non-stationary, although further analysis is needed to verify that.

The OLS model indicates the similarity of locations tends to decline with distance, while the GWR model indicates further that are regional variations in the strength and nature of this relationship. There also are differences in the localized goodness of fits (R^2), which range between 0.000 and 0.6352. Further, the parameter estimates

Table 4 Results of ordinary least squares (OLS) and geographically weighted regression (GWR)

	OLS	GWR			
		Min	Max	Mean	Median
Avdist	−0.2000***	−3.3847	0.5859	−0.5648	−0.4325
Constant	0.2089***	−0.0900	1.1998	0.3102	0.2545
R2	0.0356	0.0000	0.6352	0.1410	0.0764
Adjusted R2	0.0342	0.4402			
AIC	−1089.43	−1454.2784			

Significant at $^{*}p < 0.05$, $^{**}p < 0.01$, and $^{***}p < 0.001$

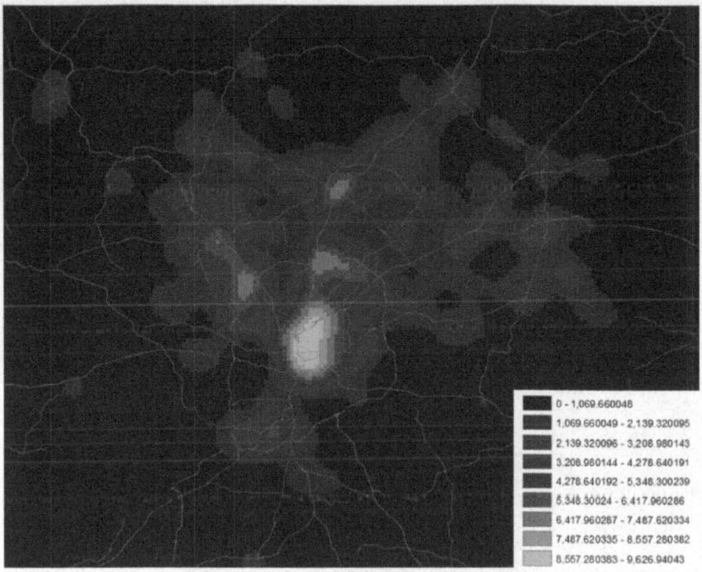

Fig. 6 Kernel density surface of un-weighted points from Fig. 1

for Avdist_i range from −3.847 to 0.56, which indicates that a distance-decay effect is not present at all locations. To visualize this, we generated kernel density surfaces of the local R^2 and the parameter estimates for Avdist_i. As a point of reference, we also generated the kernel density surface for the un-weighted distribution of "check-ins." These maps are shown in Figs. 6, 7 and 8.

From Figs. 7 and 8, one can see that the distribution of local R2s, as well as that of the parameter estimates for Avdist_i are quite different from what would be expected based on a uniform assignment of either value to the point distribution. The parts of the city where the relationship between distance and similarity of locations are the highest are in the suburbs. In these locations, the similarity of locations to other locations in the region declines with distance, with the highest distance-decay effect in the suburbs in the northwest part of the Atlanta metropolitan. In most parts of the region, there is a distance-decay effect. The exception is the center of the city, where the parameter estimates for Avdist_i are positive. The goodness of fit is also lowest in this part of the city; thus, results should be interpreted with caution.

7 Conclusion

The primary purpose of this paper is to introduce a methodology for examining the extent and nature of geographic homophily and related patterns of socio-spatial interaction at the sub-regional level. Our conclusions are consistent with what other

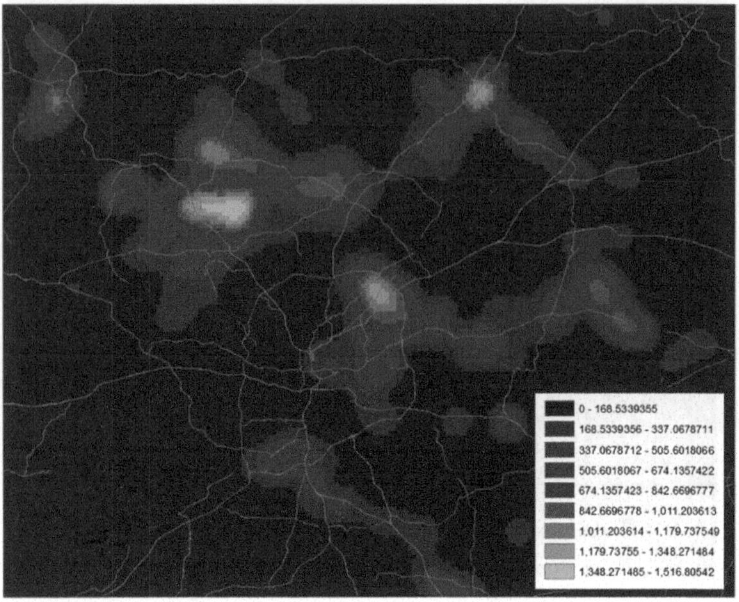

Fig. 7 Local R2's from geographically weighted regression using 100th percentile distance network

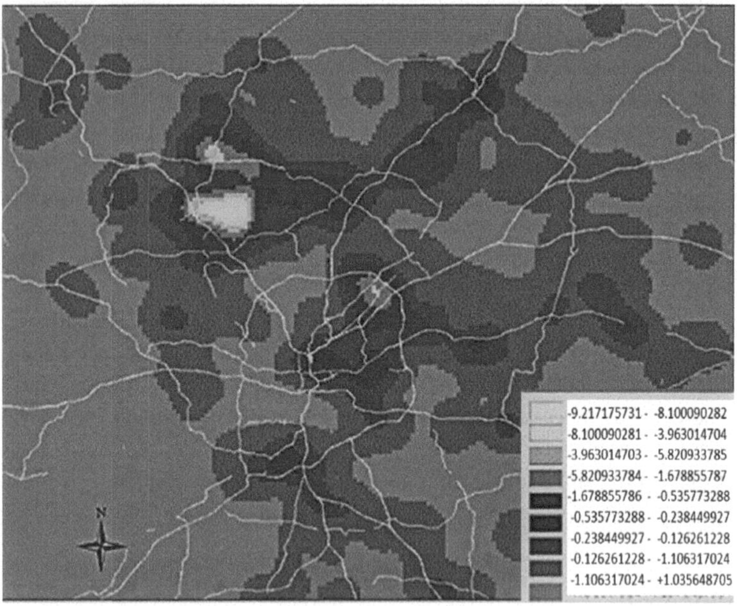

Fig. 8 Local parameter estimates for $Avdist_i$ from geographically weighted regression using 100th percentile distance network

studies have found regarding the role of space in digital networks. Specifically, distance does matter. Even though digital social networks like Gowalla are global in nature, their members tend to co-locate in space, and co-location tends to be highly localized. Moreover, there are regional variations in these patterns of association.

The analysis raises some questions that merit further examination. First, are the findings of this study specific to Atlanta, or are they generalizable to other metropolitan areas? Second, are the conclusions consistent with what we would find using other types of data, e.g., cell phone transactions? Further, to what extent are digital socio-spatial communities based on geographic homophily versus other factors, e.g., interests or demographics of group members, or even the underlying spatial configuration of land use and activities in the first place? Lastly, more research is needed to understand better the extent and nature of bias in data produced by location-sharing services and to develop techniques for managing the nuances of the bias.

References

Blondel, V. D., Guillaume, J. L., Lambiotte, R., & Lefebvre, E. (2008). *Fast unfolding of communities in large networks*. arXiv.org Physics, arXiv: 0803.0476.

Brown, C., Nicosia, V., Scellato, S., Noulas, A., & Mascolo, C. (2012a). The importance of Being placefriends: Discovering location-focused online communities. WOSN '12, August 17, 2012, Helsinki, Finland.

Brown, C., Nicosia, V., Scellato, S., Noulas, A., & Mascolo, C. (2012b, May). Where online friends meet: Social communities in location based networks. In *Sixth International AAAI Conference on Weblogs and Social Media*.

Cho, E., Myers, S. A., & Leskovec, J. (2011). Friendship and mobility: User movement in location-based social networks. In *ACM SIGKDD International Conference on Knowledge Discovery and Data Mining (KDD)*. San Diego, California, USA.

Davis, A., & Gardner, B. D. (1941). *Deep South*. Chicago: The University of Chicago Press.

Elwood, S., Goodchild, F. G., & Sui, D. Z. (2011). Researching Volunteered geographic information: Spatial data, geographic research, and new social practice. *Annals of the Association of American Geographers, 102*(3), 571–590.

Expert, P., Evans, T. S., Blondel, V. D., & Lambiotte, R. (2011). Uncovering space-independent communities in spatial networks. *Proceedings of the National Academy of Sciences, 108*(19), 7663–7668.

Fortunato, S. (2009). *Community detection in graphs*. arXiv: 0906.0612 [physics.soc-ph].

Fotheringham, A. S., Charlton, M., & Brunsdon, C. (1997a). Measuring spatial variations in relationships with geographically weighted regression. In: M. M. Fischer & A. Getis, (Eds.) *Recent developments in spatial analysis* (pp. 60–82). London: Springer-Verlag.

Fotheringham, A. S., Charlton, M., & Brunsdon, C. (1997b). Two techniques for exploring non-stationarity in geographical data. *Geographical Systems 4*, 59–82.

Freeman, L., & Duqueene, V. (1993). A note on colorings of two mode data. *Social Networks, 15*, 437–441.

Kaltenbrunner, A., Scellato, S., Volkovich, Y., Laniodo, D., Currie, D., Julemar, E. J., & Mascolo, C. (2011). Far from the eyes, close on the web: Impact of geographic distance on online social interactions. WOSN'12, August 17, 2012, Helsinki, Finland.

Lawlor, A., Coffey, C., McGrath, R., & Pozdnoukhov, A. (2012). *Stratification structure of urban habitats*. Working paper, National Centre for Geocomputation, National University of Ireland Maynooth.

McPherson, M., Smith-Lovin, L., & Cook, J. M. (2001). Birds of a feather: Homophily in social networks. *Annual Review of Sociology 27*, 415–444.

Onnela, J.-P., Arbesman, S., González, M. C., Barabási, A.-L., & Christakis, N. A. (2011). Geographic constraints on social network groups. *PLoS ONE 6*(4), e16939. https://doi.org/10.1371/journal.pone.0016939.

Pelechrinis, K., & Krishnamurthy, P. (2016). Socio-spatial affiliation networks. *Computer Communications*, *73*, 251–262.

Takheteyev, Y., Gruzd, A., & Wellman, B. (2012). Geography of Twitter networks. *Social Networks 34*, 73–81.

Volkovich, Y., Scallato, S., Laniado, D., Mascolo, C., & Kaltenbrunner, A. (2012). The length of bridge ties: Structural and geographic properties of online social interactions. *Association for the Advancement of Artificial Intelligence* (www.aaai.org).

Walsh, F., & Pozdnoukhov, A. (2011). *Spatial structure and dynamics of urban communities*. Working paper, National Centre for Geocomputation, National University of Ireland Maynooth.

Wasserman, S., & Faust, K. (1994). *Social network analysis: Methods and applications*. Cambridge, MA: Cambridge University Press.

Wellman, B. (1997). An electronic group is virtually a social network. *Culture of the Internet, 4*, 179–205.

Accessibility of Public Employment Services Within Large Urban Areas: A Spatial Approach

Patricia Suárez, Matías Mayor, and Begoña Cueto

1 Introduction

The Spanish labor market has experienced great change since the economy began to deteriorate in 2007. The unemployment rate has increased from 8% in 2007 to 24% in the second quarter of 2012, and more than 3 million jobs have been eliminated. Within this economic context, the Spanish government has undertaken several reforms aimed at increasing labor market flexibility and the likelihood of employment for job seekers through the strengthening of public employment services and active labor market policies (ALMPs). Although Law 35/2010 of 17 September 2010 regarding Urgent Measures for Labor Market Reform aimed to improve job brokering by allowing private employment agencies to compete with public employment services (PES), further reforms in both 2011 and 2012 have emphasized the need for PES to begin providing individual employment pathways to the unemployed.

In addition, it must be taken into account that regional labor disparities are rather large and persistent in Spain; hence, they must be addressed via active regional labor market policies. In 1998, the Spanish government began to decentralize public employment services among the autonomous communities (NUTS 2 regions), which were granted complete authority to enact ALMPs. Each region adopted a needs-based approach which brought in better management of the available resources and adapted employment and training programs to the features of the particular labor market and unemployed population profile.

The aims of this chapter are twofold. First, we analyze the spatial accessibility of employment offices in the largest urban areas of Spain, taking into consideration the

P. Suárez (✉) · B. Cueto
Department of Economics, Faculty of Economics and Business, University of Oviedo, Avda. Del Cristo, s/n., 33006 Oviedo, Spain
e-mail: suarezcpatricia@uniovi.es

M. Mayor
Department of Applied Economics, Faculty of Economics and Business, University of Oviedo, Avda. Del Cristo, s/n., 33006 Oviedo, Spain

J. Thill (ed.), *Innovations in Urban and Regional Systems*,
https://doi.org/10.1007/978-3-030-43694-0_4

varying size of catchment area sizes for employment offices in Spain's urban areas and the spatial distribution of urban unemployment in 2011. Second, we explore whether accessibility to employment service offices has any effect on urban unemployment rates.

The chapter defines the concept of accessibility as the potential for an individual to request employment services. As public employment services are essentially aimed at meeting the demand for job search assistance, any improvements in accessibility should increase the chance of a job seeker to move from unemployment to finding a job.

The remainder of the chapter is organized as follows. Section 2 delineates the operation of public employment services in Spain and the role of PES in the Europe 2020 Strategy (European Commission 2010b). Section 3 describes the primary urban areas in Spain, the spatial distribution of unemployment and accessibility to employment offices in 2011. Section 4 is devoted to the spatial model, in which spatial considerations have been included so as to capture the connections between large urban areas. Finally, Sect. 5 provides a conclusion to the chapter.

2 PES in Spain and the Europe 2020 Strategy

Theories about the relationship between accessibility and space originate at least as far back as with Ihlenfeldt and Sjoquist (1990). Since then voluminous literature regarding job access has emphasized the importance of space in labor markets. In this chapter, we seek to assess the ease with which an unemployed worker can access the assigned employment office, as unemployment rates are affected not only by access to jobs, but also by access to counseling, information, and training.

Since geographical and social spaces are interrelated, the importance placed upon accessibility by local and national labor market agencies is key in the performance of any labor market. This is particularly relevant in the case of Spain, where responsibility for labor market policies has been distributed across different levels of government. Autonomous communities do not have authority to enact passive labor market policies, although they do on active labor market policies. Following the decentralization of active labor market policies, regional PES have become not only job brokers but also de facto labor market regulators (Suárez and Mayor 2012). For this reason, an explicit spatial perspective is so essential in today's decentralized framework. As a matter of fact, accessibility plays an important role in policymaking, insofar as any improvement in the accessibility to an employment service office increases the likelihood of employment for its assigned unemployed workers.

The Europe 2020 Integrated Guidelines (European Commission 2010a) assign a central role to public employment services, namely to bring the 20–64 employment rate for women and men aged 20–64 to 75% by 2020. PES are largely responsible for the development of two specific guidelines: (1) increase male and female labor force participation rate, reduce structural unemployment, and promote job quality and (2) develop skilled workforces that can respond to employer needs and promote lifelong learning.

According to these guidelines, PES play an important role in activation and matching and they should, therefore, be strengthened with personalized services and active and preventive labor market measures at an early stage. That is, PES should develop the provision of tailored employment assistance, identify the individual job seeker's needs, and see to the appropriate staffing of services and programs.

In the same line, the strategy stresses the importance of matching knowledge and skill supply with present and future labor market demand. Since all PES in Europe offer training for job seekers, PES offices must work closely with employers to become knowledgeable about which skills are in demand.

The role of PES varies widely among countries and regions of Europe. In some cases, their main task for the future is to further develop service and quality standards and to better coordinate all levels in order to ensure equal services for the citizens all over the country.

Over the past ten years, labor market reforms in Spain have strengthened the role of public employment services in the development of active labor market policies. Royal Decree Law 3/2011, of 18 February, on Urgent Measures for Employability Enhancement and Reform of Active Labor Market Policies has reformed ALMPs and reinforced the role of public employment services in Spain. The aim of this law is to develop a model for providing customized attention to the unemployed, based upon individually tailored employment pathways.

The implementation of an individualized employment pathway requires intense monitoring. According to the law, job counselors should begin by assessing the employment potential of the individual and making recommendations for participation in the program based upon this information. They should periodically review whether the job seeker is complying with the pathway requirements and is suitable for re-employment. By regulation (Royal Decree Law 3/2011), the following services are to be provided at PES offices[1]:

- Initial assessment of the job seeker's profile, needs, and goals through personal interviews.
- Information about the labor market, suitable vacancies, hiring incentives, and support to entrepreneurs.
- Design, creation, and implementation of individualized employment pathways, including information about employment and self-employment. Employer and institutional contacts that may facilitate labor market insertion.
- Professional training and assessment of competencies and skills.
- Information about unemployment benefits and subsidies.

Consequently, job seeker accessibility to public employment services should be a priority of Spanish labor market policies, especially in areas with a high concentration of unemployment.

[1]Services addressed to employed people and employers are also to be provided at PES offices.

3 The Spatial Organization of Labor Markets in Large Urban Areas

3.1 Data

The definition of an urban area in this chapter is based upon the Statistical Atlas of Urban Areas (Spanish Department of Housing 2006) and the Urban Information System Web site, which in turn uses the most up-to-date information from the census and municipal register. In this study, Census 2001 and Municipal Register 2009 are used.

Accordingly, we have defined 85 large urban areas containing one or more municipalities, encompassing a total of 747 municipalities (9.2% of all Spanish municipalities). Of these 85 urban areas, 21 are composed of just one municipality; the remaining 64 urban areas are composed of more than one municipality. These urban areas contain 32 million people or 68% of the total population in Spain; in terms of land area, these urban areas account for approximately 9.6% of the total. Also, these 85 urban areas represent relevant urban agglomerations which can be used as reference points in the analysis of urban labor markets in Spain. It should be noted that within the provinces of Soria and Teruel, the 2010 population is below 50,000 inhabitants, yet we have considered them because they are provincial capitals and, therefore, serve as urban centers. Figure 1 displays municipalities which comprise the 85 large urban areas delineated for this study.

In 2011, the number of registered job seekers in Spain was 4,257,160, of whom 2,951,344 (69.3% out of total) lived in a municipality belonging to one of the 85 large urban areas. The 85 urban areas do not completely encompass the provinces in

Fig. 1 Municipalities which comprise the 85 large urban areas

which they are located. In other words, non-urban municipalities are not included in urban area delineation. For example, in the province of Huesca, only 14.5% of the unemployed workers in the province reside in the urban area of the province; among the provinces included in this study, Huesca Province has the lowest percentage of its unemployed workers residing in the urban area. This is in contrast to the autonomous community of Madrid, in which 93.6% of the unemployed workers in the province reside in the urban area—the highest percentage for any province in the study.

Among the 85 large urban areas, Sanlucar de Barrameda, Cadiz, is the urban area with the highest unemployment rate, at 26.8% followed by Elda-Petrer in Alicante at 15.7%. By contrast, the urban area of Huesca has the lowest unemployment rate (3.8%).

Figure 2 displays the ten urban areas with the highest growth in unemployment between 2006 and 2011. Castellon de la Plana, Castellon, nearly tripled the number of registered job seekers in five years. Unfortunately, this is hardly an exception; experienced a significant increase in the number of unemployed workers over the same period. In contrast, the urban areas with lower growth in unemployment over this period are shown in Fig. 3.

The growth in registered job seekers and the rising overcrowding at public employment offices call for an assessment of the accessibility to PES offices for currently unemployed workers in Spain. We need to keep in mind that the numbers of unemployed workers registered at PES offices in 2011 in some urban areas are really high. For instance, in Malaga or Madrid, the average number of unemployed workers per employment office is 11,000 and 12,500, respectively.

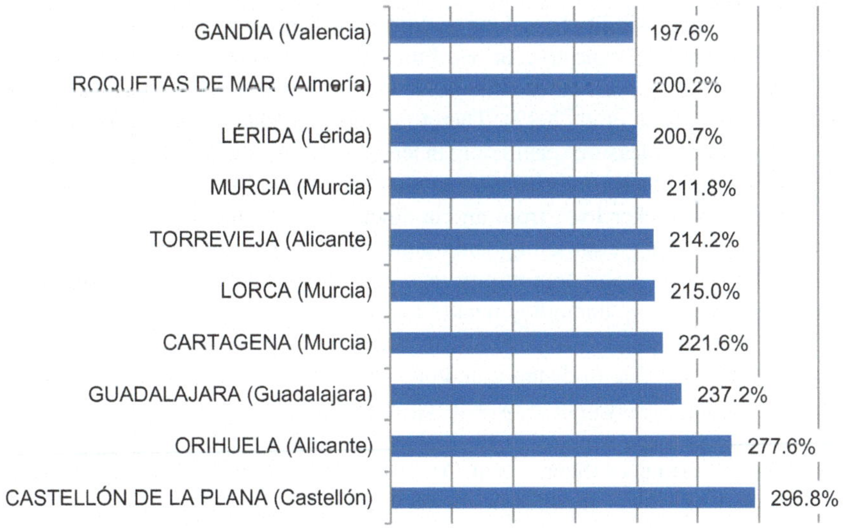

Fig. 2 Urban areas with higher unemployment growth, 2006–2011. Authors' own calculations based on SEPE data (2011)

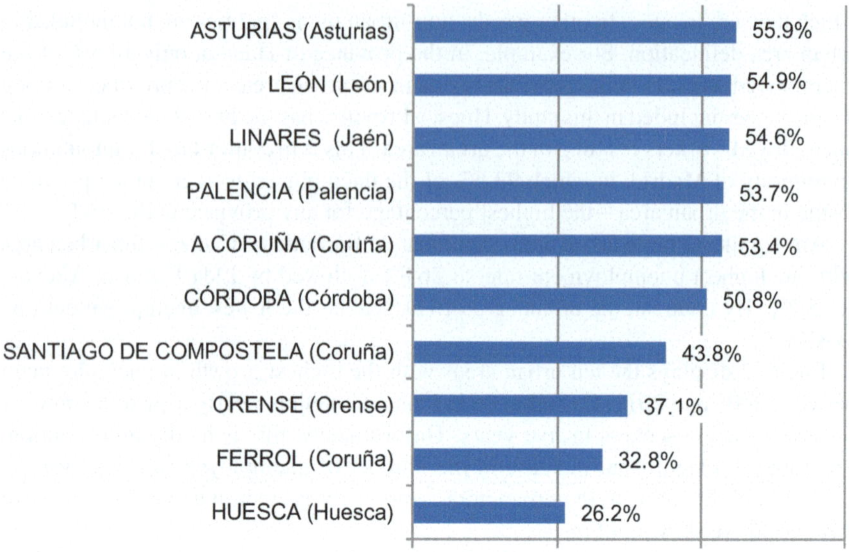

Fig. 3 Urban areas with lower unemployment growth, 2006–2011. Authors' own calculations based on SEPE data (2011)

3.2 Measuring Accessibility to Employment Services: Outputs and Results/Outcomes

The concept of accessibility has a long tradition in urban and regional sciences. Considered from a labor perspective, accessibility plays an important role in the spatial distribution of job opportunities (Ihlanfeldt 1997; Van Wee et al. 2001; Bunel and Tovar 2014; Di Paolo et al. 2017). Therefore, our first task is to compute the proposed accessibility measure specifically in terms of unemployed workers who reside in municipalities which comprise large urban areas. We propose in this chapter that the further a job seeker lives from an employment office, the less intense the job seeker's job searching, counseling, and training will be. Consequently, low accessibility to PES offices creates opportunity gaps. Not all registered jobless have the same chance of moving from unemployment to finding a job—in fact, some unemployed workers find it extremely difficult. Should unemployed workers lack access to an employment office (due to distance and/or overcrowding), they will have impeded from connecting with resources which represent their best chance for re-entering employment.

In a companion paper (Suárez et al. 2012), we examined how PES office accessibility affected 2009 local unemployment rates, without making a distinction between the different types of municipalities (urban or non-urban). This chapter differs from that approach in that it is focused upon the urban labor market structure, evaluating the impact of accessibility upon unemployment in a worsening economic situation. As mentioned above, the Spanish labor market has been disproportionately affected by the economic crisis. Unemployment has increased every month since its historical low of 1,954,984 unemployed workers in July 2006, spiraling to a new low of 4,750,867 in March 2012.

This work relies upon an accessibility measure previously developed by Suárez (2011). This takes into account the number of employment offices, together with the distance and the size of their catchment areas. Catchment areas are set by counting the unemployed assigned to an employment office, namely by adding up the number of jobless in the municipalities serviced by a given employment office. In the case of Spain, every unemployed worker is assigned to an employment office by the National Employment Authority depending upon each worker's place of residence. This measure is based on the number of employment offices per unemployed workers within a catchment area, adjusted for distance using the formula by Rich (1975) and recently used by Condeço-Melhorado et al. (2011), based upon the area of the municipalities considered $\left(d_i = \frac{1}{2}\sqrt{\frac{\text{area}}{\pi}}\right)$:

$$\text{Accessibility}_i = \left[\frac{\text{EO}_j}{\sum_{i\in j} u_i}\left(\text{e}^{-\lambda d_i}\right)\right] \tag{1}$$

where Accessibility_i is the accessibility for municipality i, EO_j is the number of employment offices in the catchment area j, and $\sum_{i\in j} u_i$ is the size of the catchment area measured as the number of unemployed workers in the municipalities i within a single catchment area. Finally, d_i is the internal distance previously defined and λ is a parameter of the distance-decay function. There is no data for trips to employment offices so we have been unable to estimate the parameters of the distance-decay function. Even though several values were used for this parameter in Suárez et al. (2012), the performance of a sensitivity analysis led us to set $\lambda = 0.10$ and $\lambda = 0.25$. Holl (2007) points out that this parameter depends on the type of activity involved. In this sense, Joassart-Marcelli and Giordano (2006) analyze the relationship between the location of one-stop centers in California and their accessibility levels establishing $\lambda = 0.25$.

Figure 4 presents the municipality level distribution of the accessibility variable for large urban areas, with $\lambda = 0.10$ and $\lambda = 0.25$, respectively.

Generally speaking, differences in accessibility values can be sizable for offices within the same urban area—while accessibility values are more comparable for office within urban centers, across the different urban areas. In other words, this suggests similar accessibility values occurring among offices located in urban centers (in terms of urban intensity) rather than similarities occurring within urban areas (in terms of proximity). The offices located in urban center are similar in their low accessibility scores, as the most crowded employment offices are located in cities; consequently, accessibility is lower.

The job seeker's need for an employment office in a smaller urban area may, in some ways, be viewed as more crucial. With smaller urban areas lacking the level of agglomeration effects of larger cities, the smaller urban area may simply have a narrower range of job types; for the job seeker in a smaller urban area, this may

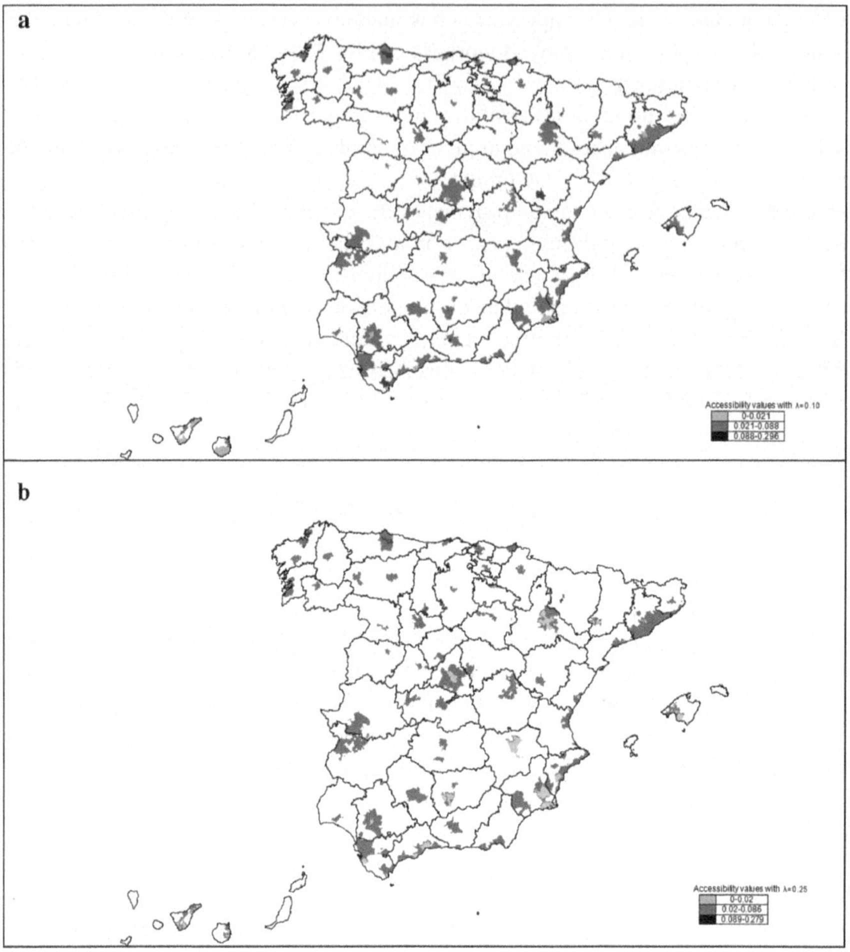

Fig. 4 **a** Accessibility to employment offices with $\lambda = 0.10$ and **b** accessibility to employment offices with $\lambda = 0.25$. Authors' own (2011)

mean lesser employment opportunities compared with Spain's primary urban areas (such as Madrid, Barcelona and Seville). Among job seekers in a smaller urban area, the particular opportunities being fewer, the PES office can in some ways attempt to compensate for the reduced opportunities by providing the essential individualized assessment, counseling, and training which will uniquely position a job seeker to take advantage of the smaller pool of fitting opportunities. Without such services, the job seeker in the smaller urban area may have fewer alternatives.

As Barca and McCann (2011) point out, we have to distinguish results/outcomes from outputs, especially in a framework that aims at a more performance-oriented regional policy. Moreover, regional policy goals should be explicitly spatial in terms of outcomes. Given the complexity of labor markets, it is essential for researchers to have available a method of analysis that can quantify the effect of an accessibility improvement upon the unemployment rate, taking into consideration existing spatial processes which may provide an outcome for said improvement, increasing or decreasing its final effect upon unemployment.

3.3 *Spatial Autocorrelation in the Spanish Labor Market*

Any study which aims to tackle both the definition and operation of regional labor markets has to consider explicitly the role that space plays as a source of economic information and as a catalyst and channel for a plethora of economic relations. Should spatial effects be overlooked, an erroneous model of the phenomenon could be produced. Several contributions have considered the spatial dimension of regional labor markets, pointing out the high degree of interdependence between local labor markets (Molho 1995; López-Bazo et al. 2002; Overman and Puga 2002). Patacchini and Zenou (2007) have analyzed the reasons for spatial dependence in local unemployment rates from a theoretical perspective; they concluded that this is mainly due to the fact that an unemployed worker may seek and find work in various areas, so spatial interactions (commuting flows) result from the mobility of the unemployed. Manning and Petrolongo (2017) propose a theoretical model of job search across space finding that distance to jobs exerts a great discouraging effect on the search efforts. When data are collected at the administrative level, spatial autocorrelation is likely to be relevant. Suárez et al. (2011, 2012) have proposed to begin the effect of the existence of PES offices upon local unemployment rate differentials and, if indeed such an effect is evidenced, to quantify such effects using a model in which the existence of spatial processes is accounted for within the unemployment variable. The predetermined location of an employment office determines its levels of accessibility and crowding, and it is precisely this difference in accessibility which explains the varying influence of PES offices on local unemployment rates.

The consideration of spatial relationships between municipalities is a long-standing problem to which a definitive solution still has not been found. The most common approach is the use of spatial weight matrices, though they can be constructed using different criteria. Notwithstanding this generalization, the case considered in this chapter is even more delicate, since two different levels of spatial relationship must be taken into account: (1) between municipalities within each of the urban areas considered and (2) between municipalities located in different urban areas.

We began by using the global Moran test (Cliff and Ord 1981) so as to analyze the existence of spatial autocorrelation in the local unemployment rates of municipalities belonging to urban areas. Table 1 summarizes the results of the global spatial autocorrelation test using various spatial weight matrices. In this exploratory analysis, we used a k-nearest neighbor matrix (with $k = 2$), which computes neighboring relationships between municipalities within the same urban area and precludes any relationship with municipalities in other urban areas. The statistic is significant, indicating

Table 1 Spatial autocorrelation tests results

	k-nearest	Distance	$1/d_{ij}$
Unemployment rate	0.601 (17.509)	0.529 (27.535)	0.543 (26.287)
Accessibility I (λ = 0.10)	0.609 (17.949)	0.478 (23.612)	0.301 (29.968)
Accessibility II (λ = 0.25)	0.585 (17.241)	0.394 (20.759)	0.459 (22.464)

Note P-values are reported in parentheses. All statistics are significant at 1% level

positive spatial autocorrelation, possibly caused by the existence of externalities. The same conclusion is reached when Moran's I is computed for the accessibility measures of these municipalities.

The first conclusion stemming from these results is that the spatial distribution of these variables is not random. Consequently, the existence of spatial processes must be taken into consideration in any model that aims to capture and explain local unemployment rate differentials and the effects of PES offices accessibility.

4 Models, Estimation Issues, and Results

Public policy may be a tool of intervention used to alleviate the problem of high unemployment nationwide, as well as to address the under-performance of particular region labor markets. Given the relationship between the aggregate unemployment rate and its implications for social welfare, the persistence of regional unemployment rate differences may be considered a major economic problem (Partridge and Rickmand 1995; Bande et al. 2008). PES offices may be critical for addressing such problems and achieving labor market objectives.

From a policy perspective of decentralization of ALMPs, as described above, we propose a method of analysis in which the effects of accessibility upon local unemployment rates may be fully assessed.

The analysis in this chapter is inspired by the debate regarding place-neutral versus place-based economic development policies, which in turn may emerge in the design of ALMPs. As Barca et al. (2012) state, the discussion has been focused on the crucial question of whether efficiency should be concentrated in the core or there is potential for growth and development in every territory (non-urban). In the case of Spain, ALMPs are highly decentralized, and regional and local authorities are responsible for their implementation. However, ALMPs are designed at the national or regional level, which prevents local factors from being considered in the development of the policies. The current location pattern of employment offices in Spain stems from political decisions made in recent decades; this suggests that priority should be placed upon delineating current urban labor markets, so that this consideration can be included in the process of siting PES offices in the future.

Differences between the roles of cities and peripheral areas in terms of economic growth have traditionally been one of the main issues in regional economics. Economic growth was traditionally discussed in terms of cities, but recent data and research have called this approach into question. Garcilazo et al. (2010) state that a synchronous increase in the potential of non-core outlying areas may significantly enhance national growth. Consequently, we have chosen to focus on urban areas not because we work under the hypothesis that they are more important economically than rural areas but simply because most job seekers are concentrated within urban areas, indicating that the areas most impactful for assessing whether PES offices are located conveniently, so that may be redesigned ALMPs accordingly, if necessary.

As explained above, local unemployment differences have been studied theoretically and empirically. Elhorst (2003) provides an excellent survey of most of the alternative models used in papers on the topic since 1985. He concludes that whichever model is used, they all result in the same reduced form equation of the regional unemployment rate.

Labor supply, labor demand, and wage-setting factors are generally used as explanatory variables in such models. Due to the high level of spatial disaggregation in this study, there is limited information available about these factors at this scale. The rates of the foreign population, males and females of working age, educational attainment, industries' employment shares, population density and urban concentration, as well as the accessibility measures previously described are used as explanatory variables in the model (see Table 2). The initial model specification is:

$$\log(u_i) = \alpha \log(A_i) + \beta X_i + e_i \quad (2)$$

where u_i is the local unemployment rate for each municipality **i**, A_i is the accessibility measure for each municipality **i**, and the matrix **X** collects the explanatory variables. Following Alonso-Villar and Río (2008) and Alonso-Villar et al. (2009), local unemployment rates have been calculated dividing the number of unemployed workers registered at PES offices in 2011 by the number of people of working age (i.e., population aged 16–64) on the 2011 municipal register. Because there are no data on the economically active population at the municipal level, local unemployment rates were calculated by dividing the number of unemployed workers registered at PES offices by the number of people of working age (i.e., population aged 16–64) on the 2009 municipal register. Even if we had had access to data on the local economically active population in the Labor Force Survey (LFS), they could hardly have been used as a reference for registered unemployment, as the active population in the LFS is calculated by adding the number of employed people to the number of jobless people, with the latter calculated in accordance with international statistical standards (International Labor Organization), which need not correspond to those in administrative-based statistics (Toharia 2005).

The initial hypothesis is that improvements in the accessibility of PES offices are associated with a reduction in the unemployment rate. Assuming that the spatial distribution of employment offices is adequate, a significant negative coefficient for A_i is expected.

Table 2 Variables in the model

Variable	Definition	Data source
Local unemployment rate (LUR)	Unemployed population/working age population (16–64)	SEPE and Municipal Register
Accessibility	Accessibility measure	Authors' own
Population density	Population/km^2	Statistical Atlas of Urban Areas
Urban concentration (UC)	Municipality population/total population in the large urban area	Statistical Atlas of Urban Areas
Foreign	Foreign population (16–64)/working age population (16–64)	Municipal Register (2009)
Female	Female population (16–64)/total female population	Municipal Register (2009)
Male	Male population (16–64)/total male population	Municipal Register (2009)
Industry[a]	Share of employment in industry	Census (2001)
Construction	Share of employment in construction	Census (2001)
Services	Share of employment in services	Census (2001)
Illiteracy[a]	% Illiteracy	Census (2001)
Primary	% Primary education	Census (2001)
Secondary	% Secondary education or vocational training	Census (2001)
Higher Education	% Higher education	Census (2001)

[a]The percentage of population with incomplete primary education and the share of employment in agriculture have been omitted to avoid multicollinearity

Equation (2) was estimated using ordinary least squares (OLS). The estimated accessibility coefficients are negative (-0.052 with $\lambda = 0.10$ and -0.062 with $\lambda = 0.25$) and statistically significant at 5% and 1% level, respectively, but the model may overlook the existence of a spatial dependence process in the local unemployment rate or in the explanatory variables.

Another issue concerns whether the accessibility variable is endogenous. Wooldridge's score test (1995) was performed to test the null hypothesis that the accessibility measure is exogenous. This test—whose instruments are geographic (municipality surface) and demographic characteristics—is more appropriate when the residuals show heteroskedasticity. The value of the statistic is $\chi_1^2 = 0.0296$, suggesting that there is no statistical evidence to reject the null hypothesis that accessibility is exogenous.

We then use the estimated residuals to calculate the Moran's statistic and determinate the existence of a spatial dependence process. The value of the Moran's statistic is 12.327 (statistically significant at 1% level), which means that there is positive spatial autocorrelation, which violates the assumptions of a linear regression model using OLS estimation. Therefore, a model which accounts for spatial processes must be used instead. The spatial econometric literature has traditionally relied on bottom-up model selection strategies based upon the Lagrange Multiplier (LM) tests (Anselin 1988) and the robust versions proposed by Anselin et al. (1996), and as a result, the spatial lag model, the spatial error model, or a combination of both has usually been selected.

From a theoretical perspective, recent contributions by LeSage and Pace (2009), Elhorst (2010), and LeSage and Fisher (2012), among others, have remarked upon the accuracy of more complete spatial models. Consequently, model specifications may use a general to specific strategy. Our model follows Manski's model (Elhorst 2010), which may be reduced to a series of simpler models by imposing restrictions on one or more parameters (such as spatial Durbin model, Kelejian-Prucha model, spatial lag model, spatial error model…). This is the expression of the Manski model:

$$\begin{aligned} y &= \rho W y + \alpha i + X\beta + W X \gamma + \varepsilon \\ \varepsilon &= \lambda W \varepsilon + u \end{aligned} \tag{3}$$

where the dependent variable y represents an n-by-1 vector of the unemployment rates in urban municipalities; the n-by-k matrix X contains k explanatory variables excluding the intercept vector, which is represented by i. The variable WY denotes the endogenous interaction effects among the dependent variables; WX denotes the exogenous interaction effects among the independent variables. $W\varepsilon$ denotes the interaction effects among the disturbance terms of the different spatial units. ρ and λ are the spatial autoregressive coefficients, and β and γ represent a k-by-1 vector of unknown parameters.

Manski's model is the most complete alternative and includes three different interaction effects: endogenous, exogenous, and correlated effects. LeSage and Pace (2009) and Elhorst (2010) have considered the simultaneous estimation of these effects and concluded that endogenous and exogenous effects are indistinguishable from each other. Should the spatial dependence in either the dependent or the explanatory variables be ignored, a problem of omitted relevant variables will arise, leading the estimators of the model coefficients to be biased and inconsistent. However, if the spatial dependence (if present) in the disturbances are ignored; only a loss in efficiency will occur. Consequently, the best option is to exclude the spatial dependence of the disturbances, which in turn leads to the spatial Durbin model (SDM) specification. This is the general expression for an SDM:

$$\boldsymbol{y} = \rho \boldsymbol{W} \boldsymbol{y} + \alpha \boldsymbol{i} + \boldsymbol{X} \boldsymbol{\beta} + \boldsymbol{W} \boldsymbol{X} \boldsymbol{\gamma} + \boldsymbol{\varepsilon} \tag{4}$$

Table 3 LM, robust LM, and LR diagnostic tests*

	LMlag	LMerr	R-LMlag	R-LMerr	$H_0 : \theta = 0$	$H_0 : \theta + \rho\beta = 0$
Accessibility I ($\lambda = 0.10$)	171.821	133.432	47.613	9.225	82.013	87.730
Accessibility II ($\lambda = 0.25$)	170.935	135.237	45.957	10.259	80.021	85.364

*All values are statistically significant at the 1% level

One strength of the SDM is that it produces unbiased coefficient estimates, even if the true data generation process is a spatial lag/error model. In addition, it does not impose a priori restrictions on the magnitude of potential spatial spillover effects. Unlike other spatial regression specifications, these spillover effects may be global or local and may also differ for different explanatory variables. In addition, the standard errors of the coefficient estimates in the SDM are correct even if the true data-generating process is a spatial error model.

Following Elhorst (2010), we have used a combination between the two selection strategies—specific to general and general to specific. Once the model is estimated using OLS, we use the LM test and its robust version to determine whether spatial lag or spatial error is a better specification. These results (summarized in Table 3) indicate that OLS is rejected in favor of any spatial alternative, so the spatial Durbin model should also be estimated and compared with the alternative models.

Since the model is estimated using maximum likelihood, a likelihood ratio (LR) test may be performed to test whether the SDM may be simplified in either the spatial lag model or the spatial error model.[2] Both hypotheses are rejected, indicating that the spatial Durbin model is the most appropriate for observing data behavior using both accessibility measures.[3]

The estimation results of the selected models are summarized in Table 4. The SDM has been estimated using maximum likelihood (ML). Although some spatial weight matrices have also been considered, we present only the results obtained using a k-nearest neighbor matrix (with $k = 5$) in Model I, a geographic contiguity matrix ($w_{ij} = 1$ if i and j share a border and $w_{ij} = 0$ in other cases) in Model II and a distance-based matrix with ($w_{ij} = d_{ij}^{-1}$ if $d_{ij} < 100$ km) in Model III.

With respect to the results, the estimate of the spatial autoregressive parameter in all models provides evidence for the existence of significant spatial effects operating through the dependent variable. This outcome confirms the initial hypothesis concerning the existence of spatial autocorrelation processes in urban unemployment rates. In addition, there are many spatially lagged independent variables that are statistically significant, which in turn shows the importance of proximity in explaining

[2]The tests follow a chi-squared distribution with K degrees of freedom.

[3]For estimation and testing purpose, we use a distance-based spatial weight matrix, where the distance cut-off is 150 km.

Table 4 Estimation results for the spatial Durbin model

	Model I (*k*-nearest)		Model II (Contiguity)		Model III (Distance)	
	Not lagged	Lagged	Not lagged	Lagged	Not lagged	Lagged
Intercept	−2.127 (−4.158)		−4.774 (−17.526)		−4.858 (−17.344)	
$\log(A_i)$	0.002 (0.091)	−0.035 (−1.372)	0.023 (1.142)	−0.061 (−2.427)	0.004 (0.199)	−0.040 (−1.537)
Density	0.015 (5.631)	−0.006 (−1.262)	0.015 (5.143)	−0.005 (−1.059)	0.013 (4.702)	−0.005 (−1.037)
UC	0.270 (8.936)	−0.164 (−2.313)	0.248 (6.745)	−0.063 (−1.494)	0.295 (8.534)	−0.140 (−2.630)
Foreign	−0.160 (−1.486)	−0.188 (−1.310)	−0.062 (−0.540)	−0.290 (−2.086)	−0.193 (−1.830)	−0.129 (−0.928)
Female	0.689 (1.872)	2.494 (3.704)	0.704 (1.859)	2.731 (4.099)	1.087 (2.833)	2.259 (3.310)
Male	1.650 (4.045)	−3.158 (−4.270)	1.289 (3.097)	−1.279 (−1.960)	1.321 (3.123)	−1.810 (−2.604)
Industry	1.552 (10.098)	−1.567 (−5.605)	1.515 (9.030)	−0.968 (−4.271)	1.176 (7.435)	−0.231 (−1.028)
Construction	1.878 (8.063)	−1.226 (−2.985)	1.950 (8.049)	−0.207 (−0.616)	1.928 (7.964)	0.143 (0.420)
Services	1.206 (8.234)	−1.060 (−3.722)	1.286 (7.724)	−0.654 (−2.913)	1.162 (7.613)	−0.011 (−0.046)
Illiterate	4.270 (5.028)	−1.364 (−0.970)	3.966 (4.567)	3.092 (2.723)	4.631 (5.356)	3.745 (3.182)
Primary	0.006 (0.030)	−0.075 (0.194)	−0.113 (0.5340)	0.950 (3.17)	−0.128 (−0.575)	1.119 (3.935)
Secondary	−0.267 (−1.464)	−0.358 (−1.199)	−0.421 (−2.313)	0.074 (0.288)	−0.497 (−2.566)	0.196 (0.731)
Higher Ed	−1.828 (−7.844)	0.545 (1.463)	−1.980 (−8.437)	1.278 (3.722)	−2.152 (−8.855)	1.391 (3.997)
LUR		0.505 (12.757)		0.367 (10.823)		0.378 (10.729)
Log-likelihood	293.275		279.857		260.307	

Notes For each model, first column includes de-estimated coefficients for the explanatory variables and the second column includes the estimated coefficients associated to the spatially lagged variables. *z*-values are in parentheses

differences in local unemployment rates. Some spatially lagged independent variables have greater coefficients—e.g., *female labor force*—suggesting that important influences may be overlooked in models that exclude neighboring municipalities (LeSage and Fischer 2008). In other cases, the estimated coefficients of the spatially lagged variables and its own variable are roughly equal in size—e.g., *illiterate*

(Model II) and *percentage of foreign population* (Model I and III)—again showing the importance of the characteristics of neighboring regions. In addition, once the spatial dependence is explicitly included in the model, the coefficient of the accessibility variable is positive and smaller but statistically non-significant, whereas the coefficient of the spatially lagged accessibility variable is negative and statistically significant.

It is essential that the SDM are interpreted properly. LeSage and Pace (2009) have shown that the standard partial derivatives interpretation of regression may not be considered valid. In this spatial regression, any change in an explanatory variable of municipality i has a ***direct impact*** on municipality i but also an ***indirect impact*** upon neighboring municipalities' $j \neq i$ throughout the system. The direct impact reflects the average response of the dependent variable to independent variables, including feedback loops that arise from impacts passing through neighbors and back to the municipality itself.[4] The indirect impact reflects the effect that any change in a municipality has on others or how changes in all municipalities affect a region—which are commonly thought of as spatial spillovers. Because the impact of changes in an explanatory variable differs across observations, LeSage and Pace (2009) propose three scalar summary measures of these varying impacts: average direct effect, average indirect effect, and average total effect.

Table 5 shows the estimated direct impact, indirect impact, and total impact upon the urban unemployment rate that resulting from changing each variable in the model, *ceteris paribus*.

In most cases, the direct effects are similar to the estimated coefficient of the non-lagged spatially variables—e.g., *illiterate* (for Models II and III). The estimated coefficient for this variable in Model I is 4.270, which is roughly equal to its direct impact, 4.364. The differences that exist between these two values are due to the feedback effects. If we analyze the results of Model I, the direct effect for the *accessibility* variable is −0.002 (though it is not statistically significant), whereas the estimated coefficient of the non-spatially lagged variable is 0.002. In all models, the direct effect of the accessibility variable is not statistically significant.

Turning to the indirect impacts, we note discrepancies between these effects and the model coefficients on the spatially lagged variables. Then, if the coefficients of the spatially lagged variables are interpreted as an indicator of the size and significance of the spatial spillover, we are making an important mistake. For example, the estimated coefficient associated with *female labor force* (in Model I) is 2.494, whereas the estimated indirect impact is 5.410, indicating the existence of positive feedback effects on the unemployment rate. Following the aim of the paper, we found that the indirect impact of the accessibility variable is negative and statistically significant in all models. This finding is key for evaluating the influence of the public employment offices for two reasons. First, statistical evidence is obtained on the existence and

[4]These arise because any municipality is a neighbor of its neighbors, so changes that impact municipality i will also impact its neighboring municipalities, which will in turn have higher-order feedback effects upon municipality i.

Table 5 Direct, indirect and total effects

	Model I (*k*-nearest)			Model II (Contiguity)			Model III (Distance)		
	Direct	Indirect	Total	Direct	Indirect	Total	Direct	Indirect	Total
$\log(A_i)$	−0.002	−0.065*	−0.067*	0.017	−0.076**	−0.059**	0.000	−0.058**	−0.058**
Density	0.016***	0.004	0.020**	0.015***	0.000	0.015**	0.013***	0.000	0.013**
UC	0.267***	−0.053	0.214	0.252***	0.041	0.293***	0.292***	−0.043	0.249***
Foreign	−0.192*	−0.511**	−0.703***	−0.099	−0.457***	−0.556***	−0.215**	−0.303**	−0.518***
Female	1.023***	5.410***	6.433***	1.056***	4.367***	5.422***	1.380***	4.002***	5.382***
Male	1.377***	−4.425***	−3.048**	1.194***	−1.179	0.015	1.177***	−1.964**	−0.786
Industry	1.459***	−1.491***	−0.032	1.466***	−0.603*	0.864***	1.200***	0.321	1.521***
Construction	1.846***	−0.527	1.319*	2.010***	0.741*	2.751***	2.023***	1.307***	3.330***
Services	1.153***	−0.857*	0.297	1.265***	−0.267	0.998***	1.210***	0.643**	1.853***
Illiterate	4.364***	1.511	5.875**	4.501***	6.639***	11.140***	5.234***	8.241***	13.474***
Primary	−0.002	−0.137	−0.139	−0.006	1.328***	1.322**	−0.011	1.604***	1.594***
Secondary	−0.325**	−0.939**	−1.264**	−0.431**	−0.117	−0.548	−0.496***	0.012	−0.484
Higher Ed	−1.873***	−0.721	−2.594***	−1.915***	0.807*	−1.108**	−2.088***	0.865**	−1.223**

*Statistical significance at the 10% level
**Statistical significance at the 5% level
***Statistical significance at the 1% level

size of this influence; second, this influence acts through the neighboring municipalities, rendering it necessary to consider all of the spatial processes embedded in the local labor market. Because both the unemployment rate and accessibility are in log form, its indirect impact (−0.076 in Model II) may be interpreted as an elasticity, so if accessibility increases by 1% in all urban municipalities, the unemployment rate decreases by 0.08%, *ceteris paribus*. This is one interpretation of the indirect impact. The second interpretation measures the cumulative impact of a change of one independent variable in municipality i averaged over all other municipalities. In the context of that interpretation, a 1% increase in accessibility in municipality i exerts a small influence in each neighboring municipality, but the cumulative effect over all municipalities is a 0.08% decrease in the unemployment rate.

Finally, the total effects and its significance are also included in Table 5. Most of the estimated impacts are significant, and the obtained sign is as expected. The impacts for the educational attainment variables are generally positive and large in magnitude for *illiterate* and *primary education* and negative and significant for *higher education*. The total impact estimates for the accessibility variable range between −0.058 and −0.067 in the proposed models.

This result implies that the actual configuration of the public employment offices in Spain exerts a significant but small influence on urban unemployment rates. Previous studies by Suárez et al. (2012) note that accessibility to employment offices has a low but a statistically significant impact when the data include all types of municipalities.

5 Results and Policy Recommendations: Looking Ahead

The aim of this chapter was to analyze the accessibility to public employment offices in the 85 large urban areas in Spain from a spatial perspective. In a recent paper, we found accessibility to have a positive effect upon reduction in unemployment (Suárez et al. 2012). As 69.3% of job seekers resided in urban areas in 2011, this chapter complements the above-mentioned paper by limiting the analysis of accessibility to PES offices to urban municipalities.

Our analysis supports the existence of spatial spillovers in urban labor markets. Consequently, the model must include this spatial process explicitly, and the results must be interpreted accordingly. Concerning the relationship between accessibility and urban unemployment rate, we find negative and significant indirect and total effects of a small magnitude, but the direct effect is not statistically significant.

Evidence from previous years makes it clear that PES offices offer essential services to the vastly growing number of unemployed; it is well known that these offices are overcrowded, which prompted our research into accessibility. Yet the results were not of the magnitude expected. In addition, we used data from 2011, the worst year for the Spanish labor market since 2007, during which PES offices were flooded beyond capacity by job seekers. Because the future prospects for the economy are rather bleak, and unemployment is expected to remain high for the foreseeable future, there is an urgent need for reforms to improve the job search assistance that unemployed workers receive at PES offices. In the future, this type of analysis may be conducted using data at a smaller scale of the actual offices in order to focus on the individual office performance.

As the analysis of spatial difference in accessibility to PES services evolves, the results should help policymakers to appropriately redesign ALMPs, where necessary, considering the urban configuration of the country. In other words, policymakers should be encouraged to consider place-based policies rather than spatially neutral policies.

Acknowledgements Funding from the Ministry of Economics and Competitiveness (Spain) ECO2016-75805-R and ECO2017-86402-C2-1-R is gratefully acknowledged.

References

Alonso-Villar, O., & del Río, C. (2008). Geographical concentration of unemployment: A male-female comparison in Spain. *Regional Studies, 42*, 401–412.

Alonso-Villar, O., del Río, C., & Toharia, L. (2009). Un análisis espacial del desempleo por municipios. *Revista de Economia Aplicada-Spain, 49*, 47–80.

Anselin, L. (1988). *Spatial econometrics methods and models*. Dordrecht, The Netherlands: Kluwer Academic Publishers.

Anselin, L., Bera, A., Florax, R., & Yoon, M. (1996). Simple diagnostic tests for spatial dependence. *Regional Science and Urban Economics, 26*, 77–104.

Barca, F., & McCann, P. (2011). Outcome indicators and targets: Towards a performance oriented EU Cohesion Policy. *DG Regio Website*. Available at https://ec.europa.eu/regional_policy/sources/docgener/evaluation/performance_en.htm. Accessed November 2011.

Barca, F., McCann, P., & Rodríguez-Pose, A. (2012). The case for regional development intervention: Place-based versus place-neutral approaches. *Journal of Regional Science, 52*, 134–152.

Bande, R., Fernández, M., & Montuenga, V. (2008). Regional unemployment in Spain: Disparities, business cycle and wage setting. *Labour Economics, 15*, 885–914.

Bunel, M., & Tovar, E. (2014). Key issues in local job accessibility measurement: Different models mean different results. *Urban Studies, 51*, 1322–1338.

Cliff, A., & Ord, J. K. (1981). *Spatial processes: Models and applications*. London: Pion.

Condeço-Melhorado, A., Gutierrez, J., & Garcia-Palomares, J. C. (2011). Spatial impacts of road pricing: Accessibility, regional spillovers and territorial cohesion. *Transportation Research Part A-Policy and Practice, 45*, 185–203.

Di Paolo, A., Matas, A., & Raymond, J. L. (2017). Job accessibility and job-education mismatch in the metropolitan area of Barcelona. *Paper in Regional Science, 96*, 91–112.

Elhorst, J. P. (2003). The mystery of regional unemployment differentials: Theoretical and empirical explanations. *Journal of Economic Surveys, 17*, 709–748.

Elhorst, J. P. (2010). Applied spatial econometrics: Raising the BAR. *Spatial Economic Analysis, 5*, 9–28.

European Commission. (2010a). *EUROPE 2020. Integrated guidelines for the economic and employment policies of the Member States*. Available at https://ec.europa.eu/eu2020/pdf/Brochure%20Integrated%20Guidelines.pdf.

European Commission. (2010b). *EUROPE 2020: A strategy for smart, sustainable and inclusive growth*. COM(2010) 2020 final, 3 December, March 2010, Brussels. Available at https://eur-lex.europa.eu/LexUriServ/LexUriServ.do?uri=COM:2010:2020:FIN:EN:PDF.
Garcilazo, J. E., Martins, J. O., & Tompson, W. (2010). *Why policies may need to be place-based in order to be people-centred*. Paris: OECD Regional Development Policy Division. Available at https://www.voxeu.org/.
Holl, A. (2007). Twenty years of accessibility improvements. The case of the Spanish motorway building programme. *Journal of Transport Geography, 15*, 286–297.
Ihlanfeldt, K. R. (1997). Information on the spatial distribution of job opportunities within metropolitan areas. *Journal of Urban Economics, 41*, 218–242.
Ihlanfeldt, K. R., & Sjoquist, D. L. (1990). Job accessibility and racial differences in youth employment rates. *The American Economic Review, 80*, 267–276.
National Statistical Institute (INE) Population Census 2001.
National Statistical Institute (INE) Municipal Register 2009.
Joassart-Marcelli, P., & Giordano, A. (2006). Does local access to employment services reduce unemployment? A GIS analysis of One-Stop Career Centers. *Policy Sciences, 39*, 335–359.
LeSage, J. P., & Fischer, M. M. (2008). Spatial growth regressions: Model specification, estimation and interpretation. *Spatial Economic Analysis, 3*, 275–304.
LeSage, J. P., & Fischer, M. M. (2012). Estimates of the impact of the static and dynamic knowledge spillovers on regional factor productivity. *International Regional Science Review, 35*, 103–127.
LeSage, J., & Pace, R. K. (2009). *Introduction to spatial econometrics*. CRC Press.
López-Bazo, E., del Barrio, T., & Artis, M. (2002). The regional distribution of Spanish unemployment: A spatial analysis. *Paper in Regional Science, 81*, 365–389.
Manning, A., & Petrongolo, B. (2017). How local are labor markets? Evidence from a spatial job search model. *The American Economic Review, 107*, 2877–2907.
Molho, I. (1995). Spatial autocorrelation in British unemployment. *Journal of Regional Science, 35*, 641–658.
Overman, H. G., & Puga, D. (2002). Unemployment clusters a cross Euro's regions and countries. *Economic Policy, 17*, 115–148.
Patacchini, E., & Zenou, Y. (2007). Spatial dependence in local unemployment rates. *Journal of Economic Geography, 7*, 169–191.
Partridge, M. D., & Rickman, D. S. (1995). Differences in state unemployment rates: The role of labor and product market structural shifts. *Southern Economic Journal, 62*, 89–106.
Rich, D. (1975). *Accessibility and economic activity: A study of locational disadvantage in Scotland*. Ph.D. Thesis, University of Cambridge.
Spain. (2011). Real Decreto-Ley 3/2011, de 18 de febrero, de medidas urgentes para la mejora de la empleabilidad y la reforma de las políticas activas de empleo. Boletín Oficial del Estado. 43, de 19 de febrero de 2011. 19240 a 19260.
Spanish Minister of Housing. (2006). Statistical Atlas of Urban Areas.
Suárez, P. (2011). *El Servicio Público de Empleo en España: ensayos desde una perspectiva regional*. Ph.D. Thesis, Universidad de Oviedo, Spain.
Suárez, P., & Mayor, M. (2012). La intermediación laboral del Servicio Público de Empleo en España: un análisis regional con los datos del SISPE. *Revista del Ministerio de Empleo y Seguridad Social, 96*, 175–194.
Suárez, P., Mayor, M., & Cueto, B. (2011). How important is access to employment offices in Spain? An urban and non-urban perspective. *Investigaciones Regionales, 21*, 119–140.
Suárez, P., Mayor, M., & Cueto, B. (2012). The accessibility to employment offices in the Spanish labour market. *Paper in Regional Science, 91*, 823–848.

Toharia, L. (2005). El desempleo en España. In V. Navarro (Ed.), *La situación social en España*. Madrid: Fundación Largo Caballero y Biblioteca Nueva.
Van Wee, B., Hagoort, M., & Annema, J. A. (2001). Accessibility measures with competition. *Journal of Transport Geography, 9*, 199–208.
Wooldridge, J. M. (1995). Score diagnostics for linear models estimated by two stage least squares. In G. S. Maddala, P. C. B. Phillips, & T. N. Srinivasan (Eds.), *Advances in econometrics and quantitative economics: Essays in Honor of Professor C. R. Rao*. Oxford: Blackwell.

Which Places Grow Faster?

An Empirical Analysis of Employment Growth Factors at the Local Level for the Spanish Economy

Diana Gutiérrez Posada, Fernando Rubiera Morollón, and Ana Viñuela

1 Introduction

Which places grow faster? How relevant are geographical factors to local employment growth? How effective can political action be? How important are size and distance to the largest cities? Do the same factors affect growth in the same way in both urban and rural areas? How relevant is the regional economic environment to the local growth? Can regional economic policies promote local economic growth over a medium- to long-run scenario?

We have these questions and many others in mind for this research. Using an empirical approach, our aim is to analyse employment growth factors in Spain at the *local* level. Most analyses of growth are usually conducted at the national or regional scale using political or administrative boundaries to delineate regions (Autonomous Communities or Provinces in the Spanish case). However, studies at the local scale are scarcer, and as of yet nonexistent in the specific case of Spain.

Several reasons explain the lack of studies on local growth. First, local growth analysis suffers from data limitations for most countries. In Europe, economic growth quantification based upon GDP figures becomes an impossible task at a higher degree of spatial disaggregation than Nomenclature of Territorial Units for Statistics Level 3 (NUTS 3) regions, and Spain is not an exception. If interested in studying

D. Gutiérrez Posada
University of Birmingham, Strategy and International Business City-REDI, Birmingham, UK
e-mail: D.gutierrez.posada@bham.ac.uk

F. Rubiera Morollón (✉) · A. Viñuela
Regional Economic Analysis Laboratory–REGIOlab, Applied Economics Department, University of Oviedo, Asturias, Spain
e-mail: frubiera@uniovi.es

A. Viñuela
e-mail: avinuela@uniovi.es

J. Thill (ed.), *Innovations in Urban and Regional Systems*,
https://doi.org/10.1007/978-3-030-43694-0_5

local growth, researchers are forced to accept employment—which figures are commonly available at a high degree of disaggregation—as a proxy for GDP. However, researchers must be aware of the different behaviours that these two variables (GDP and employment) may exhibit and the potential limitations imposed by the lack of more suitable local data.

A second reason for the scarcity of studies on local growth is that its results are at times inconsistent with the economic growth theories. For instance, some studies show how particular factors exert influence at the local level in different ways, depending on the period analysed (Massey 1995; Shearmur and Polèse 2007; Strambach 2001). Thus, in Canada, education levels were demonstrated to be strongly associated with employment growth in the 1980s, yet this variable shows no influence in the 1970s and 1990s. Blien et al. (2006), who analysed the German case focusing on the effect of diversity and industrial concentration upon local growth, show that industrial structure tends to have only a short-term effect. However, Strambach (2001) asserts on his study for the Stuttgart area that a 'top position in technological fields is no guarantee of the future competitiveness of a region'. These three studies show how any particular factor affecting local employment growth may depend upon time, but more importantly, upon the location within a country.

Aware of all this, in this research we aim to provide a better understanding of the determinants of local employment growth and the relevance of inherent characteristics of the area, as they can be affected by national/regional economic policy in a different way. A brief reappraisal of the different types of employment growth factors and a review of previous studies is offered in Sect. 2. How to translate the general ideas presented into a specific empirical model is discussed in Sect. 3. First, the concept of functional areas is briefly explained, and the election of the functional areas called *local labour market areas* (LLMs) is justified. Second, we make a revision of the data that is available at the local level. And finally, a general empirical model is proposed and then adapted to the characteristics of the Spanish economy. Section 4 presents the results for the different linear model estimated. Employment growth seems to have a strong locational component that confirms the need to estimate spatial auto-regressive models (Anselin 1988; Anselin et al. 2004), which are presented at the end of Sect. 4. The step-by-step procedure adopted in this piece of work intends to illustrate the reader into the complex and at times conflicting approaches commonly taken when tackling local economic growth. Section 5 summarises the main contributions of this chapter and suggests policy implications of our conclusions.

2 Factors of Local Employment Growth: A Reappraisal

Local employment growth and local economic development can be affected by several factors that can be classified into three groups.

The first group is comprised of factors that are difficult to measure and include substantial qualitative components, making their effects difficult to capture using a statistical approach (Doloreux et al. 2001). These factors encompass elements such as the local institutional context (Cooke et al. 2004), specific actors such as groups, organisations and communities (Galaway and Hudson 1994; Gutíerrez et al. 2019), and inter-firm dynamics and knowledge spillover (Malecki and Oinas 1999; Porter 1990). In particular, researchers in the field of innovation studies describe how certain regions have managed to develop local systems of innovation by combining these factors in particular ways (Cooke et al. 2004). Numerous case studies have described how such factors can induce employment growth at the local level; however, the nature of such studies does not enable extrapolation to general conclusions (Markusen 1999). Yet these studies are used to determine best practices which can serve as a basis for implementing policies in other regions.

The second group is comprised of factors that are also local in nature, but which can be quantitatively measured and are based upon general economic growth theory. The endowment of human capital (Florida 2002; Romer 1990), its industrial structure (Porter 1998), its relative costs (Weber 1929), and its level of diversity (Jacobs 1984; Porter 1990) are all theorised to be growth determinants. The effect of such factors upon employment growth, either at national and subnational level, is verified by statistical analyses of various sorts. Usually, a large group of regions or cities is considered, and the effect of each factor upon growth is then determined for cities or regions under study (Beckstead and Brown 2003; Florida 2002; Shearmur and Polèse 2007; Gutièrrez et al. 2018) or for nations (Barro and Sala-i-Martin 1995). Some general conclusions about *local* growth are typically drawn from this type of analysis. Florida (2002), for instance, suggests that a highly educated local workforce is conducive to local growth of employment and income. Henderson (2003) shows that local specialisation in certain industries tends to lead to employment growth in those industries. Quigley (1998) shows that for a region, a diverse economy tends to be associated with increased growth.

However, the results of the analysis of growth factors are not always conclusive and can be contradictory. For instance, as Blien et al. (2006) point out, there is an ongoing debate between researchers who defend that a diverse economy leads to growth (Jacobs 1984; Markusen 1996; Quigley 1998) and those who defend that specialisation is conducive to growth (Porter 1996). Florida's (2002) contention that, for regions, an educated workforce is conducive to growth—while to some extent compatible with human capital theory—is not supported by evidence in the Canadian case after controlling for other growth factors (Shearmur and Polèse 2007). Thus, despite the theoretical possibility of deriving conclusions about the effect of certain measurable local factors upon local growth, a general model applied to regions has proven elusive so far.

The third group is comprised of factors that are structural. By *structural,* we do not mean the industrial structure (which is considered in this study to be a growth factor of the second type), but rather the geographical and historical structures. Geographical location and, in particular, the proximity to markets (Krugman 1995; Partridge and Rickman 2006), historical trends or accidents, such as the US technological superiority (Krugman 1995; Davis and Weinstein 2002), and the centre–periphery and urban–rural divide (Parr 2001) have been suggested as having effects upon employment growth outcomes.

To the extent that there has been greater emphasis upon local economic development over the past twenty years (Martin and Sunley 1998; Parr 2001), policy makers have tended to show less interest in geographical and historical structures which generally operate at a scale larger than local (Eisinger 1988; Keating 1993). This can partly be attributed to the failure in the 1960s and 1970s of top-down policies which were aimed at balancing growth across wide geographical areas, and partly to the fact that statistical models used to describe geo-structural effects fell out of fashion over this period as well (Philo et al. 1998). They were often seen as too simplistic—unable to deal with the qualitative factors that are also important for understanding economic development—and as tending to make grandiose claims about development factors that did not bear out in practice. Moreover, these geographical or structural factors can be considered fixed in the short and medium run and are therefore unable to be affected by economic policies.

However, this does not necessarily mean that they do not affect local employment growth. As Parr (2002) points out, there are two basic variables that must be taken into consideration as *geo-structural factors*: size and location. The *size* of a country or region—commonly quantified in the field of regional economics in terms of population—defines an area as rural or urban and assists in determining the existence of agglomeration economies. Some researchers suggest that the flows of economic activity tend to favour more urbanised areas, and others state that flows *within* urban areas tend to favour larger cities. Mackay (2003) conducted a study on employment growth in the UK adopting both approaches. Coffey and Polèse (1988), Coffey and Shearmur (1996) and Polèse and Shearmur (2004) described the distribution of employment growth *across* the urban system and *between* central and peripheral areas for Canada, while in Polèse et al. (2006) a similar analysis is done for the Spanish case. All of these studies show how employment growth—particularly in strategic economic sectors such as high-order services—tends to concentrate in and around cities, and more specifically in and around large metropolitan areas. Size matters in terms of agglomeration; in addition, the possibility of the existence of *agglomeration diseconomies* should also be taken into account.

Regarding *location*, the proximity to some borders with important cross-border trade flows could be relevant, as well as the proximity to the coast as it facilitates access to ports and may even allow *door-to-door* shipping, thus reducing costs (Hummels 1999). This idea is reinforced by the fact that airports in coastal cities are normally the most important gateways to international air connections. The coast also includes spaces with a greater propensity to develop a standard type of tourism. Rappaport and Sachs (2003) studied the relevance of the coast in the US economy, finding clear correlations not only with population density, but also with productivity and growth.

As Polèse (2009) points out, not only absolute location matters because industries (and therefore economic activity and employment) are always drawn to places best suited for commerce and interaction with markets; but *proximity to size*—relative location—also matters.

Thus, spaces can be classified according to their *size* into 'metropolitan areas', 'urban areas' and 'rural areas,' and according to their *proximity to major metropolitan areas* into 'centre' and 'periphery.' This is consistent with the classic idealised landscape described in Christaller (1935), Lösch (1938) and Von Thünen (1826)—with one big marketplace at the centre (the metropolis) surrounded by some areas of smaller population size which, depending upon their respective distances to the metropolis would be categorised as either centre or periphery.

The problem lies in how these spaces can be defined in terms of being located 'close' or 'far' from the metropolis. An ingenious way of solving this dilemma was recently proposed by Partridge et al. (2008, 2009) based upon Christaller's ideas regarding the hierarchy of places, and the connection suggested by Zipf (1949) between urban size and the hierarchal position of the cities. If it is assumed that only large cities are able to offer a full range of goods and services, and if only the distance to the closest large metropolitan area (the one ranking highest in Christaller's hierarchy) is considered, this assumption neglects the fact that certain goods and services are also offered in smaller and nearer cities. One way to address this is to define a set of incremental distances to each tier (size category) of urban area. This is a way to measure the *relative* location of an area (to a large agglomeration) via a 'penalty' to access the goods and services offered there. Individuals and businesses need access to the higher-order services, urban amenities, highly qualified employment/labour and lower-cost products that are only present in highly populated places due to the presence of strong agglomeration economies.

3 An Empirical Proposal: Local Units, Databases, Variables and a Basic Linear Model

3.1 *A Starting Point for the Empirical Analysis: A Proper Definition of* Local Level

What is meant by *local* level? A number of researchers, initially in the USA in the 1960s (Fox and Kumar 1965), and then in Europe in the 1970s (Smart 1974), have devised quantitative techniques for the identification of local spatial units that were consistent with the theoretical framework of the field of regional economics. In order to delimitate these *functional economic areas* that are defined by purpose rather than by the administrative region boundaries,[1] different criteria can be utilised, such as

[1]Administratively, Spain is divided into 8106 municipalities that, excluding Ceuta and Melilla, are aggregated into 50 provinces (NUTS III level); those 50 provinces are again aggregated into seventeen Autonomous Communities or NUTS II regions of very different population sizes. The number of municipalities also varies widely within each province and ranges from 34 in Las Palmas to 371 municipalities in Burgos.

spatial uniformity, homogeneity between the units or even differentiation between them, as is the case of an urban centre surrounded by an urban fringe (Meyer 1963).

When studying local employment growth—and assuming similar behaviour for GDP growth—it is best to use a functional area that guarantees that resident workers are also generating most of the production or GDP within the area, in order to ensure that when local employment grows, so does economic growth. These functional areas that internalise the *home-to-work* journeys of their residents are called *Local Labour Market* (LLMs) in the literature. A complete explanation of the method to delineate these areas developed for the Italian National Institute for Statistics (ISTAT) can be found in Sforzi and Lorenzini (2002) and Sforzi (2012). This methodology was applied to Italy in Sforzi et al. (1997) and the same methodology was also applied to Spain by Boix and Galleto (2004), who identified 806 LLMs.

Using the 8108 Spanish municipalities as building blocks, Boix and Galleto (2004) apply an algorithm of five stages. This process highlights candidates to be the centre of a LLM, gradually adding other municipalities to generate the 806 spatial conglomerates used in this study. There are two principles underlying the algorithm which make possible to delineate functional regions (in the economic sense): labour self-containment (a minimum of 75% of employed residents work within the area) and commuting (a maximum of 25% of employed residents commute to jobs outside of the area). Therefore, a Local Labour Market encompasses within the same unit the vast majority of labour- and income-related movements and is an area with high internal homogeneity and, at the same time, high external heterogeneity (Rubiera and Viñuela 2012).

As mentioned previously, these *travel-to-work* areas have very interesting characteristics for the purpose of our analysis. Recognising that labour plays a basic role in the lives of individuals and guides their territorial behaviour with regard to the election of the municipality where they live and also work, LLMs have the advantage of including both locations within the same spatial unit. This is an important quality when using databases at a very high level of disaggregation, as some information regarding the place of work can be censored for small municipalities due to confidentiality concerns. However, this is not usually the case for information on employed population living in small municipalities. The use of LLMs as the local unit of analysis, given the availability of data on the municipality of residence of the working population, guarantees that we are considering nearly the total employment growth in the area regardless of the (unknown) municipality of work (Sforzi 2012). Moreover, a LLM describes a place that corresponds to the area where the local population develops most of its economic and social relationships; it is a place where the common interest of the local population can be identified as a whole, and therefore it could be considered an appropriate level for implementing policies at local or regional level (Parr 2008). However, the implementation of such policies requires some degree of municipal cooperation or coordination, something theoretically simple that, however, for political reasons usually confronts lots of resistance.

3.2 Second Step: Database Restrictions and Possible Variables for the Analysis

On the basis of the relationships between employment creation (Sect. 2) and the measurable local growth factors, our objective is to analyse this causal mechanism using LLMs as the *local* unit of analysis.

Despite the severe limitations on statistical information at high levels of spatial disaggregation, the Spanish Censuses, administered by the National Statistics Institute of Spain (INE), provide information on some relevant economic variables that have been identified as local growth factors. Although there are partial updates every three years, a comprehensive database only becomes available every ten years. The last two available Spanish Censuses are from 1991 and 2001.

Employment and population figures can be calculated for each of the 806 Spanish LLMs using the 1991 and 2001 Censuses. In our empirical proposal, the dependent variable will be the logarithmic employment growth (G_{emp}) in each LLM, calculated as the difference between the logarithm of the employment in 2001 and the logarithm of the employment in 1991. The LLMs initial population size (P) is used as an explanatory term. Certainly, population and employment in a local area can have a bivariate causal relationship (Freeman 2001). In this case, as many terms that will be included in the estimation are rates, we will use the logarithm of the population to preserve consistency.

The Censuses also offer data regarding the highest level of education reached by the resident population at the local level. Thus, a variable used to capture the effect of human capital endowment in the LLM, could be comprised of the percentage of the population living in a particular LLM in 1991 who holds a university degree (Ed).

Data regarding employment by type of sector is also available within the Censuses. From this data, location quotients (LQ) can be calculated for each LLM and sector. The location quotients compare the employment share in one sector in a LLM with the equivalent share at national level and therefore provide information about the specific sectors in which LLMs specialise. Moreover, following Shearmur and Polèse (2005) the degree of specialisation/diversity of the LLM in the base year 1991 (S) can be calculated via the following specialisation index:

$$S_r = \text{Ln}\left[\frac{1}{\sum_{i=1}^{n} E_i}\sqrt{\sum_{i=1}^{n}(E_i(\text{LQ}_i - 100))^2}\right] \tag{1}$$

where S_r is the specialisation index for area r; LQ_i is the location quotient of sector i for area r; and E_i is the employment in sector i for area r. Accordingly, values tend to $-\infty$ when the degree of specialisation in the LLM is identical to that of the whole national economy; the value tends to $+\infty$ as the specialisation profile of the LLM diverges from the overall Spanish specialisation (i.e. when the LLM is more specialised in one or more of the n sectors analysed).

Information on the LLMs' structural factors was collected both from the Spanish Census and the Spanish National Centre for Geographical Research (CNIG). Data regarding LLMs' population size was calculated from the municipal population data available in the Spanish Census. To calculate the linear (LD) or incremental distances (ID), we used data from the digital maps of the CNIG, which provide the longitude (X) and latitude (Y) coordinates for all Spanish municipalities.

We can classify the LLMs first according to *size* and then to *distance to size*. The first column of Table 1 presents the distribution of Spanish LLMs by population size in four tiers. The two first tiers, LLM_1 and LLM_2, correspond to the largest *metropolitan areas* or centres, to follow Christaller's nomenclature. Given the big gap in size between Madrid and Barcelona metropolitan areas and those classified as LLM_2 (with more than 500,000 but less than 2,500,000 inhabitants), we considered it appropriate to distinguish between these two levels. The next level (LLM_3) includes cities of more than 500,000 inhabitants but less than 500,000. Finally, those local labour markets with less than 50,000 inhabitants are considered rural areas (LLM_4).

Once LLMs are organised in tiers by population size, regarding *distance to size* either of the following can be considered: (i) the absolute distance (Euclidean or linear distance) to the main metropolises, in this case Madrid or Barcelona; or, (ii) the relative distance (incremental distance) to the cities of various tiers as proposed in Partridge et al. (2008, 2009).

Table 1 Distribution of LLMs by population size (1991)

Population size classification	Number of LLMs	Number of municipalities	% of total population	Category
LLM_1 > 2,500,000 inhabitants	Madrid	152	20.58%	Metropolitan areas (MAs)
	Barcelona	51		
2,500,000 inhabitants ≥ LLM_2 > 500,000 inhabitants	Valencia	52	15.15%	
	Sevilla	39		
	Bilbao	59		
	Zaragoza	95		
	Malaga	20		
	Palmas de Gran Canaria	15		
	Sta. Cruz Tenerife	17		
500,000 inhabitants ≥ LLM_3 > 50,000 inhabitants	104 LLMs	2911 municipalities	39.35%	Urban areas (UAs)
LLM_4 ≤ 50,000 inhabitants	693 LLMs	4697 municipalities	24.92%	Rural areas (RAs)
Total	806 LLM	8108 municipalities	38,871,359 inhabitants	

Source Authors' own elaboration with 1991 Spanish Census data (INE 2007)

However, as Polèse (2009) points out, the location of a LLM must be considered not only with regard to the national urban system (relative location), but also in terms of its geographic coordinates (absolute location). Therefore in order to capture the absolute location, data about LLMs includes longitude (X) and latitude (Y) coordinates, and also a variable indicating whether the LLM is located in the Atlantic Coast (AC) or in the Mediterranean and South Atlantic Coast (MC).

As a *political* factor, a dummy variable *Cap* is also included, taking the value 1 if the LLM contains an administrative capital of a province, and the value 0 otherwise. This variable captures the influence of having been identified as the administrative centre of the province, thereby concentrating a large proportion of public sector jobs and offering public services. One might wonder why this type of capital-city LLM could not be identified simply via population size, as areas including the capital city are usually the larger ones in terms of population. However, this criticism can vanish when including such as important *geo-structural* factor as *size*.

All of the variables available at LLM level that will be used in the empirical model are summarised in Table 2.

3.3 Final Step: Proposing a Simple Linear Model for the Estimation

The local growth factors (*socio-economic*, *geo-structural* and *political*), the variables which are feasible to include in this analysis subject to data constraints, and the local spatial units chosen (LLMs), in this section, the model to be estimated and the methodology will be defined. Our proposal is to estimate a set of models which become increasingly more complex, ranging from the simplest linear regressions considering only the geo-structural factors to a spatial auto-regressive methodology including all of the local growth factors presented in Table 2. In this way, we intend to quantify and come to understand the individual contribution of the different sets of growth factors. This is undertaken in an effort to provide answers to the questions raised at the beginning of this chapter, bringing us closer to a final model that identifies which LLMs grow faster and why.

The final linear model to be estimated is given by Eq. 2 if using linear distances, and Eq. 3 if using incremental distances:

$$
\begin{aligned}
G_{\text{emp}} = \alpha &+ \left[\beta_1 \text{Ed} + \beta_2 S + \beta_3 S^2 + \sum_{i=1}^{n} \beta_4^i \text{LQ}_i\right] \\
&+ \left[\gamma_1 P + \gamma_2 \text{AC} + \gamma_3 \text{MC} + \gamma_4 X + \gamma_5 Y\right] \\
&+ \delta_1 \text{LD} + \zeta_1 Cap + \varepsilon_i
\end{aligned} \tag{2}
$$

$$
G_{\text{emp}} = \alpha + \left[\beta_1 \text{Ed} + \beta_2 S + \beta_3 S^2 + \sum_{i=1}^{n} \beta_4^i \text{LQ}_i\right]
$$

Table 2 Variables and data used in the empirical approach

Variable	Term	Variable definition	Data source
Local employment growth	G_{emp}	Logarithmic employment growth: (1991–2001)	1991 and 2001 Spanish Census. Spanish National Institute of Statistics (INE)
Socio-economic growth factors			
Human capital endowment	Ed	Educational level: percentage of resident population with a university degree	Authors' own elaboration using data from the 1991 Spanish Census. Spanish National Institute of Statistics (INE)
Specialisation/diversification	S S^2	Specialisation index	Authors' own elaboration using data from the 1991 Spanish Census. Spanish National Institute of Statistics (INE)
Type of specialisation	LQ	Location quotient	Authors' own elaboration using data from the 1991 Spanish Census. Spanish National Institute of Statistics (INE)
Geo-structural factors			
Size	P	Population logarithm 1991	1991 Spanish Census. Spanish National Institute of Statistics (INE)
Absolute location	AC MC	Cantabrian and North-Atlantic Coast Mediterranean and South-Atlantic Coast	Dummy variables. Value 1 if the LLM contains a coastal municipality. 0 otherwise Spanish National Centre for Geographical Research (CNIG 2001)
	X	Longitude coordinate	Spanish National Centre for Geographical Research (CNIG 2001)

(continued)

Table 2 (continued)

Variable	Term	Variable definition	Data source
	Y	Latitude coordinate	Spanish National Centre for Geographical Research (CNIG 2001)
Relative location (linear distance)	$\mathrm{LD}_{\mathrm{LLM}_1}$	Linear distance to the highest ranked LLM	Spanish National Centre for Geographical Research (CNIG 2001)
Relative location (incremental distances)	$\mathrm{ID}_{\mathrm{LLM}_1}$ $\mathrm{ID}_{\mathrm{LLM}_2}$ $\mathrm{ID}_{\mathrm{LLM}_3}$	Incremental distances from a LLM to the nearest LLM in each tier	Spanish National Centre for Geographical Research (CNIG 2001)
Political factors			
Capital city	*Cap*	LLM containing capital of province or autonomous community	Dummy variable. Value 1 if the LLM contains the capital of the province or autonomous community. 0 otherwise

$$+\left[\gamma_1 P + \gamma_2 \mathrm{AC} + \gamma_3 \mathrm{MC} + \gamma_4 X + \gamma_5 Y\right]$$
$$+\sum_{j=1}^{n} \delta_1^j \mathrm{ID}_{ij} + \zeta_1 Cap + \varepsilon_i \quad (3)$$

In these equations, the first bracket corresponds to *socio-economic growth factors*, where Ed the percentage of population with university education in each LLM;S is the specialisation index, which is also included in a quadratic form to control for possible non-linear effects, and LQ_i are the location quotients calculated for each sector in each of the LLMs.

The second set of variables, inside the second set of brackets, corresponds to *geo-structural factors*, where P is the logarithm of the population; AC and MC are dummy variables that take the value 1 when the LLM includes a coastal municipality (Atlantic or Mediterranean coast), and X and Y are the longitude and latitude coordinates, which represent the absolute location. As for the relative location, Eq. 2 includes the Euclidean distance, LD, to the closest LLM with more than 2.5 million inhabitants; alternatively, Eq. 3 considers the incremental distance, ID, to the three tiers of LLMs, classified by size. Estimating both models with different relative location measures provides an interesting comparison. Finally, *Cap* is the dummy variable that takes a value of 1 when the area contains the capital city of a province or autonomous community.

However, before estimating these linear models (Eqs. 2 and 3), we test the *socio-economic growth factors* isolated from the *geo-structural* factors as the latter are not susceptible to policy intervention. Thus, we estimate a model that includes the socio-economic growth factors only (*A*) and another that includes only the geo-structural factors (*B*), which cannot be easily modified, if at all, in the short to medium run.

4 Employment Growth Factors in Spain: Results

4.1 The Map of Local Employment Growth in Spain

Are there a priori expectations regarding employment growth in Spain? Before analysing the local growth factors, it is useful to represent the spatial distribution of employment growth during the study period in the 806 Spanish LLMs (Fig. 1).

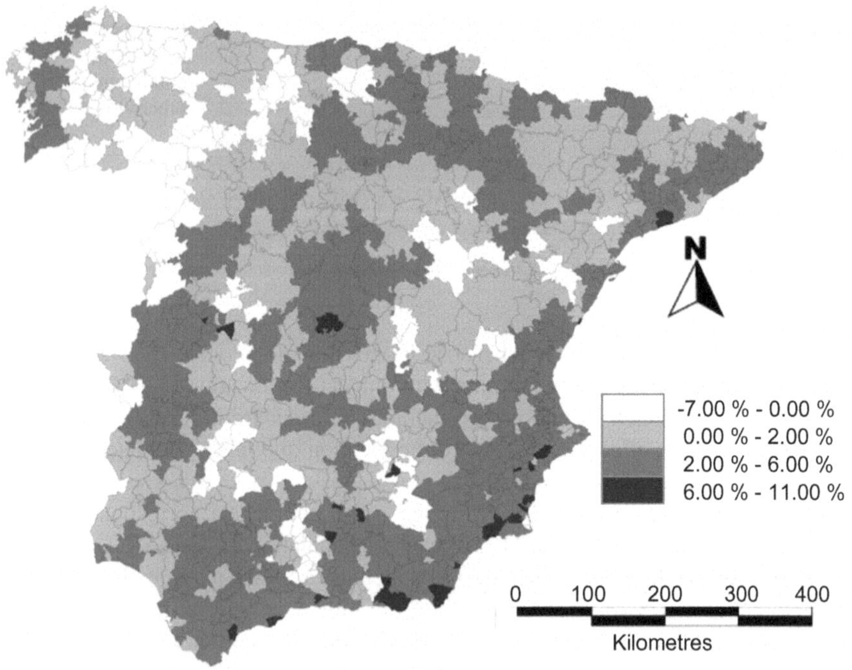

Fig. 1 Annual employment growth rates for Spanish LLMs (1991–2001). *Source* Authors' own elaboration with 2001 Spanish Census data (INE 2007); and Spanish Municipal Division map (CNIG 2001)

As expected, the Mediterranean coast comprises many LLMs above the half of the employment growth distribution, as do the areas around Madrid and Barcelona metropolitan areas. Other isolated groups can be detected also in the North and North-West. In sum, the location of faster-growing areas does not seem to be arbitrary. Our aim is either to uncover common factors responsible for this growth, and to determine the importance of attributes inherent to the LLM, such as the geo-structural factors. If in the end location (absolute and relative) explain most of their successes, then spatial dependence between the LLMs should be considered. Such a result would have important implications for the coordination of economic policies at the subnational level.

4.2 First Results with the Basic Models

As can be observed in Table 3 by comparing the results for the second and third columns, the total of all *geo-structural* and *political* factors has almost the same explanatory capability as the total of all of the *socio-economic* and *political* factors This result confirms the remarkable importance of population size and geographical location for understanding local employment growth. All the variables are then analysed jointly as proposed earlier in Eqs. 2 and 3, and the results are shown in the same table as model 2 and model 3, respectively.

In model 2, a linear distance to the nearest large metropolis will be used to represent the relative distance, while model 3 adds more information about the relative position via inclusion of the concept of incremental distances.

In relation with the absolute positions of the LLMs, as expected, the coastal LLMs grow faster than those located inland; specifically, the LLMs located along the Mediterranean coast are growing faster than the Atlantic coastal ones. However, after controlling for the rest of determinant factors, location on the Atlantic coast has more effect over growth than location on the Mediterranean coast. Growth is more intensely affected by absolute location in those territories located in the northeast of the country, as results for the *Longitude* (X) and *Latitude* (Y) coordinates indicate. The Spanish northeast is a highly urbanised area with easy access to the commercial corridor and near the French/European border.

Regarding the local economic structure, the level of specialisation is crucial to understand local employment growth: the more specialised the territory, the faster its growth. However, the sector in which the LLM is specialised also plays an important role: the Location Quotients (LQ) show positive values for Manufacturing and Construction, as well as for Public Services and Business Services. Once the location quotients are included in the model, the expected positive sign of the political factor included (*Cap*) is no significant. Being the capital city of a province or autonomous community does not positively affect growth beyond what corresponds to its concentration of public organisations and business services.

The size of the LLM measured by its population (P), and the level of education (E), both require a careful interpretation. Concerning the level of education, the

Table 3 Local employment growth: basic linear models[a,b]

	Socio-economic and political factors	Geo-structural and political factors	Equation 2 model 2 linear distance	Equation 3 model 3 incremental distances
Constant	−0.256**	0.045	−0.142	0.015
Ed (education level)	−5.393***	–	−1.348	−1.432*
S (specialisation index)	0.129***	–	0.084**	0.086**
S^2 (sqr specialisation index)	−0.017***	–	−0.011***	−0.011***
LQ values				
LQ manufacturing	0.001***	–	0.001***	0.001***
LQ construction	0.001***	–	0.001***	0.001***
LQ public services	0.002***	–	0.001***	0.001***
LQ transport and storing	0.000	–	0.000**	0.001**
LQ RS and business services	0.003***	–	0.001***	0.001***
LQ financial services	0.000	–	0.000	0.000
P (LOG population)[b]	–	0.015**	−0.019**	−0.007***
AC (Atlantic coastal) (1/0)	–	0.164***	0.112***	0.133***
MC (Mediterranean c.) (1/0)	–	0.133***	0.110***	0.099***
X (longitude)	–	−0.025***	−0.015***	−0.010***
Y (latitude)	–	0.040***	0.034***	0.033***
LD_{LLM_1}	–	−0.003	0.011	–
ID_{LLM_1}	–	–	–	−0.006
ID_{LLM_2}	–	–	–	−0.009
ID_{LLM_3}	–	–	–	−0.028**
Cap (capital) (1/0)	−0.020	0.037	0.027	0.030
Moran's I	19.575***	19.243***	14.233***	7.387***
Adjusted R^2	0.320	0.351	0.459	0.486
F-Snedecor[c]	36.580***	59.397***	41.049***	36.756***

*, **, *** significance at 10%, 5%, 1% level

[a]Dependent variable: Logarithmic growth rate of employment (Gemp)

[b]*Note* Local and spatially lagged variables are for 1991. The dependent variable, logarithmic growth rate of employment, is calculated over the period 1991–2001

[c]The F-Snedecor represents the global significance of the regression

estimations show that its effect, if significant, is not positive at the local level. This is a counterintuitive result that contradicts the human capital theory and the educationally oriented policies designed to promote growth at the regional level. Some explanations for this phenomenon were given by Pritchett (2001), who states that the negative and significant effect of higher education may be due to: (i) the creation of ill-directed cognitive skills; (ii) stronger growth of the supply of educated workers than that of the demand for them; or (iii) failure in the creation and transfer of knowledge. It is important to note that the effect of education on growth vary greatly across countries, and even within the same country depending upon the time frame under study (Shearmur and Polèse 2005).

Regarding size—as measured by the initial population living in the LLM—the a priori expectation is that a positive relationship with economic growth exists if agglomeration economies are operating. As for the Spanish case, the result for this variable is significant but *negative*. This may indicate the presence of agglomeration *dis*economies: the bigger the LLM, the higher the destruction/expulsion of employment and economic activity. According to this result, over this decade the centrifugal forces (congestion, pollution, loss of amenities which were once a draw, etc.) were superior to centripetal ones at least at the local level.

However, when considering those results in conjunction with the results regarding LLMs' relative locations, the LLMs' population sizes are probably not as important as proximity to an urban LLM—that is, to city of a minimum size. In Eq. 2, the LLMs' relative locations are quantified by linear distance to the nearest large metropolis (of over 2.5 million inhabitants) and added to the specification, while in Eq. 3 the idea of *incremental distances* is introduced. The Proximity to Madrid and Barcelona metropolitan areas—those over 2.5 million inhabitants—is not significant in explaining growth. The only variable that is significant among the LD and ID variables is the ID variable for cities of 250,000–500,000 inhabitants (LLM_3), as included in Eq. 3 The effect is positive but small, explaining only 2.8% of total variance.

4.3 A Rural–Urban Variation Over the Basic Model

The sample used in this analysis includes LLMs of very different population sizes and characteristics. Out of the 806 LLMs, there are more than 600 that can be considered rural, with a population below 50,000 inhabitants, which is a threshold commonly used in the literature. This heterogeneity in the degree of urbanisation can be masking different employment growth dynamics between rural and urban areas, which may not be affected in the same ways by the same factors—and, if so, may be affected at different levels of intensity.

To control for the size heterogeneity, we have divided the sample into two subsamples—rural and urban LLMs—and estimated the models for these two subsamples (Eq. 3/Rural and Eq. 3/Urban), measuring their relative locations via incremental distances (Table 4).

Table 4 Local employment growth: rural–urban distinction[a,b]

	Equation 3/Rural (rural LLMs only)	Equation 3/Urban (urban LLMs only)
Constant	−0.078	0.655**
Ed (education level)	−2.080**	−0.290
S (specialisation index)	0.088*	0.042
S^2 (sqr specialisation index)	−0.012**	0.006
LQ values		
LQ manufacturing	0.001***	0.000
LQ construction	0.001***	0.000
LQ public services	0.001***	−0.001
LQ transport and storing	0.001***	−0.001
LQ RS and business services	0.001***	0.001**
LQ financial services	0.001	0.001
P (LOG population)[b]	−0.022**	−0.048**
AC (Atlantic coastal) (1/0)	0.128***	0.055
MC (Mediterranean c.) (1/0)	0.153***	0.054*
X (longitude)	−0.010***	−0.002
Y (latitude)	0.033***	0.031***
$\mathrm{ID}_{\mathrm{LLM}_1}$	0.001	−0.017*
$\mathrm{ID}_{\mathrm{LLM}_2}$	−0.013	0.004
$\mathrm{ID}_{\mathrm{LLM}_3}$	−0.017***	−0.044
Cap (capital) (1/0)	–	0.073*
Moran's I	7.140***	1.794**
Adjusted R^2	0.454	0.398
F-Snedecor[c]	32.831***	4.818***

*, **, *** significance at 10%, 5%, 1% level

[a]Dependent variable: Logarithmic growth rate of employment (Gemp)

[b]*Note* Local and spatially lagged variables are for 1991. The dependent variable, logarithmic growth rate of employment, is calculated over the period 1991–2001

[c]The F-Snedecor represents the global significance of the regression

Given the low Adjusted R^2 obtained and the fact that many variables are non-significant, the growth model and the variables considered are not capable of explaining employment growth in *urban* areas. Large urban areas are assumed to undergo processes of endogenous growth and are thus less affected either by the geo-structural factors or the economic environment. However, their location along the coast is significant, although only for the Atlantic coast, explaining 5.4% of total variance. For the rural LLMs, both coastal locations (Atlantic or Mediterranean) are significant, explaining 15.3% and 12.8% of total variance, respectively. The urban areas close to the largest metropolises (Madrid and Barcelona) enjoy higher growth rates; this may be due to complementarities with higher quality goods and services only offered in the largest cities.

The ability of this particular growth model to address employment growth is better for rural LLMs (higher Adjusted R^2). In rural areas, location quotients are significant at 1% level (except for Financial Services) although each industry contribution is as low as it could possibly be, explaining only 1% of total variance. Thus, the variables which are most important to the rural model are the two coastal variables, the *Mediterranean* coastal variable being of primary importance; both coastal variables are positive and significant. *Latitude* having a positive effect, presents as the most important factor, whereas *Longitude* has a negative effect of lesser magnitude. In other words, moving towards the North and the East has a positive effect on employment growth of the incremental distance variables in Eq. 3/Rural, only one is significant—ID_{LLM_3}, representing the distance from a rural LLM to the nearest medium-sized *urban* LLM (50,000–500,000 inhabitants), and its effect is positive, though small. Interestingly, while for the rural LLMs the important thing is being located close to medium-sized LLMs, for the urban LLMs (Eq. 3/Urban) only being close to a large metropolises is significant—ID_{LLM_1} representing the distance to the nearest metropolitan area of more than 2.5 million inhabitants.

The result for *Education Level* (measured as the percentage of resident population with a University degree) requires special attention. Its effect on overall employment growth (Eq. 3) is negative and significant, while the results for the subsamples show no significance in the urban case (Eq. 3/Urban) and are negative and significant for the rural subsample (Eq. 3/Rural). This unexpected finding compels further research and analyses.[2]

Finally, the dummy *Cap* is only considered in Eq. 3/Urban, as no rural LLM (less than 50,000 inhabitants) is simultaneously the administrative capital of the province or autonomous community. Thus, when working with two subsamples the magnitude of the effect of this variable becomes significant for the urban LLMs after controlling for the rest of factors.

[2]Our intuition is that the educational level effect on overall employment growth might be biased by the effect of the rural LLMs, where having a pool of highly educated labour force does not necessarily imply that the area offers high qualified jobs and/or experiences high rates of employment growth.

4.4 Analysis with Spatial Dependence Models

Based upon results from Model 3, geo-structural factors appear to explain a large part of local employment growth. Applying a spatial analysis approach, in this section we carry out further analysis of local employment growth to extend the model and include possible spatial effects. Spatial autocorrelation or spatial dependence exists when there is a functional relationship between processes occurring in a LLM and processes occurring in proximate LLMs.

The application of these methods is justified by the values obtained via the Moran's I test in the linear models estimated in the previous section. The values resulting from the Moran's I test indicate that one should reject the null hypothesis of random spatial distribution of the dependent variable, thus corroborating the existence of spatial autocorrelation.

Specifically, the following spatial dependence models were estimated:

(i) *Spatial lag model*, considering the influence of employment growth in neighbouring LLMs and measuring this effect via a *Rho* parameter.
(ii) *Spatial error model*, introducing the spatial auto-regressive component which allows the existence of spatial dependence in the error term to be parsed from the random error through a *Lambda* parameter in the estimation.
(iii) *Spatial Durbin model*, which is a combination of the two previous models, allowing for spatial dependence in the dependent and the independent variables simultaneously.

A review of the various methods for addressing spatial auto-regressive processes can be found in the seminal works from Anselin (1988), Cliff and Ord (1981), Griffith (1988, 2003), Haining (1990) and Anselin et al. (2004). In this analysis, the three different models were estimated always using a proximity matrix *W*, calculated under the rule of the k-nearest neighbours—in this, case the ten nearest neighbours.

The estimation and comparison of the results of these three models can provide additional answers to the questions that motivated this research (in particular, which local areas grow faster and why), as they take into account the economic dynamics of surrounding areas. In other words, these models investigate the importance of not only an LLM's own economic performance, but that of neighbouring LLMs' economic performance. The results of these three models are presented in Table 5.

Here, it is worth sounding a cautionary note before proceeding with a discussion of the results. Due to the estimation methods employed, estimated coefficients cannot be interpreted in the same straightforward manner as the ordinary least squares (OLS) regression results. The interpretation is more complex in spatial econometric models due to the spatial multiplier effect and spatial connectivity—i.e. a change in an explanatory variable in LLM_i will have a direct impact on LLM_i, as well as an indirect impact on other LLMs.[3] Considering the complications associated with interpreting

[3]For a detailed discussion of these complexities, we refer the reader to Anselin (2003) or LeSage and Pace (2009).

Table 5 Local employment growth with spatial dependence[a,b]

	Spatial lag model[c]	Spatial error model[c]	Spatial Durbin model[c]	
			Factor	Spatial lag
Constant	−0.106	−0.055	0.298	–
Rho	0.508***	–	0.530***	–
Lambda	–	0.573***	–	–
Ed (education level)	−1.456**	−1.927**	−2.061**	1.031
S (Specialisation index)	0.072**	0.076**	0.068**	−0.071
S^2 (sqr specialisation index)	−0.009**	−0.009***	−0.009**	0.005
LQ values				
LQ manufacturing	0.001***	0.001***	0.001***	0.000
LQ construction	0.001***	0.001***	0.001***	0.001 **
LQ public services	0.001**	0.001***	0.001***	0.001
LQ transport and storing	0.000**	0.001**	0.001**	0.000
LQ RS and business services	0.001***	0.001***	0.001***	0.001
LQ financial services	0.000	0.000	0.000	0.000
P (LOG population)[b]	−0.015**	−0.013*	−0.014	−0.025
AC (Atlantic coastal) (1/0)	0.096***	0.124***	0.131***	−0.045
MC (Mediterranean c.) (1/0)	0.082***	0.102***	0.075***	−0.025
X (longitude)	−0.003	−0.016***	0.017	−0.015
Y (latitude)	0.016***	0.035***	0.145***	−0.125 **
ID_{LLM_1}	−0.007	−0.008	−0.012	0.010
ID_{LLM_2}	−0.002	−0.002	0.003	−0.002
ID_{LLM_3}	−0.025*	−0.046**	−0.059	0.057
Cap (capital) (1/0)	0.040	0.035	0.039	−0.039

(continued)

Table 5 (continued)

	Spatial lag model[c]	Spatial error model[c]	Spatial Durbin model[c]	
			Factor	Spatial lag
Lagrange multiplier	155.840***	164.825***		
Robust Lagrange multiplier	10.233***	19.218***		
Likelihood ratio test	98.197***	99.101***	80.493***	
Akaike information criterion (AIC)	−781.13	−782.04	−764.97	

*, **, *** significance at 10%, 5%, 1% level
[a]Dependent variable: Logarithmic growth rate of employment (Gemp)
[b]*Note* Local and spatially lagged variables are for 1991. The dependent variable, logarithmic growth rate of employment, is calculated over the period 1991–2001
[c]The W contiguity matrix was calculated using the k-nearest neighbours proximity rule (order 10)

coefficients in spatial econometric models, we instead focus on assessing whether results are reasonable, and providing a general sense of the magnitude of results.

In the *spatial lag* model, the dependent variable is introduced as an explanatory term (Rho) weighted by the contiguity matrix W. Its effect is significant and positive, which underlines the importance of the neighbours' employment growth for a given LLM. The significance of the lambda parameter in the *spatial error* model reflects unobserved interactions between LLMs through unobserved factors that are spatially correlated. However, after controlling for spatial dependence, the socio-economic and geo-structural factors are still significant and consistent with the results for the linear model (Eq. 3).

In the *spatial Durbin* model, both of the previous effects are included: on the one hand, there is the spatially lagged dependent variable, and on the other hand there are the lags of the explanatory factors. The significance levels of the variables in the spatial Durbin model are similar to those for the other spatial models. However, the spatial error model seems to explain better the spatial dependence according to the likelihood ratio test (highest value), the Akaike information criterion (lowest value) and the Lagrange multipliers (highest value).

Thus, there is evidence for positive spatial dependence: higher employment rates in contiguous LLMs are associated with higher employment growth in a particular LLM and vice versa.

5 Conclusions

The aim of the analysis developed here was to shed light upon the processes underlying local employment growth in Spain. As there is a lack of GDP data for the scale of analysis considered, this study of employment growth was proposed as a means of obtaining clues concerning local in general. Spanish Local Labour Markets (LLM) served as basic spatial units, and data from the Spanish Censuses of 1991 and 2001, and from the Spanish National Centre for Geographical Research, were used to estimate the empirical models proposed. The specification has a logarithmic employment growth rate as the dependent variable, including a compendium of socio-economic, geo-structural and political factors as explanatory terms. To account for the distance to more populous regions, we used incremental distances as an alternative to the traditional linear distance approach. We estimated nine models of increasing complexity: the first four consisted of simple linear regressions (with the gradual addition of variables); the following two estimated the most complete of the previous specifications for rural and urban subsamples; and the last three were estimated using spatial econometric methods which account for the impact of neighbouring spatial units.

When considered separately, socio-economic growth factor variables and geo-structural growth factor variables have similar explanatory power. This demonstrates that policies may be effective in stimulating employment growth to some extent—but a comparable part of the growth comes from features that cannot be affected by political decisions, namely, the geographical characteristics of an area. The geographical terms are significant, highlighting the importance of being located near the European commercial corridor (northeast) and on the coasts. Regarding the *socio-economic* variables, our results point to the importance of the industrial structure: being a specialised economy has a positive effect upon employment growth. Specialisation in sectors such as Manufacturing, Construction, Public services and Business services contribution to employment growth is significant although very little in magnitude. The negative effect of the educational level is unexpected and this result should be subject to further investigation regarding the urban/rural split in education and labour characteristics.

Population has a negative impact that may be due to agglomeration diseconomies, possibly leading to employment expulsion favouring medium-sized regions. It suggests that population size, in and of itself, does not translate into employment growth. More than size, what is relevant is the distance to size (incremental distances). The results show the significant and positive effect of being proximate to medium-sized LLMs possibly due to fact that these medium-sized cities and surrounding areas may feature some higher-level goods and services without the same agglomeration diseconomies of larger cities.

When the estimation is carried out separately for urban and rural areas, it is demonstrated that the proposed model fits the rural case better. It seems that urban areas follow a path of endogenous growth affected only by its own size and the capital condition of the city. For the rural areas, nearly all the socio-economic and geographical variables have a significant impact, except for the incremental distance variables ID_{LLM_1} and ID_{LLM_2}. The negative and significant effect of educational level in the rural case is noteworthy and may bias the results of the estimations for the complete sample, and as such, may serve as an explanation for the unexpected previous outcome for this variable; however, this must be subject to further analysis.

Spatial dependence tests confirm our thoughts about the significance of location, as the spatial distribution of employment growth does not seem to be random. The spatial lag estimation uncovers a positive relationship between our own employment growth and the employment growth of neighbouring regions, while the spatial error estimation highlights the relevance of the unobserved characteristics of neighbours. In the spatial Durbin estimation, the existence of unobserved factors interacting spatially between LLMs is confirmed.

In light of the results obtained from this analysis, answers can be formulated for the question that motivates our research, namely: *Which places grow faster?* The profile of a successful region in terms of employment growth would be represented by an area located in the northeast, preferably on the coast, with a specialised economy, located near areas with actively growing employment.

References

Anselin, L. (1988). *Spatial econometrics: Methods and models*. Boston: Kluwer Academic Publishers.

Anselin, L. (2003). Spatial externalities, spatial multipliers, and spatial econometrics. *International Regional Science Review, 26*(2), 153–166.

Anselin, L., Florax, R. J. G. M., & Rey, S. J. (Eds.). (2004). *Advances in spatial econometrics: Methodology tools and applications*. Berlin: Springer.

Barro, R. J., & Sala-i-Martin, X. (1995). *Economic growth*. Boston: McGraw Hill.

Beckstead, D., & Brown, W. M. (2003). From Labrador city to Toronto: The industrial diversity of Canadian cities, 1992 to 2002. In Statistics Canada (Ed.), *Insights on the Canadian economy* (Vol. 3). Ontario.

Blien, U., Suedekum, J., & Wolf, K. (2006). Local employment growth in West Germany: A dynamic panel approach. *Labour Economics, 13*(4), 445–458.

Boix, R., & y Galletto, V. (2004). Identificación de Sistemas Locales de Trabajo y Distritos Industriales en España. Dirección General de Política para la Pequeña y Mediana Empresa. Mimeo. Madrid.

Christaller, W. (1935). *Die Zentralen Orte in Süddeutschland*. Jena: Fischer.

Cliff, A. D., & Ord, J. K. (1981). *Spatial processes: Models and applications*. London: Pion.

CNIG. (2001). Mapas Digitales de los Municipios Españoles. Centro Nacional de Información Geográfica, Ministerio de Fomento. https://centrodedescargas.cnig.es/CentroDescargas/equipamiento/BD_Municipios-Entidades.zip. Accessed 13 Dec 2011.

Coffey, W. J., & Polèse, M. (1988). Locational shifts in Canadian employment, 1971–1981: Decentralization versus decongestion. *The Canadian Geographer, 32*(3), 248–256.

Coffey, W. J., & Shearmur, R. (1996). *Employment growth and change in the Canadian urban system*. Ottawa: Canadian Policy Research Networks.

Cooke, P., Heindenreich, M., & Braczyk, H. J. (2004). *Regional innovation systems: The role of governance in a globalized word*. London: Routledge.

Davis, D. R., & Weinstein, D. E. (2002). *Technological superiority and the losses from migration*. National Bureau of Economic Research Working Paper Series No. 8971. https://www.nber.org/papers/w8971. Accessed 3 Jan 2012.
Doloreux, D., Shearmur, R., & Filion, P. (2001). Learning and innovation: Implications for regional policy. An introduction. *Canadian Journal of Regional Science, 24*(1), 5–11.
Eisinger, P. K. (1988). *The rise of the entrepreneurial state: State and local economic development policy in the United States*. Madison: The University of Wisconsin Press.
Florida, R. (2002). *The rise of the creative class*. New York: Basic Books.
Fox, K. A., & Kumar, T. K. (1965). The functional economic area: Delineation and implications for economic analysis and policy. *Papers of the Regional Science Association, 15*(1), 57–85.
Freeman, D. G. (2001). Sources of fluctuation in regional growth. *The Annals of Regional Science, 35*(2), 266–294.
Galaway, B., & Hudson, J. (Eds.). (1994). *Community economic development: Perspectives on research and policy*. New York: Thompson.
Griffith, D. A. (1988). *Advanced spatial statistics: Special topics in the exploration of quantitative spatial series*. Dordrecht: Kluwer.
Griffith, D. A. (2003). *Spatial autocorrelation and spatial filtering: Gaining understanding through theory and scientific visualization*. New York: Springer.
Gutierrez, D., Rubiera, F., & Viñuela, A. (2019). Determinants of immigrants concentration at local level in Spain: Why size and position still matter. *Population, Space and Place* (forthcoming).
Gutierrez, D., Rubiera, F., & Viñuela, A. (2018). Ageing places in an ageing country: The local dynamics of elderly population in Spain. *Tijdschrift voor economische en sociale geografie—Journal of Economic and Social Geography, 109*(3), 332–349.
Haining, R. (1990). The use of added variable plots in regression modelling with spatial data. *The Professional Geographer, 42*, 336–344.
Henderson, V. (2003). Marshall's scale economies. *Journal of Urban Economics, 53*, 1–28.
Hummels, D. (1999). *Have international transportation cost declined?* mimeo, University of Chicago.
INE (2007) Censo de Población, 2001. Instituto Nacional de Estadística. https://www.ine.es/jaxi/menu.do?type=pcaxis&path=%2Ft20%2Fe243&file=inebase&L=0. Accessed 26 Nov 2011.
Jacobs, J. (1984). *Cities and the wealth of nations*. New York: Vintage.
Keating, M. (1993). *The politics of modern Europe: The state and political authority in the major democracies*. Aldershot. Edward Elgar.
Krugman, P. R. (1995). *Development, geography, and economic theory*. Cambridge: The MIT Press.
LeSage, J. P., & Pace, R. K. (2009). *Introduction to spatial econometrics*. Florida: Taylor & Francis.
Lösch, A. (1938). The nature of economic regions. *Southern Economic Journal, 5*(1), 71–78.
Mackay, R. R. (2003). Twenty-five years of regional development. *Regional Studies, 37*(3), 303–317.
Malecki, E., & Oinas, P. (1999). *Making connections technological learning and regional economic change*. Aldershot: Ashgate.
Markusen, A. (1996). Sticky places in slippery space: A typology of industrial districts. *Economic Geography, 72*(3), 293–313.
Markusen, A. (1999). Fuzzy concepts, scanty evidence, policy distance: The case for rigour and policy relevance in critical regional studies. *Regional Studies, 33*(9), 869–884.
Martin, R., & Sunley, P. (1998). Slow convergence? The new endogenous growth theory and regional development. *Economic Geography, 74*, 201–227.
Massey, D. B. (1995). *Spatial divisions of labour: Social structures and the geography of production*. New York: Routledge.
Meyer, J. R. (1963). Regional economics: A survey. *The American Economic Review, 53*, 19–54.
Parr, J. (2001). Regional economic planning: learning from past experience. In D. Felsenstein, R. McQuaid, P. McCann & D. Shefer (Eds.), *Public investment and regional economic development*. Edward Elgar, Cheltenham.
Parr, J. (2002). Agglomeration economies: Ambiguities and confusions. *Environment and Planning A, 34*, 717–731.

Parr, J. (2008). Cities and regions: Problems and potentials. *Environment and Planning A, 40*, 3009–3026.

Partridge, M. D., & Rickman, D. S. (2006). *The geography of American poverty: Is there a role for place based policy?* W.E Upjohn Employment Institute

Partridge, M. D., Rickman, D. S., Ali, K., & Olfert, M. R. (2008). Lost in space: Population growth in the American hinterlands and small cities. *Journal of Economic Geography, 8*, 727–757.

Partridge, M. D., Rickman, D. S., Ali, K., & Olfert, M. R. (2009). Agglomeration spillovers and wage and housing cost gradients across the urban hierarchy. *Journal of International Economics, 78*, 126–140.

Philo, C., Mitchell, R., & More, A. (1998). Reconsidering quantitative geography: The things that count. *Environment and Planning A, 30*(2), 191–201.

Polèse, M. (2009). *The wealth and the poverty of regions: why cities matters*. Chicago: University of Chicago Press.

Polèse, M., & Shearmur, R. (2004). Is distance really dead? Comparing the industrial location patterns over time in Canada. *International Regional Science Review, 27*(4), 1–27.

Polèse, M., Shearmur, R., & Rubiera, F. (2006). Observing regularities in location patterns: An analysis of the spatial distribution of economic activity in Spain. *European Urban and Regional Studies, 14*(2), 157–180

Porter, M. (1990). *The competitive advantage of nations*. New York: The Free Press.

Porter, M. (1996). Competitive advantage, agglomeration economies, and regional policy. *International Regional Science Review, 19*(1–2), 85–90.

Porter, M. (1998). *Competitive strategy: Techniques for analyzing industries and competitors*. New York: The Free Press.

Pritchett, L. (2001). Where has all the education gone? *The World Bank Economic Review, 15*(3), 367–391.

Quigley, J. M. (1998). Urban diversity and economic growth. *Journal of Economic Perspectives, 12*(2), 127–138.

Rappaport, J., & Sachs, J. D. (2003). The United States as a coastal nation. *Journal of Economic Growth, 8*, 5–46.

Romer, P. (1990). Human capital and growth: theory and evidence. *Carnegie-Rochester Conference Series on Public Policy, 32*(1), 251–286.

Rubiera, F., & Viñuela, A. (2012). From local units to economic regions in Spain. Where the agglomeration economies make sense. In E. Fernández & F. Rubiera (Eds.), *Defining the spatial scale in modern regional analysis: New challenges from data at local level*. Advances in spatial science. Springer, Berlin.

Sforzi, F., Openshaw, S., & Wymer, C. (1997). Le Procedura di Identificazione dei Sistemi Locali del Lavoro. In F. Sforzi (Ed.), *I Sistemi Locali del Lavoro* (pp. 235–242). Rome: ISTAT.

Sforzi, F., & Lorenzini, F. (2002). I Distretti Industriali. In: L´esperienza Italiana dei Distretti Industriali. Istituto per la Promozione Industriale (IPI), Roma.

Sforzi, F. (2012). From administrative spatial units to local labour market areas. Some remarks on the unit of investigation of regional economics with particular reference to the applied research in Italy. In E. Fernández & F. Rubiera (Eds.), *Defining the spatial scale in modern regional analysis: New challenges from data at local level*. Advances in spatial science. Springer, Berlin.

Shearmur, R., & Polèse, M. (2007). Do local factors explain local employment growth?: Evidence from Canada, 1971–2001. *Regional Studies, 41*(4), 453–471.

Shearmur, R., & Polèse, M. (2005). Diversity and employment growth in Canada, 1971–2001: Can diversification policies succeed? *The Canadian Geographer, 49*(3), 272–290.

Smart, M. W. (1974). *Labour market areas: Uses and definition*. Oxford: Pergamon Press.

Strambach, S. (2001). Innovation processes and the role of knowledge-intensive business services (KIBS). In K. Koschatzky, M. Kulicke, & A. Zenker (Eds.), *Innovation networks—concepts and challenges in the European perspective* (pp. 53–68). Heidelberg: Physica Verlag.

Von Thünen, J. H. (1826). *Der Isolierte Staat in Beziehung auf Landwirtchaft und Nationalökonomie*. Hamburg: Perthes.

Weber, A. (1929). *Theory of the location of industries*. (trans: Friedrich, C. J. from Weber's 1909 book). The University of Chicago Press
Zipf, G. K. (1949). *Human behaviour and the principle of least effort: An introduction to human ecology*. Cambridge: Adison Wesley Press.

A Spatial Optimization Approach for Solving a Multi-facility Location Problem with Continuously Distributed Demand

Jing Yao and Alan T. Murray

1 Introduction

Location theory is a branch of geographic inquiry concerned with analysis and siting of socio-economic activities, including land use, industrial production, central places and spatial competition (Murray 2008). Founded on such theory, location modeling has long been recognized as playing an important role in regional and urban planning, as well as other contexts, as location is an indispensable component for all human activities (Church and Sorensen 1996). Location-related decisions are fundamental in many aspects of human activity. Examples include, but are not limited to, locating retail stores, deploying switching centers in communication networks, selecting nature reserves to preserve threatened species, etc. In fact, whenever a question is posed about where to site goods and services, a location problem arises.

Facility location problems are usually concerned with determining where to site one or more facilities, subject to certain constraints in order to optimize objectives (Brandeau and Chiu 1989). Of particular interest, here is a minimization problem—the multi-facility Weber problem with continuously distributed demand or simply the continuous multi-Weber problem. The Weber problem is a classic facility location model tasked with identifying a site for a single facility in continuous space in order to minimize the total transportation costs from the facility to a set of fixed demand points (Weber 1909). The multi-facility extension of the Weber problem requires multiple facilities to be sited (Cooper 1963) and remains a challenge to solve (see Rosing 1992; Righini and Zaniboni 2007). When the complexity of continuously distributed

J. Yao (✉)
School of Social and Political Sciences, Urban Big Data Centre, University of Glasgow, Glasgow G12 8RZ, UK
e-mail: Jing.Yao@glasgow.ac.uk

A. T. Murray
Department of Geography, University of California at Santa Barbara, Santa Barbara, CA 93106, USA
e-mail: amurray@ucsb.edu

J. Thill (ed.), *Innovations in Urban and Regional Systems*,
https://doi.org/10.1007/978-3-030-43694-0_6

demand is added, the continuous multi-Weber problem arguably becomes one of the most difficult facility location problems to solve optimally; there is a need for good heuristics to be developed in order to address this challenge.

Continuous demand is often encountered in planning and analysis, reflecting more generally the distribution of people, risk, danger, vegetation, businesses, retail markets and customers. Typically, the continuous demand is abstracted as a finite set of discrete points in order to facilitate model formulation, as well as reduce computational expense (Miller 1996; Ouyang and Daganzo 2006; Francis et al. 2009). Though attractive for those purposes, point-based simplifications can lead to significant errors and uncertainty in analyses due to the loss of spatial detail (Murray and O'Kelly 2002; Murray et al. 2008; Francis et al. 2009; Alexandris and Giannikos 2010; Cromley et al. 2012; Yao and Murray 2013). An alternative is to describe continuous demand as a surface defined by a mathematical function (Drezner 1995; Gastner and Newman 2006; Brimberg et al. 2008; Murat et al. 2010). The issue, however, is that the exact demand distribution/function is never known or given with certainty (Yao and Murray 2013). Therefore, effectively dealing with continuously distributed demand in facility location remains a great challenge. To this end, the intent of this chapter is to address continuous demand representation in the continuous multi-Weber problem.

When facilities can be located anywhere and demand is continuously distributed, model formulation as well as its solution becomes mathematically complicated. Traditionally, the facility location problems have been solved by operations research techniques like linear and integer programming. In recent years, solution capabilities have been significantly enhanced by spatial optimization approaches incorporating geographic information systems (GIS) functions and methodologies (Church 2002; Wei et al. 2006; Church and Murray 2009; Matisziw and Murray 2009; Murray 2010; Cromley et al. 2012). GIS is used to facilitate data input and visualization, yet GIS is arguably even more valuable and meaningful when its use is integrated with modeling and solution processes (Curtin et al. 2010; Murray 2010).

The aim of this chapter is to develop an approach that combines GIS and optimization methods for solving the continuous multi-Weber problem. The next section reviews relevant research in this area. The problem specification is given in Sect. 3. Spatial representation in GIS is covered in Sect. 4. This is followed by a detailed description of the proposed method in Sect. 5 and empirical results in Sect. 6. Finally, the chapter ends with discussion and conclusions.

2 Background

As one of the first location problems formally posed, the Weber problem involves placing a single facility anywhere in space (continuous space) to serve a finite set of demand points in order to minimize total transportation costs. It has been extensively investigated and continues to be of interest since first proposed in the seventeenth century (Wesolowsky 1993; Drezner et al. 2002). The classic context is siting a factory in order to minimize the transportation costs to acquire raw materials and distribute products but could as well involve problems such as locating a fire station or hospital in order to minimize average response time. As such problems—and a vast number of extensions—can be addressed by the widely applicable Weber model, it has attracted much interest. Many location models can, in fact, be formally connected to the Weber problem (Drezner et al. 2002).

One extension of interest here is replacing discrete demand with continuous demand, usually referred to as the continuous Weber problem (Drezner 1995; Fekete et al. 2005). In this context, the Weber problem becomes a stochastic optimization problem with the goal of minimizing average distance to the demand area from the facility. The most straightforward method for dealing with a demand area is to represent it by a single point, such as a centroid (Bennett and Mirakhor 1974). Given the well-known concerns and limitations of data aggregation, the research has employed knowledge from computational geometry to evaluate average travel distance (Drezner 1995; Carrizosa et al. 1998; Fekete et al. 2005) or its bounds (Carmi et al. 2005; Abu-Affash and Katz 2008; Puerto and Rodríguez-Chía 2011). Such approximations of average distance are relatively straightforward for regular shapes but can be quite complex in other cases. Finally, some efforts have focused on numerical methods for the problem solution. For example, the Weiszfeld algorithm, a popular choice for solving the Weber problem, has been adapted for dealing with continuous extensions (Drezner and Wesolowsky 1980; Franco et al. 2008).

Another relevant extension considers multiple facilities instead of a single facility, known as the multi-Weber problem (Cooper 1963). The distinct feature of this extension is that it concerns siting multiple facilities simultaneously in continuous space with discrete demand assigned to its closest facility. The multi-Weber problem has challenged generations of researchers because the objective function is neither concave nor convex, making it difficult to find the global minima. Since the early ALTERNATE heuristic developed by Cooper (1964), many heuristics have been proposed, such as Tabu search (Brimberg and Mladenović 1996), genetic algorithms (Houck et al. 1996; Salhi and Gamal 2003), variable neighborhood decomposition (Brimberg et al. 2006), the improved ALTERNATE approach by a "transfer follow-up" (Brimberg and Drezner 2013) as well as hybrid approaches combining several heuristics (Drezner et al. 2015, 2016; Drezner and Salhi 2017). The benefit of such approaches is that they can solve large problems quickly as well as provide a good initial solution for exact approaches. In contrast, exact or optimal methods are constrained by application problem size (Rosing 1992; Righini and Zaniboni 2007). A detailed survey of exact methods and heuristic approaches can be found in Brimberg et al. (2008).

When extension involves both continuous demand and multiple facilities, the problem becomes extremely complicated, as facilities may be sited anywhere and demand varies across space. Algorithms have been proposed for the simplest assumption of a uniform demand distribution (Marucheck and Aly 1981; Drezner 1986). Work dealing with other probabilistic distributions of demand can be found in Rao and Varma (1985) and Altınel et al. (2009). To facilitate the demand allocation process, some researchers have utilized Voronoi partitions of the demand area (see Suzuki and Okabe 1995; Gastner and Newman 2006). More recently, Murat et al. (2010)

applied an approximate line search method to find a single facility location for each Voronoi polygon, but this can be computationally expensive due to a prioritization of the allocation process.

Compared to the continuous Weber problem and the multi-Weber problem, the extension of both multiple facilities and continuous demand has drawn less attention. It is worth noting that the current solution approaches for the continuous multi-Weber problem are problem-specific, inexact or computationally inefficient. Therefore, the problem of addressing multiple facilities and continuous demand represents an unresolved issue, with great potential for use and application. Given the inherent spatial nature of the problem and the availability of advanced spatial techniques linked to GIS, spatial optimization approaches represents the potential for improved problem solutions that exploit knowledge of distributed demand. This avoids the problematic assumptions which are characteristic of the existing solution approaches to the continuous multi-Weber problem.

3 Problem Specification

The continuous multi-Weber problem involves continuously distributed demand served by several facilities that may be located anywhere in continuous space. To begin, the mathematical formulation of the classic Weber problem is detailed. Then, two related extensions—the continuous Weber problem and the multi-Weber problem—are given. The formulation of the continuous multi-Weber problem is then presented.

3.1 The Weber Problem

Given demand represented by discrete, aggregated points, the objective function of the Weber problem is to minimize the sum of weighted distance to the sited facility over the finite set of demand points. This is equivalent to minimizing average distance to demand. The decision variables are the location of the facility. Consider the following notation:

i = index of demand points
(x_i, y_i) = location of demand point i
n = number of demand points
w_i = weight associated with demand point i
(X, Y) = facility location decision.

The Weber problem is as follows (Wesolowsky 1993):

$$\text{Minimize} \quad \sum_{i=1}^{n} w_i \sqrt{(X - x_i)^2 + (Y - y_i)^2} \tag{1}$$

The goal is to find the best location (*X*, *Y*) in continuous space so that the total weighted distances from the demand points to the sited facility is minimized. It should be noted that the distance function in Eq. 1 is defined by the Euclidean metric.

3.2 *Continuous Weber Problem*

As mentioned previously, the Weber problem (Eq. 1) can be extended in many ways. Previous extensions include models considering other types of distance metrics, negative weights associated with demand, or multiple objective functions, among others (Drezner et al. 2002).

An important extension is the explicit accounting for continuous variation of demand over space, rather than assuming aggregate demand points. This extension is associated with significant additional complication, as representation of a continuous surface is quite complex. Consider the following additional notation:

$g(x, y) =$ function of demand at point (x, y)
$d(x, y, X, Y) = \sqrt{(X - x)^2 + (Y - y)^2}$
$R =$ region of demand.

Discrete demand at a priori defined locations, w_i, are replaced by a function corresponding to demand at any location (x, y). Further, the distance function reflects this change as well. Formal specification of the continuous Weber problem follows (Church and Murray 2009):

$$\text{Minimize} \quad \iint\limits_{(x,y)\in R} g(x, y) d(x, y, X, Y) \mathrm{d}x\mathrm{d}y \tag{2}$$

The objective (Eq. 2) involves a double integral over the demand area instead of the sum in the Weber problem, Eq. 1. The distance, $d(x, y, X, Y)$, remains measured as a Euclidean metric, but the objective function now accounts for continuously distributed demand using $g(x, y)$, causing the calculation to be far more complex. The reason it is more complex is that in order to apply this model, $g(x, y)$ must first be mathematically defined. As previously alluded to, defining such mathematical functions usually involves approximation procedures, since the actual demand distribution cannot be known (or replicated) with certainty in practice.

Though more complicated in its formulation and calculation, the continuous Weber problem (Eq. 2) remains equivalent to Eq. 1, minimizing the average travel distance to the sited facility (Carrizosa et al. 1998).

3.3 Multi-Weber Problem

Another extension to the Weber problem considers multiple facilities. Thus, decisions regarding where each facility should be sited must also consider which demand is assigned to which facility. This makes the problem a location and allocation problem, termed a location–allocation problem. Consider the following additional notation:

$j =$ index of facilities
$p =$ number of facilities to be located
$(X_j, Y_j) =$ location of facility j
$$\Psi_{ij} = \begin{cases} 1 \text{ demand point } i \text{ is served by facility } j \\ 0 \text{ otherwise} \end{cases}.$$

Now there are decision variables for each of the facilities p to be sited. Further, allocation variables to assign demand to each facility are introduced. The formulation of the multi-Weber problem can be expressed as follows (Cooper 1963, 1964):

$$\text{Minimize} \quad \sum_{i=1}^{n}\sum_{j=1}^{p} w_i \Psi_{ij} \sqrt{(X_j - x_i)^2 + (Y_j - y_i)^2} \tag{3}$$

$$\text{Subjectto} \quad \sum_{j=1}^{p} \Psi_{ij} = 1 \quad \forall i \tag{4}$$

The objective (Eq. 3) remains as the minimization of total weighted distances from the assigned facility. The constraint in Eq. 4 specifies that each demand point is served by exactly one facility.

This model is more complicated than the Weber problem (Eq. 1) since it not only considers a facility location (X_j, Y_j) but now also accounts for allocation of demand points among several facilities, indicated by Ψ_{ij}.

3.4 Continuous Multi-Weber Problem

Combining the above extensions—the continuous Weber problem and the multi-Weber problem—results in a very challenging facility location problem. The variation of the Weber problem in this case concerns both continuous demand and multiple facilities, referred to as the continuous multi-Weber problem. Consider the following additional notation:

$d(x, y, X_j, Y_j) = \sqrt{(X_j - x)^2 + (Y_j - y)^2}$
$R_j =$ sub-region of demand served by facility j.

The discrete allocation variables, Ψ_{ij}, are now replaced by the variables R_j which define a portion of the demand region to be served. More specifically, this is the area served by facility (X_j, Y_j) and includes $(x, y) \in R_j$ reflecting demand assignment to the facility. The continuous multi-Weber problem follows:

$$\text{Minimize} \quad \sum_{j=1}^{p} \iint_{(x,y)\in R_j} g(x, y)d\left(x, y, X_j, Y_j\right)\mathrm{d}x\mathrm{d}y \tag{5}$$

$$\text{Subject to} \quad \bigcup_j R_j = R \tag{6}$$

$$R_j \cap R_{j'} = \Phi \forall j, j' \text{ where } j' \neq j \tag{7}$$

The objective (Eq. 5) includes both a double integral over each sub-region R_j served by facility j and a sum over the entire demand area characterized by a demand distribution $g(x, y)$. Similar to the previous models, this is equivalent to minimizing average weighted distance from each facility to its assigned demand points. In addition to seeking the optimal facility locations, the model must simultaneously determine the optimal allocation as well and this is reflected in the constraints (Eqs. 6 and 7). The constraint in Eq. 6 ensures that the union of sub-regions R_j must be comprised of the total demand region R and the constraint in Eq. 7 requires that no sub-region may overlap.

Essential here is making the distinction between a discrete and continuous representation of demand across space. To this end, this chapter will investigate how GIS can accommodate the continuous distribution of demand, as well as how this might facilitate structuring and solving the multiple facility location problem associated with such a demand representation.

4 Spatial Representation

The appropriate representation of geographic space has long been a concern in the field of GIScience and spatial analysis (Miller and Wentz 2003; Goodchild and Haining 2004). In GIS, discrete-object and continuous-field are two common ways to conceptualize geographic space (Worboys and Duckham 2004; Longley et al. 2011). Discrete objects usually refer to features, objects, or conceptual divisions with distinct boundaries, such as parcels, building footprints, or sidewalks, while continuous fields are often utilized to describe phenomena with indefinite bounds such as pollution, elevation and precipitation, where a value exists at every location across the area of interest. Though many more advanced concepts have been proposed in recent years to describe complex spatial phenomena (Cova and Goodchild 2002; Yuan 2001), the object/field views remain the basis for geographic representation (Goodchild et al. 2007).

Given the geographic nature of facility location problems, spatial representation is a crucial consideration. It is necessary in any facility location model to appropriately represent facilities and demand in geographic space. Of particular interest here is continuous demand, which, as previously discussed, can be conceived of as an object or a field. Representation by discrete points based upon the object perspective has been popular in location modeling for the past several decades, and this is largely attributable to the simplification of models and computation (Miller 1996; Francis et al. 2009). However, significant errors can result from data aggregation, if the underlying demand is in fact continuous in nature (Murray and O'Kelly 2002; Murray et al. 2008; Francis et al. 2009; Alexandris and Giannikos 2010; Cromley et al. 2012).

Due to theoretical as well as practical limitations associated with discrete representations of demand, interest in location modeling has recently shifted to addressing continuous representations of demand. There are primarily two approaches that have been considered in location models thus far. One option is to address continuous demand using an assumed exact surface defined by a mathematical function (Carrizosa et al. 1998; Murat et al. 2010). One advantage of modeling continuous demand in this way is that, in the model, demand is readily known for every location (x, y) in continuous space. In addition, solution approaches may be possible that exploit and benefit from the specific mathematical properties of these theoretical functions. The issue, however, is that such functions are never known with certainty due to the lack of complete information about the actual underlying demand distribution. This is clearly a problem.

As an alternative, the continuous surface may be approximated and then validated in a model. This is typically performed using sample data, in an initial or practice run analysis. There are a number of techniques in GIS that can be used to fit a continuous surface, including inverse distance weighting (IDW), natural-neighbor, trend surface, kriging and so on. These are generally known as spatial interpolation methods (Longley et al. 2011). The goal of spatial interpolation is to estimate attribute values at unobserved locations using the known values collected/measured at a finite set of locations (Cressie 1993). Consider the following notation:

$g(x, y)$ = true attribute value at location (x, y)
$\hat{g}(x, y)$= estimated attribute value at location (x, y)
$\varepsilon(x, y) = \left|g(x, y) - \hat{g}(x, y)\right|$.

The intent of spatial interpolation is to estimate $\hat{g}(x, y)$, necessarily making it an approximation of $g(x, y)$. Clearly it is desirable to minimize total estimation errors associated with an approximated function. This may be stated formally as follows (Yao and Murray 2013):

$$\text{Minimize} \quad E = \iint\limits_{(x,y)\in R} \varepsilon(x, y)\mathrm{d}x\mathrm{d}y \tag{8}$$

It is well recognized that estimation error $\varepsilon(x, y)$ is unavoidable and results from inaccurate sample data, assumptions inherent in interpolation approaches, as well as other sources (Lam 1983; Cressie 1993). Thus, errors and uncertainty exist in any derived approximation surfaces. That is, $\varepsilon(x, y) > 0$ for most locations in the model's representation of a continuous surface, regardless of the interpolation technique employed—though there may be differences in the magnitude of error, depending upon the interpolation method chosen.

If such fitted surfaces are used in further spatial analysis, cumulative errors can be introduced in the findings and results obtained (Wood and Fisher 1993; Miller and Wentz 2003; Oksanen and Sarjakoski 2005). In the case of continuous demand in facility location problems, though errors caused by point-based abstraction have been well recognized, issues associated with the representation of continuous surfaces are not well understood. Yao and Murray (2013) explicitly demonstrated error in the obtained facility location when the continuous Weber problem is considered. Addressing continuous representation in the multi-facility case remains a major challenge.

5 Solution Approach

As discussed above, it is extremely unlikely in practice that continuous demand could be represented using spatial interpolation or a mathematical function without any associated error. Rather than pursuing an exact representation, asymptotic theories are often employed in practice to obtain results. As Lee (2004) notes, there is a choice of two different frameworks which can encompass an asymptotics application, depending upon the particular basis; if based upon a growing observation region, the application is termed increasing domain asymptotics, whereas infill asymptotics (Cressie 1993), or fixed-domain asymptotics (Stein 1999), refers to a basis in a fixed and bounded region with increasingly dense observations (Lee 2004, p. 1900). Such theories are widely applied in the field of spatial analysis (Lee 2004; Zhang and Zimmerman 2005; Furrer et al. 2006; Zhu and Zhang 2006). The framework adopted here is infill (fixed-domain) asymptotics. The underlying principle is that the estimation error in spatial interpolation tends to zero as the sample size approaches infinity. In other words, when the number of sample points, n, approaches infinity, the difference between the true attribute value and the estimated attribute value can be so small as to be negligible and can be formally expressed as follows:

$$\lim_{n \to \infty} \hat{g}(x, y) = g(x, y) \tag{9}$$

From Eq. 8 this implies:

$$\lim_{n \to \infty} E = 0 \tag{10}$$

The infill asymptotic theory implications for facility location problems reliant on continuous demand are that it is possible to improve continuous demand approximation by increasing the number or density of sample points for a bounded region. Similar conclusions are also proven by Francis and Lowe (2011).

The proposed heuristic approach for solving the continuous multi-Weber problem is therefore based upon the discrete approximation for continuous demand, exploiting the above asymptotic theory in an innovative way. Although discrete approximations are often found lacking in their representations of continuous surfaces—due to the amount of error produced by such representations—infill asymptotic theory suggests that increasing the density of (adding) sample points in a bounded area improves the approximation; as the number of sample points approaches infinity, it is theorized that the estimation error approaches zero (Eq. 10). The assumption is that better approximation of continuous demand can be obtained if $n \rightarrow \infty$, which indicates that a solution to the multi-Weber problem with higher demand point densities would approach the optimum of the continuous multi-Weber problem. In practice, however, a finite set of demand points must suffice; it remains to be seen whether such a representation enables a good approximation of the actual continuous distribution to be achieved. In one respect, the discrete point approximation is less restrictive compared to other alternatives, as the demand points can be identified/collected based upon known characteristics of the underlying continuous distribution. Point density can be systematically increased to improve the accuracy of a discrete representation.

The proposed heuristic solution procedure for the continuous multi-Weber problem is given in Fig. 1, in which the parameters are defined as follows:

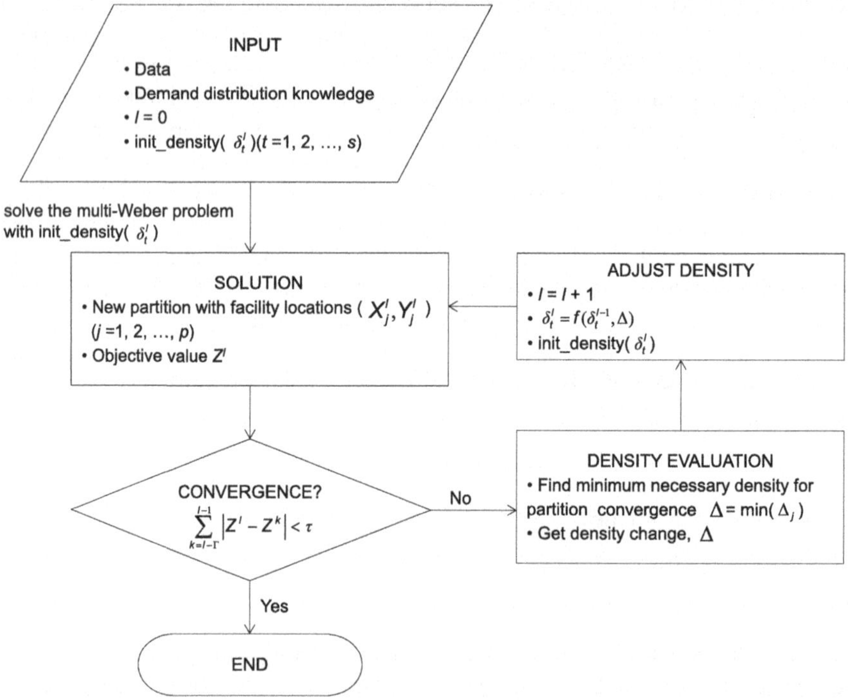

Fig. 1 Solution procedure for the continuous multi-Weber problem

init_density = initial demand point density
l = demand point layer
p = number of facilities
t = index of sub-areas in the demand region
s = number of sub-areas in the demand region
δ_t^l = demand point density in sub-area t
Γ = number of layers to consider in examining convergence of objective
τ = tolerance used to determine convergence of the objective value
Δ_j = change in demand point density for facility j service area.

In Fig. 1, the process begins with INPUT, where initialization occurs. Using the data and knowledge of the underlying continuous distribution, an approximation layer l is generated. The representation in the approximation layer is systematically enhanced as n increases (Fig. 1). Considering the spatial variation of demand, the study area can be divided into a set of sub-areas (in total s) with demand point density δ_t^l. That is, if the demand has non-uniform distribution, it is possible to keep dividing the whole area into smaller sub-areas until the demand variation within each sub-area is negligible or no demand distribution information on the finer scale can be obtained. Thus, each sub-area has a unique density. Taking non-uniformly distributed continuous demand as an example, possible knowledge regarding the underlying distribution can be the proportion of demand density among the sub-areas within which the demand distribution can be considered uniform. This information is important when generating new demand layers because we can retain spatial variation of demand densities by keeping this proportion constant. Furthermore, the partition of sub-areas can affect the optimal facility location. Again, this relies upon the knowledge of the underlying demand distribution, such as the spatial unit on which the demand density is defined. Conversion of the continuously distributed variable into discrete points can be performed with GIS software like ArcGIS.

Once a demand layer l is approximated, the multi-Weber problem is then solved, as presented in Fig. 1. The algorithm applied for the multi-Weber problem is ALTERNATE, a heuristic proposed originally by Cooper (1964). Assuming that p facilities are to be sited, the study area is therefore divided into p adjacent partitions, one for each facility, so that all demand points are served (note that it is possible that $p \neq s$). Next, the Weber problem is solved to find the optimal facility location for each partition. The two steps are repeated until no further improvement in the objective can be achieved. The final solution is the set of facility locations (X_j^l, Y_j^l) for layer l, as well as the allocation of each demand point to its nearest facility. Additionally, an objective value representing the minimized average distance to the closest facility, Z^l, is obtained.

The next step in Fig. 1 is the assessment of convergence (the decision point labeled CONVERGENCE?). The evaluation is based upon the last Γ layers, calculated as the sum of the absolute differences between the objective value of the current layer Z^l and previous layers, Z^k where $0 < k < l$. Specifically, the convergence measure is calculated as $\sum_{k=l-\Gamma}^{l-1} \left| Z^l - Z^k \right|$, which is then compared to a given tolerance threshold, τ. Usually, τ is a small value representing the relative location accuracy of sited facilities. The definition of τ relies upon the spatial context under study. The value can be a few meters for facilities like retail stores. If τ is larger, convergence is assumed because the facility location changes little as denser demand points are used. In other words, little improvements in facility location accuracy can be achieved when the convergence condition is satisfied, so the solution procedure terminates (END). As a result, the solution derived from layer l is the approximate solution for the continuous multi-Weber problem with average distance to the nearest facility Z^l.

Otherwise, the process in Fig. 1 proceeds to the DENSITY EVALUATION process. In this stage, the Weber problem is solved for each partition to seek a sufficient point density for the discrete representation. The intent of this step is to better approximate the continuous surface. Suppose $(\Delta_1, \Delta_2, \ldots, \Delta_p)$ are the changes in the point densities that are necessary to obtain solution convergence for different partitions, the minimum value $\Delta = \min(\Delta_1, \Delta_2, \ldots, \Delta_p)$ is used as the density change for the entire layer in ADJUST DENSITY.

Based upon change density Δ and the point density of the current layer, Fig. 1 indicates that a new demand layer is generated by adjusting the demand density for each sub-area, keeping the proportion of point densities among all sub-areas fixed. Detailed in Fig. 2 is a description of how each demand layer is generated. Suppose a demand area consists of two sub-areas and the proportion of demand density is 2:1. In Fig. 2, the first demand layer ($l = 1$) has the lowest demand density, and the number of demand points in the two sub-areas is 4 and 2, respectively. In the second ($l = 2$) and third ($l = 3$) layers, the number of discrete points is increased to 8 and 4, and 12 and 6, respectively. Other layer instances are created in a similar manner. This ensures that the demand points in each new demand layer reflect the proportion of point densities among sub-areas. The new demand point layer is then used as input for the next iteration of the solution process. This process is repeated until convergence is achieved.

In summary, the proposed approach uses a systematic discrete point-based approximation of demand, each time moving closer toward asymptotic convergence. At each iteration, the best solution is found and comparison is made to previous representations. Upon convergence, the best facility locations (X_j, Y_j) are found that minimize average travel distance Z for the continuous multi-Weber problem. The approximation of $g(x, y)$ using a series of demand point layers results in a series of solutions, where $Z^l \to Z$ as $n \to \infty$ (recall n is the number of demand points). Again, this

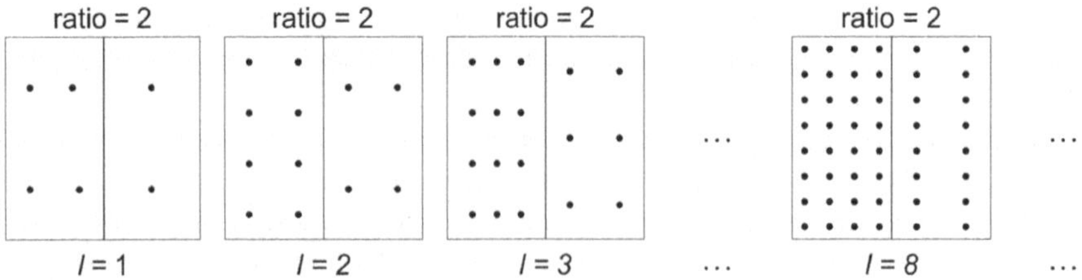

Fig. 2 Process of demand layer generation

is based upon asymptotic theory, but in practice, convergence is achieved for a reasonable n. Finding this reasonable n is accomplished through the proposed heuristic process given in Fig. 1.

6 Empirical Results

The developed approach outlined in Fig. 1 to solve the continuous multi-Weber problem was implemented using the Python[1] programming language integrating an Open Source GIS library, Shapelib,[2] to process geographic data. Analysis was performed on a personal computer (Mac OS X system) running a 3.06 GHz, Intel Core 2 Duo processor with 4 GB, 800 MHz memory.

To assess the performance of the proposed heuristic, two types of demand distributions are considered: uniform and non-uniform. The former represents the situation where continuous demand is uniformly distributed over a region, and the latter reflects the situation where demand varies across space.

6.1 Uniform Demand

The data used here are the Census tracts in Oregon, USA, as shown in Fig. 3, including two demand regions. The areas covered are 19.373 km^2 and 2.574 km^2 for Region A and Region B, respectively. It is assumed that the continuous demand is uniformly distributed over both regions, with the same initial point density that there are two demand points per km^2. Specifically, equally spaced points are used; this spacing of point features can be obtained using automated functions available in GIS. Four facilities ($p = 4$) are sited. The termination criterion is based upon the last five consecutive data layers. Specifically, the parameters are $\Gamma = 5$ and $\tau = 2$. That is, the last consecutive 5 layers are evaluated and the tolerance level is 2 m. They are used for experimental purpose but also can be considered to be able to generate quite accurate solutions in most actual facility location applications. Of course, other parameters can be employed which are relied upon the application context in reality.

Using the proposed solution procedure (Fig. 1) with the above parameters, the best facility locations derived from each iteration are illustrated in Fig. 3 as green dots. Suppose Ω is the set of total data layers generated in reaching convergence, then for solution, here, we have $|\Omega| = 94$. Thus, every four points in different groups in Fig. 3 correspond to the facility locations derived from a data layer l, where $l \in \Omega$. It can be seen that the four groups of facility locations are naturally clustered, with a few points scattered around the group centers. It is not surprising because as indicated by the infill asymptotic theory, the facility locations are approaching the

[1]https://www.python.org/.

[2]https://shapelib.maptools.org/.

a Region A

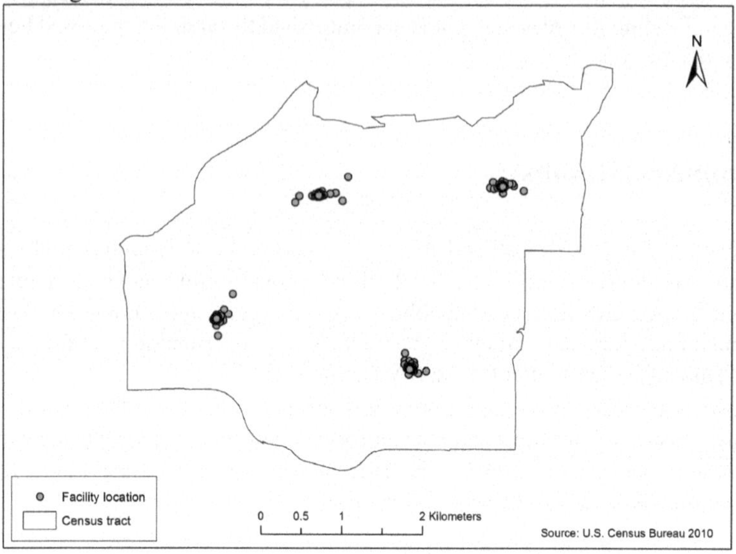

b Region B

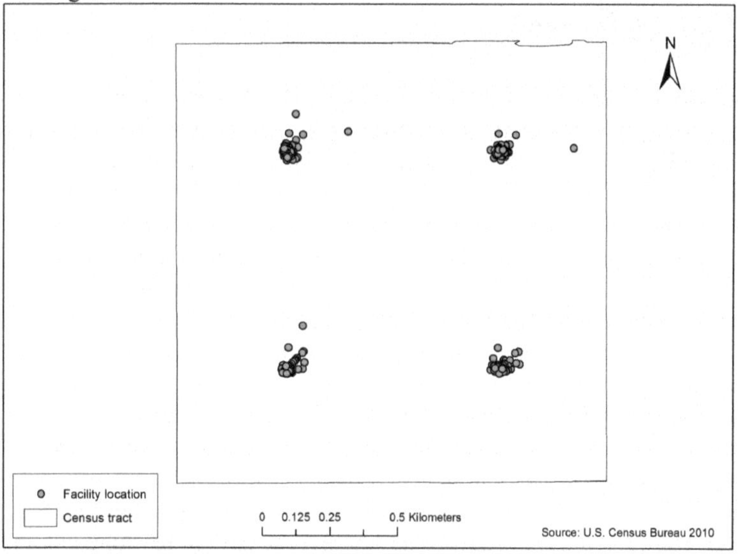

Fig. 3 Solution for the study region with uniform continuous demand: **a** Region A, **b** Region B

optimal solution as the demand point density increases. A more detailed description of spatial distribution, as well as the distance between consecutive approximated locations of one facility group displayed in Fig. 3 is provided in Fig. 4. The numbers in the top portion of Fig. 4a correspond to the indices of data layers, *l*. It can be observed that the approximated facility locations for the first several demand layer

a

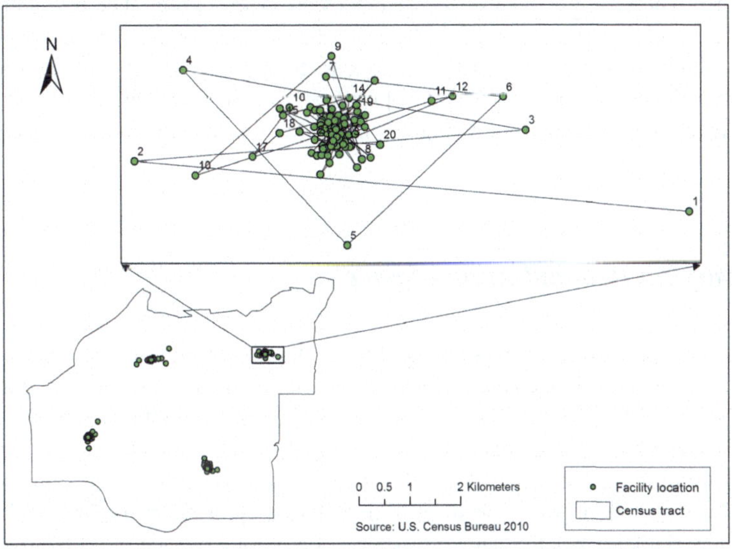

b

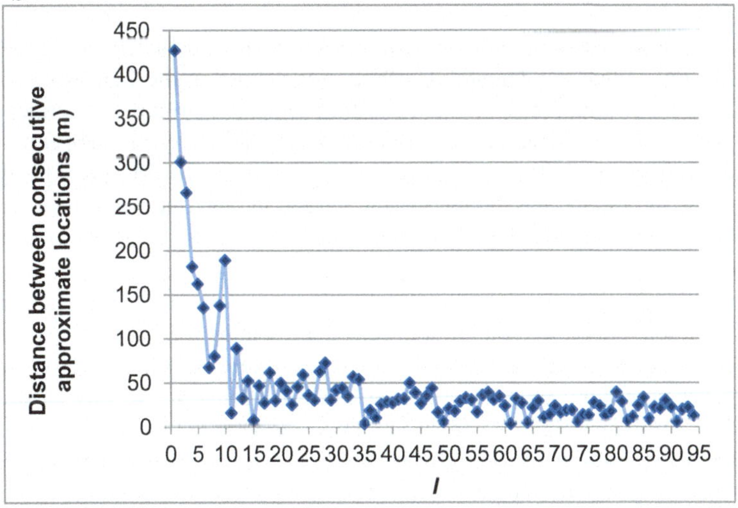

Fig. 4 Detailed solution information for Fig. 3a: **a** facility locations, **b** distance between consecutive approximate locations (m)

approximations are far away from the others. The facilities geographically converge as the point densities increase, indicating that little improvement can be obtained in the solution.

This pattern is more apparent in Fig. 4b, which presents the distance between the two facility locations obtained from layer l and $l-1$, given $l \in \Omega$ and $l > 1$. The greatest distance between the facilities is as high as 430 km. This value quickly declines with the increase of point densities. There is much less variation in distance when $l > 12$; most of the distances are within 50 m. The implication is that $\lim_{l\to\infty} \sqrt{(X_l - X_{l-1})^2 + (Y_l - Y_{l-1})^2} = 0$. The solution from the multi-Weber problem tends to be changing little and appears to approach the theoretical optimal solution of the continuous multi-Weber problem. The distance variation for the last several layers is less than 30 m, which is undoubtedly reasonably accurate for siting facilities in practice.

6.2 Varying Demand Across Space

The second application concerns non-uniformly distributed continuous demand over space. This involves 55 Census tracts with varying demand. The smallest tract has an area of 0.325 km^2 and the largest area is 23.495 km^2; the total size of the study area is 209.968 km^2. The demand density among the tracts varies from 224 unit/km^2 to 9493 unit/km^2 as shown in Fig. 5b. "Unit" can have different meanings (e.g. people for population or m^3 for air pollution) based upon the real application context. Figure 5a describes the demand using a continuous surface fitted from spatial interpolation using the kriging method. An alternative, discrete representation uses demand points. Although demand density varies across the study area, the demand density is considered to be uniform within each tract, as depicted in Fig. 5c. The reason for this is that the demand distribution is defined at the Census tract level, and there is no information available regarding demand distribution at a finer level. Again, the method described in Fig. 1 is applied after the discretization of continuous demand in this context. The specific parameters used are: $p = 4$, init_density $= 1$ point/km^2, $\Gamma = 5$ and $\tau = 8$. It should be noted that the discrete points here are used to approximate the actual demand distribution and their densities are not necessarily the same.

The solution is presented in Fig. 6e. Similar to the results in Fig. 4a, the best facility locations are clustered in four groups, representing the sites derived for each facility from all data layers Ω. In total, 20 layers are generated ($|\Omega| = 20$) when the convergence criterion is satisfied. The spatial distribution of the facility locations in each group is also investigated, with distance between consecutive locations obtained from the iterative solving procedure displayed in Fig. 6a–d. A clear convergence pattern can be observed. The geographical location differences for all four location groups quickly decrease to within 50 m after only a few iterations, then significantly decrease when convergence is achieved. For example, in Fig. 6c, the final distance is less than 10 m, which is very accurate for actual facility planning.

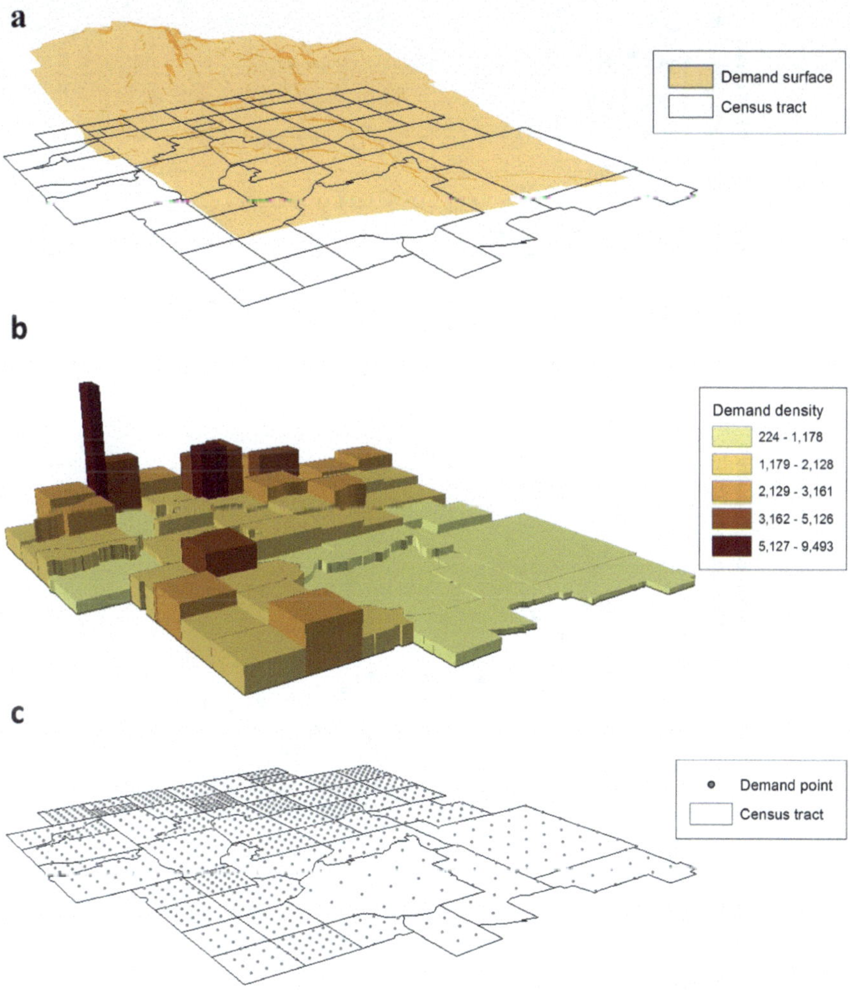

Fig. 5 A region with varying demand: **a** demand surface, **b** actual demand, **c** discrete demand point

7 Discussion and Conclusions

Continuous demand modeling is commonly required for regional planning concerned with siting facilities to provide social services for underlying demand. However, an appropriate representation of continuous demand in a digital environment is challenging and remains a key issue in facility location modeling. Traditional mathematical functions and surfaces fitted through spatial interpolation inevitably introduce error and uncertainty into the results of such analyses. With a basis in infill asymptotic theory, this research developed a spatial optimization approach integrating GIS functionality and optimization techniques to solve the continuous multi-Weber problem

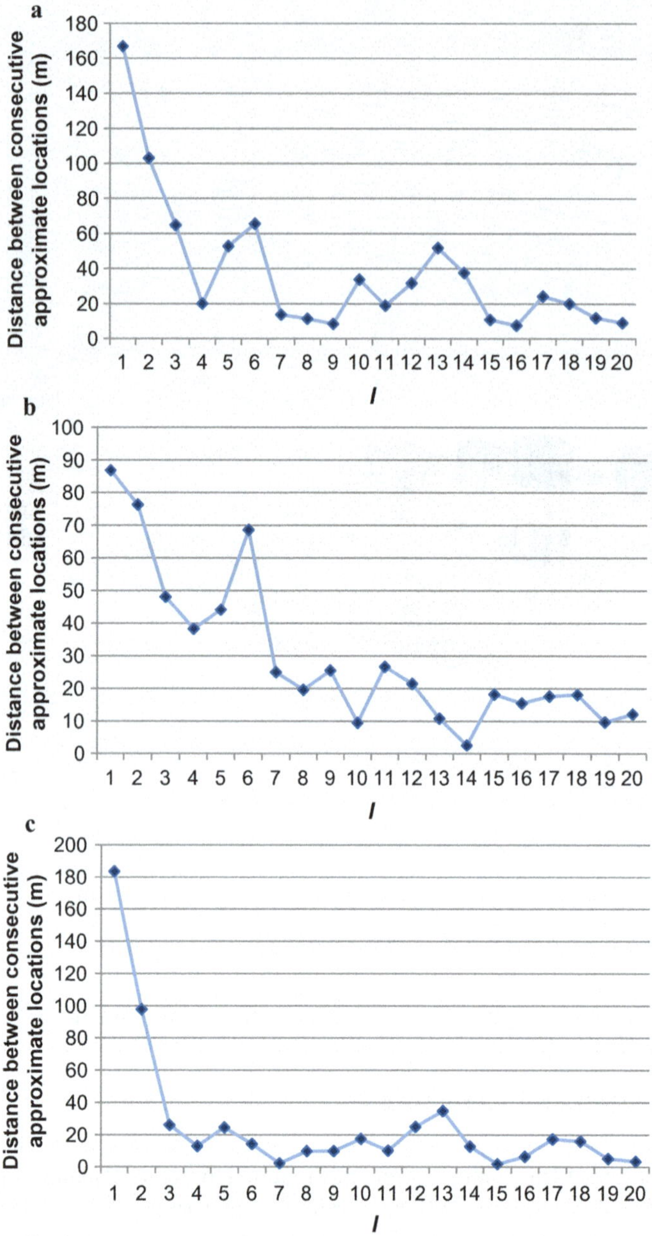

Fig. 6 Solution for the study region with varying continuous demand: **a** distance between consecutive approximate locations (m) for the facilities in cluster a (**e**), **b** distance between consecutive approximate locations (m) for the facilities in cluster b (**e**), **c** distance between consecutive approximate locations (m) for the facilities in cluster c (**e**), **d** distance between consecutive approximate locations (m) for the facilities in cluster d (**e**), **e** facility locations

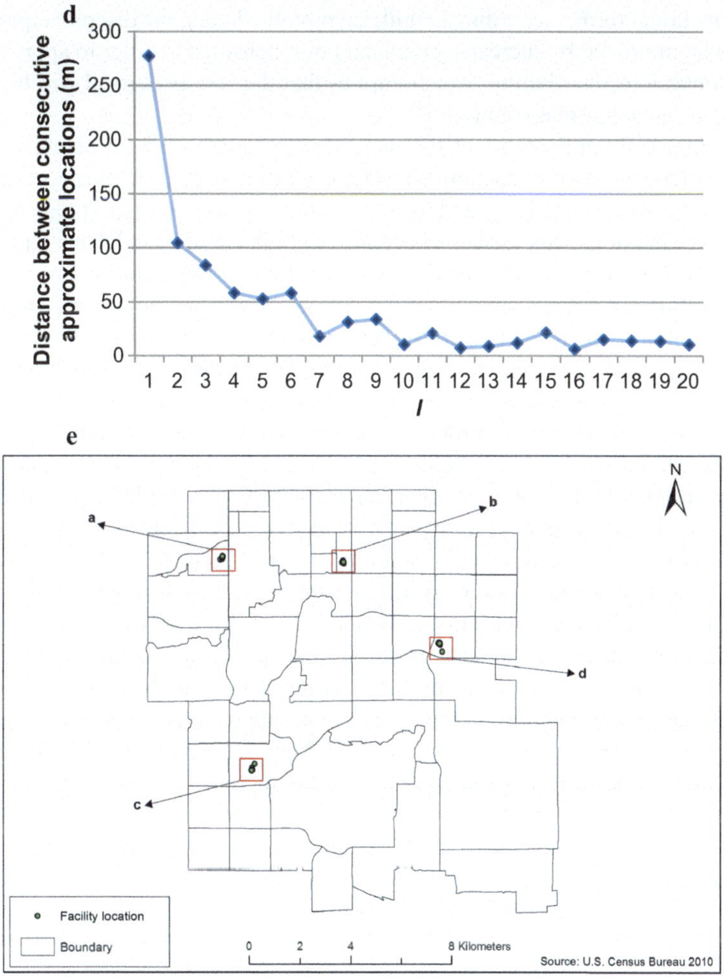

Fig. 6 (continued)

in a way that reduces representation error and conforms to what is actually known about the demand region.

The essential challenge here is how to represent continuous demand. It is well known that discrete-object and continuous-field are two primary models that are widely applied in GIS to represent geographic space. Though the surface models based upon the continuous-field representation can reflect the continuous nature of the underlying demand, such surface models defined by mathematical functions or fitted by spatial interpolations are subject to significant errors. The discrete point approximation employed in this research may be a better option, both in terms of error minimization as well as conceptually, in that there are less unrealistic assumptions about functional form. According to infill asymptotic theory, the discrete representation can be improved by increasing demand point densities in order to approach the actual continuous distribution, which implies that the results derived will be closer to the theoretical optimal solution.

The empirical applications of the methodology proposed here demonstrated a consistent tendency upon reaching convergence terms of both facility location and average travel distance (Figs. 4 and 6). Though the solution changes among different point approximations, this variation becomes negligible as demand density increases sufficiently. For example, the distance between the first two solution layers was as large as 430 m but became smaller as the density of demand points increased, ultimately less than 10 m at solution convergence. In other words, higher accuracy in the facility locations is achieved by using more dense demand points. The convergence of solutions demonstrated the effectiveness of the proposed method.

There are two aspects of the proposed approach which deserve further exploration. One is the tolerance (τ) value for solution convergence assessment. This value largely depends on the spatial context of a particular study. For example, if the intent is to locate a facility like a public library or a retail store, a tolerance level of 10 m for facility locations from consecutive solutions might be considered sufficient. The second issue is that the ALTERNATE algorithm is used when solving the multi-Weber problem given an approximate demand layer l. ALTERNATE is employed in this research because of its broad application and ease of implementation. Of course, other solution approaches for the multi-Weber problem could be applied.

Given the spatial nature of facility location problems, it is necessary and crucial to incorporate GIS into the solution methodology. Incorporating GIS functionality, this chapter developed a spatial optimization approach to solve the continuous multi-Weber problem, addressing continuous demand representation. Results from empirical applications showed the effectiveness of the proposed method, and its general applicability to support planning and decision-making processes.

Acknowledgements This work is supported by the National Natural Science Foundation of China (Grant 41201117).

References

Abu-Affash, A. K., & Katz, M. J. (2008). Improved bounds on the average distance to the Fermat-Weber center of a convex object. In *Proceedings of the 20th Annual Canadian Conference on Computational Geometry*, Montreal, Canada, August 13–15.

Alexandris, G., & Giannikos, I. (2010). A new model for maximal coverage exploiting GIS capabilities. *European Journal of Operational Research, 202*(2), 328–338.

Altınel, I. K., Durmaz, E., Aras, N., & ÖzkIsacIk, K. C. (2009). A location-allocation heuristic for the capacitated multi-facility Weber problem with probabilistic customer locations. *European Journal of Operational Research, 198*(3), 790–799.

Bennett, C. D., & Mirakhor, A. (1974). Optimal facility location with respect to several regions. *Journal of Regional Science, 14*(1), 131–136.

Brandeau, M. L., & Chiu, S. S. (1989). An overview of representative problems in location research. *Management Science, 25*, 645–674.
Brimberg, J., & Drezner, Z. (2013). A new heuristic for solving the p-median problem in the plane. *Computers & Operations Research, 40*(1), 427–437.
Brimberg, J., Hansen, P., & Mladenović, N. (2006). Decomposition strategies for large-scale continuous location-allocation problems. *IMA Journal of Management Mathematics, 17*, 307–316.
Brimberg, J., Hansen, P., Mladenović, N., & Salhi, S. (2008). A survey of solution methods for the continuous location-allocation problem. *International Journal of Operations Research, 5*, 1–12.
Brimberg, J., & Mladenović, N. (1996). Solving the continuous location-allocation problem with Tabu search. *Studies in Locational Annals, 8*, 23–32.
Carmi, P., Har-Peled, S., & Katz, M. J. (2005). On the Fermat-Weber center of a convex object. *Computational Geometry, 32*(3), 188–195.
Carrizosa, E., Muñoz-Márquez, M., & Puerto, J. (1998). The Weber problem with regional demand. *European Journal of Operational Research, 104*, 358–365.
Church, R. L. (2002). Geographic information systems and location science. *Computers & Operations Research, 29*, 541–562.
Church, R. L., & Murray, A. T. (2009). *Business site selection location analysis and GIS*. New York: Wiley.
Church, R. L., & Sorensen, P. (1996). Integrating normative location models into GIS. In P. Longley & M. Batty (Eds.), *Spatial analysis: Modeling in a GIS environment 1996*. Cambridge, UK: GeoInformation International.
Cooper, L. (1963). Location-allocation problems. *Operations Research, 11*, 37–52.
Cooper, L. (1964). Heuristic methods for location-allocation problems. *SIAM Review, 6*, 37–53.
Cova, T. J., & Goodchild, M. F. (2002). Extending geographical representation to include fields of spatial objects. *International Journal of Geographical Information Science, 16*, 509–532.
Cressie, N. (1993). *Statistics for spatial data* (Revised ed.). New York: Wiley.
Cromley, R. G., Lin, J., & Merwin, D. A. (2012). Evaluating representation and scale error in the maximal covering location problem using GIS and intelligent areal interpolation. *International Journal of Geographical Information Science, 26*(3), 495–517.
Curtin, K., Hayslett-McCall, K., & Qiu, F. (2010). Determining optimal police patrol areas with maximal covering and backup covering location models. *Network and Spatial Economics, 10*(1), 125–145.
Drezner, Z. (1986). Location of regional facilities. *Naval Research Logistics Quarterly, 33*, 523–529.
Drezner, Z. (1995). Replacing discrete demand with continuous demand. In Z. Drezner (Ed.), *Facility location: A survey of applications and methods. Springer Series in Operations Research* (Chap. 2, pp. 33–42). New York: Springer.
Drezner, Z., Brimberg, J., Mladenović, N., & Salhi, S. (2015). New heuristic algorithms for solving the planar p-median problem. *Computers & Operations Research, 62*, 296–304.
Drezner, Z., Brimberg, J., Mladenović, N., & Salhi, S. (2016). New local searches for solving the multi-source Weber problem. *Annals of Operations Research, 246*(1), 181–203.
Drezner, Z., Klamroth, K., Schöbel, A., & Wesolowsky, G. O. (2002). The Weber problem. In Z. Drezner & H. Hamacher (Eds.), *Facility location: Applications and theory* (pp. 1–36). Berlin: Springer.
Drezner, Z., & Salhi, S. (2017). Incorporating neighborhood reduction for the solution of the planar p-median problem. *Annals of Operations Research, 258*(2), 639–654.
Drezner, Z., & Wesolowsky, G. (1980). Optimal location of a demand facility relative to area demand. *Naval Research Logistics Quarterly, 27*, 199–206.
Fekete, S. P., Mitchell, J. S. B., & Beurer, K. (2005). On the continuous Fermat-Weber problem. *Operations Research, 53*, 61–76.
Francis, R. L., & Lowe, T. J. (2011). Comparative error bound theory for three location models: Continuous demand versus discrete demand. *Top*. https://oi.org/10.1007/s11750-011-0244-2.
Francis, R. L., Lowe, T. J., Rayco, M. B., & Tamir, A. (2009). Aggregation error for location models: Survey and analysis. *Annals of Operations Research, 167*, 171–208.

Franco, C. V., Rodríguez-Chía, A. M., & Espejo-Miranda, I. (2008). The single facility location problem with average-distances. *Top, 16*, 164–194.

Furrer, R., Genton, M. G., & Nychka, D. (2006). Covariance tapering for interpolation of large spatial datasets. *Journal of Computational and Graphical Statistics, 15*(3), 502–523.

Gastner, M. T., & Newman, M. E. J. (2006). Optimal design of spatial distribution networks. *Physical Review E, 74*(1), 016117.

Goodchild, M. F., & Haining, R. P. (2004). GIS and spatial data analysis: Converging perspectives. *Papers in Regional Science, 83*, 363–385.

Goodchild, M. F., Yuan, M., & Cova, T. J. (2007). Towards a general theory of geographic representation in GIS. *International Journal of Geographical Information Science, 21*, 239–260.

Houck, C. R., Joines, J. A., & Kay, M. G. (1996). Comparison of genetic algorithms, random restart and two-opt switching for solving large location-allocation problems. *European Journal of Operational Research, 20*, 387–396.

Lam, N. S. (1983). Spatial interpolation methods: A review. *The American Cartographer, 10*(2), 129–149.

Lee, L.-F. (2004). Asymptotic Distributions of quasi-maximum likelihood estimators for spatial autoregressive models. *Econometrica, 72*(6), 1899–1925.

Longley, P. A., Goodchild, M. F., Maguire, D. J., & Rhind, D. W. (2011). *Geographic information systems and science* (3rd ed.). Chichester: Wiley.

Marucheck, A. S., & Aly, A. A. (1981). An efficient algorithm for the location-allocation problem with rectangular regions. *Naval Research Logistics Quarterly, 28*, 309–323.

Matisziw, T. C., & Murray, A. T. (2009). Siting a facility in continuous space to maximize coverage of continuously distributed demand. *Socio-Economic Planning Sciences, 43*, 131–139.

Miller, H. J. (1996). GIS and geometric representation in facility location problems. *International Journal of Geographical Information Science, 10*, 791–816.

Miller, H. J., & Wentz, E. A. (2003). Representation and spatial analysis in geographic information systems. *Annals of the Association of American Geographers, 93*(3), 574–594.

Murat, A., Verter, V., & Laporte, G. (2010). A continuous analysis framework for the solution of location-allocation problems with dense demand. *Computers & Operations Research, 37*, 123–136.

Murray, A. T. (2008). Economic geography: Location theory. In R. Kitchin & N. Thrift (Eds.), *International encyclopedia of human geography*. Oxford: Elsevier.

Murray, A. T. (2010). Advances in location modeling: GIS linkages and contributions. *Journal of Geographical Systems, 12*, 335–354.

Murray, A. T., & O'Kelly, M. E. (2002). Assessing representation error in point-based coverage modeling. *Journal of Geographical Systems, 4*, 171–191.

Murray, A. T., O'Kelly, M. E., & Church, R. L. (2008). Regional service coverage modeling. *Computers & Operations Research, 35*, 339–355.

Oksanen, J., & Sarjakoski, T. (2005). Error propagation of DEM-based surface derivatives. *Computers & Geosciences, 31*, 1015–1027.

Ouyang, Y., & Daganzo, C. F. (2006). Discretization and validation of the continuum approximation scheme for terminal system design. *Transportation Science, 40*, 89–98.

Puerto, J., & Rodríguez-Chía, A. M. (2011). On the structure of the solution set for the single facility location problem with average distances. *Mathematical Programming, 128*(1–2), 373–401.

Rao, J. R., & Varma, N. H. (1985). Stochastic multi-facility minisum location problem involving Euclidean distances. *Operations Research, 22*, 232–240.

Righini, G., & Zaniboni, L. (2007). A branch-and-price algorithm for the multi-source Weber problem. *International Journal of Operational Research, 2*(2), 188–207.

Rosing, K. E. (1992). An optimal method for solving the (generalized) multi-Weber problem. *European Journal of Operational Research, 58*, 414–426.

Salhi, S., & Gamal, M. D. H. (2003). A genetic algorithm based approach for the uncapacitated continuous location-allocation problem. *Annals of Operations Research, 123*, 203–222.

Stein, M. (1999). *Interpolation of spatial data: Some theory for kriging. Springer Series in Statistics*. New York: Springer.

Suzuki, A., & Okabe, A. (1995). Using Voronoi diagrams. In Z. Drezner (Ed.), *Facility location: A survey of applications and methods 1995* (pp. 103–118). New York: Springer.

Weber, A. (1909). Uber den standort der Industrien. Tubingen. (C. J. Fredrich, Trans., 1929, Theory of the location of industries). Chicago: University of Chicago Press.

Wei, H., Murray, A. T., & Xiao, N. (2006). Solving the continuous space p-center problem: Planning application issues. *IMA Journal of Management Mathematics, 17*, 413–425.

Wesolowsky, G. O. (1993). The Weber problem: History and perspectives. *Location Science, 1*, 5–23.

Wood, J. D., & Fisher, P. F. (1993). Assessing interpolation accuracy in elevation models. *IEEE Computer Graphics and Applications, 13*(2), 48–56.

Worboys, M. F., & Duckham, M. (2004). *GIS: A computing perspective* (2nd ed.). Boca Raton: CRC Press.

Yao, J., & Murray, A. T. (2013). Continuous surface representation and approximation: Spatial analytical implications. *International Journal of Geographical Information Science, 27*(5), 883–897.

Yuan, M. (2001). Representing complex geographic phenomena with both object- and field-like properties. *Cartography and Geographic Information Science, 28*, 83–96.

Zhang, H., & Zimmerman, D. L. (2005). Towards reconciling two asymptotic frameworks in spatial statistics. *Biometrika, 92*(4), 921–936.

Zhu, Z., & Zhang, H. (2006). Spatial sampling design under the infill asymptotic framework. *Environmetrics, 17*(4), 323–337.

Some Consequences of Ignoring Relocations in the Cost–Benefit Analysis of Transportation Infrastructure Investments

David Philip McArthur, Inge Thorsen, and Jan Ubøe

1 Introduction

Cost–benefit appraisals of investments in transportation infrastructure are often based on estimates of traffic demand on the relevant road links. A correct appraisal of the investments requires that the estimation of the demand is made on a sound basis. In this chapter, commuting is the only trip purpose to be studied. Predictions of commuting flows are typically based on a demand curve estimated within a doubly-constrained modelling framework, which means that the spatial distribution of jobs and workers (households) is assumed to be given. This may be a reasonable assumption in some cases, for instance, if the investments mainly affect the long-distance traffic on a particular transport corridor. In many cases, however, changes in the transportation network will affect the location decisions of local households and employers. In the long run, this may induce significant changes in the spatial distribution of jobs and people. In such cases, predictions based on a doubly-constrained modelling framework would be seriously biased. The main motivation of this chapter is to study the importance and the character of such a bias.

Relocation effects certainly represent a possible source of prediction errors in calculating induced traffic and welfare benefits in traditional cost–benefit appraisals of investments in road transportation infrastructure. In addition, the spatial distribution of jobs and households may itself be a target for regional policy. One important

D. P. McArthur (✉)
University of Glasgow, Glasgow, UK
e-mail: david.mcarthur@glasgow.ac.uk

I. Thorsen
Western Norway University of Applied Sciences, Haugesund, Norway
e-mail: inge.thorsen@hvl.no

J. Ubøe
Norwegian School of Economics, Bergen, Norway
e-mail: jan.uboe@nhh.no

J. Thill (ed.), *Innovations in Urban and Regional Systems*,
https://doi.org/10.1007/978-3-030-43694-0_7

aspect to discuss is, for instance, whether improvements in the road network may contribute to preventing depopulation of peripheral areas. Predictions of such effects are therefore both interesting from a regional policy point of view, and as input to a cost–benefit analysis.

The first challenge is to specify an appropriate modelling framework. The approach to be followed in this chapter is to study potential relocation effects and the resulting prediction bias in a spatial general equilibrium modelling framework. The literature offers some very interesting operative spatial general equilibrium models. One successful set of models is the spatial computable general equilibrium (SCGE) models, originating from the 'new economic geography'; see, for instance, Fujita et al. (1999). This tradition of spatial general equilibrium models normally adopts a macroscopic perspective of the geography. The models are designed for a multiregional setting, with relatively large regions, ignoring spatial interaction of commuting and shopping, and intraregional disparities of, for example, labour and housing market characteristics.

Another tradition of spatial equilibrium modelling is large-scale models for urban planning. Such models are designed for metropolitan areas and offer a very detailed description of spatial interaction and location characteristics of the local geography. This involves, for instance, traffic assignment problems, mode choice in transport, and several aspects of urban land use. Many specifications and applications can be found in the literature, after this kind of modelling had its renaissance at the end of the 1980s (Boyce 1988). One well-known example of a large-scale model is the UrbanSim model; see Waddell et al. (2003).

Studying regional policy issues in, for example, a typical west Norwegian region calls for specifying a spatial dimension in between the large-scale models for metropolitan areas and the multiregional spatial general equilibrium models of the 'new economic geography'. Congestion, mode choice, and urban environmental and land-use problems are not very relevant, but flows of commuting and shopping should be accounted for in explaining and predicting changes in the location pattern.

A nontechnical description of the modelling framework is provided in Sect. 2. Section 3 introduces a simple 5-node imaginary geography, that is used in the numerical experiments to follow. Relocation effects of varying the distance from a peripheral zone to the central business district are studied in Sect. 3.1, while Sect. 3.2 focuses on prediction errors of ignoring relocation effects in cases where the travel distance is reduced by $\frac{1}{3}$. Section 3.3 demonstrates that the prediction errors are sensitive to the parameter representing the distance deterrence in intraregional moving decisions. The discussion is taken into a standard microeconomic cost–benefit analysis in Sect. 4, while Sect. 5 addresses environmental issues, represented by the possibility of negative externalities related to induced traffic from improvements in the road network. Finally, some concluding remarks are offered in Sect. 6.

2 A Nontechnical Description of the Modelling Framework

This chapter utilises the spatial equilibrium model presented and developed in McArthur et al. (2014). The core of the model centres on the definition of equilibrium, involving intraregional migration and commuting flows corresponding to a specific spatial distribution of jobs and workers between the zones of a region. To reach an operational model specification, we introduce a set of reasonable hypotheses on the spatial behaviour of firms and households. Hence, we concentrate on the spatial dimension of the supply and demand for labour and ignore many other aspects, for instance, related to heterogeneity of jobs and workers.

Consider first the demand for labour at specific locations. Like most spatial general equilibrium models, the model to be presented incorporates the core elements of economic base modelling. This means that the model distinguishes between two types of firms. The activity of local sector firms is determined by demand arising from within the region, while production in basic sector firms is determined by factors unrelated to intraregional demand.

2.1 *Basic Sector Firms, Local Innovativeness and Competitiveness*

As indicated in the left upper part of Fig. 1, the number of basic sector jobs at a specific location depends on local innovativeness and competitiveness. This may reflect agglomeration economies or agglomeration diseconomies, the wage level, entrepreneurial spirit, transport costs, the availability of qualified workers, etc. No attempt is made to explicitly account for local variations of innovativeness and competitiveness in the current version of the model to be presented, and the spatial distribution of basic sector jobs is treated exogenously in the model.

2.2 *Local Sector Firms: The Spatial Shopping Pattern*

The spatial distribution of local sector jobs reflects the residential location pattern and the spatial shopping behaviour of the households/customers. Retailing is the dominating local sector activity. As in most other forms of spatial interaction, there is a distance deterrence effect in shopping travel patterns. In economic base theory, this can be used to support an assumption of proportionality between local sector employment and the population of a specific geographic area. It can be argued, however, that such an assumption of proportionality is unreasonable at the level of spatial aggregation considered in this chapter.

Hence, an alternative, more appropriate, approach is used. The basic idea is that consumer shopping behaviour results from a trade-off between price savings and

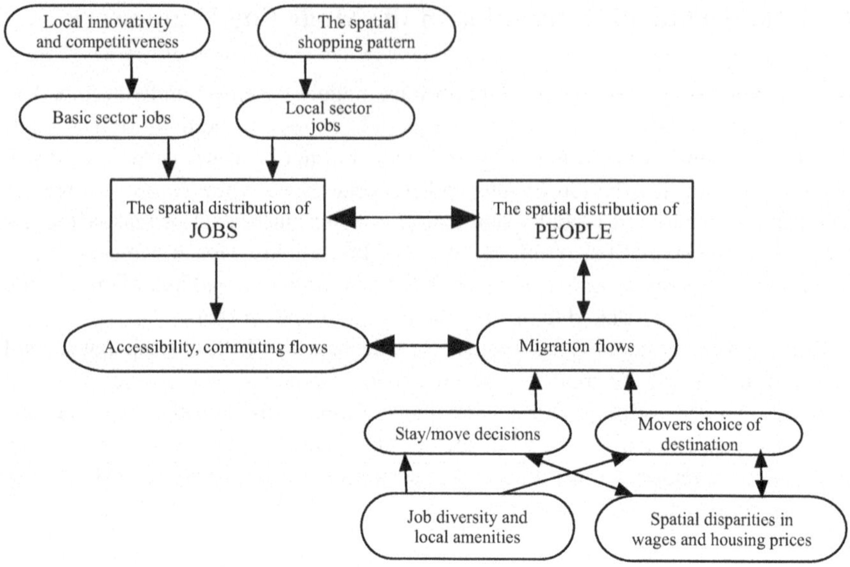

Fig. 1 Basic mechanisms in the modelling framework

transport costs. Gjestland et al. (2006) derived some results regarding this trade-off. The more realism added to assumptions on the distribution of price savings, product range, shopping frequency and the valuation of time, the closer Gjestland et al. (2006) come to a smooth, concave function between the frequency of shopping locally and the distance from the shopping centre offering favourable prices.

A next step in deriving a spatial pattern of local sector activities is by recognising that economies of scale, transportation costs and agglomeration benefits allow firms in a central location to offer goods and services at a lower price than firms located in more peripheral locations. Agglomeration benefits explain why some types of local sector activities will largely be concentrated in a centre. Administrative services often locate in the centre, giving rise to agglomeration benefits which in turn attract more activity. At the same time, businesses in many cases choose to locate in the same area because consumers often perceive it to be beneficial if they can satisfy their demand for several goods and services with one shopping trip.

In other words, the potential for price savings pulls shopping towards urban centres, while transport costs contribute to explain why customers do some of their shopping close to where they live. For an illustration of the outcome of the relevant trade-off, assume that the region has just one centre. This assumption can be easily relaxed, into cases with many centres of various sizes. Consider next the number of shop employees per resident is measuring the local sector density (LSD). The trade-off is then represented by a pattern corresponding to the graph in Fig. 2. Low prices explain a very high local sector density in the centre of the region. A significant proportion of the shopping trips emanating from a suburb will, as a rule, be directed

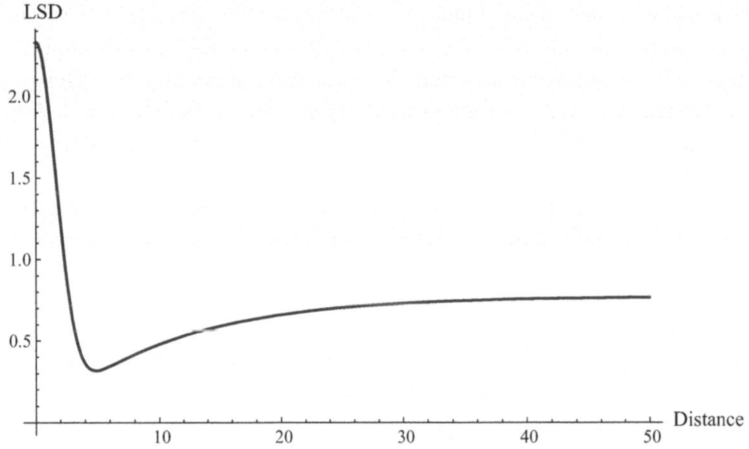

Fig. 2 Spatial distribution of local sector jobs in a monocentric geography

towards the regional centre, since customers here can benefit from low prices in the centre at relatively low transport costs. For zones which lie at a long distance from the centre, virtually all shopping will take place within the zone. As illustrated in Fig. 2, the local sector density will be high in the regional centre, low in suburbs, and it will be approaching the regional average as the distance from the centre increases. Gjestland et al. (2006) found empirical support for such an intuitively appealing pattern from observations of Norwegian regions.

The level of local sector density at a centre reflects the central place system of the region, and the importance and dominance of a centre can be argued to be a decreasing function of the average distance to potential customers outside the centre. The appendix explains how this reasonable hypothesis and the relationship illustrated in Fig. 2 are made operational in the spatial general equilibrium model.

The discussion so far means that the intraregional distribution of (local sector) jobs reflects the residential location pattern. At the same time, it makes sense to assume that residential location decisions are influenced by the job opportunities within a reasonable commuting distance. As indicated in Fig. 1, this means that the spatial distribution of jobs and people is interdependent. This is the fundamental mechanism in economic base modelling, which will be addressed later in this section.

2.3 The Decision to Stay or Move from a Residential Site

The residential location choice can be considered to result from a two-step decision process. First, a household decides whether to stay or move from the current residential site. Second, households moving have to choose between alternative locations. Consider first the diagonal elements of a matrix of transition, migration, probabilities.

One hypothesis incorporated into our model is that the probability of remaining in a zone is positively related to the labour market accessibility of the zone. This is consistent with the findings from Swedish microdata (Lundholm 2010; Eliasson et al. 2003), while Van Ham and Hooimeijer (2009) find a similar result for the Netherlands. The explanation is that labour market accessibility allows greater flexibility and can generally be seen as a favourable attribute for a residential location.

It is easy to find examples of sparsely populated rural areas where unemployment is close to zero and out-migration of the working population is high (McArthur et al. 2012). An important reason for this is that the probability of finding an appropriate job in a peripheral area is low, and known, causing workers to migrate out of the area or drop out of the labour force. The point is that when labour market accessibility is below some critical level, it is the local balance between the demand and supply of labour which determines the probabilities of staying in a zone. This effect is represented by a relationship between the tendency to move and labour market accessibility in the model.

In this chapter, accessibility is represented by a measure for generalised distance to all other zones of the geography. Each zone is weighted by the number of jobs, adjusted for the competition for jobs, measured by the number of jobs as a proportion of the local number of job seekers. In addition, the weights involve a distance deterrence function that places a relatively high weight on destinations which lie within a short distance from the residential location.

Finally, the measure of generalised distance is combined with information on the local labour market situation in the function that determines the probability that workers move from a specific zone. Assume that a zone has high unemployment. If this zone is centrally located in the region, with a low value of generalised distance, many workers will choose to commute rather than move from their current residential location. If, however, the generalised distance is long, then migration will be a more frequent spatial interaction response to high unemployment, defining a process towards a situation with a balance in the local labour market.

2.4 Spatial Equilibrium and Migration Flows Between Different Zones

The impact of job opportunities and the labour market situation is accounted for through the diagonal elements in the matrix of migration probabilities. The migration between different zones is modelled through the introduction of a search strategy where a worker evaluates destinations successively outwards over the network. The worker will move to the first place where the conditions are 'satisfactory'. Options further out in the network will then not be evaluated. Hence, an absorption effect is introduced, analogously to the basic idea in the theory of intervening opportunities (Miller 1972). This further means that the probability of moving decreases as the worker evaluates alternatives which lie progressively further out in the network.

Another central hypothesis within the regional science literature is that distance limits spatial interaction. Accounting for the absorption effect and the distance deterrence effect forms a symmetric matrix, that is next normalised into a migration probability matrix. This matrix is then used to find the equilibrium solution for the system. An equilibrium spatial population pattern is reached if implementing the migration probabilities leaves the distribution of population between zones unchanged (Nævdal et al. 1996).

2.5 *Job Diversity, Local Amenities, Spatial Wage Disparities and House Prices*

We do not incorporate job diversity, local amenities, spatial wage disparities and house prices into this version of this model. We include them in Fig. 1 to illustrate where they would fit into the model.

2.6 *The Relationship Between the Spatial Distribution of Jobs and People, an Economic Base Multiplier Process*

The spatial distribution of jobs is linked to the spatial distribution of people through labour market accessibility, and the simultaneity between commuting flows and migration flows. The economic base mechanism represents the more direct link between the location of jobs and people in Fig. 1. As mentioned above, location decisions of local sector firms reflect the shopping behaviour and the location pattern of the households demanding the goods and services being offered. At the same time, workers employed in local sector firms tend to prefer a residential location close to the firm. Assume increased basic sector activity in a zone. This causes a rise in labour demand, attracts labour to the zone and increases the demand for goods and services produced in the local sector. This creates further demand for labour and initiates a positive growth cycle, known in the literature as an economic base multiplier process.

The equilibrium modelling approach in this research accounts for different kinds of interdependencies in a simultaneous treatment of location decisions made by firms and households. This can be argued to be a preferred approach to introducing a specific causality on the employment–population interaction. According to Hoogstra et al. (2011), the nature of this causality differs across space and time. They carried out a statistically supported literature review ('meta-analysis') and found that the empirical evidence is highly inconclusive on the jobs-people direction of causality, although most result point towards 'jobs follow people'.

3 Prediction Errors in a Simple 5-Node Imaginary Geography

Consider the very simple, 5-node, network illustrated in Fig. 3. In this imaginary geography, the central business district (CBD) is assumed to be located in zone B. The distances from the CBD to the other zones are indicated in the figure. The suburban zone D is located 5 km from zone B, while the zones C and E are located within a reasonable commuting distance from the CBD: 30 and 20 km, respectively. In Fig. 3, zone A appears to be a peripheral rural location, 80 km from the CBD. The distance between the zones A and B, d_{AB}, will be systematically varied to study the impact of this distance on the equilibrium employment and population in zone A. Reductions in distances may be due to investments in road infrastructure, for instance, by removing the effect of topographical barriers through the construction of tunnels and/or bridges. Alternatively, d_{ij} can be interpreted as travelling times, and reduced travelling times can result from improved road standards, higher speed limits and more efficient traffic management.

Finding an equilibrium spatial distribution of population and workers of course calls for a parametrisation of the model. The parameters are defined in the appendix, which provides a technical model presentation. The parameter values chosen for the standard case of the numerical experiments are presented in Sect. A.7.

3.1 *Relocation Effects of Variations in the Distance from the Central Business District*

Assume that the central business district has a concentration of basic sector jobs, while non-CBD basic sector jobs are evenly spread between the zones A, C, D and E; $E^b_B = 10{,}000$, $E^b_A = E^b_C = E^b_D = E^b_E = 3000$. The equilibrium solution following

Fig. 3 A 5-node network of zones

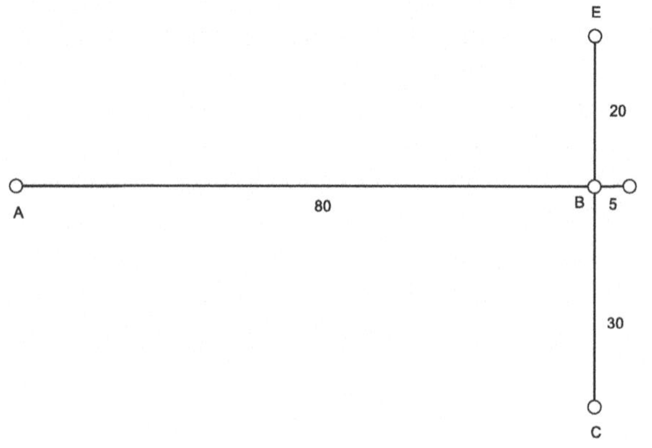

from the model depends on the characteristics of the network. In this section, we focus on the distance from zone A to the CBD, d_{AB}. Figure 4 offers an illustration of how the equilibrium employment and population in zone A changes when d_{AB} is systematically varied from 5 to 100 km.

Consider the first part (a) of Fig. 4, corresponding to a value of 0.25 of the parameter β, representing the elasticity of migration flows with respect to distance. For very low values of d_{AB}, zone A appears as a suburban zone, with a high population and a relatively low number of local sector jobs. The zone is attractive for commuters. It has a relatively low number of local sector jobs, since households living here tend to do their shopping in the CBD rather than locally. Notice from the figure that increasing d_{AB} has no unambiguous effect on population and employment in zone A. There are two forces, pulling the equilibrium in different directions. First, increasing d_{AB} makes zone A less attractive as an origin of commuting. On the other hand, people do more of their shopping locally when the distance to the CBD increases, making the zone more attractive for local sector activities. In Fig. 4a, the effect through labour market accessibility on population dominates for variation of distance in the intervals $d_{AB} < 20$ and $d_{AB} > 30$, while the effect of increased local sector employment tends to dominate for $15 < d_{AB} < 30$. Compared to the other cases in the figure, however, population and employment are not substantially affected by variations in distance.

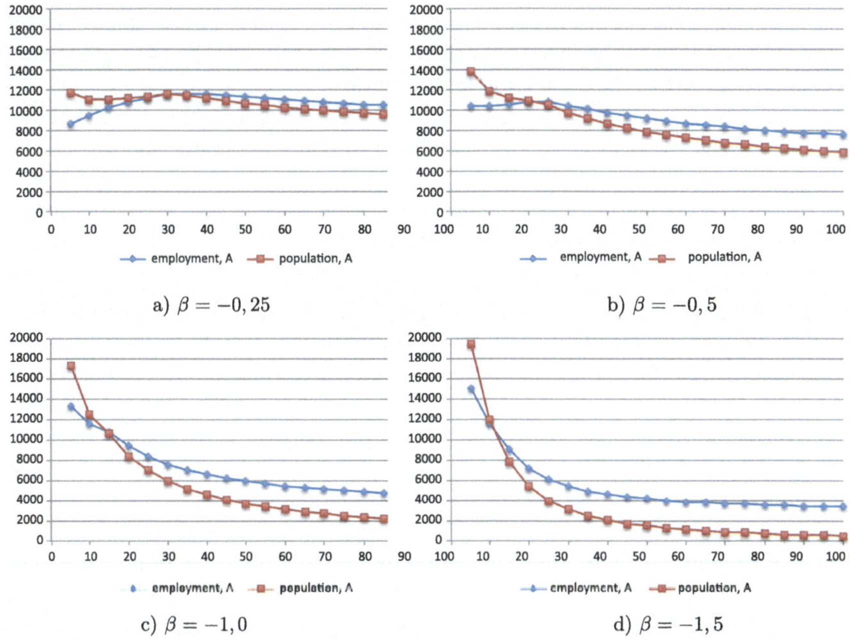

Fig. 4 Equilibrium employment and population in zone A, corresponding to variations in d_{AB}, and to different values of the distance deterrence in migration (β)

The other cases in Fig. 4 correspond to higher negative values of the distance deterrence in migration (β), i.e. where migration is more deterred by distance. In the literature, estimates of this elasticity typically range between -1.0 and -3.0; see, for example, Schwartz (1973). It can be argued, however, that distance deterrence in migration depends on the geographical scale that is considered. The value of $\beta = -0.25$ corresponds to a hypothesis that internal moves within a region are not very sensitive to distance. Figure 4 demonstrates that the impact of variations in distance depends heavily on the value of this parameter. Keep in mind that the employment curves in the four parts of Fig. 4 represents the local sector employment plus the number of basic sector jobs.

According to the four parts of Fig. 4, equilibrium population and employment of zone A are a lot more sensitive to variations in distance in cases with strong residential site preferences, that is high negative values of β. If zone A is located close to the CBD and the rest of the system, workers living here can take advantage of a high labour market accessibility, without moving a long distance from their preferred residential location. Hence, this makes zone A popular as a residential location, especially in cases with a strong distance deterrence effect in local migration.

In a case where the negative value of β is high, the inhabitants have strong location preferences for their current residential area, and this is detrimental if the zone is peripherally located in the region. This contributes to explaining rural depopulation, even in cases where there are good prospects of receiving job offers in the zone. For high values of d_{AB}, zone A offers an unfortunate combination of low labour market accessibility and a location in a long distance from the preferred residential location for a large majority of the population in the region.

Consider next transportation infrastructure investments, reducing the travel distance from zone A to the CBD by one-third. In one experiment, the distance is reduced from 60 to 40 km, while the distance is reduced from 30 to 20 km in another experiment. The model is used to predict relocation effects, and the results of the numerical experiments are presented in Table 1.

Assume first that relocation effects are measured by absolute changes, that is the changes in the total number of jobs and people in zone A. According to Table 1, the absolute changes resulting from transport innovations tend to be most dramatic in the case where zone A is initially located relatively close to the central area of the region. In this case, zone A is predicted to experience a reduced population and employment if $\beta = -0.25$, while the zone is predicted to experience substantial increases in population and employment if intraregional moves are severely deterred by distance.

If workers are not very concerned with distance in making their intraregional moving decisions, an equilibrium solution results where the population is relatively evenly distributed between the zones A, C, D and E. In such a case, the character of the equilibrium solution is less sensitive to variations in distance than in cases where workers are more reluctant to move a long distance from their current residence. This is the reason why the model predicts more substantial relocation effects in the cases with high negative values of the relevant distance deterrence parameter, as reported

Table 1 Relocation effects of reduced distance from zone A to the CBD

	$\beta = -0.25$		$\beta = -0.5$		$\beta = -1.0$		$\beta = -1.5$	
	E_A	L_A	E_A	L_A	E_A	L_A	E_A	L_A
$d_{AB} = 60$	10,997	10,280	8677	7281	5445	3135	3884	1133
$d_{AB} = 40$	11,522	11,153	9716	8670	6575	4584	4601	2052
$d_{AB} = 30$	11,576	11,602	10,400	9743	7626	5941	5412	3093
$d_{AB} = 20$	10,725	11,152	10,819	10,944	9428	8413	7195	5403
$\Delta E(60 \rightarrow 40)$	525		1039		1130		717	
$\%\Delta E(60 \rightarrow 40)$	4.8		12.0		20.8		18.5	
$\Delta L(60 \rightarrow 40)$	873		1389		1449		919	
$\%\Delta L(60 \rightarrow 40)$	8.5		19.1		46.2		81.1	
$\Delta E(30 \rightarrow 20)$	−851		419		1802		1783	
$\%\Delta E(30 \rightarrow 20)$	−7.4		4.0		23.6		32.9	
$\Delta L(30 \rightarrow 20)$	−450		1201		2472		2310	
$\%\Delta L(30 \rightarrow 20)$	−3.9		12.3		41.6		74.7	

in Table 1. The table also reveals a tendency that relative relocation effects are more dependent on the value of β than on the initial location of zone A.

Measured in percentage changes, the relocation effects are not very sensitive to the initial location of zone A. Relative changes are also, however, strongly dependent on the value of the distance deterrence in migration; relocation effects are considerably stronger for high negative values of β. Notice also from Table 1 that the balance between people/workers and jobs is more disturbed if the road transport innovation is introduced in a case where zone A is located close to the CBD. The problem to be addressed in the sections to follow is how such relocation effects affect the predicted increase in commuting flows resulting from improvements in the road network. What is the impact of ignoring relocation effects in traditional cost–benefit analyses of investments in transport infrastructure?

3.2 *Relocation Effects, Prediction Errors and the Initial Position of Zone A*

As stated in the introduction, ignoring relocation effects potentially introduces errors in the prediction of commuting flows. In this section, the focus is on how the prediction bias is depending on the initial location of the zone that benefits from improvements in the road transportation network. Commuting flows in the system are predicted by a doubly-constrained gravity model; see Sect. 6. Consider once again variations in the distance between zone A and the CBD. Results of numerical experiments are presented in Table 2.

Table 2 reports induced commuting in two cases. One case is starting from a situation where $d_{AB} = 60$, the other is taking $d_{AB} = 30$ as a starting point. The model is next used to predict induced commuting flows corresponding to a reduction in travel distance of $\frac{1}{3}$. This corresponds to a reduction of distance of 40 and 20 km in the two cases, respectively. The upper part of the table refers to experiments where the spatial distribution of jobs and people is assumed to be given, unaffected by changes in the location of zone A relative to the CBD. The lower part of the table refers to experiments that account for the predicted changes in the row and column sums of the commuting flow matrix, that is for the changes in the spatial distribution of jobs and people that was predicted by the spatial equilibrium model and discussed in the preceding section. The location pattern corresponding to $d_{AB} = 60$ and $d_{AB} = 30$ is of course the same in the two parts of the table.

The results in Table 2 are based on a value of the distance deterrence parameter in migration of $\beta = 1.0$. It follows from part c) of Fig. 4 that zone A has a very low population if $d_{AB} = 60$ in such a case, and that the number of (basic sector) jobs is considerably higher. This high local net supply of jobs is reflected in the predicted commuting flows; according to Table 2, a lot more workers are commuting into zone A than from zone A in the case where $d_{AB} = 60$. Notice that the induced commuting flows to and from zone A are equal in the upper part of the table, ignoring rounding

Table 2 Responses in commuting flows to changes in the travel distance (d_{AB}) in a case with $\beta = 1.0$

	$\sum_j T_{Aj}$	$\sum_i T_{iA}$	$\sum_j T_{Aj} + \sum_i T_{iA}$
Ignoring relocation effects			
$d_{AB} = 60$	25	2335	2360
$d_{AB} = 40$	322	2633	2955
$d_{AB} = 30$	1583	3268	4851
$d_{AB} = 20$	2989	4674	7663
Induced commuting (60 → 40)	297	298	595
Induced commuting (30 → 20)	1406	1406	2812
% Induced commuting (60 → 40)	1188.0	12.8	25.2
% Induced commuting (30 → 20)	88.8	54.1	58.0
Accounting for relocation effects			
$d_{AB} = 60$	25	2335	2360
$d_{AB} = 40$	492	2483	2975
$d_{AB} = 30$	1583	3268	4851
$d_{AB} = 20$	4022	5036	9058
Induced commuting (60 → 40)	467	148	615
Induced commuting (30 → 20)	2439	1768	4207
% Induced commuting (60 → 40)	1868.0	6.3	26.1
% Induced commuting (30 → 20)	154.1	54.1	86.7

Note T_{ij} = the number of commuters from zone i to zone j
$\sum_j T_{Aj}$ is commuting from zone A, while $\sum_i T_{iA}$ is commuting into zone A

errors. This follows as a consequence of keeping constant the row and column sums of the commuting flow matrix.

The two cases in Table 2 refer to equal percentage changes in travel distance, but the reduction in physical distance is two times larger in the case where $d_{AB} = 60$ initially. This does not necessarily mean that the costs involved in this case are two times higher than the costs of reducing the travel distance from 30 km to 20 km. There may be higher costs involved in road infrastructure investments closer to the centre of the geography.

Notice first that the induced commuting in absolute terms are a lot higher in the case where zone A is initially located close to the CBD. With one exception, this also applies in relative terms. The exception is commuting from zone A, which is very low prior to the reduction in travel distance. Hence, even a moderate increase in the number of commuters appears as a huge relative increase.

It can be argued that ignoring relocation effects leads to an underprediction of the induced commuting flows resulting from improvements in the road infrastructure. Transport innovations are offering new options that workers will take advantage of by choosing more preferred combinations of job and residence location. Intuitively, this leads to more commuting. To some degree, this represents a hypothesis that is supported by the results of the numerical experiments reported in Table 2.

Consider first the reduction of travel distance from 60 to 40 km. It follows from Table 1 that this induces a marked increase in the number of workers that are living in zone A. This explains why ignoring relocation effects leads to an underprediction in the number of commuters from zone A, as reported in Table 2. At the same time, however, more jobs in zone A are now occupied by local workers, explaining why the relocation effects actually contribute to reduce the induced commuting to zone A. In total, ignoring relocation effects in the case where $d_{AB} = 60$ initially leads to a very marginal underprediction of commuting flows involving zone A; according to Table 2, the induced commuting is 26.1% in the case where relocation effects are accounted for and 25.2% in the case where they are ignored.

According to the results reported in Table 2, the prediction errors from ignoring relocation effects are substantially higher, both in absolute and relative terms, in the case where zone A is located closer to the CBD.

3.3 Prediction Errors and the Distance Deterrence in Moving

It was demonstrated in Sect. 3.1 that relocation effects of changes in the transportation infrastructure are sensitive to the distance deterrence in intraregional moving decisions. Hence, prediction errors of ignoring relocation effects should also be expected to be sensitive to the value of β. This is confirmed by the simulation results illustrated in Fig. 5. The curves in the figure refer to errors of ignoring relocation effects in predicting total commuting flows to and from zone A. Adjusted for the fact that the scaling on the vertical axis differs in the two parts of the figure, and it is very apparent that prediction errors are a lot larger in part b) of the figure, corresponding to the case where the negative value of β is highest. This is consistent with the finding in Sect. 3.1 that the equilibrium spatial distribution of jobs and residences is not very sensitive to variations in distance for low negative values of β.

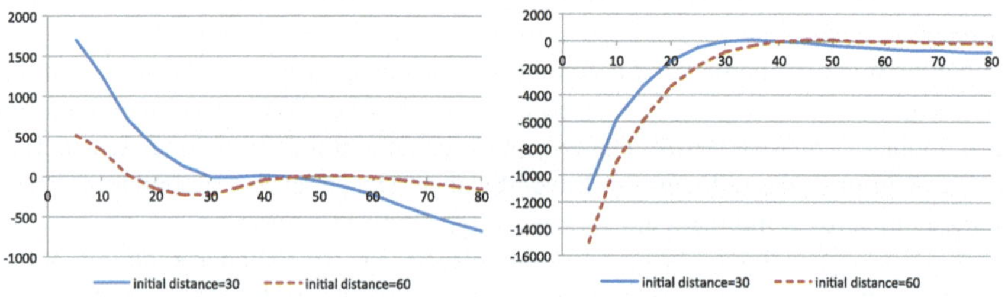

a) $\beta = -0.25$ b) $\beta = -1.0$

Fig. 5 Prediction errors of ignoring relocation effects in predicting total commuting flows to and from zone A

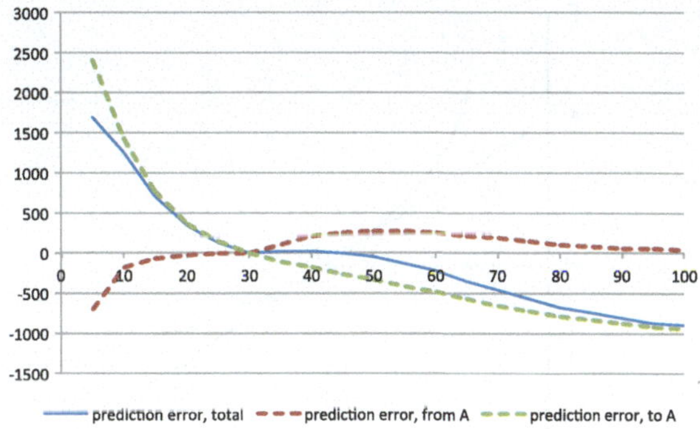

Fig. 6 Prediction errors of commuting flows to and from zone A

Figure 5 also illustrates the finding from Sect. 3.2 that the prediction bias from ignoring relocation effects is most severe in the case where the initial location of zone A is close to the centre of the geography. If zone A is initially located in a distance of 60 from the CBD, huge improvements in the transportation infrastructure are required to generate significant relocation effects, causing prediction errors. In the case with $\beta = 1.0$ and $d_{AB} = 30$ initially, commuting flows might be substantially underpredicted even for moderate reductions in the travelling time from zone A to the CBD.

One striking difference between the two parts of Fig. 5 is that ignoring relocation effects might actually lead to an overprediction of induced commuting flows in the case with a low negative value of β. This might at a first glance seem to be a counter-intuitive result. As mentioned above, the reduced travel distance makes zone A more attractive as an origin for commuting flows, and workers can take advantage of new options by choosing more preferred combinations of job and residence location. Remember, however, from part (a) of Fig. 4, that a reduced d_{AB} might lead to reduced employment in zone A, since the households make more of their shopping in the CBD. This is reflected in Fig. 6, where the prediction errors are split into flows to and from zone A. The figure clearly demonstrates that incoming commuting flows are overpredicted when the relocation of local sector jobs are ignored. It is important to consider predictions of total commuting flows as a net outcome of in- and out-commuting. As expected, ignoring relocation effects lead to an underprediction of commuting flows from zone A, also in the case with a low negative value of β. In Fig. 6, however, the effects originating from shopping decisions are dominating.

4 Welfare Calculations

When deciding whether to undertake a transportation investment, it is important to understand the costs and benefits. In this section, focus will be placed on the estimation of the direct benefits flowing to users. A standard microeconomic framework can be used to asses the change in welfare resulting from a change in the transportation

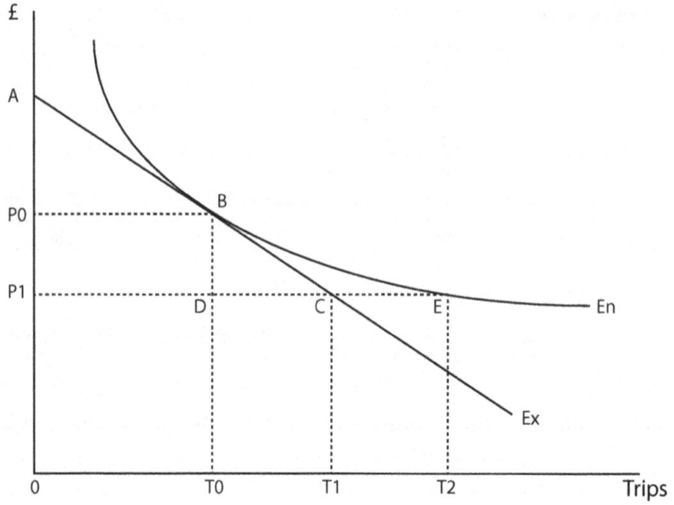

Fig. 7 Demand for trips across a link as a function of the generalised cost of travel

network. In order to proceed, an estimate of the demand function is required. An example demand curve for trips between two locations is shown in Fig. 7.

Figure 7 depicts a situation where the generalised cost of travel across a link is given by P_0. Looking at the demand curve labelled ‘Ex’, the demand for trips is T_0. The consumer surplus for these road users is given by the area $AB\,P_0$. Assume now that an investment is made which reduces the generalised cost of travel to P_1. From a welfare perspective, two effects which must be considered. Firstly, the users who made the T_0 trips can now do so at a lower cost. This increases their consumer surplus by the amount $P_0BD\,P_1$. The additional effect is the *induced demand*, i.e. the increase in demand from T_0 to T_1. These users were not willing to travel at the previous price, but are willing to do so at the new price. The consumer surplus for these users is lower than that for the previous users. The consumer surplus from the new users is given by the area $BC\,D$. The total change in consumer surplus caused by the change is therefore given by $P_0BC\,P_1$.

An important assumption invoked in the construction of the demand curve ‘Ex’ illustrated in Fig. 7 is that all other factors remain equal. One of these factors assumed to remain fixed is the location of workers and firms. However, firms and workers will often relocate in response to a change in the transportation network, as has been demonstrated. In such a case, changes in the generalised cost of travel will result in a shift in demand rather than a movement along the same demand curve. Whether the shift will be outward or inward will depend on the preferences of the agents populating the geography as well as the geography itself.

Consider again a price of P_0, where the demand for trips is T_0. If the generalised cost is lowered to P_1, we would expect the demand for trips to rise to T_1 if the locations of firms and workers are fixed. If, however, the locations are allowed to vary, movement along the demand curve labelled 'En' is observed. Demand would therefore increase from T_0 to T_2 in response to the reduction of the generalised cost of travel. Failure to account for relocation effects will result in an underestimation of the change in welfare by the area *BCE*. The total welfare change when accounting for relocation effects is therefore *BDE*.

This effect can be explored in our model. Two different demand curves for the demand for trips between zones A and B will be constructed. The distance will be systematically varied from 1 to 80 km to construct these curves. The change in welfare resulting from a reduction in the distance between A and B from 40 to 20 km will be estimated. In order to calculate welfare, this distance will be converted to a generalised cost of travel. To do this, the UK's Automobile Association's (The AA) estimate of the variable cost of motoring.[1] The variable costs amount to 13.92 pence per km. An estimate of the value of time is required. Firstly, we assume travel at a speed of 60 km/h, so that 1 km is the equivalent of 1 min. We use the value of time spent commuting from the UK's Department for Transport's (DfT) Transport Appraisal Guidance (TAG)[2] inflated to 2012 GBP. This gives the cost of one minute of 11.73 pence. The total cost of commuting one km is therefore £0.26.

One theoretical point which should be noted is that we work with the Marshallian demand curve which accounts only for substitution effects and not income effects. This means that consumer surplus is measured rather than the more theoretically appealing measures of compensating variation (CV) or equivalent variation (EV) which can be derived from the Hicksian demand curve. In practice, consumer surplus is by far the most used measure. Partly this is because of the CV and EV measures are more difficult to calculate. The CS will usually lie somewhere between the CV and EV measures. When income effects are small, as is usually the case with transport projects, the measures will provide similar results De Jong et al. (2007).

Demand curves under the assumptions of fixed and variable locations for firms and workers and firms are presented in Fig. 8. We begin at a situation where the distance between A and B is 40 km, corresponding to a cost of £10.26, and a total demand of 2975 one-way trips per day. This corresponds to 1.19 million trips per year if workers work a 200 day year and make a return trip on each of these days. We now consider the impact of a reduction in distance to 20 km, or a generalised cost of £5.13.

If fixed locations are assumed, then a rise in demand to 6783 one-way trips per day is predicted, a change of 3808. If, however, firms and workers are allowed to move in response to the change, a rise in demand to 9058 is predicted, a rise of 6083. This means an assumption of fixed locations underestimates demand by 76%.

The estimated change in the consumer surplus can be obtained by integrating the demand function over the interval of the price change, i.e. 10.26–5.13. We use

[1] See www.theaa.com/resources/Documents/pdf/motoring-advice/running-costs/petrol2012.pdf (Retrieved 11/09/2012) will be used. We assume a petrol engine and that the car's original purchase price was £14,000 to £17,000. Parking and toll charges are excluded from the cost.

[2] www.dft.gov.uk/webtag/documents/expert/pdf/u3_5_6-vot-op-cost-120723.pdf (Retrieved 11/09/2012).

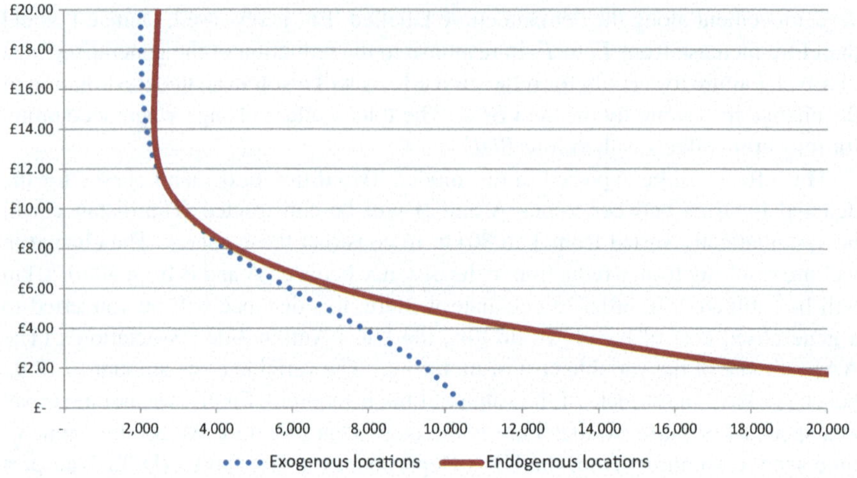

Fig. 8 Demand for trips between zones A and B as a function of the generalised cost under the assumptions of a fixed and a variable locations of firms and workers

Simpson's rule to numerically integrate the demand function. Assuming workers and firms have fixed locations, the change in consumer surplus is estimated to be £23,681 per half-day for all road users. If workers and firms are allowed to relocate, then the consumer surplus is estimated to be £26,873. This is a difference of £3192 or 13.5%. On an annual basis, this amounts to an underestimation of £1,276,808.

In the example presented here, ignoring relocation effects resulted in an underestimation of consumer surplus of 13.5%. However, under different conditions, an under- or overprediction may be observed. One important parameter is the sensitivity of workers to distance in making migration decisions. The example is shown in Fig. 8 assumed a distance deterrence parameter of $\beta = 1$. If we follow the same procedure as above but change β to a value of 0.2, we get an overestimation of 0.2% by assuming fixed locations of firms and workers.

It is unsurprising that making workers less sensitive to distance when considering where to live has this effect. When workers do not care about distances, shortening distances has little effect on their decisions. A range of under- and over-predictions is therefore possible. Careful modelling which takes into account location decisions is therefore important in evaluating the effect of changes in infrastructure.

5 Road Improvements and Aggregate Commuting

The welfare calculations that were made in the previous section did not account for the possibility of negative externalities related to induced traffic resulting from improvements in the road network. From an environmental point of view, it is often

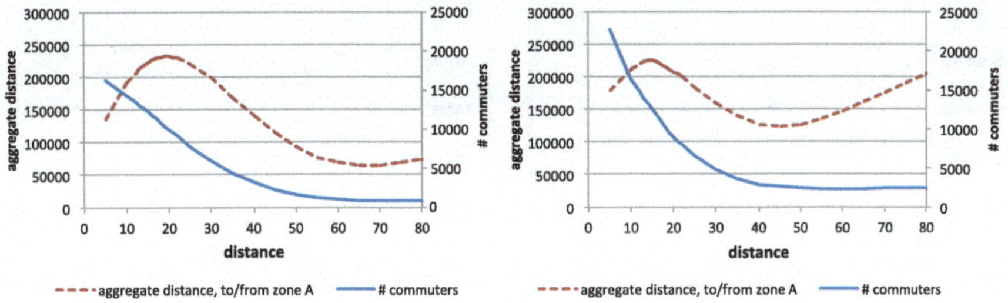

a) $\beta = -0.25$ b) $\beta = -1.0$

Fig. 9 Impact of d_{AB} on the number of commuters and the aggregate commuting distance to and from zone *A*, for two values of the parameter representing distance deterrence in intraregional moving decisions

claimed that investments in road infrastructure, in general, generate more traffic and, consequently, more emissions and pollution. This is obviously an interesting perspective in a welfare evaluation of such investments. Figure 9 illustrates how the total number of commuters to and from zone *A* and the aggregate commuting distance to and from zone *A* depends on the distance between the zone and the regional centre, d_{AB}. No attempt has been made to account for the possibility that the average speed and emissions per vehicle might depend on the number of road users, due to congestion problems. The parameter representing the distance deterrence in intraregional moving decisions has a value of -0.25 in the left part of the figure, and -1.0 in part b) of the figure.

Notice first that the number of commuters to and from zone *A* increases to a high level if the distance to the regional centre, d_{AB}, is short. This increase is more modest, however, in the case where the moving decisions of workers are not very sensitive to distance. Once again, this reflects the finding in Sect. 3.1 that the equilibrium spatial distribution of jobs and residences is less sensitive to variations in distance for low negative values of β. Notice also, from the right part of Fig. 9, that the number of commuters is very insensitive to variations in distances over 45 in the case with $\beta = 1.0$. It also follows from the figure that reductions in distance, for $d_{AB} > 45$, induce reduced aggregate commuting distance. The dominating effect is that the average commuter to and from zone *A* travels a shorter distance as a consequence of road improvements.

Assume next that the distance d_{AB} is initially shorter than approximately 15–20 km, prior to investments in the road network. It follows from Fig. 9 that reductions in distance then lead to a reduced aggregate distance of commuting to and from zone *A*. The explanation is that shorter average commuting distance, combined with the effect of relocation effects, more than outweighs the effect of an increased number of commuters. The reduction in aggregate commuting distance is especially distinct in the case with a low negative value of β, reflecting, for instance, that the increased number of commuters is more modest in this case.

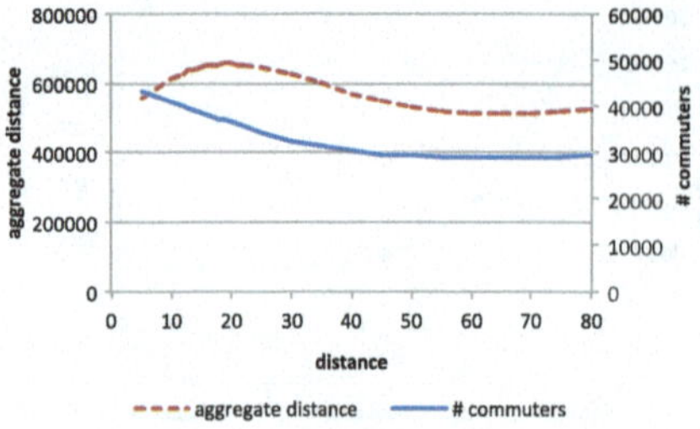

a) $\beta = -0.25$

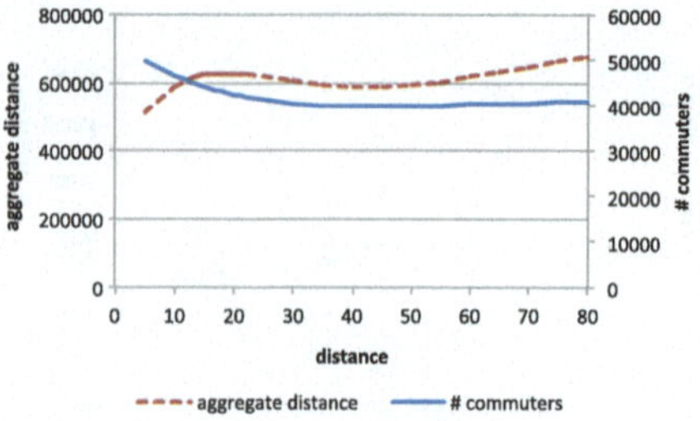

b) $\beta = -1.0$

Fig. 10 Impact of d_{AB} on the number of commuters and the aggregate commuting distance in the geography, for two values of the parameter representing distance deterrence in intraregional moving decisions

In principle, this reduction in aggregate commuting on a specific link does not necessarily provide an environmental argument in favour of road investments. The reduction may have its counterpart in increased traffic on other links in the system. Figure 10 illustrates how the total number of commuters and the aggregate system-wide commuting distance depend on the distance between zone A and the regional centre, d_{AB}. According to the figure, improvements in the road network may actually cause reductions in the aggregate commuting distance, despite the fact that the number of commuters increases.

6 Conclusions

It has been the aim of this chapter to draw attention to a potentially important source of error in estimating the benefits arising from changes in transportation infrastructure. This error occurs when it is assumed that firms and workers will not relocate in response to a change in infrastructure. In this chapter, a model is constructed for analysing this problem on an intraregional scale.

It was shown, using our model, that assuming fixed locations could result in either an under- or overestimation of the traffic on the road link under analysis. The direction and magnitude of the error will depend on the preferences of agents in the geography and the spatial structure of the region. One of the key factors is the sensitivity of people's migration decisions to distance. When workers are insensitive to distance, there is very little relocation in response to a change in the infrastructure. The reasoning is that distance receives a low weight in people's decisions about where to live. When people are sensitive to distance, the opposite was true.

The spatial structure of the region also turned out to be important. Adjusting distances for settlements closer to the CBD had a much larger impact than making the same changes at a larger distance from the CBD. This is logical, given that thresholds generally exist for how far people are willing to commute, or how far they are willing to travel to go shopping. The configuration of settlements within a region will therefore also affect any relocation response to the change in the infrastructure.

We can therefore say, based on our numerical experiments, that there are two factors which ought to raise particular concern about using a naive model assuming fixed locations. Firstly, when workers are sensitive to distance regarding their location decisions we would expect larger errors when erroneously assuming fixed locations. The second factor we can identify is when the change in infrastructure is taking place close to the CBD. As the distance to the CBD declines, and as workers become more sensitive to distance, the prediction errors can be expected to rise.

In terms of the cost–benefit analysis of an infrastructure project, a number of important effects have been identified. A failure to adequately deal with location modelling may deny us, ex ante, information about outcomes which are potentially important in the decision-making process. For instance, if we are concerned about equity between urban and rural areas, we may favour policies which encourage people to live in more peripheral areas. Here, the location effects are important as an end, rather than simply a means.

Perhaps the most important aspect of neglecting relocation effects is the effect on the estimated demand for a new or improved road. Relocation may result in higher or lower demand than would be expected if locations were assumed to be fixed. We have shown an example where the benefit was underestimated by 13.5% due to an underestimation of demand. Such an error may make the difference between the project going ahead or not.

Incorrect predictions of demand will also prove problematic when estimating the expected environmental costs of a change in infrastructure. As has been shown, the environmental effects can be complex. A shorter journey time may increase demand, but reduce the overall distance driven by all drivers. Understanding how location decisions will respond to a change in infrastructure is important in understanding the environmental impact. A project which may appears to have a negative impact when using a naive model may in fact have a positive impact when taking relocation effects into account.

Appendix

A. A Technical Presentation of the Model

This appendix provides a technical presentation of the mechanisms represented in Fig. 1. The spatial distribution of basic sector firms is considered to be exogenously given, which means that aspects of local innovativity and competitiveness are not explicitly accounted for. In this version of the model, we further ignore the possibility that migration decisions are affected by job diversity and local amenities, while housing prices and wages are assumed to be exogenously given.

A.1 Basic and Local Sector Firms; the Economic Base Multiplier

Total employment in zone i (E_i) is defined to be the sum of basic sector employment (E_i^b) and local sector employment (E_i^l) in the zone:

$$E_i \equiv E_i^b + E_i^l \tag{1}$$

Let $\boldsymbol{L}$ be a vector representing a given residential location pattern of workers, while T_{ij} is the probability that a worker lives in zone i and works in zone j. Hence, $\boldsymbol{T} = [T_{ij}]$ represents the commuting matrix in the geography, and by definition:

$$\boldsymbol{TE} = \boldsymbol{L} \tag{2}$$

The spatial distribution of local sector activities reflects both the spatial residential pattern and the spatial shopping behaviour. Assume that the number of workers living in a zone is proportional to the number of residents/consumers in the zone, and let C_{ij} be the number of local sector jobs in zone i which are supported by shopping from worker living in zone j. Hence, $\boldsymbol{C} = [C_{ij}]$ is a shopping matrix, and the spatial distribution of employment in local sector activities is given by:

$$\boldsymbol{E}^l = \boldsymbol{CL} \tag{3}$$

Given that the inverse of the matrix $(\boldsymbol{I} - \boldsymbol{TC})$ exists, it follows from Eqs. (1), (2) and (3) that:

$$\boldsymbol{L} = (\boldsymbol{I} - \boldsymbol{TC})^{-1}\boldsymbol{TE}^b \tag{4}$$

$$\boldsymbol{E}^l = \boldsymbol{C}(\boldsymbol{I} - \boldsymbol{TC})^{-1}\boldsymbol{TE}^b \tag{5}$$

These solutions capture the economic base multiplier process: people attract local sector activities, while local sector employment opportunities attract workers (see Sect. 2). As mentioned above, the spatial distribution of basic sector activities ($\boldsymbol{E}^b$) will be considered exogenous in the model, while the other variables ($\boldsymbol{C}$, $\boldsymbol{E}^l$, $\boldsymbol{T}$, $\boldsymbol{L}$) are represented by a set of equations representing shopping, commuting, location and migration decisions of households and firms. In the next two subsections of this appendix, we consider how residential location and migration decisions reflect spatial disparities in the labour market situation as well as characteristics of the spatial structure and the road transportation network.

A.2 Interzonal Migration Flows and Spatial Equilibrium

In modelling migration probabilities, Nævdal et al. (1996) introduced a nice trick to facilitate construction of Markov chains. The construction uses a symmetric matrix $\boldsymbol{Q} = \{Q\}_{i,j=1}^{N}$, where all the elements ($Q_{ij} \geq 0, i, j = 1, 2, \ldots, n$) are dependent on the characteristics of the geography. The transition matrix $\boldsymbol{M} = \{P_{ij}\}_{i,j=1}^{N}$ is given by:

$$P_{ij} = \frac{Q_{ij}}{\sum_{k,k \neq j} Q_{kj}} \quad i, j = 1, 2, \ldots, N \tag{6}$$

Nævdal et al. (1996) showed that any assumption about the coefficients Q_{ij} can be interpreted as an assumption about migration flows in the equilibrium state. As a next step, Nævdal et al. (1996) introduced some network characteristics which are symmetric between zones and which are relevant in explaining the relevant kind of spatial interaction. For a connected network with fixed Q_{ij}-s, the construction produces regular Markov chains. Nævdal et al. (1996) showed that the equilibrium condition is given by the eigenvector:

$$\boldsymbol{L} = \begin{bmatrix} \frac{\sum_{i,i \neq 1} Q_{i1}}{1-P_{11}} \\ \cdot \\ \cdot \\ \cdot \\ \frac{\sum_{i,i \neq 1} Q_{iN}}{1-P_{NN}} \end{bmatrix} \tag{7}$$

Let $\alpha_i = 1 - P_{ii} \neq 0, i = 1, 2, \ldots, N$ be the probability that a person will not stay in zone i within the given time frame. In this subsection, we focus on internal migration flows, and the diagonal elements α_i of the migration probability matrix are assumed to be given. Assume next a strategy where a migrant evaluates destinations successively outwards over the network and moves to the first zone where the conditions are 'satisfactory'. In addition, we introduce a simplifying assumption of constant absorption, defined by the absorption parameter s:

$$s = \frac{\text{Probability of moving to } (m+1)\text{-th neighbours}}{\text{Probability of moving to } m\text{-th neighbours}}, \qquad \text{constant in } m$$

As an operational assumption accounting for the impact of both distance and absorption the migration flows between zone i and its m-th neighbour will be proportional to $\frac{s^m}{d_{ij}^{\beta}}$, where β is a distance deterrence parameter. The step parameter n defines the maximum transition length, i.e. the transition probability is zero between neighbours of order greater than n. The symmetric $\boldsymbol{Q}$-matrix derived from this procedure defines the transition matrix $\boldsymbol{M}$ by (6). As an illustration, the transition matrix for a simple linear three-node system is given by:

$$\begin{bmatrix} 1-\alpha_1 & \frac{s\alpha_2}{d_{21}^{\beta}}\left(\frac{s}{d_{12}^{\beta}}+\frac{s}{d_{32}^{\beta}}\right) & \frac{s^2\alpha_3}{(d_{12}+d_{23})^{\beta}}\left(\frac{s}{d_{23}^{\beta}}+\frac{s^2}{(d_{12}+d_{23})^{\beta}}\right) \\ \frac{s\alpha_1}{d_{21}^{\beta}}\left(\frac{s}{d_{21}^{\beta}}+\frac{s^2}{(d_{21}+d_{32})^{\beta}}\right) & 1-\alpha_2 & \frac{s\alpha_3}{d_{23}}\left(\frac{s}{d_{23}^{\beta}}+\frac{s^2}{(d_{12}+d_{23})^{\beta}}\right) \\ \frac{s^2\alpha_1}{(d_{21}+d_{32})^{\beta}}\left(\frac{s}{d_{21}^{\beta}}+\frac{s^2}{(d_{21}+d_{32})^{\beta}}\right) & \frac{s\alpha_2}{\left(\frac{s}{d_{21}^{\beta}}+\frac{s}{d_{32}^{\beta}}\right)d_{32}^{\beta}} & 1-\alpha_3 \end{bmatrix}$$

In our chapter, the coefficients will be state-dependent, i.e. the Q_{ij}-s are functions of $\boldsymbol{E}$ and $\boldsymbol{L}$. In that case, the equilibria are no longer unique, but the interpretation in terms of the strength of migration flows in the equilibrium state remains valid; see Nævdal et al. (1996). A standard application of Brouwer's fixed-point theorem gives that equilibria always exist in the state-dependent case.

A.3 The Decision to Stay or Move from a Zone

It is a central hypothesis in the model that the decision to stay or move from a zone depends on the labour market accessibility of the zone. Labour market accessibility is introduced by a measure of generalised distance, rather than, for example, by a gravity-based Hansen accessibility measure (Hansen 1959). The generalised distance from zone i is given by:

$$d_i = \sum_{j \neq i} \frac{W_j}{\sum_{k \neq i} W_k} d_{ij} \tag{8}$$

Labour market accessibility is of course not just a matter of distances, the weight W_i represents the size of alternative job destinations. The size, and thickness, of a potential destination is assumed to be represented by the number of jobs; $W_j = E_j, j = 1, 2, \ldots, N$, defining d_i as the average Euclidean distance to potential employment opportunities in the geography. In a spatial labour market context, however, it can be argued that potential destinations within a reasonable commuting distance should be put more weight on than more distant destination alternatives. This is done through the introduction of a distance deterrence function $D(d_{ij})$, that places a relatively high weight on destinations which lie within a short distance from the residential location:

$$W_j = E_j(1 - D(d_{ij})) \tag{9}$$

The distance deterrence, and, hence, the weights are parameterised by d_∞, d_0 and μ in the following logistic expression:

$$D(x) = \frac{1}{1 + e^{-k(x - x_0)}}, \quad x_0 = \frac{1}{2}(d_0 + d_\infty), \quad k = \frac{2\log(\frac{1}{\mu} - 1)}{d_\infty - d_0} \tag{10}$$

d_∞ is the upper limit for how far workers, as a rule, are willing to commute on a daily basis, d_0 is the lower limit (internal distance) where people are insensitive to further decreases in distance, while μ captures friction effects in the system. The values of x_0 and k are given to satisfy the conditions $D(d_0) = \mu$ and $(1 - D(d_\infty)) = \mu$. If , e.g. $\mu = 0.05$, this means that the function will fall to 5% of its value outside the range where $d_0 \leq x \leq d_\infty$. Glenn et al. (2004) gave a microeconomic and geometric justification for the use of such a function.

Finally, the definition of generalised distance also accounts for the competition for jobs at alternative locations (Liu and Zhu 2004; Shen 1998), represented in the model by the proportion of the total number of job seekers in each potential destination, $\frac{E_j}{L_j}$:

$$W_j = E_j(1 - D(d_{ij}))\frac{E_j}{L_j} \tag{11}$$

The definition of generalised distance is included in the diagonal elements of the migration matrix, reflecting workers spatial interaction response to an unfortunate local labour market situation ($L_i > E_i$). A high value of d_i [and $D(d_i)$] means that the migration decisions are very sensitive to the local labour market situation. On the other hand, a high local unemployment does not in itself bring about a significant out-migration from zones in highly accessible labour market location (low d_i), with an excellent commuting potential. This is captured by the following specification of α_i:

$$\alpha_i = \alpha_i(L_i) + D(d_i)\max\left\{\rho\left(\frac{L_i - E_i}{L_i}\right), 0\right\} \tag{12}$$

Here, the parameter ρ reflects the speed of adjustment to an unfortunate labour market situation, towards a situation with a balance in the local labour market, $L_i = E_i$.

A.4 The Spatial Distribution of Local Sector Employment

It is reasonable to assume that local sector activities in a whole region (E_r^l) are proportional to population in the region (L_r):

$$E_r^l = \sum_i^n E_i^l = b\sum_i^n L_i = bL_r \qquad b > 0 \tag{13}$$

where b is the proportion parameter. Let the spatial distribution of local sector employment be represented by $\frac{E_i^l}{L_i}$, that is the number of shop-employees per resident at location i. Assume, as a simplification, a monocentric region, offering agglomeration benefits for local sector firms and price savings for households in shopping. Shopping decisions then results from a trade-off between price savings and transport costs.

Transportation costs provide an incentive for local sector firms to decentralize in order to cater for local demand. The trade-off between transport costs and potential price savings plays a central role in Gjestland et al. (2006), providing a theoretical base in favour of the hypothesis that the frequency of shopping locally is a smooth, concave, function of the Euclidean distance from the CBD. In our chapter, we assume that there is only one CBD and define the local sector density by:

$$\text{Local sector density} = \frac{E^l}{L}(\text{distance to CBD}) \tag{14}$$

$$= R_\infty(1 - \exp[-\beta_{\text{CBD}} \cdot \text{distance to CBD}]) \tag{15}$$

$$+ C \cdot \exp[-(\gamma \cdot \text{distance to CBD}/d_{\text{dispersion}})^2] \tag{16}$$

The only free parameter is β_{CBD} which controls the decay in the local sector density curve. The other parameters are defined as follows:

$$R_\infty = \frac{\sum_{i=1}^N E_i^l}{\sum_{i=1}^N L_i} \quad \text{(average local sector density in the system as a whole)} \tag{17}$$

$d_{\text{dispersion}}$ is the spatial extension of the CBD, $\gamma = \sqrt{-\ln[\kappa]}$ forces the effect of the second term (16) down to $\kappa\%$ of its peak value at the boundary of the CBD. Given values for β_{CBD}, R_∞ and $d_{\text{dispersion}}$, C are chosen such that the integral

$$\int_0^{d_{\text{dispersion}}} \text{Local sector density}(r) \cdot \frac{2r}{d^2_{\text{dispersion}}} \cdot L_{\text{CBD}} dr = E^l_{\text{CBD}} \tag{18}$$

The spatial distribution of local sector activities reflects the net effect of the price savings resulting from agglomeration forces and the transport costs of shopping in the CBD rather than locally.

A.5 Commuting Flows

In the model to be used in this chapter, the location of workers and jobs is assumed to be fixed, calling for a doubly-constrained version of the gravity model. It is well known that a doubly-constrained gravity model is equivalent to the multinomial logit model; see Anas (1983) for details. This means that the model can be derived from random utility theory. The doubly-constrained gravity model incorporates set of balancing constraints, representing an assumption of a given spatial distribution of jobs and households. The following model specification ensures that the column sums of the predicted commuting flow matrix equal the total number of jobs at the corresponding destinations, and that each row sum equals the number of workers residing in the corresponding zone:

$$T_{ij} = A_i O_i B_j D_j e^{(-\beta_{\text{gravity}} d_{ij})} \tag{19}$$

$$A_i = \left[\sum_j B_j D_j e^{(-\beta_{\text{gravity}} d_{ij})} \right]^{-1} \tag{20}$$

$$B_j = \left[\sum_i A_i O_i e^{(-\beta_{\text{gravity}} d_{ij})} \right]^{-1} \tag{21}$$

where

T_{ij} is the number of commuters from origin i to destination j
O_i is the observed number of commuting trips originating from zone i
D_j is the observed number of commuting trips terminating in zone j
d_{ij} is the travel time from origin i to destination j.

A_i and B_j are the balancing factors which ensure the fulfilment of the marginal total constraints: $\sum_j T_{ij} = O_i$ and $\sum_i T_{ij} = D_j$.

A.6 An Iterative Process Towards Spatial Equilibrium

To initiate the iterative process, we begin with more or less random random initial values for employment and population ($\boldsymbol{E}_0 = \boldsymbol{E^l}_0 + \boldsymbol{E^b}_0$ and $\boldsymbol{L}_0$). These values are fed into a state-dependent migration matrix $\boldsymbol{M}$ and adjusted to fit a local sector density curve, which is then iterated until we find a fixed point $\boldsymbol{L}$, which represents the equilibrium solution for population (workers), i.e. that $\boldsymbol{ML} = \boldsymbol{L}$, and equilibrium values fitting the local sector (jobs) $\boldsymbol{E^l}$ to the local sector density curve.

A.7 Parameter Values Chosen for the Numerical Experiments

Absorption effects are ignored in the very simple transportation network (Fig. 3), $s = 1$. The distance deterrence in internal migration flows is represented by an elasticity of $\beta = -1.0$, unless else is stated. The logistic distance deterrence function involved in determining the decisions to stay or move from a zone is specified by $d_\infty = 80$, $d_0 = 5$, and $\mu = 0.05$, while the speed of adjustment to an unfortunate labour market situation is given by $\rho = 1$. The form of the local sector density function is given by $\kappa = 0.05$ and a spatial extent of the CBD of $d_{\text{dispersion}} = 4$. Estimated commuting flows reflect a distance deterrence parameter of $\beta_{\text{gravity}} = 0.07$.

References

Anas, A. (1983). Discrete choice theory, information theory and the multinomial logit and gravity models. *Transportation Research Part B: Methodological, 17*(1), 13–23.

Boyce, D. (1988). Renaissance of large-scale models. *Papers in Regional Science, 65*(1), 1–10.

De Jong, G., Daly, A., Pieters, M., & Van Der Hoorn, T. (2007). The logsum as an evaluation measure: Review of the literature and new results. *Transportation Research Part A: Policy and Practice, 41*(9), 874–889.

Eliasson, K., Lindgren, U., & Westerlund, O. (2003). Geographical labour mobility: Migration or commuting? *Regional Studies, 37*(8), 827–837.

Fujita, M., Krugman, P., & Venables, A. (1999). *The spatial economy: Cities, regions and international trade*, Vol. 213. Wiley Online Library.

Gjestland, A., Thorsen, I., & Ubøe, J. (2006). Some aspects of the intraregional spatial distribution of local sector activities. *The Annals of Regional Science, 40*(3), 559–582.

Glenn, P., Thorsen, I., & Ubøe, J. (2004). Wage payoffs and distance deterrence in the journey to work. *Transportation Research Part B: Methodological, 38*(9), 853–867.

Hansen, W. G. (1959). How accessibility shapes land use. *Journal of the American Institute of planners, 25*(2), 73–76.

Hoogstra, G., van Dijk, J., & Florax, R. (2011). Determinants of variation in population-employment interaction findings: A quasi-experimental meta-analysis. *Geographical Analysis, 43*(1), 14–37.

Liu, S., & Zhu, X. (2004). An integrated GIS approach to accessibility analysis. *Transactions in GIS, 8*(1), 45–62.

Lundholm, E. (2010). Interregional migration propensity and labour market size in Sweden, 1970–2001. *Regional Studies, 44*(4), 455–464.

McArthur, D., Encheva, S., & Thorsen, I. (2012). Exploring the determinants of regional unemployment disparities in small data sets. *International Regional Science Review, 35*(4), 442–463.

McArthur, D., Thorsen, I., & Ubøe, J. (2014). Employment, transport infrastructure, and rural depopulation: A new spatial equilibrium model. *Environment and Planning A, 46*(7), 1652–1665.

Miller, E. (1972). A note on the role of distance in migration: Costs of mobility versus intervening opportunities. *Journal of Regional Science, 12*(3), 475–478.

Nævdal, G., Thorsen, I., & Ubøe, J. (1996). Modeling spatial structures through equilibrium states for transition matrices. *Journal of Regional Science, 36*(2), 171–196.

Schwartz, A. (1973). Interpreting the effect of distance on migration. *The Journal of Political Economy*, pp. 1153–1169.

Shen, Q. (1998). Location characteristics of inner-city neighborhoods and employment accessibility of low-wage workers. *Environment and planning B: Planning and Design, 25*(3), 345–365.

Van Ham, M., & Hooimeijer, P. (2009). Regional differences in spatial flexibility: long commutes and job related migration intentions in the Netherlands. *Applied Spatial Analysis and Policy, 2*(2), 129–146.

Waddell, P., Borning, A., Noth, M., Freier, N., Becke, M., & Ulfarsson, G. (2003). Microsimulation of urban development and location choices: Design and implementation of urbansim. *Networks and Spatial Economics, 3*(1), 43–67.

A New Approach for User Benefit Assessment by a LUTI Model: The Case of Madrid

Yang Wang, Floridea Di Ciommo, and Andrés Monzón

1 Introduction

Transport policies (such as road pricing, public transport improvement and parking management) have been applied to help mitigate traffic congestion and decrease negative external impacts (e.g., pollution, noise, traffic safety) in many cities throughout Europe. However, these transport policies are mostly deemed as regressive policy.

In this context, social equity effects are important for at least two reasons (Wangsness et al. 2017). First, the magnitude of the redistribution between different groups of users can be so large that it could dwarf the net benefit of the transport policy implementation. In some real-world applications, the distributional effects may be so much larger than the efficiency gains that congestion charging may not be seen as worthwhile. Second, the regressiveness of a transport policy (such as congestion pricing) depends on the quality of the alternative available modes, the revenue recycling and on the design of the transport policy schemes (Eliasson and Mattsson 2006; Martens and Di Ciommo 2017).

- The equity analysis of transportation policies is related to the complexity and wide impact of those policies on mobility, land use and the environment (Levinson 2010; Venables 2017). Meanwhile, land use transport interaction models are probably the most complex quantitative models used in any aspect of urban and regional

Y. Wang (✉)
Vectio Traffic and Transport Engineering, Calle Orense 58, 7 A, 28020 Madrid, Spain
e-mail: yang.wang@vectio.com

F. Di Ciommo
CambiaMO/Changing Mobility, Calle Duque de Fernán Núñez 2, 1°, 28012 Madrid, Spain
e-mail: Floridea.diciommo@cambiamo.net

A. Monzón
TRANSyT, Universidad Politécnica de Madrid, Calle del Profesor Aranguren, 3, 28040 Madrid, Spain
e-mail: andres.monzon@upm.es

J. Thill (ed.), *Innovations in Urban and Regional Systems*,
https://doi.org/10.1007/978-3-030-43694-0_8

planning (Wegener 2014; Geurs and Van Wee 2004). The LUTI model described in this paper is a strategic, dynamic and integrated urban land use and transport model (Metropolitan Activity Relocation Simulator, MARS) which has been calibrated for the Madrid metropolitan area (MMA) to model and assess several transport policies.

This paper aims to assess equity issues in terms of user benefits (including travel time and trip costs) testing different TDM polices. The impact analysis is developed using the following steps:

- Step 1: assess the problems of the study area, select possible policy instruments and design transport policy scenarios.
- Step 2: improvement of MARS model including a modified social welfare function (WF).
- Step 3: assessment of the user benefits using a social equity and transport network efficiency approach.

A wide range of transport policy instruments are available for use in the planning context of the Madrid metropolitan area. By studying the mobility situation and socioeconomic status of the MMA, and by considering the restrictions of the MARS model, several TDM policies have been selected as policy packages to reduce congestion and internalize traffic externalities.

The current methodology for evaluating transport projects, notably cost–benefit analysis (CBA), has certain limitations regarding equity. In the existing literature and practice, one way of considering equity in urban transport issues is the use of consumer surplus. However, from a purely welfare economics point of view; the richer population groups have a higher value of utility. If we adopt this conventional approach, projects that primarily serve higher income groups will score better in CBA than transport investments that primarily serve lower income groups. High income groups could receive more transport infrastructure, as they have a higher weight in the welfare function and receive the infrastructure benefits (Martens 2011; Martens et al. 2012).

A social welfare function is modified by including costs and benefits to both users and operators of the transport system, but also costs and benefits to the government, as well as what are termed "externalities costs." The commonly used consumer surplus variable neglects the difference of the utilities and weight between low income and high income travelers (Pfaffenbichler and Shepherd 2002; Guzman et al. 2013; Di Ciommo and Lucas 2014). Actually, the conventional consumer surplus formulation gives a higher value to the upper income class. If the income differences between transport users are minimal, the old approximation approach of the consumer surplus may work; if the difference is greater, it needs to calibrate the utility functions using a compensating weight. Consequently, we adopt a CW represented by the inverse of household income level (Jara 2007).

Based on this modified social welfare function, the social equity effects on user benefit as determined by the generalized travel costs are estimated in different policy scenarios, in terms of time period and household income classes.

This paper is organized as follows: After the introduction, the second section defines the case study of the Madrid metropolitan area and describes three scenarios, including the simulation of transport policies by the MARS model. After that, it modifies the user benefit measure in the social welfare function. The results in terms of social equity and traffic system efficiency are presented, respectively, in the fourth section. Finally, the conclusions make the point about the user benefits variations using the consumer surplus with or without the compensative weight.

2 Selecting TDM Policy Instruments to Define the Scenarios

The selection of an appropriate policy instrument for MMA depends on the main objective of decision makers, the general mobility situation and other requirements of the studied area. Therefore, it firstly reviews the challenges which MMA faces and then considers the relative merits of the general transport policy schemes to choose which one is best suited to the case study area.

2.1 The Case Study: Madrid Metropolitan Area

The Madrid metropolitan area is located in the center of Spain with a population of slightly more than six million people and covers an area of 8030 km^2 (CTRM 2006). Table 1 shows the basic information of MMA. The metropolitan road network includes three orbital highways (e.g., M30, M40 and uncompleted M50), eight freeways and four tolled radial highways (Fig. 1).

The average speed of traffic in the area inside of M40—which is also the majority of MMA—is 24.84 km/h. The lowest average speed is 9.57 km/h in the city center. The average speed of traffic of the whole MMA is 23.31 km/h, which shows a critical congestion problem in this area.

Table 2 lists data about car ownership rates (i.e., 529 per 1000 inhabitants), which is the highest rate of motorization among Spanish regions. The mobility household survey developed for the Madrid metropolitan area shows that from 1996 to 2004, the number of mechanized trips increased by 52%, whereas the population increased

Table 1 Basic information of MMA (*2005 ATP, INE)*

Population (10^3)	Area (km^2)	Density (hab/km^2)	Work places	Unemployment	Average family size	GDP (€)
6,008	8,030	748	2,878	6.4	2.9	27,279

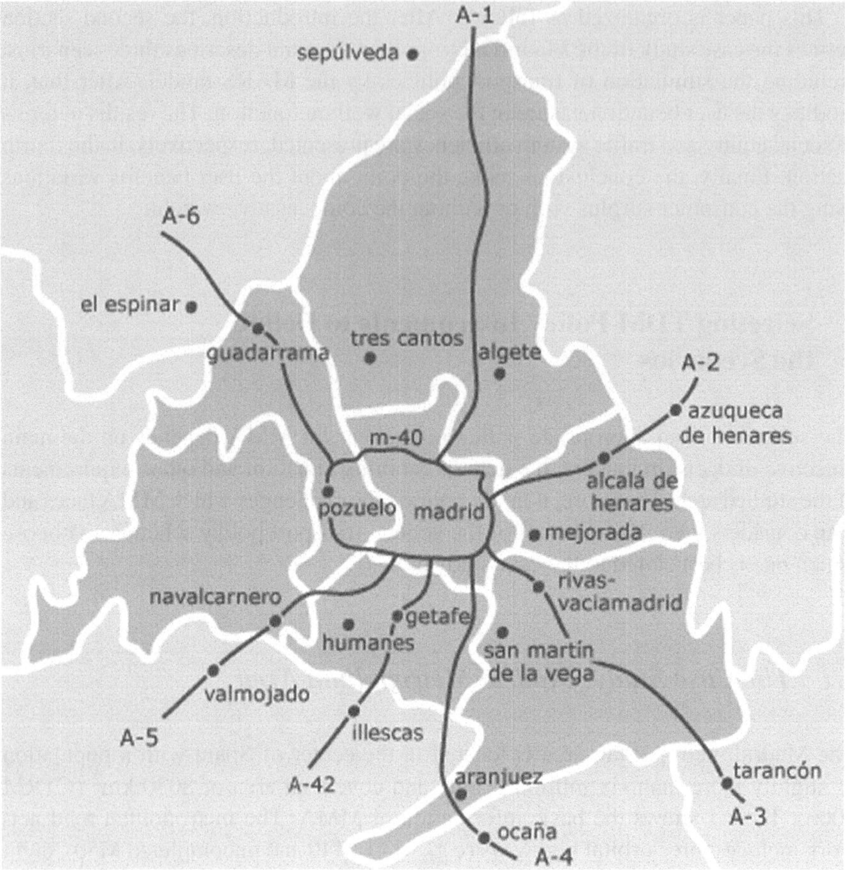

Fig. 1 Map of Madrid region

Table 2 Data of transportation in MMA (*Resource* National Statistics Institute & Public Transport Associations of Spain, 2005)

Car per head	Modal split (PT/PC[a])	No. of autobus lines	Velocity of bus (km/h)	Urban bus frequency[b] (min)	MMA bus frequency[c] (min)	Number of bus	No. of parking place
0.529	31/69	441	13.80	8.6	15.7	3,091	20,758

[a]PT: public transport; PC: private car

[b]Average frequency in peak hour of urban bus (minute)

[c]Average frequency in peak hour of metropolitan bus (minute)

by only 14% (CTRM 2006). From the table, one can also see the public transport services provided in MMA. Even if the public transport system in Madrid is one of the most efficient, the sprawled urban form makes it difficult to offer efficient public transportation services outside of the Madrid municipality. In addition, the good quality of urban roads makes the use of the private vehicles more efficient in terms of time (Di Ciommo et al. 2011).

Nevertheless, the suburbanization of both residence and employment has prompted new transport mobility trends in the MMA, increasing congestion, especially along the metropolitan rings of M40 in peak hours.

Moreover, the Spanish Strategy of Sustainable Development (Gobierno de España 2002) has drafted a long-term perspective aimed at producing a more coherent society in terms of rational use of common resources and one that is more equitable and cohesive in terms of land use. In this context, transportation policy makers require the appropriate policy instruments to achieve this goal.

2.2 *Transport Policy Instruments and Scenario*

This part explains in detail three general transport policy instruments corresponding to scenarios: cordon charge, parking charge and public transport improvement. The designed scenarios within the policy instruments are explained as below. Table 3 summarizes all four scenarios corresponding to each policy scenario.

2.2.1 Scenario 0: Do Nothing

The "Do Nothing" scenario is needed to quantify the impact of these policies. This scenario includes only the infrastructures built in 2004 and will not consider any intervention during the 30-year evaluation period (2004–2034). Under the base scenario, the net social welfare is set to zero. The optimal toll value obtained in Alternative

Table 3 Scenario summary

Scenario	Policy instrument	Applied area	Implement level
Scenario 0	Do nothing	–	–
Scenario 1	Cordon toll	On M40	1.5 €/vehicle
	Parking charge	Inside of M40	2.5 €/vehicle
Scenario 2	Public transport (PT) frequency increasing	All the network	100%
Scenario 3	Combination of cordon toll+	On M40	1.5 €/vehicle
	Parking charge+	Inside of M40	2.5 €/vehicle
	Public transport frequency increasing	All the network	100%

Scenario 1, which maximizes the net present value (NPV) of the objective function proposed, is the surplus of the social welfare compared to Scenario 0.

2.2.2 Scenario 1: Cordon Toll and Parking Charge Scenario

Road pricing involves charging motorists directly for driving on a particular road or in a particular area. It can reduce traffic congestion on a particular roadway or area, particularly if it is implemented as part of a comprehensive transport management program—for example, alongside transit improvements and parking charge programs (Vickrey 1955).

For the current study, a toll is applied within the cordon of the Madrid metropolitan area, which alleviates congestion along the orbital highway of M40. The toll scheme is implemented only for passenger vehicles. The baseline year for this study is 2004, and the toll ring scheme is assumed to be implemented in 2007. It is charged during peak hours with a fixed pricing of 1.5 euros/vehicle. In order to manage the optimization problem, this work adopts a linear policy profile with single-variable policy instrument values as suggested by Pfaffenbichler and Shepherd (2002).

Besides that, parking charges have been used to control the demand for space that has appeared in high-density areas but has gradually been extended to areas with different uses and almost any size (Parry and Bento 2001). As a result, parking management is one of the most effective tools to control the attractiveness of private car travel and, through it, circulatory congestion.

In this case, the implementation of parking charges within the area inside of M40 is designed to discourage car use in some inner metropolitan areas. For practical reasons, it applies a uniform rate of 2.50 euros/vehicle for all of the kernels of Madrid, which must be paid by all users without exception.

2.2.3 Scenario 2: Public Transport (PT) Improvement Scenario

In the MARS model, public transport is divided into two categories: road services with normal vehicular traffic (urban bus and interurban bus) and rail service (metro and light rail). This scenario aims to assess the gain of user benefit by increasing bus frequency of public transit services by 100% on all roads inside of M40 (see Fig. 1). Applying PT frequency instruments is also intended to provide an alternative mode for travelers.

Higher frequency of public transport services implies higher operation costs provided by operators. Thus, the frequency improvement starts from the base operating costs of the bus and rail and calculates the increase or decrease of operating costs as the variation. Implementation of this measure occurs throughout Madrid and gradually, year by year, until reaches the optimal value at the end of the evaluation period.

2.2.4 Scenario 3: Integrated Policies Scenario

An integrated policies scenario combines a road pricing and parking charge scenario on one side, and public transport improvement scenario on another side (e.g., a combination of Scenario 1 and Scenario 2). It aims to examine the impact of integrating the two different policies: restricting the use of vehicles and improving public transport services. The improvement of public transport services increases the modal shift from a more expansive private vehicle transport mode to cheaper public transport mode. Hence, social welfare will be improved by the combined policies scenario.

3 Proposed Methodology

3.1 Metropolitan Activity Relocation Simulator for Madrid

We adopt a strategic, dynamic LUTI model (Metropolitan Activity Relocation Simulator, MARS) (Pfaffenbichler et al. 2008) to define and assess transport policies to feed into the objective function for the optimization process.

MARS includes a transport model to simulate travel behavior of the population, a housing location model, a workplace location model and fuel consumption and emission model. It includes representation of three composite modes: private vehicle (driver and passenger), public transport (bus and rail) and non-motorized modes (cycling and walking). MARS uses the concept of a constant time budget (Levinson and Kumar 1994) and assumes that commuting trips are inelastic (Emberger et al. 2008). Table 4 shows the main characteristics of the MARS for Madrid.

The MARS model is based on a concept of causal loop diagrams (CLDs) and is used to explain the variables of cause-and-effect relations in the model. In the case of car usage, for example, Fig. 2 shows some of the CLDs (balancing and reinforcing causal loops) that drive the behavior of commuting car trips from one zone to another. Similar CLDs could be constructed for other modes and for non-commuting trips (Pfaffenbichler et al. 2008). Changes in the price of fuel charged to the end user influence the attractiveness of car use. This level of attractiveness can also be modified by other CLDs that take into account changes in speed (time) due to a different level of congestion (loop L2). Changes in accessibility of transport affect, in the long term, residential and workplace locations, which may have an impact on the attractiveness of car use as well (De la Hoz et al. 2010).

This simple causal loop structure and user-friendly software help to understand the modeling approaches used. Transport model and optimization procedures are interdependent. The transport policy assessed in the MARS model is redefined in an optimization process through feedback from the results obtained. MARS can model the transport and behavioral responses to several demand and supply-side instruments. MARS assumes that land use is not a constant but is rather part of a SD that is influenced by transport infrastructure (Shepherd et al. 2006). The interaction

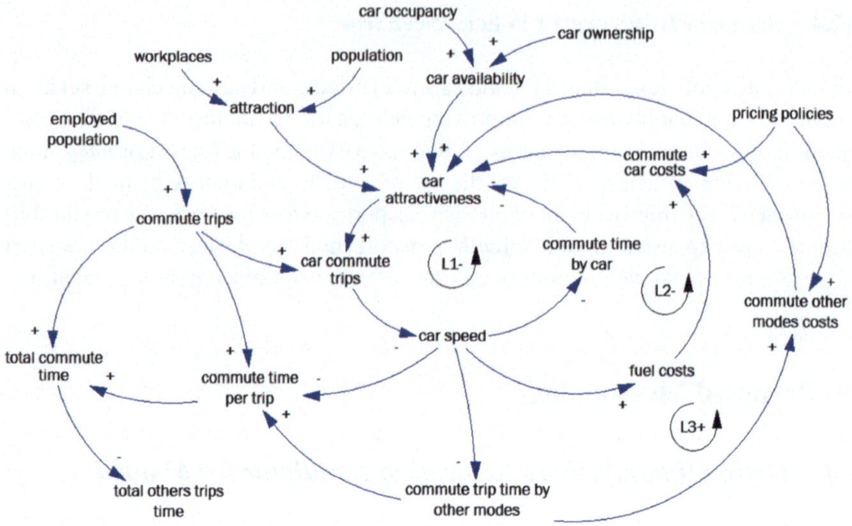

Fig. 2 Commute car trips causal loop diagram

Table 4 Characteristics of MARS Model

Model feature	MARS
Number of zones	90 zones: usually administrative boundaries
Modes of travel	Car, public transport, slow
Congestion effects	OD-specific speed-flow curves for commute trips. No speed effect for other trips (assumed to be in the off-peak)
Generalized costs	In-vehicle time, money, access/egress, parking search time, wait times, change times
Journey purposes	Commute, other
Time periods	Peak, off-peak
Levels of car ownership	0; 1
Demographics/household categories	Average household size, employed residents, cars per head, average income per zone
Route choice	No
Mode/destination choice	Simultaneous
Time of day choice	No
Demand response	Commute trips inelastic. Constant time budget
Land use response	Yes

process is modeled using time-lagged feedback loops between the transport and land use sub-models over a 30-year period. The basis of the stock-flow diagram for the expanded SD model is reproduced in Pfaffenbichler et al. (2008).

In terms of transport policies which can be modeled, MARS was developed as a way to investigate strategic issues over a long-term period. A multimodal CBA must be based on output from the land use and transport model, where choices as to the destination and mode are open to travelers.

3.2 Appraisal Framework

Social equity and sustainable transport efficiency are identified as the primary objectives for this work. The objectives are converted into a welfare function, which ranks all possible policy combinations with respect to their contribution to the objectives. Figure 3 shows the logical structure of the appraisal framework used in this work. The most important part is to calculate consumer surplus using the old and new approach within the welfare function.

Social welfare depending on a transport system is defined as the sum of social benefits of users, operators and government. In this paper, our focus is user benefit measures. Generally, the calculation of transport user benefits is based on the conventional consumer surplus theory, where the benefits are obtained from the variations of travel time, operating costs and mobility patterns. Equation 1 shows the social welfare function that includes the cost of users, operators, government and environmental externalities in general.

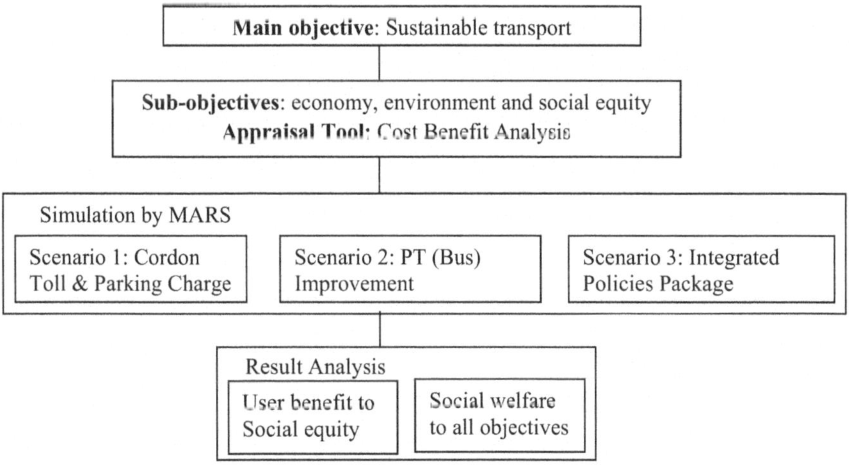

Fig. 3 Structure of appraisal framework

$$\Delta W(U) = \int_0^n \sum_{ij} \frac{1}{(1+r)^t} \cdot \left[\Delta CS_{ij}(t) + \Delta O_{ij}(t) + \Delta G_{ij}(t) + \Delta E_{ij}(t)\right] \mathrm{d}t \quad (1)$$

where

$\Delta W(U)$	the change in social welfare compared to base case scenario
ΔCS_{ij}	customer surplus changing from zone i to j
ΔO_{ij}	operators benefit changing from zone i to j
ΔG_{ij}	government benefit changing from zone i to j
ΔE_{ij}	external costs from zone i to j
r	social discount rate [%]
t	time (year)
n	number of years (from 0 to 30).

3.2.1 Consumer Surplus CS_{ij}

It is well known that the consumer surplus from a change in travel times and/or travel costs should in general be calculated at the level of origin–destination pairs.

$$\begin{aligned} \Delta CS &\approx \frac{1}{2} CW_i \sum_i \sum_j \sum_m \left(T_{ijm}^{E_0} + T_{ijm}^{E_1}\right) \cdot \left(G_{ijm}^{E_0} - G_{ijm}^{E_1}\right) \\ &\approx \frac{1}{2} CW_i \sum_i \sum_j \sum_m \left(T_{ijm}^{E_0} + T_{ijm}^{E_1}\right) \cdot \left(C_{ij}^{E_1} + t_{ij}^{E_1} * VOT_i - C_{ij}^{E_0} + t_{ij}^{E_0} * VOT\right) \end{aligned} \quad (2)$$

where

CW_i	compensating weight in zone i
T_{ij}^k	trips between i to j in scenario k
G_{ij}^k	travel generalized cost from i to j in scenario k
C_{ij}^k	total travel cost including charging and operation cost from i to j in scenario k
t_{ij}^k	travel time from i to j in scenario k
VOT	value of time (10.45 €/h during the peak hour and 5.7 €/h during non-peak hour).

CW_i, compensating weight represents a weight to compensate the user utility for different levels of user income. It is calculated by the inverse of the household income in origin zones as shown in Eq. 3.

$$CW_i = \frac{9600}{I_i} \quad (3)$$

where

I_i	household income in zone i

9600 coefficient to transfer monthly household income to household income in minute.

Consumer surplus (i.e., the difference between user utility and its cost) implies a higher consideration for high income users. Actually, consumer surplus increases with the income of users. In this context, a compensating parameter is adopted in order to consider more equally the consumer surplus value of different income groups of users. As we see in Eq. 2, the consumer surplus is calculated by adopting a parameter equal to the value of the inverse of average household income (I_i). By this method, the different economic level of users is included in the equation, and then, the final value is more concrete to the real user utility. This is the first paper including the compensating weight function, which is more focused on user benefit evaluation than on assessment of all social welfare function. By using the compensating weight welfare function, the MARS model processes the simulation of three transport policies scenarios.

Travel cost C_{ij}^k includes the cost of charging (like a congestion toll or parking charging) and operation costs, like shows in Eq. 4.

$$C_{ij}^k = F_{ij}^k + P_{ij}^k \tag{4}$$

where

F_{ij}^k operation cost from zone i to j in scenario k.
P_{ij}^k cost of road charging from zone i to j in scenario k.

3.2.2 Operator Costs O_{ij}

The operators' cost is calculated by Eq. 5.

$$\Delta O_{ij} = \sum_t \sum_{ijm} T_{ijm}^1 \cdot \left(\tau_{ijm}^1 + c_{ijm}^1\right) - T_{ijm}^0 \cdot \left(\tau_{ijm}^0 + c_{ijm}^0\right) \tag{5}$$

τ_{ij} consists of two parts, one of which is revenue from private vehicles in the case of an urban toll or parking charge fee; the other part is revenue collected from public transport services, e.g., trip tickets.

C_{ij} represents the cost of administration, operation and maintenances cost, energy-consuming cost, etc.

3.2.3 Government Costs G_{IJ}

Equation 6 represents the government's benefits or losses resulting from changes in the revenue fuel tax (ΔF_{ij}) and changes in the cost of road maintenance (ΔM_{ij}).

$$\Delta G_{ij}(t) = \sum_{t}\sum_{ijm}\left(\mathrm{T}_{ij}^{1} - \mathrm{T}_{ij}^{0}\right) \cdot \left(\Delta \mathrm{F}_{ij} + \Delta \mathrm{M}_{ij}\right) \tag{6}$$

3.2.4 Externalities Costs E_{ij}

Equation 7 represents the value of externalities, associated with greenhouse gas emissions (ΔGHE_{ij}), air pollution (ΔP_{ij}) and safety (ΔS_{ij}).

$$\Delta E_{ij}(t) = \sum_{t}\sum_{ijm}\left(T_{ij}^{1} - T_{ij}^{0}\right) \cdot \left(\Delta GHE_{ij} + \Delta P_{ij} + \Delta S_{ij}\right) \tag{7}$$

4 Analysis of the Results

The simulation has been executed by including the compensating weight (CW_i) into the consumer surplus calculation. Table 5 shows the changes of the total user benefit as compared to the do nothing scenario. The value is calculated using and non-using compensating weight and is given in terms of three periods of time. The changes of the total user benefit show an overview of how the compensating weight influences the user benefit. It can be seen that when the user benefit is negative like in Scenario 1, the user benefit value obtained by using CW_i is around 14% lower than without CW_i. However, in the case of positive user benefit, like in Scenarios 2 and 3, the value calculated using CW_i is 10% higher than without CW_i, that is, consistent with the user benefit including compensating weight, which could more equally estimate the value of user benefit for different income users.

With respect to policy implication, there are two findings. Firstly, the highest user benefit is obtained from the public transport improvement scenario. Since the bus frequency increases, more travelers shift to public transport mode, which leads to lower travel costs. The user-detrimental policy measure in the current study is the scenario with cordon toll and parking charge. It mainly affects users by increasing travel costs. Secondly, according to the user benefit changes by time, travelers lose more and more when the cordon toll and parking charges are implemented; however,

Table 5 Total user benefit changes comparison with respect to time period (€ millions)

Total user benefit	Without CW_i			With CW_i		
Time period	S1–S0	S2–S0	S3–S0	S1–S0	S2–S0	S3–S0
Short term (5 year)	−48.72	550.02	502.49	−56.62	497.06	441.90
Middle term (10 year)	−50.03	568.05	519.52	−57.16	517.61	462.13
Long term (15 year)	−56.25	551.44	496.70	−62.36	503.10	442.50

the highest user benefit in Scenarios 2 and 3 is achieved in the middle term, not long term. It can be explained that travelers gain more when the public transport is improved, but the long-term benefit is reduced due to the increase in travel time for public transport users due to traffic increasing.

The following results of user benefit changes are given in terms of each scenario and based on the time period as well as household income. Travelers are divided into three groups based on their household income level. For low income users who earn less than 1850 euros/month, it estimates 0.48×10^6 households. Middle income users are estimated to be 1.12×10^6 households who earn between 1850 and 2500 euros/month. For high income users who earn more than 2500 euros/month, there are estimated 0.36×10^6 households. Therefore, the middle income users are the majority in this user benefit assessment.

It also lists the variation in percentage by comparing the user benefit value **without** CW_i and **with** CW_i. It aims to compare user benefit changes in social equity dimension and also to illustrate potential advantage by including the CW into user benefit calculation.

S1: Scenario of Cordon Toll and Parking Charge

Scenario 1 includes the road charge policy of a cordon toll on the orbital highway M40 and a parking charge of 2.5 €/vehicle in peak hours for all kinds of vehicles. The results in terms of user benefits variation are presented in Table 6. The main findings are:

1. The user benefit changes are all **negative** in the case of the cordon toll and parking charge scenario, regardless of the household income level. However, when CW_i is included, low income users lost 5.3 times more in the long term than high income users. It means that a road pricing scheme without any revenue redistribution would be very regressive to low income travelers.
2. Measuring the user benefit variation by percentage, the lower income users lost 28%, middle income users lost around 6% and high income users gain 33% more by adopting *CW*. So, it means in the case of implementing a user-detrimental policy measure (such as road pricing), it is more harmful for the low income group. By including the *CW*, the results in terms of user benefits are closer to the real user utility. It can be seen that low income users lose much more benefit in terms of travel cost and trip time than middle and high income users. With *CW*, the difference of user detriment between high and low income users is enlarged.

S2: Scenario of Public Transport Improvement

In Scenario 2, 100% of bus services are improved during peak hours by increasing the operators' costs. It increases a large number of benefits for travelers, as shown in Table 7. More findings are listed below:

1. There is a large increase of user benefit in short term and long term in the public transport improvement scenario. Comparing different levels of household income, high income users achieve around 60% more benefit than the low income

Table 6 User benefit changes in respect to time period and households income (€ million)

Income level	Low income (≤1850 €/month)			Middle income (1850 ≤ x ≤ 2500)			High income (≥2500 €/month)		
Time period	Short term (5 year)	Middle term (15 year)	Long term (30 year)	Short term (5 year)	Middle term (15 year)	Long term (30 year)	Short term (5 year)	Middle term (15 year)	Long term (30 year)
Without	−33.6	−32.9	−32.9	−9.0	−9.1	−11.7	−6.1	−8.1	−11.6
With	−43.0	−42.0	−42.1	−9.6	−9.7	−12.2	−4.0	−5.4	−8.0
With/without (%)	−28	−28	−28	−7	−7	−4	+34	+33	+31

Table 7 User benefit changes in respect to time period and households income (€ million)

Income level	Low income (≤1850 €/month)			Middle income (1850 ≤ x ≤ 2500)			High income (≥2500 €/month)		
Time period	Short term (5 year)	Middle term (15 year)	Long term (30 year)	Short term (5 year)	Middle term (15 year)	Long term (30 year)	Short term (5 year)	Middle term (15 year)	Long term (30 year)
Without	124.3	130.0	121.0	133.9	149.2	156.3	291.8	288.8	274.1
With	159.6	166.9	155.3	130.5	146.1	153.7	207.0	204.6	194.1
With/without (%)	28	28	28	−3	−2	−2	−29	−29	−29

users with *CW*. This result may owe to a great mode shift from private vehicle to public transport, particularly for low income users; thus, high income users gain benefit from the decrease in travel time.

2. Considering the impact of incorporating the *CW*, low income users receive 28% more benefit than without *CW*, middle income users see only a slight difference and high income users lose 29% benefit in both short and long term. It means that where the value of user benefit is positive, the benefit value without *CW* is overestimated for high income users. Besides that, user benefit with *CW* reduces the difference between high and low income users. High income users gain more than 2.2 times the benefit compared to low income users without *CW*, and in the case of including *CW*, the difference is 0.25 times higher.

S3: Integrated Policy Package (S1 + S2)

Scenario 3 is the combination of two scenarios: a cordon toll and parking charge (S1) and an increase in bus frequency (S2). A number of researchers have proven (Parry and Bento 2001; Pucher 1983) that providing an alternative transport mode encourages travelers to shift their travel mode from private vehicle to public transport mode. The findings of this scenario are included in Table 8 and analyzed in the following items:

1. The impact of user benefit changes for users of different income levels is similar to Scenario 2, which is that all transport users benefit from the integrated policy measure, though high income users gain more benefit than low income users. The user benefit increase is less in S3 than S2, due to the implementation of a cordon toll and parking charges.

   Besides that, the benefit for high income users is reduced in the long term (from 199.4 to 186.3). Because the toll charge maintains a fixed price for all period, road congestion would not continue to decrease in longer term. Therefore, high income users, who are mostly users of private cars, will receive less benefit over a long period of time.
2. Seeing the difference in the benefit value with and without *CW*, the percentages in the third row are very close, as they are in Scenario 2. It signifies that the estimation of user benefit without compensating weight is also overestimated for higher income users. With *CW*, the value is three times higher for high income users than low income users, and without *CW*, the difference reduces to 0.6 times higher.

Some Policy Implications

Incorporating compensating weight into the consumer surplus calculation avoids overestimating user utility for high income users, as can be seen in Tables 6, 7 and 8. Higher income travelers are normally more willing to pay to save their travel time. The new consumer surplus calculation considering this effect provides a revised result, which is particularly useful for analyzing social impacts.

Table 8 User benefit changes in respect to time period and households income (€ million)

Income level	Low income (≤1850 €/month)			Middle income (1850 ≤ x ≤ 2500)			High income (≥2500 €/month)		
Time period	Short term (5 year)	Middle term (15 year)	Long term (30 year)	Short term (5 year)	Middle term (15 year)	Long term (30 year)	Short term (5 year)	Middle term (15 year)	Long term (30 year)
Without	91.7	98.2	89.1	125.0	140.3	144.8	285.8	281.1	262.8
With	117.9	126.1	114.5	121.0	136.6	141.7	203.1	199.4	186.3
With/without (%)	29	28	29	−3	−3	−2	−29	−29	−29

In the case of road pricing implementation (S1), low income users lose more and high income users lose less, as opposed to failing to consider *CW*. It indicates the high regressiveness of a road pricing policy and sheds light on the level of social inequity by the new consumer surplus with compensating weight.

5 Conclusions

This paper has investigated the use of three different policy instruments (e.g., road pricing, parking charge and public transport frequency improvement) and the implications for the performance of urban transport policies in terms of user benefit.

The paper presents a modified social welfare function accounting for a compensating weight calculated using the household income level. This welfare function could calculate the benefits for users, transport operators, government and social externalities. However, this first paper with the new welfare function focuses on the different groups of user benefits. The simulation is applied to the case study of the Madrid metropolitan area. Based on a dynamic CBA framework, costs and benefits generated by three different policy schemes have been assessed.

By using the modified welfare function, the result analysis shows that road pricing leads to a negative consumer surplus if there is not a redistribution of revenue from the road and parking charges. In the long term, road pricing may appear like a regressive policy, as high income users gain more benefit than low income users. Public transport improvement scenario brings more positive benefits for all users and seems to be less a regressive policy. The combined policies scenario produces a higher level of benefits for low, middle and high income users than Scenario 1. Nevertheless, the level of benefit is higher for high income users.

By incorporating the compensating weight, in the case of implementing a user-detrimental policy (like S1), the low income users lose more and high income users lose less than without considering this coefficient. It proves the high regressiveness of a road pricing policy and sheds light on the social inequity level by the new consumer surplus with compensating weight. When all users benefit from a policy measure implementation, such as in S2 and S3, the high income users benefit less than by the conventional consumer surplus calculation. The user benefit result with compensating weight also reduces the differences for high and low income users, which shows a public improvement policy is the least regressive policy measure in terms of social equity.

References

CTRM. (2006). EDM'04: Encuesta domiciliaria de movilidad en día laborable.

De la Hoz, D., Guzmán, L. A., Pfaffenbichler, P., & Shepherd, S. (2010). *Fuel tax levels necessary to achieve the agreed reduction targets of* CO_2 *emissions. The Case of Madrid*. In *Highway and urban environment*, pp. 377–385

Di Ciommo, F., Monzon, A., Valdes, C., & Wang, Y. (2011). *A methodology to evaluate a toll-ring implementation in terms of metropolitan equity. The case of Madrid Metropolitan Area*. In Transportation Research Board 90th Annual Meeting. No. 11-0750.

Di Ciommo, F., & Lucas, K. (2014). Evaluating the equity effects of road-pricing in the European urban context—The Madrid metropolitan area. *Applied Geography, 54*, 74–82.

Eliasson, J., & Mattsson, L. G. (2006). Equity effects of congestion pricing: Quantitative methodology and a case study for Stockholm. *Transportation Research Part A: Policy and Practice, 40*(7), 602–620.

Emberger, G., Shepherd, S. P., & May, A. D. (2008). The effects of appraisal methodology on the specification of optimal urban transport strategies. *International Journal of Sustainable Transportation, 2*(1), 58–75.

Geurs, K. T., & Van Wee, B. (2004). Land-use/transport interaction models as tools for sustainability impact assessment of transport investments: Review and research perspectives. *European Journal of Transport and Infrastructure Research, 4*(3), 333–355.

Gobierno de España. (2002). Estrategia Española de desarrollo sostenible.

Guzmán, L. A., Di Ciommo, F., & de la Hoz, D. (2013). Integration of congestion pricing and intertemporal preference rate in social welfare function. *Transportation Research Record: Journal of the Transportation Research Board, 2346*(1), 1–12.

Jara Diaz, S. (2007). *Transport Economic Theory*, pp. 75–106.

Levinson, D. (2010). Equity effects of road pricing: A review. *Transport Reviews, 30*(1), 33–57.

Levinson, D. M., & Kumar, A. (1994). Integrating feedback into the transportation planning model: Structure and application. *Transportation Research Record, 1413*, 70–77.

Martens, K. (2011). Substance precedes methodology: On cost–benefit analysis and equity. *Transportation, 38*(6), 959–974.

Martens, K., & Di Ciommo, F. (2017). Travel time savings, accessibility gains and equity effects in cost–benefit analysis. *Transport Reviews, 37*(2), 152–169.

Martens, K., Golub, A., & Robinson, G. (2012). A justice-theoretic approach to the distribution of transportation benefits: Implications for transportation planning practice in the United States. *Transportation Research Part A: Policy and Practice, 46*(4), 684–695.

Parry, I. W., & Bento, A. (2001). Revenue recycling and the welfare effects of road pricing. *The Scandinavian Journal of Economics, 103*(4), 645–671.

Pfaffenbichler, P. C., & Shepherd, S. P. (2002). A dynamic model to appraise strategic land-use and transport policies. *European Journal of Transport and Infrastructure Research, 2*(3/4), 255–283.

Pfaffenbichler, P., Emberger, G., & Shepherd, S. (2008). The integrated dynamic land use and transport model MARS. *Networks and Spatial Economics, 8*(2–3), 183–200.

Press, W. H. (2007). *Numerical recipes: The art of scientific computing* (3rd ed.). Cambridge University Press.

Pucher, J., Markstedt, A., & Hirschman, I. (1983). Impacts of subsidies on the costs of urban public transport. *Journal of Transport Economics and Policy*, 155–176.

Shepherd, S. P., Zhang, X. N., Emberger, G., Hudson, M., May, A. D., & Paulley, N. (2006). Designing optimal urban transport strategies. The role of individual policy instruments and the impact of financial constraints. *Transport Policy, 13*(1), 49–65.

Venables, A. J. (2017). *Incorporating wider economic impacts within cost-benefit appraisal*. ITF Roundtable Reports Quantifying the Socio-economic Benefits of Transport, 109.

Vickrey, W. (1955). Some implications of marginal cost pricing for public utilities. *The American Economic Review*, 605–620.

Wangsness, P. B., Rødseth, K. L., & Hansen, W. (2017). A review of guidelines for including wider economic impacts in transport appraisal. *Transport Reviews, 37*(1), 94–115.

Wegener, M. (2014). Land-use transport interaction models. *Regional science* (pp. 741–758). Berlin, Springer.

City Land Use and Rent Dynamics with Location Externalities and Zoning Regulations: A Dynamic Spatial General Equilibrium Model

Wenjia Zhang, Kara M. Kockelman, and Jean-Claude Thill

1 Introduction

Land-use market in reality is never lassie faire, filled with externalities. Land use or location externalities are broadly recognized as negative, and zoning regulations are needed to correct the corresponding inefficiency. For example, the first type of land-use externalities comes from publicly provided goods (Tiebout 1956), and zoning regulation is effective here to sustain a community size that is optimal for using public property, like avoiding overcrowding. Exclusionary zoning is also effective to prevent "free riders," low-income outsiders who seek to live near neighbors with higher housing consumption than themselves. Without excluding these free riders, some people will pay less on property taxes even though they share the same benefits from public goods as those paying more (Hamilton 1975). The second type of location externalities comes from the potential adverse effects related to the proximity of incompatible land uses. Land-use regulation can be seen as a tool to correct these negative externalities, protect property rights, and enhance the system of nuisance law (Clawson 1971; Ellickson 1973; Fischel 1985). On the other hand, an increasing number of recent studies has recognized the external benefits of land-use mixture, and

W. Zhang (✉)
School of Urban Planning & Design, Peking University Shenzhen Graduate School, 518055 Shenzhen, Guangdong, China
e-mail: zhangwj@pkusz.edu.cn

K. M. Kockelman
Department of Civil, Architectural and Environmental Engineering, The University of Texas at Austin, Austin, TX, USA
e-mail: kkockelm@mail.utexas.edu

J.-C. Thill
Department of Geography & Earth Sciences, University of North Carolina, Charlotte, NC, USA
e-mail: jean-claude.thill@uncc.edu

School of Applied Economics, Renmin University of China, Beijing, China

J. Thill (ed.), *Innovations in Urban and Regional Systems*,
https://doi.org/10.1007/978-3-030-43694-0_9

in these studies, regulations could zone out mixed neighborhoods and land use that are desired by the market (Levine 2010). Exclusionary and low-density zoning could lead to inefficient urban sprawl. It is thus important to regard location externalities and regulations in the analysis of city land-use markets.

While many empirical studies have investigated the effects of regulation policies on land-use dynamics and housing affordability (e.g., Glaeser and Gyourko 2003; Quigley and Rafael 2005), this article rather focuses on how to enhance land-use models by internalizing location externalities and regulations. Relying on the discussion of modeling and policy scenarios, we also explore how mixed land-use planning and zoning regulation can affect land-use and rent dynamics in simulated contemporary cities.

Although many land-use models have emerged in different disciplines, including economics, planning, geography, and regional science (see Wegener's 2004, 2014 reviews), most of them overlook or fail to address the externalities of mixed land-use and zoning regulation. Among them, spatial equilibrium models (SEMs) and agent-based (microsimulation) models (ABMs) are widely discussed and applied in planning practice. Both models have their own merits and deficiencies, and recent progress illuminates how to incorporate the advantage of one model into another (Irwin 2010). For example, the lack of market mechanism is a major critique of ABMs. Many research efforts have included market interaction and rent-biding mechanisms in (e.g., Parker and Filatova 2008; Magliocca et al. 2011; Filatova et al. 2009a; Zhou and Kockelman 2011). On the other hand, several recent SEMs reflect more spatial heterogeneity and transitional dynamics (e.g., Anas and Liu 2007; Martínez and Henríquez 2007; Jin et al. 2013), as done in ABMs. This chapter attempts to enrich the literature by developing a dynamic SEM and focusing on the behavioral and policy implications of added land-use complexity and dynamics.

Measuring land-use externalities requires land-use details. While land-use representation has improved in recent SEMs, such models still do not reflect a fine characterization of land uses. In theoretical urban economic models, the monocentric model endogenizes residential lot size (or housing size) and distance to the workplace in residents' utility functions, in order to solve for the spatial distribution of residential densities (Alonso 1964; Brueckner 1987). Non-monocentric models can simulate an additional land-use feature, employment density (Fujita and Ogawa 1982; Lucas and Rossi-Hansberg 2002; Zhang and Kockelman 2016a; Zhang and Kockelman 2016b), by recognizing that firms often prefer locations closer to each other. Such agglomeration effects generate different technology benefits across locations. In applied SEMs, urban spatial structure is often organized and represented by zones, and thus, more land-use characteristics can be considered. For example, some models allow for different building types (or land-use types) and access to daily goods and services (measured via time and money costs) (Anas and Liu 2007).

Many empirical studies find that land-use or built environment attributes affect people's activity and travel choices of households and business. These land-use characteristics are often summarized as three *D*s: density, diversity, and design (Cervero and Kockelman 1997), later extended to five *D*s, by adding distance to transit and destination accessibility (Ewing and Cervero 2001), and then seven *D*s, by adding demand management and demographics (Ewing et al. 2010). Such land-use characteristics are regularly included in residential mobility studies and the hedonic analysis of property values (e.g., Song and Knaap 2004; Löchl and Axhausen 2010). Thus, it is important to include more land-use characteristics in applied SEMs, to avoid mis-estimation of local travel decisions, land-use patterns, and community welfare.

In addition, the evaluation of externalities and zoning regulation need to consider the dynamics of urban built environments, which is often ignored by SEMs. Many SEMs are static equilibrium models (e.g., monocentric models): They assume that market-clearing processes simultaneously resolve in one shot, and external factors and shocks are absent. To address such limitations, the dynamic SEM developed here emphasizes land-use complexity and dynamics. The starting point is Anas and Liu's (2007) zone-based computable general equilibrium model called "RELU",[1] for Regional Economy and Land Use. In RELU, a consumer's utility is associated with his/her home neighborhood's land-use features, including home floor space (the inverse of residential density) and access to workplace, as well as daily goods and services. In RELU, a firm's output is a function of floor space and the access to the intermediate inputs from basic industries. RELU also summarizes other land-use information and zonal features into an exogenous variable, representing the constant "inherent" attractiveness of each zone to consumers and firms. In addition, RELU endogenously models the dynamics of real estate development and treats developers as having perfect foresight and thus as able to perfectly predict future asset prices (e.g., looking forward one year). The RELU model is thus a stationary dynamic equilibrium model, in which all the exogenous variables exhibit no change over time.

Land-use details and dynamics in the model proposed in this extension of RELU come from three key factors that better reflect the reality of many urban areas. The first is a change of demographics and zonal attractiveness, which are exogenously given. Many cities in the USA have been experiencing falling household sizes and population growth, which affect the present and future housing, neighborhood, and community preferences (Nelson 2006, 2013). The second feature relates to building stock conversions. Unlike RELU, our model assumes that building stocks evolve, changing from year to year; they do not stay constant. The third feature is the endogenous change of locational (zone-based) externalities. Here, we define two types of positive location externalities that affect households and firms, respectively. The "externality" affecting households' residential location choices is assumed to be land-use diversity (in the form of land-use mixing and job-population ratios), and the externality affecting firm location choices is an innovation-based agglomeration economy. These externalities are evolved in a dynamic context, due to the relocation of households and firms; over time, they tend to stimulate new relocation and redevelopment. Notably, zoning regulations set a limit to building stock conversions

[1] An updated version, RELU-TRAN2, is developed in Anas and Hiramatsu (2012). When compared to RELU-TRAN, RELU-TRAN2 adds the choice of vehicle fuel economy into consumers' utility functions and thus internalizes people's gasoline use.

and affect the formation of locational externalities, thus influencing the dynamics of land-use market and urban form.

This dynamic SEM is calibrated in the metropolitan city of Austin, Texas, with 38 zones, and is used to explore changes in land-use and rent dynamics from year 2010 to 2035. The applications are based on four scenarios, each one with different land-use preferences and zoning regulations in place. The following three sections introduce the model's specification, calibration, solution algorithms, and simulation results, respectively. The paper concludes with a discussion of findings.

2 The Model

2.1 Spatial and Temporal Context of the City

Let us consider a city that is divided into N_z model zones, representing districts of the considered region. Land use in the city is categorized into N_r types of residential use (e.g., low- and high-density single- and multi-family residential use), N_f types of land use for firms (e.g., low- and high-density commercial and industrial uses), and N_o types of other uses, including land use for civil, transportation, and open-space functions. Thus, there are in total $\left(N_{\mathrm{lu}} = N_r + N_f + N_o\right)$ types of land use. The land used for residences and firms is endogenously determined, while the amounts used for other functions are exogenously given and will stay constant if no specific regulations or policies leading to land-use change are imposed.

Urban subsystems evolve at different rates (Wegener, 2004). For example, land uses and transport networks change relatively slowly, while household location choices, wages, and rents are modified at a much faster pace. To reflect some of this diversity, two time scales are used here (Fig. 1). The first scale, representing relatively slow change, is indexed by T, while the second scale, representing faster change, is indexed by t. Along the long-term scale, new construction and demolition are finished at the end of period $T - 1$ and realized by households at the beginning of period T. Thus, households are assumed to update their understanding of the land-use diversity of their neighborhood at the beginning of period T, relying on the changed land-use stocks. During period T, both the land stocks and the households' understandings of neighborhood diversity stay constant, until a new update occurs at the beginning of the next time period $T + 1$. Meanwhile, technology diffuses locally at the beginning of period T. During period T, firms determine their investments in innovation, leading to the adoption of new technology at the end of period T. In the second scale, each period T is divided into several time steps, from t_T to $t_T + \mathcal{T}$. Households in each time step t_T decide whether to move and/or change jobs and where to relocate. Meanwhile, job distribution, goods prices, land rents, asset prices, and wages are altered and adjusted by the market in each time step, until they reach market equilibria.

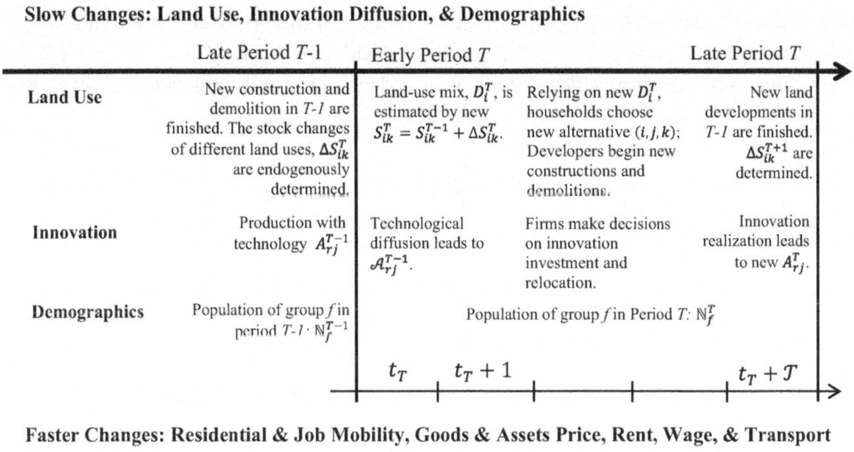

Fig. 1 Model dynamics

2.2 *Households*

While the RELU model categorizes households (or consumers) based on their skill level only, the present model develops a more detailed typology of households, relying on both households' lifecycle stage and their skill levels. Compared to skill levels, the household's stage in the lifecycle is anticipated to be more influential on their housing and neighborhood preferences. For example, Nelson (2013) recognized three types of households that may have different housing and neighborhood preferences, including starter-home, peak-demand, and downsizing households. Specifically, the starter-home households (whose household heads are under 35 years of age) often have higher demand for homes with smaller floorspaces, or townhouses, condominiums and apartments. The peak-demand households (35–64 years old), who have growing families and need more space, often prefer larger-lot single-family housing. The downsizing households (over 65 years old) likely no longer need large homes and thus may prefer smaller housing units. Also, while the share of the population with different skill levels may not change much in the future, the share of households in different life stages may significantly change in the future. Nelson (2013) predicted that the starter-home and downsizing households will account for about 84% of the new housing market from 2010 to 2035; these types of households have a stronger preference for mixed-use, walkable, amenity-rich neighborhoods, and multi-family housing types, than peak-demand households do.

Households in the city are thus subdivided into n_h types relying on their life stage (e.g., starter-home, peak-demand, and downsizing). Every household activity is assumed to be performed by a single representative household member, which is a potential worker with a level of working skill s ($s = 1, \ldots, N_s$). In total, there are $n_{hs}(= n_h \times n_s)$ household types. The number of households Ω_{hs}^{T} of type-hs in the city and its development over the coarse time scale T ($T = T_0, T_1, \ldots$) are exogenously given. In each time step t ($t = t_T, t_{T+1}, \ldots, t_{T+T}$) of time period T, each household type-hs choosing zone i ($i = 1, \ldots, n_z$) for residence, zone j ($j = 1, \ldots, n_z$) for workplace, and housing building type-k ($k = 1, \ldots, n_r$) will generate a flow utility, $U^t_{ijk|\text{hs}}$, as follows:

$$U^t_{ijk|\text{hs}}\left(C^t, q^t, D^T_i\right) = \alpha_{\text{hs}} \ln\left(\sum_{\forall z} \iota_{z|ijhs}\left(C^t_z\right)^{\eta_{hs}}\right)^{1/\eta_{hs}} + \beta_{hs} \ln q^t$$
$$+ \xi_{hs} f\left(D^T_{i1}, D^T_{i2}\mathbb{A}^T_i\right) + I_{ijk|hs} + \varepsilon^t_{ijk|hs} \tag{1}$$

where

C^t_z is the quantity of retail goods the consumer purchases from zone z, in time step t;
q^t is the size of floor space in the chosen housing type-k in zone i, in time step t;
D^T_{i1}, D^T_{i2} are the endogenously determined variables of land-use mix and job-population ratio, respectively, representing the locational externalities in zone i in period T;
$\mathbb{A}^T_i$ is a vector of exogenous local amenity variables of zone i in period T;
$I_{ijk|hs}$ is the exogenous inherent attractiveness of the residence–workplace-housing choice (i, j, k);
α_{hs}, β_{hs} are the elasticities of utility with respect to the retail goods and housing floor space (which are constant over time) and $\alpha_{hs} + \beta_{hs} = 1$;
ξ_{hs} equals 1 if households have preference on neighborhood land-use mixture and 0 if not, and
$\varepsilon^t_{ijk|hs}$ is the random error term of choice (i, j, k).

The utility function shown in Eq. (1) is similar to that used in the RELU model. One major difference is that Eq. (1) introduces the land-use mix variable as a proxy for the location externality and local amenity of residential zones, thus better depicting land-use complexity. Specifically, the vector of local amenities $\mathbb{A}^T_i$ can include variables representing the natural advantage or disadvantage of each location (such as proximity to lakes and rivers, and site topography), open space, school quality, public transit infrastructure, and other civil and cultural facilities. The formation and evolution of a neighborhood's land-use diversity is a dynamic process. Figure 1 illustrates the dynamics defined in this paper. The land-use diversity of zone i during period T is assumed to be a function of land stocks of various land-use types formed at the beginning of period T, S^T_{ik}:

$$D^T_{id} = f_D\left(S^T_{i1}, \ldots, S^T_{iN_{\text{lu}}}\right) \text{ and } S^T_{ik} = S^{T-1}_{ik} + \Delta S^T_{ik} \tag{2}$$

Type-hs households currently living in zone i and dwelling type-k and working in zone j in period $t - 1$ will have two choice alternatives in time step t:

1. continue living in zone i and dwelling type-k and working in zone j and obtain a one-time-step utility $U^t_{ijk|h}$.
2. change i, j, and/or k at the beginning of period t to (i', j', k'), $(i, j, k) \notin \{(i', j', k')\}$. In the current period, $t - 1$, the household pays all associated relocation costs, including moving and search costs (financially and physiologically), $\mathcal{U}^{t-1}_{\text{RL}}$. If households relocate only their residences, the relocation costs $\mathcal{U}^{t-1}_{\text{RL}}$ are assumed to relate less to their new residence than to a function of land rents of neighborhoods they are living in, i.e., R^{t-1}_{ik}.

The forward-looking households would maximize their expected utilities from time step t_T with a utility discount rate, μ, by making a sequence of residence-workplace-building type decisions $\{(i, j, k)\}^{t_T+T}_{t_T}$, under a budget constraint on income and time, in each time step t in period T. The optimization problem is:

$$\max_{\forall[(i,j,k)]^{t_T+T}_{t_T}} E\sum_{t=t_T}^{t_T+T} \mu^{t-t_T} U^t_{ijk|hs}\left(C^t, q^t, \mathcal{U}^{t-1}_{\text{RL}}, \varepsilon^t_{ijk|hs}\right) \tag{3}$$

subject to the following budget constraint:

$$\sum_{\forall z} \mathcal{P}^t_{z|ijhs}\left(p^t_{rz}, w^t_{js}, g^t_{iz}, G^t_{iz}\right)C^t_z + q^t R^t_{ik} = \mathcal{M}^t_{ijhs}\left(w^t_{js}, \mathcal{W}^t_{hs}, g^t_{iz}, G^t_{iz}\right) \tag{4}$$

where

$p^t_{n_f z}$ is the price of outputs from four producer types r (i.e., agriculture, retail, construction, and service sectors) produced in zone z in time step t;

w^t_{js} is the hourly wage rate paid for labor of household type-s in zone j in time step t;

$\mathcal{W}^t_{hs}$ is the non-wage annual income per household that belongs to household type-hs in time step t;

g^t_{iz} is the round-trip monetary cost per person trip from zone i to z in time step t;

G^t_{iz} is the round-trip travel time per person trip from zone i to z in time step t;

$\mathcal{P}^t_{z|ijhs}$ is the full delivered price of a retail good z for a type-hs household residing in i and working in j in time step t, which is a function of $p^t_{n_f z}$, w^t_{jhs}, g^t_{iz}, G^t_{iz}; and

$\mathcal{M}^t_{ijhs}$ is the full income of household type-hs residing in zone i and working in zone j, which is a function of w^t_{jhs}, $\mathcal{W}^t_{hs}$, g^t_{iz}, G^t_{iz}.

The one-period optimization problem represents that the current decisions of households are not only made on the basis of the current-time-step utility but also on future-steps utility. Assuming that the behavior of a household demonstrates perfect foresight, the decision-making outcome at the end of each period would fully reflect the future, and a household can be modeled as looking forward one period at a time (e.g., a household's current decision in period t will be affected by their expected utility in time step $t + 1$ but not affected by those after the time step $t + 1$). This assumption makes the optimization problem tractable and solvable. Thus, the lifetime expected utility can be represented by the value function in Eq. (5), which obeys the Bellman Eq. (1957):

$$V_{ijk|hs}^{t} = \max_{d_{ijk}}\left(v_{ijk|hs}^{t} + \varepsilon_{ijk|hs}^{t}\right), \tag{5}$$

where

$$v_{ijk|hs}^{t} = u_{ijk|hs}^{t} + E\left\{\max\left[U_{ijk|h}^{t+1}\left(u_{ijk|hs}^{t+1}, \varepsilon_{ijk|hs}^{t+1}\right), U_{i'j'k'|h}^{t+1}\left(u_{i'j'k'}^{t+1}, \varepsilon_{ijk|hs}^{t+1}\right)\right.\right.$$
$$\left.\left. -\mathcal{U}_{RL}^{t}; (i, j, k) \notin \left\{(i', j', k')\right\}\right]\right\}. \tag{6}$$

The first part of the RHS in Eq. (6), $u_{ijk|hs}^{t}$, represents the realization component of the utility function in period t, while the second part represents the expected utility maximization in period t by choosing (or not choosing) to relocate. Assuming that the idiosyncratic error term $\varepsilon_{ijk|hs}^{t+1}$ is distributed as an *i.i.d.* Type 1 extreme value term, Eq. (6) can be written as follows:

$$v_{ijk|hs}^{t} = u_{ijk|hs}^{t} + \mu \ \ln\left\{\exp\left(u_{ijk|hs}^{t+1}\right) + \sum_{\forall d_{i'j'k'} \neq d_{ijk}} \exp\left(u_{i'j'k'}^{t+1} - \mathcal{U}_{\mathrm{RL}}^{t}\right)\right\}. \tag{7}$$

Solving Eq. (5) under budget constraint (4), one can derive the direct utility function $\bar{U}_{ijk|hs}^{t} = \bar{u}_{ijk|hs}^{t} + \varepsilon_{ijk|hs}^{t}$ at the optimized choices for floor space and retail quantities produced. $\hat{u}_{ijk|hs}^{t}$ is thus calculated as follows:

$$\bar{u}_{ijk|hs}^{t} = \alpha_{hs} \ln \alpha_{hs} + \beta_{hs} \ln \beta_{hs} + \ln \mathcal{M}_{ijhs}^{t}$$
$$- \beta_{hs} \ln R_{ik}^{t} + \frac{\alpha_{hs}(1 - \eta_{hs})}{\eta_{hs}} \ln\left(\sum_{\forall z} \iota_{z|ijhs}^{\frac{1}{1-\eta_{hs}}} \left(\mathcal{P}_{z|ijhs}^{t}\right)^{\frac{\eta_{hs}}{\eta_{hs}-1}}\right) + I_{ijk|hs}^{t}. \tag{8}$$

In each period T, the model assumes that the city evolving over the time steps t_T to t_{T+T} will reach a stationary-state general equilibrium. Let $\bar{v}_{ijk|hs}^{T}$ be the *stationary-state* value function in period T:

$$\bar{v}_{ijk|hs}^{T} = \bar{u}_{ijk|hs}^{T} + \mu \ \ln\left\{\exp\left(\bar{u}_{ijk|hs}^{T}\right) + \sum_{\forall i'j'k' \neq ijk} \exp\left(\bar{u}_{ijk|hs}^{T} - \mathcal{U}_{\mathrm{RL}}^{T}\right)\right\}. \tag{9}$$

Given that $\varepsilon_{ijk|hs}^{t+1}$ follows an *i.i.d.* Gumbel distribution, the stationary-state choice probability in period T is of a multinomial logit form:

$$P_{ijk|hs}^{T} =$$

$$\frac{\exp\left(\lambda_{hs}\bar{u}^T_{ijk|hs}\right)\left[\exp\left(\bar{u}^T_{ijk|hs}\right)+\sum_{\forall i'j'k'\neq ijk}\exp\left(\bar{u}^T_{i'j'k'|hs}-\mathcal{U}^T_{RL}\right)\right]^{\lambda_{hs}\mu}}{\sum_{\forall(a,b,c)}\left\{\exp\left(\lambda_{hs}\bar{u}^T_{abc|hs}\right)\left[\exp\left(\bar{u}^T_{abc|hs}\right)+\sum_{\forall i'j'k'\neq ijk}\exp\left(\bar{u}^T_{i'j'k'|hs}-\mathcal{U}^T_{\mathrm{RL}}\right)\right]\right\}^{\lambda_{hs}\mu}},$$

$$\sum_{\forall(i,j,k)} P^T_{ijk|hs} = 1. \tag{10}$$

If one ignores the model's relocation disutility term (i.e., $\mathcal{U}^T_{\mathrm{RL}} = 0$) and the exogenously and endogenously changing variables (of land-use mix and population) between time points, the household-side model is the same as the model in RELU.

2.3 *Firms*

The model assumes that a firm's decision of how much to innovate in the current period T is affected by the technological spillover of other firms and can affect a firm's future investment decisions in technological innovation (Fig. 1). This setup has been put forth by Desmet and Rossi-Hansberg (2014), who modeled spillovers and agglomeration externalities in an endogenous growth model based on an abstract space. This type of dynamic mainly stems from the changing endogenous agglomeration externalities that arise from knowledge spillover varying over space (across locations) and between periods.[2] Agglomeration economies and dynamics of this sort are so far not been considered in *applied* land-use and transportation models, though the agglomeration economies from knowledge spillover and proximity to people (rather than intermediate goods) have become increasingly important in understanding the location choices of firms and workers (Glaeser 2010).

There are $\mathcal{R}$ types of industries, including agriculture, manufacturing, business, and retail. Firms thus can be categorized in $\mathcal{R}+2$ types, by adding construction and demolition firms[3]. The production function of the type-r ($r = 1, \ldots, \mathcal{R}+2$) firm with output X_{rj} in zone j in period T is shown in Eq. (11):

$$X^T_{rj} = \left(A^T_{rj}\right)^{\gamma} F\left(K^T_{rj}, L^T_{hs|rj}, B^T_{k|rj}, Y^T_{rj}\right), \tag{11}$$

where

A^T_{rj} is the technology level of type-r firm in zone j;
K^T_{rj} is the capital used as an input in production by type-r firm in zone j;
$L^T_{hs|rj}$ is labor of skill group s used as an input in production by type-r firm in zone j;

[2]Other sources of agglomeration externalities are endogenized in the model, as they are in the RELU model (Anas and Liu, 2007), including those that come from reducing the costs of moving intermediate goods over space and those that come from reducing the costs of accessing workers (via commuting costs).

[3]RELU has a more detailed category of construction and demolition firms than used here, based on different building types.

$B_{k|rj}^T$ is floor space of type-k ($k = n_r + 1, \ldots, n_k$) used as an input in production by type-r firm in zone j;

Y_{rj}^T is the intermediate input in production by type-r firm in zone j; and

γ is the elasticity of productivity in zone j.

As shown in Fig. 1, technology diffuses between businesses between time periods. This diffusion h is assumed to be local and to decline exponentially with distance (Desmet and Rossi-Hansberg 2014). Let A_{rj}^{T-1} be the technology used in type-r firms in zone j in period $T-1$. In the next period T, the type-r firms in zone j have access to (but do not necessarily use) technology $\mathcal{A}_{rj}^T$:

$$\mathcal{A}_{rj}^T = \max_{\forall i}\left\{\exp(-\delta g_{ij}) A_{ri}^{T-1}\right\}. \tag{12}$$

Assuming a type-r firm in zone j can access the new technology $\mathcal{A}_{rj}^T$ at the beginning of period T, this firm may decide to invest in it with a probability $\theta_{rj}^T \le 1$ at cost $Z\left(\theta_{rj}^T, w_{js}^T\right)$. Once the investment is realized, the firm has a probability θ_{rj}^T to obtain an innovative technology that incurs a productivity gain and a probability of $(1-\theta_{rj}^T)$ to obtain no effect on productivity. Thus, A_{rj}^T is the expected technology level during the period T, conditional on $\mathcal{A}_{rj}^T$, as follows (Desmet and Rossi-Hansberg 2014):

$$\begin{aligned} A_{rj}^T\left(\theta_r^T, \mathcal{A}_{rj}^T\right) &= E\left(\text{innovation}|\mathcal{A}_{rj}^T,\right) + E\left(\text{no effect}|\mathcal{A}_{rj}^T\right) \\ &= \frac{\sigma_r \theta_{rj}^T}{\sigma_r - 1}\mathcal{A}_{rj}^T + \left(1-\theta_{rj}^T\right)\mathcal{A}_{rj}^T = \left(\frac{\theta_{rj}^T}{\sigma_r - 1} + 1\right)\mathcal{A}_{rj}^T, \quad \text{for } \sigma_r > 1. \end{aligned} \tag{13}$$

Firms maximize the expected present value of profits with discount factor φ. The optimization problem of a type-r firm in zone j at time T is therefore:

$$\begin{aligned} &\max_{\left\{K_{rj}^T, L_{s|rj}^T, B_{k|rj}^T, Y_{rj}^T, \theta_{rj}\right\}_{T_1}^{\infty}} E\sum_{T=T_1}^{\infty} \varphi^{T-T_1}\left\{p_{rj}^T\left[A_{rj}^T\left(\theta_r^T, \mathcal{A}_{rj}^T\right)\right]^{\gamma} F\left(K_{rj}^T, L_{hs|rj}^T, B_{k|rj}^T, Y_{rj}^T\right)\right. \\ &\left. -\rho K_{rj}^T - \sum_{s=0}^{S} w_{js}^T L_{s|rj}^T - \sum_{k=0}^{S} R_{jk}^T B_{k|rj}^T - \sum_{r'=1}^{\mathcal{R}-1}\sum_{j'=0}^{N_z}\left(p_{r'j'}^T + \vartheta_{r'} g_{j'j}^T\right) Y_{r'j'|rj}^T - Z\left(\theta_{rf}^T, w_{rj}^T\right)\right\}, \end{aligned}$$

subject to a target output X_{rj}^T given by the production function (11). Here, p_{rj}^T is the output price of a type-r firm at zone j in period T. w_{js}^T is the hourly wage rate paid for labor of household type-s in zone j in period T. R_{jk}^T is the rent of type-k land use in zone j in period T. $Y_{r'j'|rj}^T$ is the intermediate input in production from type-r firm in zone j to type-r' firm in zone j'. ρ is the yearly interest rate.

Here, the function of innovation cost is defined as in Desmet and Rossi-Hansberg (2014):

$$Z(\theta_r^T, w_{rj}^T) = w_{rj}^T\left(\phi_r^T + \psi_r^T \frac{1}{1-\theta_r^T}\right), \quad \text{for } \psi_r^T > 0. \tag{14}$$

Since the firm's optimal dynamic innovation decisions maximize the current period profits, maximizing the expected present value of discounted lifetime utility would be equivalent to maximizing the current utility in each period T. Supposing that $\hat{K}_{rj}^T, \hat{L}_{s|rj}^T, \hat{B}_{k|rj}^T, \hat{Y}_{rj}^T, \hat{\theta}_{rf}^T$ is the optimal solution to the current profit maximization problem in each period T, then the indirect price function is derived as follows:

$$p_{rj}^T = \mathcal{P}\left(\hat{K}_{rj}^T, \hat{L}_{s|rj}^T, \hat{B}_{k|rj}^T, \hat{Y}_{rj}^T, \hat{\theta}_{rf}^T | \mathcal{A}_{rj}^T\right) = \mathcal{P}\left(w_{js}^T, R_{jk}^T, p_{rj}^T | g_{ij}^T, \mathcal{A}_{rj}^T\right). \tag{15}$$

2.4 Land Developers

Following the same rationale as that used in RELU, land developers are modeled as looking forward one year at a time. In the model, the developers can perfectly foresee the capital gains of two types of investment decisions: construction (keeping the land undeveloped versus constructing a type-k building) and demolition (keeping the land use unchanged versus demolishing an existing building). In addition, the investment decisions pertaining to land use are closely related to citywide real estate policies and land-use regulation. For example, many US metropolitan areas have zoning ordinances that typically limit building heights and lot coverage, in favor of building low-density, single-use neighborhoods. Some high-density and mixed-use neighborhoods thus are excluded or "zoned-out" under such zoning regulations (Levine 2010). To model such zoning effects, we define an alternative set $\mathbb{Z}_i$ that includes the building types that are allowed in the modeled zone i under the zoning regulations. Even though some building types (e.g., high-density multi-family housing) are desired by the market, they are absent in neighborhood i if they are not included in the alternative set $\mathbb{Z}_i$ (e.g., a zoning ordinance does not permit the development of high-density multi-family housing in zone i).

For the construction decision, given a perfectly competitive market, the expected profits of the investors after collecting the rent on vacant land at the beginning of the year and paying property taxes equal zero:

$$\begin{aligned} &E\left\{\max\left[\pi_{i00}\left(\Upsilon_{i0}^T\right), \pi_{i0k}\left(\Upsilon_{ik}^T, p_{\mathcal{R}+1}^T, \mathbb{C}_{i0k}^T\right); k = 1, \ldots, n_k \& k \in \mathbb{Z}_i\right]\right\} \\ &+ R_{i0}^T - \frac{1}{1+\rho}\tau_{i0}\Upsilon_{i0}^T = 0 \end{aligned} \tag{16}$$

and

$$E[\max(\pi_{i00}, \pi_{i0k})] = \frac{1}{\Phi_{i0}} \ln \sum_{k=0}^{n_k} \exp(\Phi_{i0}\pi_{i0k})$$

where

$E[\max(\pi_{i00}, \pi_{i0k})]$ is the expectation of vacant land development to a type-k building, known from the logit calculus with dispersion coefficient Φ_{i0};

$\pi_{i00}\left(\Upsilon_{i0}^{T}\right)$ is the expected capital gain of keeping the land undeveloped and a function of the stationary-state market price of vacant land as an asset in zone i, Υ_{i0}^{T};

$\pi_{i0k}\left(\Upsilon_{ik}^{T}, p_{\mathcal{R}+1}^{T}, \mathbb{C}_{i0k}^{T}\right); k = 1, \ldots, n_k$ and $k \in \mathbb{Z}_i$ is the expected capital gain of constructing a type-k building on the vacant land and a function of the stationary-state market price of type-k land, Υ_{ik}^{T}, construction price, $p_{\mathcal{R}+1}^{T}$, and nonfinancial costs of constructing a type-k building, $\mathbb{C}_{i0k}^{T}$. Notice that the condition $k \in \mathbb{Z}_i$ reflects the zoned-out effect on development and investment decisions;

τ_{i0} is the property tax rate;

ρ is the real interest rate.

Similarly, for the demolition decision, the total profits of the investors in type-k real estate are zero after accounting for the expected rental profits and property tax:

$$E\left\{\max\left[\pi_{ikk}\left(\Upsilon_{ik}^{T}, \mathbb{C}_{ikk}^{T}\right), \pi_{ik0}\left(\Upsilon_{i0}^{T}, p_{\mathcal{R}+2}^{T}, \mathbb{C}_{ik0}^{T}\right)\right]\right\} + E\left\{\max\left[r_v\left(\mathbb{V}_{ik}^{T}\right), r_o\left(R_{ik}^{T}, \mathbb{O}_{ik}^{T}\right)\right]\right\} - \frac{1}{1+\rho}\tau_{i0}\Upsilon_{ik}^{T} = 0 \tag{17}$$

and

$$E[\max(\pi_{ikk}, \pi_{ik0})] = \frac{1}{\Phi_{ik}} \ln\left[\exp(\Phi_{ik}\pi_{ikk}) + \exp(\Phi_{ik}\pi_{ik0})\right]$$

$$E[\max(r_v, r_o)] = \frac{1}{\Psi_{ik}} \ln\left[\exp(\Psi_{ik}r_v) + \exp(\Psi_{ik}r_o)\right]$$

where

$E[\max(\pi_{ikk}, \pi_{ik0})]$ is the expectation of land conversion from type-k use to vacant land, known from the logit calculus with dispersion coefficient Φ_{ik};

$\pi_{ikk}\left(\Upsilon_{ik}^{T}, \mathbb{C}_{ikk}^{T}\right)$ is the expected capital gain of keeping a type-k building, a function of Υ_{ik} and the nonfinancial costs of keeping the building in zone i, $\mathbb{C}_{ikk}^{T}$;

$\pi_{ik0}\left(\Upsilon_{i0}^{T}, p_{\mathcal{R}+2}^{T}, \mathbb{C}_{ik0}^{T}\right)$ is the expected capital gain of demolishing an existing type-k building, a function of Υ_{i0}^{T}, demolition costs, $p_{\mathcal{R}+2}^{T}$, and the nonfinancial costs of demolition in zone i, $\mathbb{C}_{ik0}^{T}$;

$\mathrm{E}[\max(r_{\mathrm{v}}, r_{\mathrm{o}})]$ is the expectation of rental profit from a unit amount of floor space, known from the logit calculus with dispersion coefficient Ψ_{ik};

$r_{v}\left(\mathbb{V}_{ik}^{T}\right)$ is the expected gain of keeping the type-k building vacant, a function of the costs of vacancy in zone i;

$r_{o}\left(R_{ik}^{T}, \mathbb{O}_{ik}^{T}\right)$ is the expected gain of keeping the type-k building occupied, a function of the rent and the maintenance costs for renters in zone i, $\mathbb{O}_{ik}^{T}$.

2.5 Market Clearing Within Each Period

The model of urban development assumes that the markets of products, labor, and real estate rental are clearing in each period T. First, when the three markets of basic industrial products (e.g., agricultural, manufacturing, and business) are clearing, the aggregate output of type-r ($r = 1, \ldots, \mathcal{R} - 1$) basic industry in zone i X_{ri}^{T} can be used as an intermediate input to any other type-r' ($r' = 1, \ldots, \mathcal{R} + 2$) industries in zone i' or exported outside the modeled city, $\mathbb{E}_{ri}^{T}$.

$$\sum_{r'=1,\ldots,\mathcal{R}+2}\ \sum_{i'=1,\ldots,N_z} Y_{ri\rightarrow r'i'}^{T} + \mathbb{E}_{ri}^{T} = X_{ri}^{T},\quad \forall r = 1, \ldots, \mathcal{R} - 1 \tag{18}$$

Under the condition of product market clearing, the aggregate output of the retail industry equals the aggregate demand of retail goods:

$$\sum_{\forall hs} \mathbb{N}_{hs} \sum_{\forall i',j,k} P_{i'jk|hs}^{T} C_{i|i'jk}^{T} + \mathbb{E}_{\mathcal{R}i}^{T} = X_{\mathcal{R}i}^{T}. \tag{19}$$

The equiliium outputs of the construction and demolition industries are calculated as follows:

$$X_{\mathcal{R}+1,i}^{T} = \sum_{\forall k\in=_i} m_k S_{i0}^{T} Q_{i0k}\left(\Upsilon_{i0}^{T}, \Upsilon_{i1}^{T}, \ldots, \Upsilon_{iN_z}^{T}\right). \tag{20}$$

and

$$\mathrm{X}_{\mathcal{R}+2,i}^{\mathrm{T}} = \sum_{\forall \mathrm{k}=1,\ldots,\mathrm{n_k}} S_{ik}^{T} Q_{ik0}\left(\Upsilon_{i0}^{T}, \Upsilon_{ik}^{T}\right), \tag{21}$$

where Q_{i00}, Q_{i0k}, and Q_{ik0} are the probabilities of keeping land undeveloped, developing the vacant land to a type-k building ($k \in \mathbb{Z}_i$, i.e., building types permitted by zoning ordinances), and demolishing a type-k building ($k = 1, \ldots, n_k$), respectively:

$$Q_{i0k}\left(\Upsilon_{i0}^T, \Upsilon_{i1}^T, \ldots, \Upsilon_{iN_z}^T\right) = \begin{cases} \dfrac{\exp\left[\frac{1}{1+\rho}\Phi_{i0}\Upsilon_{i0}^T\right]}{\exp\left[\frac{1}{1+\rho}\Phi_{i0}\Upsilon_{i0}^T\right] + \sum_{\forall k \in \mathbb{Z}_i} \exp \Phi_{i0}\left[\frac{1}{1+\rho}\left(\Upsilon_{ik}^T - p_{\mathcal{R}+1,i}\right)m_k - \mathbb{C}_{i0k}^T\right]}, & \text{if } k = 0 \\ 0, & \text{if } k \notin \mathbb{Z}_i \\ \dfrac{\exp \Phi_{i0}\left[\frac{1}{1+\rho}\left(\Upsilon_{i0}^T - p_{\mathcal{R}+1,i}\right)m_k - \mathbb{C}_{i0k}^T\right]}{\exp\left[\frac{1}{1+\rho}\Phi_{i0}\Upsilon_{i0}^T\right] + \sum_{\forall k \in \mathbb{Z}_i} \exp \Phi_{i0}\left[\frac{1}{1+\rho}\left(\Upsilon_{ik}^T - p_{\mathcal{R}+1,i}\right)m_k - \mathbb{C}_{i0k}^T\right]}, & \text{if } k \in \mathbb{Z}_i \end{cases} \tag{22}$$

$$Q_{ik0}\left(\Upsilon_{i0}^T, \Upsilon_{ik}^T\right) = \frac{\exp \Phi_{ik}\left[\frac{1}{1+\rho}\left(\frac{\Upsilon_{i0}^T}{m_k} - p_{\mathcal{R}+2,i}\right) - \mathbb{C}_{ik0}^T\right]}{\exp \Phi_{i0}\left[\frac{1}{1+\rho}\Upsilon_{ik}^T - \mathbb{C}_{ikk}^T\right] + \exp \Phi_{ik}\left[\frac{1}{1+\rho}\left(\frac{\Upsilon_{i0}^T}{m_k} - p_{\mathcal{R}+2,i}\right) - \mathbb{C}_{ik0}^T\right]}. \tag{23}$$

Second, when the real estate rental markets are clearing, the demands for residential and commercial floor space need to equal their supplies in each zone i, respectively:

$$\sum_{\forall hs} \mathbb{N}_{hs}^T \sum_{\forall j} P_{ijk|hs}^T b_{ijk|hs}^T = S_{ik}^T \frac{r_o\left(R_{ik}^T, \mathbb{O}_{ik}^T\right)}{r_v\left(\mathbb{V}_{ik}^T\right) + r_o\left(R_{ik}^T, \mathbb{O}_{ik}^T\right)}, \quad k = 1, \ldots, n_r \tag{24}$$

$$\sum_{\forall hs} B_{k|ri}^T = S_{ik}^T \frac{r_o\left(R_{ik}^T, \mathbb{O}_{ik}^T\right)}{r_v\left(\mathbb{V}_{ik}^T\right) + r_o\left(R_{ik}^T, \mathbb{O}_{ik}^T\right)}, \quad k = n_r + 1, \ldots, n_k. \tag{25}$$

Third, the labor market clearing also requires that the annual demand and supply for the labor hours of skill-hs groups need to be equal:

$$\sum_{r=1}^{\mathcal{R}+2} L_{hs|rj}^T = \mathbb{N}_{hs}^T \sum_{\forall i,k} H_{ijf}^T P_{ijk|f}^T. \tag{26}$$

2.6 Transitional Dynamics

From period T to $T + 1$, the land stocks of type-k buildings and the production technology level of type-r firms at zone i will change endogenously, and the population counts of type-hs residential groups are given exogenously. For example, the construction and demolition activities are assumed to be finished at the end of period T, and the land stocks are updated at the beginning of period $T + 1$, as follows:

$$S_{ik}^{T+1} = \begin{cases} S_{i0}^{T} Q_{i00} + X_{\mathcal{R}+2}^{T}, & \text{if } k = 0 \\ S_{ik}^{T} - S_{ik}^{T} Q_{ik0}, & \text{if } k \notin \mathbb{Z}_i \\ S_{ik}^{T} - S_{ik}^{T} Q_{ik0} + m_{ik} S_{i0}^{T} Q_{i0k}, & \text{if } k \in \mathbb{Z}_i \end{cases}. \tag{27}$$

Equation (26) shows that the amount of vacant land zoned for type-k building in zone i in period $T + 1$ equals the amount of the undeveloped vacant land kept in period T and the demolished building. For land-use types excluded by zoning regulation in period T, the new land stocks in period $T + 1$ may decrease due to possible demolition activities. For "zoned-in" land-use types, the new stocks equal the old stocks plus the new construction and minus demolition stocks in the previous period. These settings differ from those in the RELU model. The model here assumes that building stocks change incrementally over time, while the RELU model solves for equilibrium building stocks so that the construction stocks equal the demolition stocks in the long term. This setting of building stock conversion in the present mode allows for modeling the effects of policy intervention.

The changes in stocks of different buildings lead to new zone-based land-use characteristics, such as land-use mix. Here, we use the index of land-use mix entropy that is widely used in the planning field to measure the zone-based land-use mixture, D_{i1}^{T+1}:

$$D_{i1}^{T+1} = -\sum_{\ell=1}^{\mathbb{L}} \mathcal{L}_{i\ell}^{T} \ln \mathcal{L}_{i\ell}^{T} / \ln \mathbb{L} \tag{28}$$

where $\mathcal{L}_{i\ell}^{T}$ $(\ell = 1, 2, \ldots, \mathbb{L})$ represents the proportion in type-ℓ land use area in total land area. Notice that the land-use area is not equivalent to the floor space outcomes, S_{ik}^{T}, but can be calculated from them. In the following simulation, we recognize six types of land use in a zone ($\mathbb{L} = 6$), including single-family, multi-family, industrial, commercial, open space, and civil uses. Among them, the land areas of open space and civil uses are exogenously given, while all the others are calculated by S_{ik}^{T} and the FAR m_{ik}.

Meanwhile, as shown in Fig. 1, the technology levels of type-r firms at zone i (Eq. 12) are assumed to be updated at the beginning of period $T + 1$, due to innovation diffusion (Eq. 11) and the firms' investment in innovation during period T (Eq. 13). Both the transitions in technology level and land-use characteristics can affect the wage levels, product and asset prices, and land rents, leading to new zone-based job-housing ratios, D_{i2}^{T+1}:

$$D_{i2}^{T+1} = \sum_{\forall h,s} \mathbb{N}_{hs}^{T} \sum_{\forall i',k} P_{i'ik|hs} \Big/ \sum_{\forall h,s} \mathbb{N}_{hs}^{T} \sum_{\forall j,k} P_{ijk|hs}. \tag{29}$$

3 Calibration and Simulation

The model is calibrated and applied in the Texas capital metropolitan area, including 38 multiple listing service (MLS) neighborhoods (covering the City of Austin and Travis County that surrounds it) as modeling zones and four outer zones (representing four counties near Travis County and functionally connected to it). The MLS neighborhoods have been defined based on real estate traditions, school zones, zip codes, housing stock consistencies, and natural boundaries (like rivers). Figure 2 shows the geographical distribution of the 38 MLS neighborhoods in the urban core (12 zones), inner suburbs (16 zones), and outer suburbs (10 zones).

The model is calibrated on the basis of the 2005–2010 period, and the starting parameters are mainly calibrated using 2008 land-use data from the 2005 City of Austin travel diary and OD data from Capital Area Metropolitan Planning Organization, demographic data from the 2010 US population census, and estimated population projection data (until 2050) from the Texas Data Center (TDC). While these data sets cannot fully support the parameter calibration for the model here, some parameters (e.g., firmographies) are derived from the existing literature (e.g., Anas and Rhee 2006; Zhou and Kockelman 2011; Desmet and Rossi-Hansberg 2014) and come from empirical estimates. In each policy scenario, the simulation includes five time periods (from 2010 to 2035) defined by 5-year increments.

The applied model considered here consists of nine population groups: three lifecycle stages (defined by the household head) across three skill levels. The numbers of households (or housing units) in each of these groups are exogenously given and estimated using data from the 2010 Census and the TDC's population projections data (through 2050). Here, we use the population projection of the Travis County by age

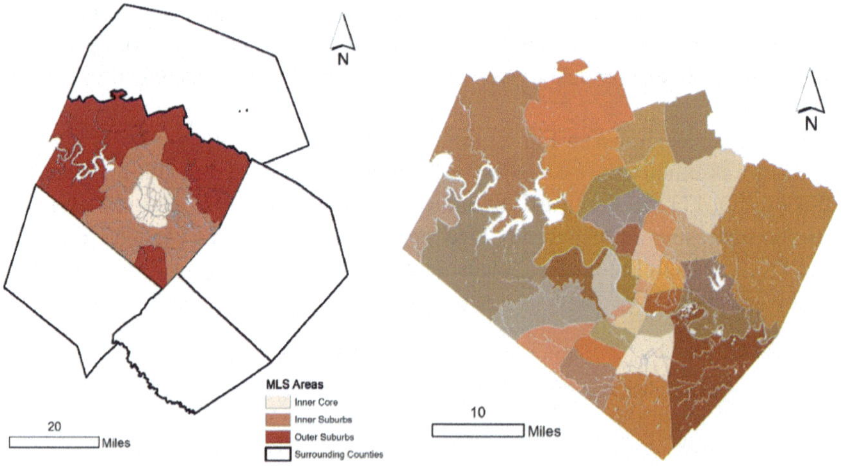

Fig. 2 Austin, Texas' 38 MLS areas and surrounding counties as outer zones (left) and divisions of MLS areas (right)

Table 1 Household groups in three lifecycle stages over the simulation horizon

	2010	2015	2020	2025	2030	2035
Total households/housing units	404,467	435,282	478,385	518,615	554,731	586,992
% increase of total households compared to 2010 (%)		7.62	18.28	28.22	37.15	45.13
% increase of total households in five years (%)		7.62	9.90	8.41	6.96	5.82
Starter home (15–34 old) [% of total households]	135,381 [33.47]	133,822 [30.74]	129,538 [27.08]	128,848 [24.84]	139,315 [25.11]	150,059 [25.56]
Peak-demand (25–64 old) [% of total households]	221,714 [54.82]	243,520 [55.95]	273,595 [57.19]	295,049 [56.89]	302,606 [54.55]	308,843 [52.61]
Downsizing (≥65 old) [% of total households]	47,372 [11.71]	57,940 [13.31]	75,252 [15.73]	94,718 [18.26]	112,811 [20.34]	128,090 [21.82]

Note These numbers are estimated using data from 2010 Census and population projection data from Texas Data Center

from 2015 to 2035 under the one-half 2000–2010 migration (0.5) scenario. Assuming the average household size for 2015–2035 is the same as 2010, we estimated the corresponding total number of households as shown in Table 1. The estimation shows that the total number of households will increase from 404,467 in 2010 to 586,992 in 2035, generating a net increase of 45.13% and a yearly increase rate of 1.5%. The population growth rates peak at around 2020: The five-year increase rates rise from 7.62% in 2010–2015 to 9.9% in 2015–2020 then fall to 5.82% in 2030–2035. Table 1 also depicts the overall household counts for these three lifecycle stages. The shares of starter-home households (with household heads up to 34 years old) and peak-demand households (35–64 years old) are anticipated to decrease, while the share of downsizing households (older than 65 years) will almost double, from 2010 to 2035. In addition, we define four types of residential buildings (low- and high density single-family and multi-family uses) and calculate the occupied and vacant land stocks and floorspace based on the future zoning maps obtained from the City of Austin (COA 2010).

Figure 3 describes the algorithm used to solve for 1110 within period equations. The population numbers $\mathbb{N}_{hs}^{T}$ are exogenously given at the beginning of each period. The variables S_{ik}^{T}, D_{i1}^{T}, D_{i2}^{T},A_{rj}^{T} are given at the starting period and calculated at later periods based on corresponding updated information from prior periods. Within each period, the endogenous variables, such as product prices and output levels, land rents, wages, and property values and rents, are solved recursively to clear product, labor, and real estate markets. The Newton–Raphson algorithm is used recursively to find the fixed point solutions of these endogenous variables. The run time for finding such spatial equilibria within a period on a standard personal computer ranges from 10 to 20 min, depending on the initial values used.

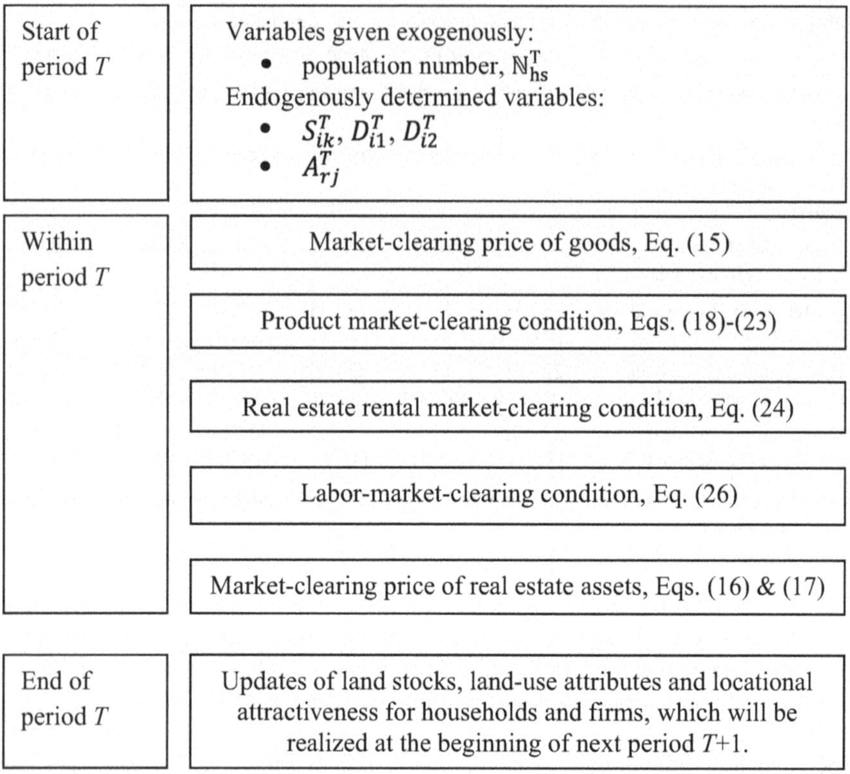

Fig. 3 Solution process within a single period T

4 Land-Use and Rent Dynamics Under Four Scenarios

This section compares the land-use, housing demand, and rent dynamics from 2015 to 2035 every five years, under four scenarios with different assumptions. The first scenario (S1) assumes that the household groups have diverse preferences for housing size but no preference (i.e., $\xi_{hs} = 0$ in Eq. (1)) for a neighborhood with mixed-use features. For example, the peak-demand group's ($\beta_{hs} = 0.35$) utility elasticity of housing size is higher than that of the starter home ($\beta_{hs} = 0.3$) and downsizing ($\beta_{hs} = 0.32$) groups. The second scenario (S2) assumes that the household groups have varying preferences for both housing size and a neighborhood with mixed-use features (including land-use mixture entropy index[4] and job-to-housing ratios), i.e., $\xi_{hs} = 1$. By comparing S1 and S2, one can determine how future demographic trends may affect city land-use and housing demand. The third and fourth scenarios (S3 and

[4]Six types of land use are included: single-family residential, multi-family residential, commercial, industrial, civil, and open space. Among them, the shares of civil and open-space land uses are exogenous. Suppose the proportions of different land uses are ϱ_i, $i = 1 \ldots 6$, the land-use mix entropy index is calculated as $\sum_{i=1}^{6} \varrho_i \ln \varrho_i / \ln 6$.

S4) add a low-density zoning regulation to S1 and S2, respectively. This low-density zoning regulation is assumed to prevent high-density and multi-family residential developments in the 10 MLS neighborhoods that are part of Austin's outer suburbs starting in 2010.[5] By comparing S1 with S3, and S2 with S4, one can examine how the supply constraints on high-density development affect land use, housing demand, rents, and property values.

5 Land-Use Dynamics Associated with Changes in Demographics and in Zoning Regulations

Figures 4 and 5 show the trends of population and employment densities in the Austin area from 2015 to 2035 under scenarios S1, S2, and S3. These figures clearly indicate that city land-use dynamics are closely connected with people's changing preference for various land-use features, changing demographics, and changing land-use supply as affected by land-use regulations and planning. These changing preferences can be either exogenously given or endogenously determined and probably cannot lead to a stationary dynamic spatial equilibrium even in the long term, especially when location externalities on the consumption and production sides and land development policies exist and vary over time and across space.

First, we compare the land-use dynamics under scenarios 1 and 2. In S1, the sole source of urban dynamics is population growth that is set exogenously to the simulation model. The simulation results show that the household and employment densities across most of the 12 inner core neighborhoods significantly increase from 2015 to 2035 (Figs. 4a and 5a). In S2, where residents are assumed to prefer living in neighborhoods that have a more mixed-use character (thus introducing the location externality as another source of dynamics), as shown in Figs. 4b and 5b, future population and job locations appear more centralized in and around the inner urban core than under the baseline case depicted by S1. Table 2 summarizes the land-use, population, and employment differences across the inner core, inner suburban, and outer suburban neighborhoods under S2 versus S1. These findings suggest that a rising demand for mixed-use environments resulting from a stronger preference for such development among residents may increase core population and levels, while lowering them in the suburbs, while improving land-use diversity in the suburban areas at the same time.

Second, we examine the "zoned-out" effects by comparing the land-use dynamics before and after low-density zoning regulations in the outer suburban areas by comparing S3 to S1. Here, the land-use regulation can be regarded as an exogenous constraint on urban development. The comparison of Fig. 4a, c appears to show that

[5]For zone i in the outer suburbs, $\mathbb{Z}_i = \{$low-density single-family housing with FAR $= 0.2\}$, and for zone i in the urban core and inner suburbs, $\mathbb{Z}_i = \{$low-density single-family housing with FAR $= 0.2$, high-density single-family. housing with FAR $= 0.4$, low-density multi-family housing with FAR $= 0.6$, high-density multi-family housing with FAR $= 0.8\}$

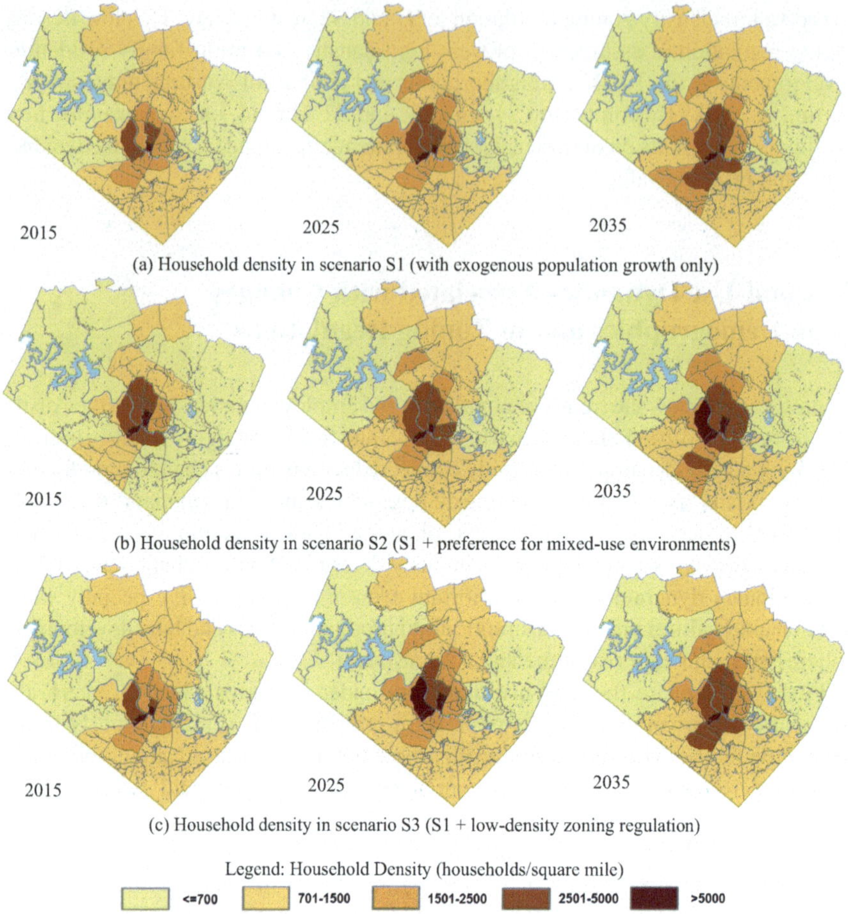

Fig. 4 Differences in household density over time (year 2015–2035), across scenarios S1, S2, and S3

such a zoning regulation may increase urban population densities at the early stage of the simulation time horizon but will not greatly affect the density distribution over the longer term. In contrast, the zoning regulation appears to have more significant effects on the spatial distribution of employment densities (Fig. 5a, c). Table 3 provides a summary of land-use, employment and population change when zoning regulation is established. When households have no mixed-use preference, at the early stage (2015–2020), the low-density zoning regulation tends to centralize more households in the urban core and in the inner suburban areas and decrease population in the outer suburbs. At later stages of development (2025–2035), both urban and outer suburban household counts fall, as these households move to the inner suburban area. Meanwhile, many potential employment opportunities would be zoned-out by such a regulation, especially in the outer suburban areas. But such regulations may restrict

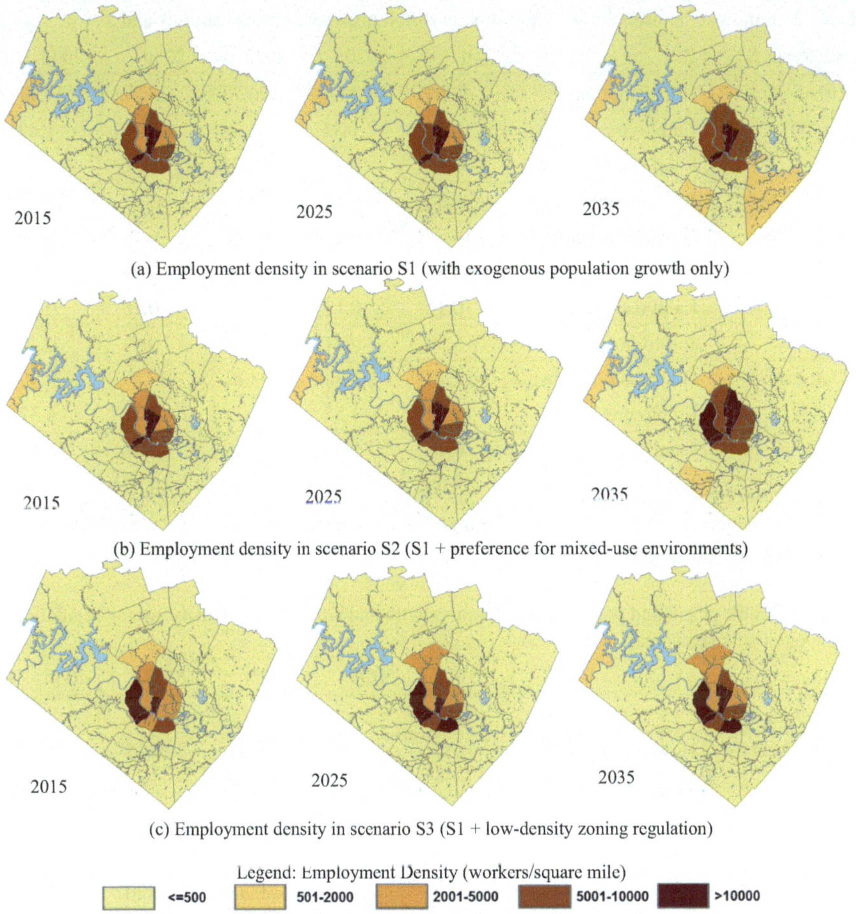

Fig. 5 Differences in job density over time (year 2015–2035), across scenarios S1, S2, and S3

firms to locate within the urban core and the inner suburb and increase employment in these two areas. In summary, the predicted demographic trends suggest that the low-density zoning regulation may encourage population decentralization alongside employment centralization in the long term, causing citywide job-housing mismatch and urban sprawl.

If demand for smaller housing units and for mixed-use neighborhoods rises but their supply is constrained by land-use regulations, do these trends aggravate urban sprawl? To answer to question, we simulate city growth under S4 and compare again S2. These simulations yield some mixed results (Table 3). Households seem to still centralize in the urban core, although the shares of households in the inner suburbs grow from 2015 to 2035. The employment distribution also shows a centralization tendency. Compared to S2, more jobs in the inner suburbs will move to the urban core than from the outer suburbs. Though high-density residences are regulated by

Table 2 Land-use, household and employment differences between scenarios 1 and 2

S2 versus S1	%	2015	2020	2025	2030	2035
Urban core	Household number	45.69	22.66	26.73	23.25	24.11
	Employment	1.88	1.32	1.38	1.33	1.43
	Land-use mix	−0.88	−0.04	−0.54	0.26	−0.31
	Job-to-housing ratio	−36.38	−19.69	−28.60	−23.02	−25.87
Inner suburbs	Household number	−14.87	−6.61	−7.70	−6.56	−6.80
	Employment	−4.32	−2.19	−3.30	−2.99	−3.32
	Land-use mix	0.15	0.12	−0.15	0.16	0.06
	Job-to-housing ratio	5.98	0.11	0.84	0.24	0.40
Outer suburbs	Household number	−12.77	−7.34	−8.81	−7.99	−8.40
	Employment	−4.72	−4.55	−4.11	−4.18	−4.37
	Land-use mix	4.77	5.84	5.97	6.28	6.36
	Job-to-housing ratio	5.08	2.37	5.05	4.87	5.28

Note The proportions are calculated as a rate of change in each time period: (value under S2 – value under S1)/value under S1

Table 3 Employment and population differences with and without low-density zoning regulations in the outer suburban areas

%		2015	2020	2025	2030	2035
S3 versus S1						
Urban core	Household number	0.37	0.16	−0.02	−0.12	−0.20
	Employment	7.75	7.61	7.48	7.57	7.69
Inner suburbs	Household number	0.28	0.28	0.27	0.22	0.18
	Employment	−1.53	−2.58	−3.18	−3.36	−3.49
Outer suburbs	Household number	−0.64	−0.49	−0.34	−0.20	−0.09
	Employment	−34.44	−35.96	−37.11	−37.57	−37.91
S4 versus S2						
Urban core	Household number	2.79	2.18	2.24	2.31	2.27
	Employment	1.19	1.53	1.69	1.65	1.56
Inner suburbs	Household number	−0.42	0.20	0.38	0.52	0.59
	Employment	−5.91	−7.36	−7.51	−7.54	−7.34
Outer suburbs	Household number	−2.78	−2.34	−2.73	−2.94	−3.03
	Employment	−0.45	−0.94	−2.01	−1.79	−1.46

Note The proportions are calculated as a rate of change in each time period: (value under S3 (or S4) – value under S1 (S2)/value under S1 (S2)

the zoning ordinances, the mixed-use demand may increase the supply of mixed-use neighborhoods with job-housing balance, thus leading to relatively matching trends of population and employment distribution. These findings suggest that when the real estate market realizes residents' preferences for mixed-use neighborhoods, the negative sprawling effects of land-use regulation may be mitigated.

6 Trends of Housing Demand and Rent

Table 4 shows the projected trends of housing demand, rent, and property price from

Table 4 Changes in housing demand, rent, and property prices from 2015 to 2035 under scenario 1

	2015 (no. of housing units)	2020 (% change in 2015–20)	2025 (% change in 2020–25)	2030 (% change in 2025–30)	2035 (% change in 2030–35)
Housing demand					
Low-density single-family	246,041	10.02	8.50	6.81	5.69
High-density single-family	90,922	9.80	8.33	7.17	5.98
Low-density multi-family	74,581	9.70	8.23	7.08	5.92
High-density multi-family	23,739	9.69	8.27	7.42	6.18
Housing rents					
Low-density single-family	4.20	8.60	6.51	−0.21	−0.50
High-density single-family	5.10	9.73	8.11	0.53	−0.80
Low-density multi-family	7.50	8.63	6.32	−0.35	−0.30
High-density multi-family	10.81	10.64	9.14	1.30	−0.49
Property prices					
Low-density single-family	178.75	10.12	8.59	6.87	5.73
High-density single-family	97.79	9.71	8.24	7.10	5.93
Low-density multi-family	82.36	9.58	8.12	7.01	5.87
High-density multi-family	142.53	9.38	8.03	7.30	6.09

2015 to 2035 in scenario 1. The projection of total housing demand basically fits the change in total number of households as estimated in Table 1. When the demographic change is the only dynamic factor, the growth rates of low-density single-family (LDSF) housing units are higher than those of other housing types before 2025. However, later during the simulation period, the demand for high-density multi-family and single-family housing increases at a faster rate across the urban region as a whole. Table 5 compares the trends of housing demand under each of the four scenarios. By comparing S1 and S2, one can find that the demand for LDSF housing decreases when the mixed-use preference is realized in the market; it continues to do so over the entire planning horizon. The demand for other housing types will rise, with the demand for high-density multi-family (HDMF) homes increasing the most. The effects of zoning regulation on housing demand seem much smaller than the realization of mixed-use preference. At the early periods, the constraint on high-density development will lower the demand for LDSF. But, in the long term, such a land-use regulation will increase LDSF demand and lower other housing demands. After comparing S2 versus S1 and S4 versus S2, we find that the low-density zoning may mitigate the decreasing trends of LDSF housing demand.

In S1, the housing rents associated with all four building types is expected to increase initially and drop later on, while real property price will keep increasing from 2015 to 2035, though the growth rate will slow down (Table 4). Table 6 compares the housing rent trends between S2 and S1 and S3 and S1. Contrary to the expected

Table 5 Percentage differences in housing demand across paired scenarios

Housing types (%)	2015	2020	2025	2030	2035
S2 versus S1					
Low-density single-family	−6.99	−4.27	−4.81	−4.44	−4.61
High-density single-family	9.39	5.22	5.96	5.23	5.48
Low-density multi-family	3.57	2.77	3.38	3.38	3.53
High-density multi-family	25.25	15.71	16.66	15.38	15.61
S3 versus S1					
Low-density single-family	−0.20	−0.10	−0.02	0.03	0.07
High-density single-family	0.28	0.20	0.12	0.06	0.01
Low-density multi-family	0.19	0.08	−0.03	−0.09	−0.14
High-density multi-family	0.35	0.06	−0.15	−0.27	−0.36
S4 versus S2					
Low-density single-family	−1.25	−1.02	−1.08	−1.14	−1.14
High-density single-family	1.24	1.13	1.34	1.44	1.47
Low-density multi-family	1.01	0.87	0.78	0.80	0.76
High-density multi-family	2.84	2.43	2.39	2.48	2.46

Note The proportions are calculated as a rate of change in each time period. For example, the numbers contrasting S3 versus S1 are calculated as (housing variables of S3 − variables of S1)/variables of S1

Table 6 Comparisons of average housing rent trends

Housing types (%)	2015	2020	2025	2030	2035
S2 versus S1					
Low-density single-family	−1.31	5.51	3.07	4.89	3.92
High-density single-family	16.97	27.30	26.22	28.76	27.79
Low-density multi-family	76.04	82.70	74.64	77.68	75.61
High-density multi-family	−2.41	8.91	4.93	5.87	4.87
S3 versus S1					
Low-density single-family	−3.01	−0.24	1.94	2.58	3.13
High-density single-family	19.61	22.87	25.08	25.42	26.21
Low-density multi-family	−3.75	0.34	4.64	6.72	7.21
High-density multi-family	−4.26	−1.79	−0.24	−0.35	0.02

Note The proportion numbers are calculated by change rate. For example, the numbers in S3 versus S1 are calculated as (landuse variables of S3 − variables of S1)/variables of S1

trend for housing demand, the demand for mixed-use neighborhood will significantly increase the rents of low-density multi-family (LDMF) and high-density single-family (HDSF) housing. The zoning regulation will raise the HDSF housing rent most. These findings suggest that the supply constraint on high-density development may raise the housing rents of such high-density housing.

7 Conclusions

This paper developed a dynamic spatial equilibrium model (SEM) to compare changes in land-use patterns, housing demand, and rents longitudinally under different scenarios of demographic changes (agent preferences) and policy interventions (zoning restrictions). As a case study, results are simulated over a 20 year period for the Austin, Texas metropolitan area. When compared to the existing dynamic SEMs (e.g., Anas and Liu 2007; Martínez and Henríquez 2007), this new model introduces more land-use details and more drivers of dynamics for land-use change. For example, the specification tracks not just different housing sizes and access attributes, but also several location externalities (e.g., land-use diversity, job-housing balance, and production externalities emerging from innovation diffusion) that affect agent (household and firm) decisions. In addition, the model allows for three dynamics that affect spatial choice, including exogenously provided demographic details, building stock conversion (as constrained by zoning regulations), and endogenously evolving location externalities. These modeling enhancements help respond to criticisms from many agent-based modelers (e.g., Simmonds et al. 2013) and demonstrate the ability of applied SEMs to reflect more realistic land-use complexity and urban dynamics.

The scenario analyses mainly explore the effects of demographic trends, land-use preferences, and low-density zoning regulations on the dynamics of land use, housing demand, and rents, and their related welfare implication. Simulation results suggest that people's rising demand for mixed-use neighborhoods may improve land-use diversity in suburban areas and lower demand for low-density single-family housing across a region. Low-density zoning regulations in Austin's outer suburbs may lead to citywide job-housing imbalances and urban sprawl (with population decentralizing and jobs potentially centralizing) while raising high-density housing rents. But such regulations do not appear to affect housing demand much, especially in the longer term. When the existing low-density zoning regulations cannot be changed (in the short term), promotion of mixed-use development may increase households' preference for mixed-use environments and thus moderate tendencies toward more excessive urban sprawl.

The new model has been found to be powerful at assessing urban futures under emerging socio-demographic trends and possible land development regulations by governments. Several modeling limitations still merit further exploration. First, further simulation analyses should discuss the effects of transitional costs (e.g., residential moving costs) and innovation diffusion on scenario results described above. More sensitivity analyses are also needed, to support the realistic land-use policy analysis. Second, this paper does not quantify welfare effects (or their distribution) across different scenarios. Ideally, future research will extend these calculations to provide efficiency information and welfare outcomes of various land-use policies (including changes to zoning regulations and subsidies for alternative developments).

References

Alonso, W. (1964). *Location and land use: Toward a general theory of land rent*. Cambridge: Harvard University Press.

Anas, A., & Hiramatsu, T. (2012). The effect of the price of gasoline on the urban economy: From route choice to general equilibrium. *Transportation Research Part A: Policy and Practice, 46*(6), 855–873.

Anas, A., & Liu, Y. (2007). A regional economy, land use, and transportation model (RELU-Tran). *Journal of Regional Science, 47*(3), 415–455.

Anas, A., & Rhee, H. J. (2006). Curbing excess sprawl with congestion tolls and urban boundaries. *Regional Science and Urban Economics, 36*(4), 510–541.

Bellman, R. (1957). *A Markovian decision process* (No. P-1066). Santa Monica, California: Rand Corporation. Available at http://www.dtic.mil/cgi-bin/GetTRDoc?Location=U2&doc=GetTRDoc.pdf&AD=AD0606367.

Brueckner, J. K. (1987). The structure of urban equilibria: A unified treatment of the Muth-Mills model. *Handbook of Regional and Urban Economics, 2*, 821–845.

Cervero, R., & Kockelman, K. (1997). Travel demand and the 3Ds: Density, diversity, and design. *Transportation Research Part D: Transport and Environment, 2*(3), 199–219.

Clawson, M. (1971). *Suburban land conversion in the United States*. Baltimore: The Johns Hopkins Press.

Desmet, K., & Rossi-Hansberg, E. (2014). Spatial development. *American Economic Review, 104*(4), 1211–1243.

Ellickson, R. C. (1973). Alternatives to zoning: Covenants, nuisance rules, and fines as land use controls. *University of Chicago Law Review, 40*(4), 681–781.
Ewing, R., & Cervero, R. (2001). Travel and the built environment: a synthesis. *Transportation Research Record, 1780,* 87–114.
Ewing, R., Greenwald, M., Zhang, M., Walters, J., Feldman, M., Cervero, R., et al. (2010). Traffic generated by mixed-use developments: Six-region study using consistent built environmental measures. *Journal of Urban Planning and Development, 137*(3), 248–261.
Filatova, T., Parker, D., & Van Der Veen, A. (2009a). Agent-based urban land markets: Agent's pricing behavior, land prices and urban land use change. *Journal of Artificial Societies and Social Simulation, 12*(1). Available online http://jasss.soc.surrey.ac.uk/12/1/3.html.
Fischel, W. A. (1985). *The economics of zoning laws: A property rights approach to american land use controls*. Baltimore: Johns Hopkins University Press.
Fujita, M., & Ogawa, H. (1982). Multiple equilibria and structural transition of non-monocentric urban configurations. *Regional Science and Urban Economics, 12*(2), 161–196.
Glaeser, E. L. (Ed.). (2010). *Agglomeration economics*. Chicago, IL: University of Chicago Press.
Glaeser, E. L., & Gyourko, J. (2003). The impact of building restrictions on housing affordability. *FRBNY Economic Policy Review, 2,* 21–29.
Hamilton, B. W. (1975). Zoning and property taxation in a system of local governments. *Urban Studies, 12*(2), 205–211.
Irwin, E. G. (2010). New directions for urban economic models of land use change: incorporating spatial dynamics and heterogeneity. *Journal of Regional Science, 50*(1), 65–91.
Jin, Y., Echenique, M., & Hargreaves, A. (2013). A recursive spatial equilibrium model for planning large-scale urban change. *Environment and Planning B: Planning and Design, 40*(6), 1027–1050.
Levine, J. (2010). *Zoned out: Regulation, markets, and choices in transportation and metropolitan land use*. London, UK: Routledge.
Löchl, M., & Axhausen, K. W. (2010). Modelling hedonic residential rents for land use and transport simulation while considering spatial effects. *Journal of Transport and Land Use, 3*(2), 39–63.
Lucas, R. E., & Rossi-Hansberg, E. (2002). On the internal structure of cities. *Econometrica, 70*(4), 1445–1476.
Magliocca, N., Safirova, E., McConnell, V., & Walls, M. (2011). An economic agent-based model of coupled housing and land markets (CHALMS). *Computers, Environment and Urban Systems, 35*(3), 183–191.
Martínez, F. J., & Henríquez, R. (2007). A random bidding and supply land use equilibrium model. *Transportation Research Part B: Methodological, 41*(6), 632–651.
Nelson, A. C. (2006). Leadership in a new era: Comment on "planning leadership in a new era". *Journal of the American Planning Association, 72*(4), 393–409.
Nelson, A. (2013). The resettlement of America's suburbs. *Planning Theory & Practice, 14*(3).
Parker, D. C., & Filatova, T. (2008). A conceptual design for a bilateral agent-based land market with heterogeneous economic agents. *Computers, Environment and Urban Systems, 32,* 454–463.
Quigley, J., & Rafael, S. (2005). Regulation and the high cost of housing in California. *American Economic Review Papers and Proceedings, 95*(2), 323–328.
Simmonds, D., Waddell, P., & Wegener, M. (2013). Equilibrium versus dynamics in urban modelling. *Environment and Planning B: Planning and Design, 40*(6), 1051–1070.
Song, Y., & Knaap, G. J. (2004). Measuring urban form: Is Portland winning the war on sprawl. *Journal of the American Planning Association, 70*(2), 210–225.
Tiebout, C. M. (1956). A pure theory of local expenditures. *The Journal of Political Economy*, 416–424.
Wegener, M. (2004). Overview of land-use transport models. *Handbook of Transport Geography and Spatial Systems, 5,* 127–146.
Wegener, M. (2014). Land-use transport interaction models. In *Handbook of Regional Science* (pp. 741–758). Berlin: Springer.

Zhang, W., & Kockelman, K. M. (2016a). Optimal policies in cities with congestion and agglomeration externalities: Congestion tolls, labor subsidies, and place-based strategies. *Journal of Urban Economics, 95,* 64–86.
Zhang, W., & Kockelman, K. M. (2016b). Congestion pricing effects on firm and household location choices in monocentric and polycentric cities. *Regional Science and Urban Economics, 58,* 1–12.
Zhou, B., & Kockelman, K. (2011). Change in land use through microsimulation of market dynamics. *Transportation Research Record, 2255,* 125–136.

An Agent-Based Model of Wealth Inequality with Overlapping Generations, Local Interactions, and Intergenerational Transfers

Gunawan Wicaksono and Yuri Mansury

1 Introduction

Regional science and urban economics are disciplines tasked with the examination of spatial inequality and the human interactions that produce uneven patterns. Studies of regional and urban systems are thus beset by the complexity of social interactions. This chapter examines how the strength of interactions with local neighbors affects human capital investments and the distribution of wealth in a decentralized economy. Drawing on the framework of complex systems studies, we propose an agent-based model in which local interactions can tilt the attitude of adaptive individuals favorably toward education, putting the economy on a path of output expansion accompanied by rising inequality. Such trajectory is largely consistent with the empirics of growth and inequality.

Among most of the advanced economies, the long-run growth of output has been more or less steady despite the few intense episodes of major recessions (Gordon 2013; IMF 2019; OECD 2013). Steady output expansion alone, however, cannot tell us whether growth is widely shared, or is accruing to a privileged group of wealthy people, as has been the case recently. Indeed, inequality has increased sharply in the last three decades (Howell 2013; Heckman and Krueger 2003) as evident by the rising gap between the poor and the rich (Krueger et al. 2010; OECD 2015; Saez and Zucman 2016). Inequality may be the result of national or even global economic forces, but it clearly has important spatial dimension. In the USA, for example, Census Bureau data reveal that the largest cities are among the most unequal places vis-à-vis the rest of the country (Berube 2014).

G. Wicaksono
John F. Kennedy School of Government, Ash Center for Democratic Government and Innovation, Harvard University, Cambridge, MA 02138, USA

Y. Mansury (✉)
Department of Social Sciences, Illinois Institute of Technology, Chicago, IL 60616, USA
e-mail: ymansury@iit.edu

J. Thill (ed.), *Innovations in Urban and Regional Systems*,
https://doi.org/10.1007/978-3-030-43694-0_10

Since the seminal work of Kuznets (1955), scores of researchers have contributed to the theoretical and empirical studies of the growth-inequality nexus. A few have explored the role of local interactions and neighborhood feedback in changing economic stratification in their analyses. Previous studies, however, have typically emphasized *income* inequality, often neglect the role of intergenerational transfers (e.g., Becker et al. 2018; Durlauf 1996), or represent space in a very abstract manner (e.g., Bénabou 1996). Space is of course central in urban economics and regional science. Intergenerational transfers, on the other hand, are one of the primary drivers of persistent inequality. Clark (2014), for example, argues that the only sure path to success is essentially to be born to the right parents.

This chapter seeks to make modest contributions to the literature. First, as far as we know, our study is among the first to explore the impact of the strength of local interactions on wealth inequality in an agent-based model with intergenerational transfers. We do so by implementing a general equilibrium model with known analytical solutions, allowing the baseline agent-based simulations to be validated. We then investigate how aggregate economic performance and inequality respond to changes in the strength of local interactions. Finally, we show that neighborhood effects impinge on the efficacy of policies that aim to generate an equitable distribution of wealth.

There are many ways to become rich, but essentially there are two main channels: either through one's own work or through pure chance. If we consider only legal means, the former includes starting a business or investing in financial assets, while the latter includes winning a lottery or through inheritance (a chance event as children cannot choose parents). Conventional wisdom views wealth inequality as an outcome of either unequal inheritance or labor market choices (i.e., the decision to work in industries that require certain skills but award higher pay). For example, Piketty (2011) argues that employment and inheritance are substitutes, in that one can become wealthy either by working hard in lucrative industries *or* by inheriting substantial gift from wealthy parents. It can be argued, however, that inheritance complements labor market outcomes when access to good jobs depends on educational achievements that only the wealthy can afford. This is the view that we espouse in this chapter.

A catch-22 situation arises when the incomes of less affluent families are stagnating in the absence of a more educated workforce, but a more educated workforce is not achievable without an increase in the incomes of moderate- and low-income families. We therefore endeavor here to develop a theoretical model of persistent inequality driven by the interplay of human capital investments, local interactions, and intergenerational transfers. To that end, we extend the overlapping generations model of human capital investments by Galor and Zeira (1993, hereafter GZ). An overlapping generations model, as Samuelson (1958) introduced and reviewed here, describes a demographic structure where parents and offspring live together during at least one period. Parents then die at the end of the shared period, while the youth live on to the next period and beget their own children before they, too, die. Intergenerational transfers occur during the time together, motivated by the parents' desire to secure a better future for offspring. The dynamic then repeats with a different, younger set of agents, in perpetuity.

GZ show how the ancestral distribution of endowments coupled with costly investments in human capital can produce a bipolar distribution of wealth. The main message is that the poor and the rich can continue to coexist (economic stratification will not be altered) even in the very long run, if the initial distribution is sufficiently heterogeneous. To this overlapping generations model, we add geography, which allows us to focus upon the role of local interactions in perpetuating the inequality between poor and rich *neighborhoods*, while maintaining the internal consistency of a general equilibrium framework.

Numerous studies have documented the relationship between human capital and economic status. The notion of human capital investments as a source of inequality was originally formalized by Mincer (1958) and developed further by Becker (1964). Since then, it has been the tradition in labor economics to describe the diversity of individual earnings as a function of schooling.[1] The micro-level empirics are generally consistent with the macro-level data. Nomura (2007), for example, shows using cross-country data that initially the poor benefit the most from education in terms of the improvement in living standards. Credit constraints, however, limit the poor's ability to enroll in school, and this preserves, if not worsens, aggregate inequality. This strand of human capital literature, however, rarely links economy-wide inequality to *spatial* patterns.

It is the studies on agglomeration economies (Rosenthal and Strange 2001) that produce evidence of the tendency for skilled workers to cluster in cities. The starting point of this research is the hypothesis that the interactions of skilled workers—a proxy for the stock of human capital—in a crowded urban environment increase the returns to education for all, which in turn attracts even more skilled workers. That is, there is something about local interactions in close proximity among city workers that promote the accumulation of human capital in both formal and informal settings. Econometric analyses have provided an overwhelming support for this hypothesis (e.g., Glaeser and Mare 2001; Moretti 2004; Roca and Puga 2017). Compelling as it is, however, the agglomeration economies literature is generally silent about the role of intergenerational transfers (i.e., gifts from parents to offspring, both *post mortem* and in vivo) in generating spatial inequality.

Intergenerational transfers come into play when the ability to finance children's schooling depends upon the parents' wealth. Das (2007) shows that descendants of the poor are more likely to end up in poverty than the offspring of the rich. The key insight here is that initial wealth distribution determines future distribution through intergenerational transfers, which may include both inheritance and gifts (e.g., parents' "purchase" of offspring's education). Left to their own devices in Das' theoretical work, offspring are severely challenged in the quest to move to higher income stratum. It is parents' insecurity about their children's future that motivates bequests according to Fan (2006). In the presence of credit constraints, bequests also provide offspring with the necessary resources to invest in human capital.

[1] See the extensive survey in Card (1999).

Nakajima and Nakamura (2009) show that indeed the poor's lack of resources aggravates inequality further because of the hampered ability to invest in education. Unequal access to human capital investments then creates a persistent gap between the rich and the poor (Andergassen and Nardini 2007). Only by lowering education costs can inequality diminish in the long run through the opportunities the former provides for the economically disadvantaged.

Equal opportunity to education alone, however, does not guarantee an egalitarian distribution of wealth. Social milieu also matters. Bowles et al. (2014) show that sorting in social networks can also cause persistent inequality. That is, segregation by class (e.g., the rich interacting only with the rich) increases the likelihood for the poor to remain trapped in poverty. The role of proximity is examined in Mookherjee et al. (2010a, b), which advances a theoretical model showing the role of the educated in inspiring others in the neighborhood to also invest in education. The theoretical importance of neighborhood effects has been confirmed by the empirical studies. Corcoran et al. (1990) and Sampson (2012), for example, documented the strong association between community characteristics and the future economic status of offspring. On a different but related front, Stewart et al. (2007) reveal the impact of neighborhood composition of younger generations on their educational aspirations.

In addition to neighborhood effects, other crucial features of inequality dynamics have also been identified in the literature. Among the most enduring ones is the persistence of economic status across generations. Solon (2002) reviews the evidence and shows the correlation between parents' and offspring's income to be at least 0.34 for Germany, the UK, and the USA. As expected, the higher-order correlations (say, between offspring and great-grandparents) diminish over time but remain statistically significant through generations (Solon 2018). The theoretical model of Becker and Tomes (1979) explains the inertia as the combined effects of the genetic and cultural transmission of traits that contribute to economic success. The superior education attained by children from wealthy families then compounds this process of economic inheritance further. Social networks appear to be an attenuating force, however, as neighborhoods that are more cohesive in terms of the strength of community involvement are also those with higher probabilities for a child of low-income families to rise to the top of the income distribution (Chetty et al. 2014).

The tendency for individuals with similar attributes, such as class or race, to cluster in space is another prominent feature of urban inequality. Schelling (1971) attempts to explain how neighborhoods become segregated using a checkerboard model that is now considered the antecedent to modern agent-based models. Recent studies, however, suggest that US cities today are more racially integrated than the beginning of the twentieth century (Glaeser and Vigdor 2012; Walker 2018). But the decline in racial segregation has not caused the end of *economic* segregation. On the contrary, the uneven sorting of households by income class has been steadily increasing since the 1970s (Jargowsky 1996; Reardon et al. 2018). Economic segregation refers to the uneven sorting of households by income. If the average income of one's neighbors somehow affects one's own economic fortune, then economic segregation is expected to worsen existing inequality. Reardon and Bischoff (2011) found that economic segregation indeed amplifies the influences of local interactions, and this widens the gap between the wealthy and the poor.

The empirics motivate the present study to introduce direct interactions and neighborhood effects to GZ's general equilibrium framework. In theoretical studies of wealth inequality, space is often represented by a single parameter designed to capture the extent of local interactions. The present study distinguishes itself from these studies in proposing a spatially-explicit model where agents are situated in a two-dimensional grid lattice. Wealth heterogeneity implies different abilities to invest in human capital. At the same time, agents learn through direct interactions whether their neighbors have the financial resources to pursue investments in human capital. Thus, if there is a sufficient number of individuals with educational potential around, those without such potential may be persuaded to adapt (and to attend school) in order to conform to the social milieu, and vice versa.

The introduction of space and local interaction agents renders the equilibrium distribution of wealth analytically intractable. Simulations are thus the only resort. It is to simulate the emerging spatial patterns of production and inequality in a decentralized economy that we exposit an agent-based model.[2] Computational models have been used to explain emerging patterns in urban and regional systems with heterogeneous agents (Mansury and Gulyas 2007; Parker and Filatova 2008). Here, we present a new bottom-up approach showing how results change depending upon the degree of local interaction and neighborhood effects.

In addition to formalizing the roles of neighborhoods, we also use the model to highlight the connections between cross-sectional inequality and intergenerational transmission of economic status. As will be shown below, a variant of our model captures intergenerational transmission by positing a positive statistical association between parents' economic success and the offspring's.

The agent-based model produces results consistent with the empirical evidence for a trajectory characterized by sustained output expansion and persistent inequality. Understanding the drivers of growth and change allows us to consider three policy instruments aimed to promote equal opportunity, namely those that lower borrowing rates, reduce education costs, and target subsidies to agents who otherwise cannot afford education. On the one hand, governments have promoted borrowing rate policies, such as the US Direct Loans Program, which offer students loans at an interest rate considerably lower than market terms. On the other hand, the USA has one of the most expensive higher education systems in the world (Usher and Medow 2010), with education costs reaching over 50% of median income. It would be of interest to simulate the equity implications of policies aimed to make education more affordable, as the Scandinavian countries seem to have accomplished. The third instrument we look at is targeted subsidies, which take the form of, for example, the Pell Grant offered to students from low-income families. As we will show below, our agent-based model

[2] ABM is also known as Agent-based Computational Economics (ACE) (Tesfatsion 2005). Tesfatsion stressed the importance of ACE in dealing with complicated micro-behavior in real world (e.g., asymmetric information, imperfect information, and multiple equilibria).

allows us to examine the differential impact of these instruments on inequality and growth, demonstrating the potential application of our model to policy analysis.

Clearly, a highly stylized model like ours should not be taken literally. To gain insight into how individual choices and interactions perpetuate inequality, we abstract away from the socioeconomic details of a real urban system and also from much of the detail of the diffusion of neighborhood effects. Thus, for example, the model describes production in terms of the dichotomy between skilled and non-skilled workers, abstracting away from intermediate degree of diversity or finer earning categories within types. Nevertheless, we have preserved what is essential for a spatial model of growth and inequality. We think the empirical evidence summarized in this section and its connections with theories hold out the promise that continuing research, using richer models that extend ours here, will improve our understanding of the interplay between human capital, intergenerational transfers, and local interactions.

The rest of the chapter proceeds as follows. Section 2 lays out the analytical framework for the numerical simulations. Following that, Sect. 3 details the implementation of the agent-based algorithm. Section 4 outlines the various scenarios that we consider in this chapter, and Sect. 5 reports the simulation results. We close in Sect. 6 with concluding remarks.

2 Mathematical Model

The model extends GZ with differences in several important respects. First, our agent-based model is numerical and spatial, and it treats workers as a discrete variable. In addition, we allow neighboring agents to interact directly, which is the primary value added of agent-based simulations in which geography is incorporated. Last but not least, we explicitly model the labor market, allowing wages to be numerically determined and calibrated for the agent-based simulations. The numerical implementation of the agent-based model is detailed in Sect. 3. This section describes mathematically the micro-foundations of production, human capital investments, and worker behavior.

We begin by reviewing GZ, which employs an overlapping generations model in which every agent lives for two periods. In the first period of life, agents are referred to as the Young who must decide whether to pursue education or to work as unskilled labor. It is assumed that the Young cannot be in school and be employed at the same time. In the second period of life, agents are referred to as the Old who consume output produced and leave bequests to their offspring. As in GZ, every parent gives birth to exactly one offspring, an assumption also adopted here for computational convenience.

The Young's decision whether to stay in school or be employed as unskilled labor is determined by how much they wish to consume and bequeath when they become old. In the beginning of their second period of life, the Old give birth to the next generation of the Young, again exactly one offspring per parent. Old agents who chose to invest in human capital when they were young can now gain employment as skilled workers in the second period of their life. As they become parents, the Old transfer bequests to their offspring, and then consume the remaining wealth. This cycle continues perpetually.

The economy produces a single commodity employing both types of workers albeit in different sectors, namely skilled and unskilled worker sectors, denoted by the subscripts s and n, respectively. Each sector utilizes a distinct production recipe in manufacturing output. Specifically, the production function for the sector employing unskilled workers is linear:

$$Y_{\mathrm{n}} = \phi L_{\mathrm{n}}. \tag{1}$$

where Y_{n} is the output of the non-skilled worker sector, ϕ the sector's constant marginal product of labor, and L_{n} the number of unskilled workers employed. By contrast, the production function for the skilled labor sector is neoclassical[3]:

$$Y_{\mathrm{s}} = AK_{\mathrm{s}}^{\beta} L_{\mathrm{s}}^{1-\beta} \tag{2}$$

where Y_{s} is the output of the skilled worker sector, K_{s} the physical capital, L_{s} the number of skilled workers, A a positive constant, and β a positive fraction. More specifically, the parameter A represents the level of technology that is Hicks neutral (Hicks 1932) in the sense that a change in technology does not change the marginal productivity of capital relative to labor's. The parameter β on the other hand represents the share of output accruing to capital in the skilled worker sector (Chiang 1984). Equations 1 and 2 together represent the total output for the same homogenous commodity, to be consumed by utility maximizing agents (individuals) in the second period of their lives. Before proceeding further, we need to identify a commodity that will serve as the unit of account in which all prices are denominated. This is because money has no role in the standard general equilibrium framework. In such a framework, every transaction occurs only when two agents want a commodity that the other agent possesses (see, e.g., Kiyotaki and Wright 1998). We simply assume here that agents agree to set the consumption good as the *numeraire* and, without loss of generality, fix its price as $P = 1$.

We depart from GZ by introducing neighborhood effects that influence agents' educational choices through local interactions. Social interactions reveal private information, allowing younger agents to observe whether their local neighbors are skilled-worker prospects. These observations are then taken into account in the Young's human capital decisions. We assume a simple rule of engagement in which a weighted average is calculated to capture the dependence on neighborhood effects of agent a's tendency to invest in human capital:

[3]The formulation (known as Cobb-Douglas) of the production function belongs to the general class of concave functions that GZ use in a more abstract setting. Because our model is computational, we need to specify a functional form so that the algorithm can compute the numerical values of output. The Cobb-Douglas formulation is chosen because it remains the most ubiquitous form of production function in today's theoretical and empirical analyses of output growth (Felipe and Adams 2005).

$$\pi_{\mathrm{a}} = \theta\varphi_{-\mathrm{a}} + (1-\theta)\varphi_{\mathrm{a}} \tag{3}$$

where $0 \leq \theta \leq 1$ is a constant parameter that represents the weight for neighborhood effects, φ_{a} is based on the agent's internal economic calculations (($\varphi_{\mathrm{a}} = 1$) if education is economically viable, 0 otherwise), and $\varphi_{-\mathrm{a}}$ is based on neighbors' economic potentials. The parameter θ thus measures the strength of local interactions. Local interactions are strongest when $\theta = 1$, representing the upper bound where only neighborhood effects matter. The other limiting case is $\theta = 0$ as in GZ, where local interactions are absent. In what follows, we will sometimes refer to the Young with the economic potentials to become skilled workers as "educational prospects" or simply "prospects."

Equation 3 is inspired by Wilson (1987), which has documented the persistent underinvestments in human capital among the poor isolated in inner-city neighborhoods. Here, educational choices depend on whether π_{a} exceeds a critical level—exogenously given π_{min}—which, if surpassed, will lead to agents investing in human capital. The exception to this rule is when wealth would fall below zero if agents end up investing in education that they could not afford because of their neighbors' influence. In this case, agents would revert back to the default choice of working as unskilled workers in both periods of their lives.

Agents consume only when old. The satisfaction that agents get from consumption (and bequest) is described by a log-linear utility function (GZ Eq. 3, p. 38):

$$U_{\mathrm{a}}(c_{\mathrm{a}}, b_{\mathrm{a}}) = \alpha \ln c_{\mathrm{a}} + (1-\alpha) \ln b_{\mathrm{a}} \tag{4}$$

where c_{a} and b_{a} denote agent a's consumption and bequest, in terms of both the aggregate commodity and $0 < \alpha < 1$, the share parameter. Agents are rational in the sense that they choose the division of wealth between consumption and bequest that maximizes welfare. The log-linear formulation implies that agents allocate a fixed proportion α of their lifetime wealth M (defined below) to own consumption, and leave the rest for their offspring.

The lifetime wealth of an agent depends on the skills possessed by the agent, as well as whether the agent borrows to finance education. Opting out of education would lead to agents working as unskilled workers in both periods of life, earning

$$M_{\mathrm{n}} = (1+r)(x+w_{\mathrm{n}}) + w_{\mathrm{n}} \tag{5}$$

where M_{n} denotes the unskilled workers' lifetime wealth, w_{n} the unskilled workers' wage, x the inheritance they receive from their parents, and r the interest rate on the first period's savings. That is, agents save the wages earned and the bequests inherited when young, which would then yield interests that become available for consumption in the second period of life.

On the other hand, those who choose to invest in human capital and can afford to self-finance their education would have lifetime wealth as follows:

$$M_{\mathrm{s}} = (1+r)(x-h) + w_{\mathrm{s}} \qquad \text{for } x \geq h \tag{6}$$

where M_s denotes the skilled workers' lifetime wealth, h the cost of education, and w_s the skilled workers' compensation. The idea here is that agents forgo work when they were young in order to acquire skills, and their inheritance, x, is large enough that they can afford the cost of education, h, without borrowing from the financial market. The acquired skills would then allow educated agents to earn the lucrative wage, $w_s \gg w_n$, as they enter the second period of their life.

The third type of lifetime wealth calculation corresponds to agents forced to borrow should they decide to invest in human capital because the cost of education exceeds their inheritance. Their lifetime wealth as skilled workers would then be as follows:

$$M_s^* = (1+i)(x-h) + w_s \qquad \text{for } x < h \tag{7}$$

where i is the interest rate on the loan, $x - h$, that borrowing agents would have to repay when they become old. Because $x < h$, this type of agents would have to borrow to help cover the cost of education. They would do so when they were young as long as the skilled workers' compensation, w_s, is large enough for lifetime wealth to exceed that of unskilled workers, $M_s > M_s^* > M_n$. Note that Eqs. 5, 6, and 7 are obtained by adding consumption to bequest in GZ's (p. 39) Eqs. 9, 11, and 13, respectively.

Since agents are rational and utility maximizing, they make choices (i.e., whether to invest in human capital or not) that yield the highest lifetime wealth. Having made educational choices, agents then decide how to split wealth between own consumption and bequests when they become old. It can be shown that the log-linear utility function implies that the optimal bequest and consumption—based solely upon economic calculations—would be (see Kreps 2012):

$$\begin{aligned} b_a(x) &= (1-\alpha)M \\ c_a(x) &= \alpha M \end{aligned} \tag{8}$$

How do local interactions affect education decisions? The spatial structure of our model corresponds to a two-dimensional grid lattice, where every agent is surrounded by and interacts with, at most, eight adjacent neighbors (the so-called Moore neighborhood). Given this structure, the decision whether to invest in human capital depends on: (i) the neighborhood composition of a young agent in terms of the educational potential of adjacent neighbors and (ii) the strength of local interactions. Thus, for example, when the strength of neighborhood effects is at the maximum ($\theta = 1$), education decisions depend exclusively on the neighbors' economic potential according to the calculations in Eqs. 5–7. Furthermore, if there are enough prospects to persuade an agent—more than the threshold number of neighbors, TD, who are expected to invest in education based purely on economic grounds—then the agent would pursue education, as long as it results in positive lifetime wealth. Note the bounded rationality inherent in agents making choices based upon neighbors' *potential*—i.e., without knowing whether neighbors will actually pursue education. This may be interpreted as agents anticipating local peers to receive significant inheritance from their well-to-do parents and use it to finance education. But some economically viable candidates eventually opt out of school if interactions are strong *and* there are insufficient prospects in *their* neighborhood. As the Young in the model make the final decisions simultaneously, we assume that they are unable to observe their neighbors' actual decisions.

For concreteness, consider once again the maximum dependency ($\theta = 1$) example where, in addition, agents need to be persuaded by at least two prospective neighbors (TD = 2) in order to pursue education. Now consider Agent C who has three neighbors: A, B, and D (see Fig. 1a). Following GZ, we assume the existence of two discrete types of agents. Though the present study does not subdivide types further into finer earning categories, we could certainly do so in future extensions of the base model discussed here. Let *Type 1* ("1" in black circle) agents be those for which a skilled job yields wealth higher than unskilled employment, and let *Type 0* ("0" in white circle) be the opposite. Suppose the economic calculations are such that education is feasible for every agent in that neighborhood (Fig. 1b). In this case, all agents end up investing in human capital—*not* due to their internal economic calculations, but because every agent has at least two skilled-worker *prospects* as neighbors. But suppose instead that education is not economically feasible for Agents A and B (Fig. 1c); this information could lead to Agent C making the opposite decision, choosing not to pursue education. If neighborhood effects are sufficiently strong, Agent C could become a worker in the unskilled sector (Fig. 1d) because, from the agent's perspective, that is where the overwhelming majority of neighbors will work.

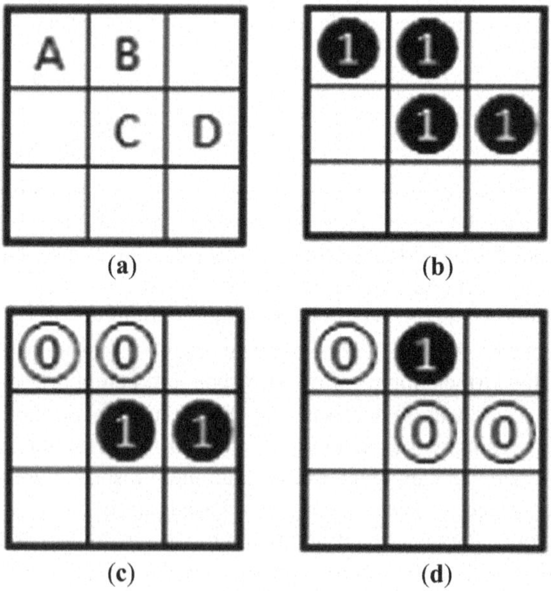

Fig. 1 Neighborhood composition and human capital decisions with maximum interaction ($\theta = 1$)

Agent C would then follow the anticipated career path of peers, in order to conform to the social milieu.

2.1 Theoretical and Computational Aspects of Wages

We discuss next the derivation of wages within the formal system of our theoretical model. In this analysis, wages are determined as follows. The unskilled workers' wages, w_n, are considerably more tractable mathematically because the production function (see Eq. 1) implies a linear relationship between price and the wages, $w_n = p\phi$. As output has been chosen as the numeraire with fixed price $P = 1$, the wages for unskilled workers are fully determined. We discuss here the initialization of skilled workers' wages, w_s, which, theoretically, is determined endogenously.[4]

We assume that initially there are L_n unskilled workers and no skilled workers. Thus, $L = L_n$, and when unskilled sector production begins at time $t - 1$, aggregate output can be determined based upon Eq. 1. To initialize production activity in the skilled sector, agents must first acquire the necessary skills by investing in human capital before they can be employed as skilled workers. For agents that choose to invest in education, how much will they be compensated as employees of the skilled sector? The question is a challenge for numerical analysis because human capital investments and the compensation are both endogenously determined: Wages depend on the supply of skilled workers and vice versa. Fortunately, the following proposition helps to resolve the chicken-and-egg aspect of the market for skilled workers, allowing skilled workers' wages to be calibrated at a level that guarantees the coexistence of unskilled and skilled employment.

Proposition 1 *The long-run wages for skilled workers converge almost surely to a level at which an agent would be indifferent between investing and not investing in human capital. It can be shown that this wage satisfies the following equation:*

$$\text{Threshold } W_s = (2 + r)w_n + (1 + i)h \tag{9}$$

The proof is by contradiction, but it can be easily demonstrated by numerical simulations. Note the "almost surely" convergence is evident from wages in our simulations that settle down in the threshold $w_s$after only a few iterations.

By Proposition 1, in the long run the compensation for skilled workers is guaranteed to be higher than (more than twice) the unskilled workers' wages, w_n, as evident by the right-hand side of Eq. 9. The marginal product of skilled workers (derived from Eq. 2) then determines how many agents, L_s, would be employed in the skilled

---

[4]GZ assume endogenous wages for *unskilled* workers for their purely mathematical treatment. For a numerical model like ours, fixing unskilled wages while permitting skilled ones to vary as a "free" variable allows the simulations to be initialized much more efficiently.

workers' sector at the threshold wage w_S. But how and when does production in the skilled workers' sector begin?

In our agent-based model, we initialize production in the skilled sector at $t = 2$ based on the following mechanism. At $t = 2$, agents from the first period become Old, who give birth and bequest to a new Young generation. The Young then must decide whether to invest in human capital or be employed in the non-skilled sector based on how attractive the skilled sector wage, w_S, is. At the beginning of $t = 2$, however, the skilled workers' wage is indeterminate because the supply of skilled workers is non-existent. The supply of skilled workers is non-existent, however, because their wage has yet to be determined. This simultaneity problem is what we refer to as the chicken-and-egg aspect of the labor market.

Fortunately, Proposition 1 allows us to initialize w_S at the threshold level that is attractive to the Young because it yields lifetime wealth that exceeds that of unskilled workers. At $t = 2$, a few from the pool of Young agents that the agent-based algorithm randomly selects are offered the skilled workers' wages. This is the only stage where the random selection is applied to initialize the population of skilled workers. It is important to note that, given the usual downward-sloping demand schedule for skilled labor implied by Eq. 2, the threshold w_S can only support a limited number of agents. Greater *supply* of skilled workers than the level that is consistent with w_S will push skilled workers' wage below the threshold. Consequently, only a limited number of agents are offered the skilled workers' wages.

It turns out that the threshold wage defined in Proposition 1 is also the wage that sustains the long-run coexistence of skilled and non-skilled workers. At a level lower than the threshold w_S, skilled production gradually vanishes as skilled workers' wage is too low to cover the cost of education for future generations. On the other hand, initializing w_S at the threshold results in rapid convergence to a stable trajectory in which unskilled and skilled workers coexist.

To help intuition, Fig. 2 illustrates the connection between wages and aggregate output. When wages in the skilled sector are lower than the threshold W_S, in

Fig. 2 Skilled workers' wages and aggregate output

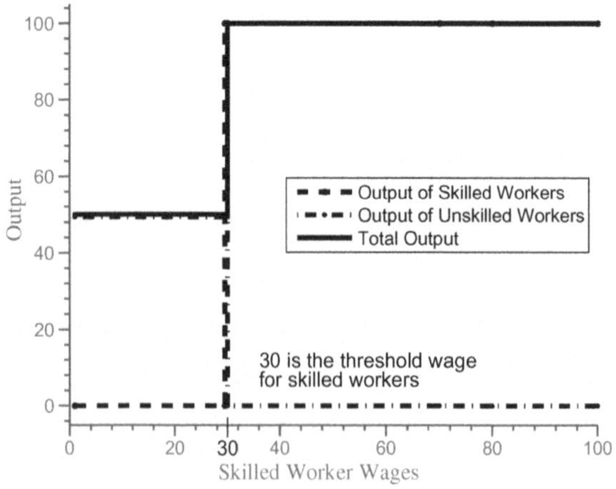

equilibrium all workers choose unskilled sector employment. Conversely, all agents invest in human capital when skilled workers' wages are persistently higher than the threshold. Unskilled and skilled workers coexist only when wages are such that the Young who did not receive bequests from their parents are indifferent between the two sectors.

We perform numerical simulations in which skilled workers' wages are calibrated according to Proposition 1, in order to prevent the modeled economy from degenerating into a homogeneous equilibrium. The next section details the agent-based algorithm.

3 Algorithm Implementation

We implement the agent-based algorithm with the Recursive Porous Agent Simulation Toolkit (Repast)[5] library. Before running the scenarios of interest, we set the model parameters at values that would ensure that the simulation produces outcomes in which both skilled and unskilled workers coexist. This is a necessary step because initial explorations suggest that skilled and unskilled workers coexist only within a narrow range of exogenous variables and parameter values. Skilled workers' wages have already been calibrated based upon Proposition 1. Table 1 lists the values of the additional variables and parameters calibrated in the agent-based simulations. According to GZ (Eq. 7, p. 41), the relationship between the interest rate, r, and the borrowing rate, i, is determined by the equation $i = (1 + br)/(b - 1)$ with $b > 1$ to ensure positive borrowing rates. To emulate the real world at the time this study is written, we set interest rate and borrowing rate below 10%, with the latter being higher than the former. Furthermore, the values listed in Table 1 for the consumption

Table 1 Fixed parameter values in the agent-based simulations

Parameters	Value
Share parameter in skilled labor production function, β	0.7
Technology parameter, A	30
Wages of unskilled workers	10
Interest rate, r	0.01
Borrowing rate, i	0.05
Maximum bequest, x	100
Total number of workers, L	100
Consumption parameter, α	0.7
Minimum threshold to invest in education, $\pi_{\min}$	0.5

[5]We use Repast library version 3.1, developed by North et al. (2006). Repast is a free open source toolkit and available at Repast homepage at https://repast.sourceforge.net/repast_3/index.html (Road 2004).

share parameter, α, and the labor share parameter in the production function, β, are chosen to ensure the coexistence of skilled and unskilled labor at positive wages. In addition, the cost of education, h, is fixed at 20.

Having initialized the parameters, the following algorithm is implemented:

1. First, we derive the labor demand schedules for both sectors from the production functions (see Eqs. 1 and 2). The demand schedules relate wages to incremental output for both the skilled and unskilled labor sectors. Aggregate supply is the combined output produced in the two sectors.
2. Each agent is then initialized with three binary variables (each taking the value of either one or zero), which represent, respectively, φ_{a} (agents' own internal economic calculation), $\varphi_{-\text{a}}$ (the educational potential of agents' neighbors), and the tendency to invest in education, π_{a}. The sequence of determining the value of each term is determined as follows. Agents first determine whether they are prospective skilled workers by comparing the lifetime wealth from an investment in education (Eqs. 6 or 7) with that from a career as an unskilled worker (Eq. 5). The binary variable φ_{a} is set $= 1$ if an agent is a prospect, $= 0$ otherwise.
3. In scenarios in which we allow neighborhood effects, agents then observe whether their nearest neighbors are skilled-worker prospects. Since the spatial structure is a square grid lattice, each agent has a maximum of eight neighbors (the so-called Moore neighborhood structure). We consider the case in which agents make educational choices depending on the number of neighbors, TD, who are expected to invest in education based purely on economic grounds (i.e., without taking neighborhood effects into account). Thus, for example, TD $= 1$ means that an agent would invest in human capital if there is at least one prospect for a future skilled worker in the neighborhood.
4. If the number of prospects $\geq$ TD, then $\varphi_{-\text{a}}$ takes the value of 1, otherwise $=$ 0. The tendency to invest in human capital, π_{a}, can then be calculated for every agent based on Eq. 3. Agents for which π_{a} exceeds the minimum threshold, π_{min}, would actually invest in human capital.
5. The computation of agents' wealth and bequest in the subsequent period follows directly from current period's education choices. In order to avoid negative wealth, a restriction is imposed on agents' choices so that an investment in human capital occurs only if lifetime wealth is not negative.[6]
6. We consider three different setups:

    a. Setup 1 corresponds to a society where initially *no* agent pursued education. Subsequently, higher wages are offered to randomly selected agents, some of which would be motivated to invest in human capital. The wages for prospective skilled workers are calibrated according to Proposition 1. Absent neighborhood effects, $\theta = 0$, Setup 1 produces trajectories of output and inequality reported in GZ.

[6]This restriction is necessary in order to prevent the economy from spiraling out of control.

b. In Setup 2, the probability for an agent to be offered the higher wage is higher if the parent is a skilled worker. This allows education choices to be correlated across generations.
c. Setup 3 combines Setups 1 and 2, and in addition all agents offered the higher wages (which they accept) which are clustered in one neighborhood. This setup is designed to capture the notion of class segregation. Thus here, unlike in the other setups, the spatial distribution of wealth will strongly affect the future educational status of the younger generations.

7. For each setup, we examine the results obtained from running the simulations with differential neighborhood effects (i.e., where the weights for the strength of local interactions, θ in Eq. 3, are varied).

The flowchart below (Fig. 3) summarizes the logic of the agent-based algorithm.

4 Scenarios

We consider the following scenarios corresponding to three alternative policy changes and compare the equilibrium number of skilled workers.

- A reduction in borrowing cost achieved by lowering the loan rates, which will make education accessible to a greater number of agents.
- A direct reduction in the cost of education for all agents in the economy while leaving the loan rates unchanged.
- Targeted subsidy, which focuses on agents who otherwise would not have invested in human capital. For simplicity, we provide direct subsidy by adjusting the cost of education using a parameter that takes the value between zero and one. This mechanism implies a transfer payment equals the difference between the subsidized and the full price of education.

In the targeted subsidy simulation, the wealth equation for agents whose bequest falls short of the education cost is modified to include a subsidy parameter, ξ. Thus, in this scenario, the wealth of skilled workers who otherwise would have to borrow in order to finance their education would be:

$$M_s(x) = (1 + i)\ (x - (1 - \xi)h) + w_s \tag{10}$$

where ξ is a fraction of education cost and takes a positive value that is less than one. Since h is the cost of education, ξh is effectively the transfer payment received by low-bequest agents. The partial subsidy thus reduces the burden of education costs, enabling agents to earn higher lifetime wealth. We use Eq. 10 to investigate the impact on human capital investments of different subsidy levels.

Comparing alternative wealth distributions in a large number of agents is potentially a daunting task. We use the Gini index, which is zero with complete equality and one when a single person has everything and everyone else nothing, to facilitate

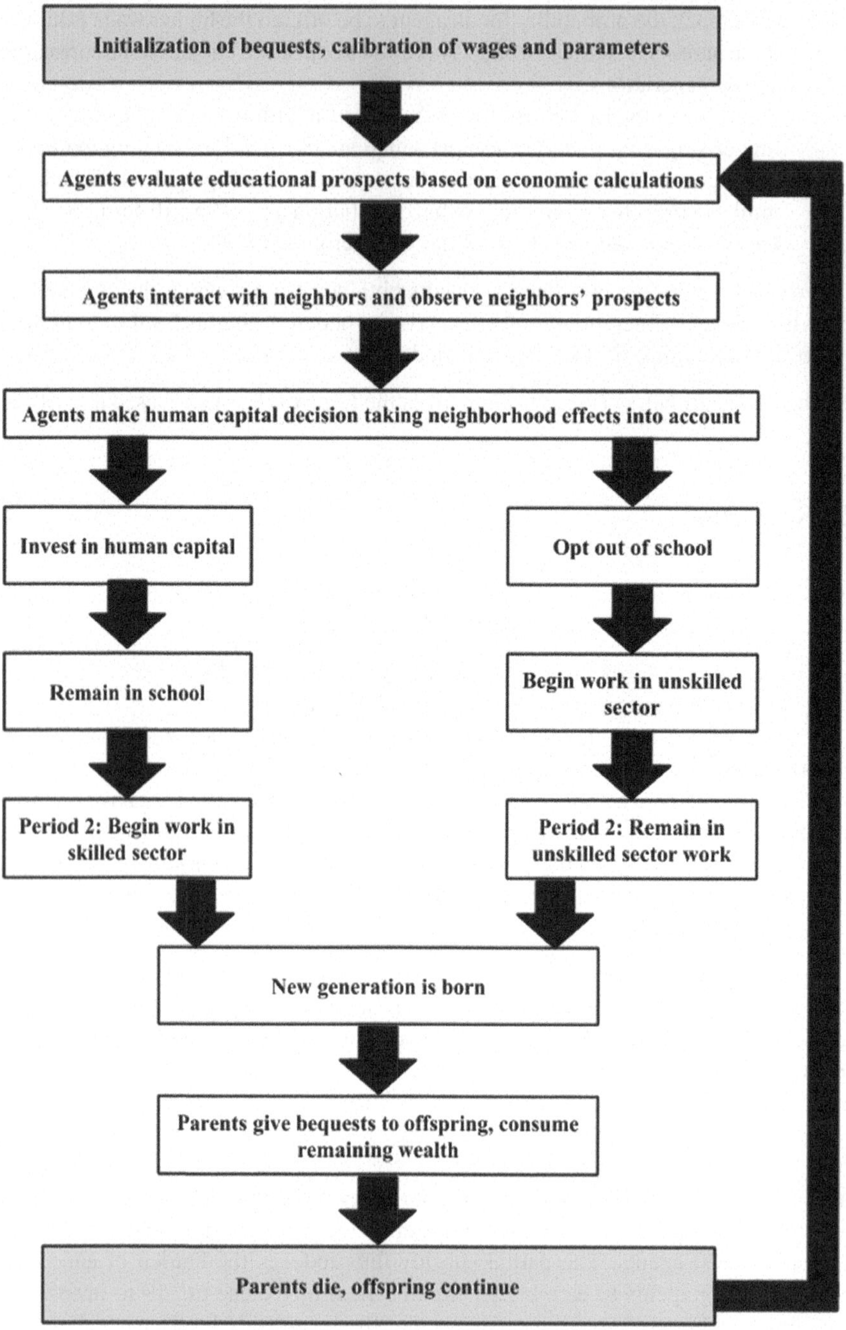

Fig. 3 Logic of the agent-based algorithm

comparisons of aggregate inequality across scenarios. The index is a summary (i.e., economy-wide) measure of inequality and has been widely used for distributional and policy analyses.[7]

5 Results

For each combination of parameter values and initial setup, we run Monte Carlo simulations with five different random seeds for each specific parameter configuration to control for random fluctuations. Each configuration is run for 1000 periods in order to ensure that the system reaches steady state. As described in Fig. 3, every period corresponds to the time where parents (the Old)'s second period of life overlaps with their offspring (the Young)'s first period of life. The results reported here correspond to a system populated by 100 agents that live for two periods, in which the borrowing rate, i, is fixed at 0.05. Exactly 100 offspring are born to the parents that consume, make bequests, and then die in the second period of life.

We validate the agent-based algorithm by replicating the analytical results of GZ, in which direct interactions are absent. Validation of the algorithm (i.e., verifying the logic of the codes) is accomplished by matching the results from the baserun (non-spatial) model with the results reported in GZ. If the algorithm's logic is valid with respect to the theoretical model, then the results should match as the baserun model is simply the computational analog of GZ's mathematical approach. In particular, our agent-based simulations successfully reproduce the long-run divergence of wealth between skilled and unskilled workers starting from a random-uniform distribution of initial wealth (GZ p. 41).

We discuss first the sway of stronger interactions as a result of an increase in the magnitude of neighborhood effects, as captured by the parameter θ (Eq. 3). Note that $\theta = 0$ represents the no-interaction case, where agents make decisions based purely on economic considerations. As Setup 2 produces results that are very similar to those from Setup 3, to save space Setup 2's results are not shown here. Figure 4 displays the relationship between interactions and human capital investments (measured by the number of skilled workers) over 1000 periods in Setup 1. Note that Fig. 4 plots 10-period averages in order to reduce short-term fluctuations and highlight the differential trends.

The figure indicates that human capital is highest when agents assign equal weights ($\theta = 0.5$) to neighbors' influence and pure economic calculations, respectively, and lowest when agents consider only the former ($\theta = 1$). Intermediate levels are observed when neighborhood effects are positive but weak ($0 < \theta < 0.5$), or somewhat strong ($0.5 < \theta < 1$). At $\theta = 0.5$, moderate interactions succeed in significantly raising the propensity of the Young to pursue education. The results thus suggest that the highest level of human capital is obtained when neighborhood effects are neither too strong nor too weak.

[7] See Sen (1997) for historical perspectives and mathematical treatment.

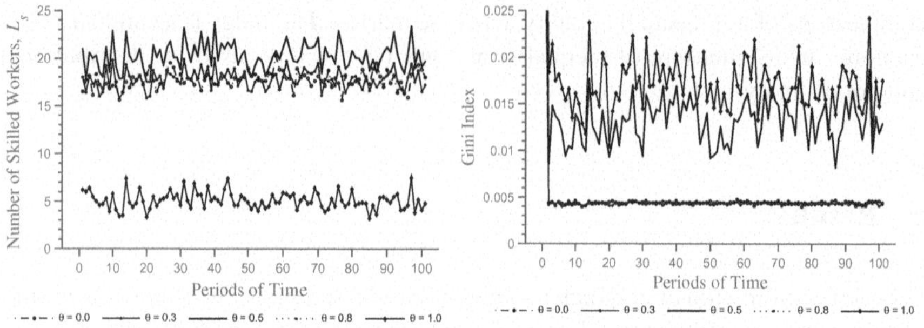

a. Society's level of education, measured by the number of skilled workers over 1,000 periods

b. Distribution of wealth, measured by Gini index over 1,000 periods

Fig. 4 Time series of **a** the number of skilled workers, and **b** the distribution of wealth, various θ's, Setup 1

Inequality, however, is highest when local interactions are very strong ($\theta \geq 0.8$), as the Gini indices suggest in Fig. 4b. Conversely, the distributions of wealth are most equal (and the Gini indices the lowest) when neighborhood effects are completely absent or weak ($\theta = 0.0$–0.3). In a nutshell, inequality is highest when neighborhood effects are strongest. It appears that a positive association characterizes the relationship between inequality and the strength of local interactions.

Figure 5 reinforces the findings in a cross section of equilibria at $t = 1000$ with TD fixed $= 4$. As Fig. 5a confirms, the steady-state relationship between the strength of interactions and human capital investments remains nonlinear—positive when neighborhood effects are weak, but becomes negative as soon as θ exceeds 0.5. When neighborhood effects are weak to moderate ($\theta < 0.5$), increasing the strength of interactions would expand output (Fig. 5a) but has social repercussions as inequality

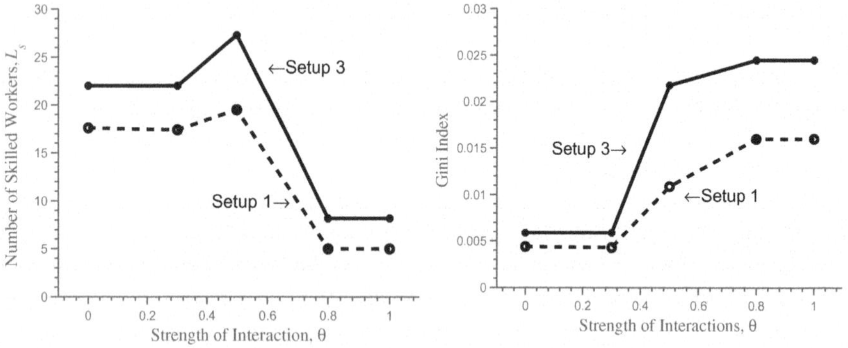

a. Society's level of education, measured by the number of skilled workers

b. Distribution of wealth, measured by Gini index

Fig. 5 Relationship between the strength of interactions and **a** society's level of education and **b** the distribution of wealth, various setups, $t = 1000$

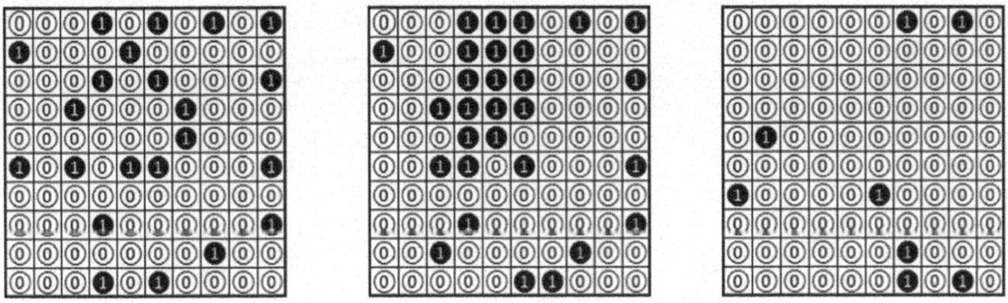

a. Without interactions, $\theta = 0$

b. With interactions, $\theta = 0.5$

c. With interactions, $\theta = 0.8$

Fig. 6 Spatial patterns of neighborhood segregation between skilled ("1" in black circle) and unskilled workers ("0" in white circle), various levels of interactions

rises (higher Gini index). Thus, stronger interactions appear to create a tradeoff between higher output and inequality when neighbors are only weakly interacting with each other. Stronger interactions (up to $\theta = 0.5$) lead to not only higher human capital investments, which promote output, but also a *less* equal distribution of wealth. The tradeoffs disappear when neighborhood effects are strong. Starting from $\theta = 0.5$, increasing the strength of interactions would decrease output (Fig. 5a) and raise inequality (Fig. 5b) at the same time. There are thus no tradeoffs when neighbors are already strongly interacting with each other. An even stronger neighborhood effect in this case will lead to a contraction of output and an increase in inequality. It is also noteworthy from Fig. 5 introducing class segregation and path dependence (Setup 3) brings about a higher level of human capital investments for a given strength of local interactions, as indicated by the upward shift of the plots. Neither segregation nor path dependence, however, change the qualitative nature of the relationships among output, inequality, and interaction strength.

Figure 6 shows the spatial patterns that emerge from Setup 3 under various interaction levels, θ, with TD fixed = 4. The no-interactions case ($\theta = 0$) leads to about twenty-two percent of workers being educated. At a moderate level of concern for neighbors' prospects ($\theta = 0.5$), the proportion of skilled workers is the highest, about thirty percent of the total. When agents are highly intertwined ($\theta = 0.8$), skilled workers fall precipitously to less than 10% of the entire labor force. Results from the other setups show very similar patterns.

Keeping the interaction level fixed at $\theta = 0.5$, an increase in the minimum number of prospects in the neighborhood required for adaptation, TD, leads to declining human capital investments (Fig. 7a) across all setups. This is intuitive. Agents are less likely to pursue education when doing so requires a stronger presence of "positive influences" to do so in the neighborhood. The upside of this decline is a fall in wealth inequality as evident in the lower Gini indexes (Fig. 7b).

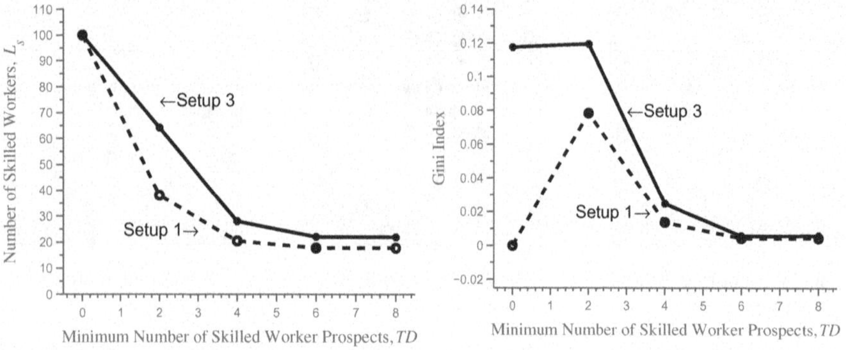

a. Society's level of education, measured by the number of skilled workers

b. Distribution of wealth, measured by Gini index

Fig. 7 Relationship between the minimum number of skilled-worker prospects (TD) and **a** society's level of education and **b** the distribution of wealth, various setups

5.1 Scenario Analysis

We turn now to the analysis of three alternative policies: (A) a reduction in borrowing rates; (B) a reduction in education costs; and (C) targeted subsidies to agents who otherwise would not have invested in human capital. Here, the simulations focus on Setup 1, where higher wages are randomly offered and class segregation is absent.[8] For this purpose, we set TD = 4 in the simulations that produce Figs. 8, 9, and 10. Figure 8a shows that lower borrowing rates in scenario A are associated with higher

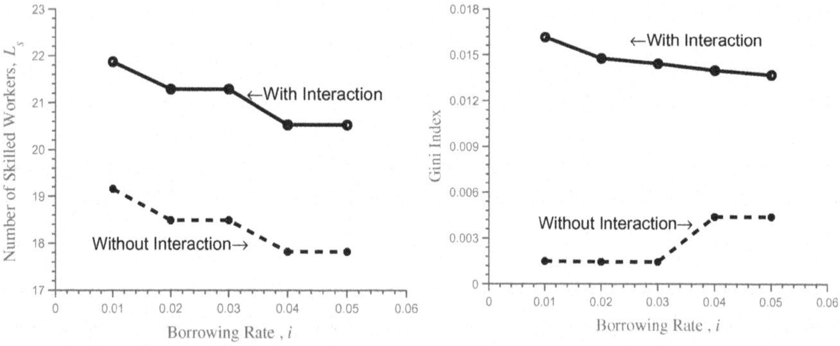

a. $\theta = 0$ and $\theta = 0.5$ for the without and with interaction cases, respectively

b. $\theta = 0$ and $\theta = 0.5$ for the without and with interaction cases, respectively

Fig. 8 Scenario A: borrowing rates and **a** society's level of education and **b** the distribution of wealth, various degrees of interactions

[8]It turns out that results of the subsidy scenario analysis are qualitatively very similar under Setups 2 and 3.

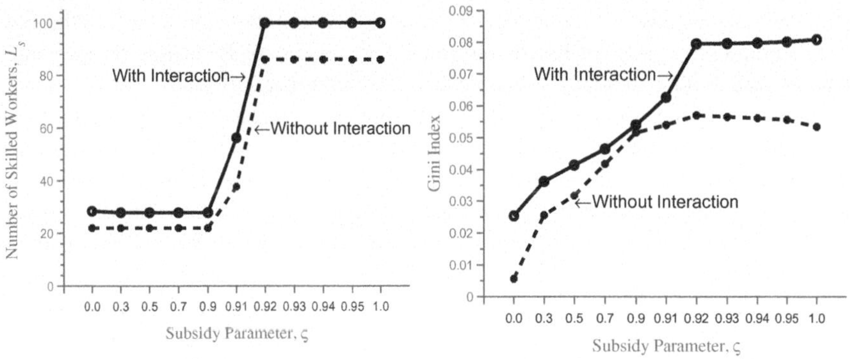

a. $\theta = 0$ and $\theta = 0.5$ for the without and with interaction cases, respectively

b. $\theta = 0$ and $\theta = 0.5$ for the without and with interaction cases, respectively

Fig. 9 Scenario C: targeted subsidies and **a** society's level of education and **b** the distribution of wealth, Setup 1

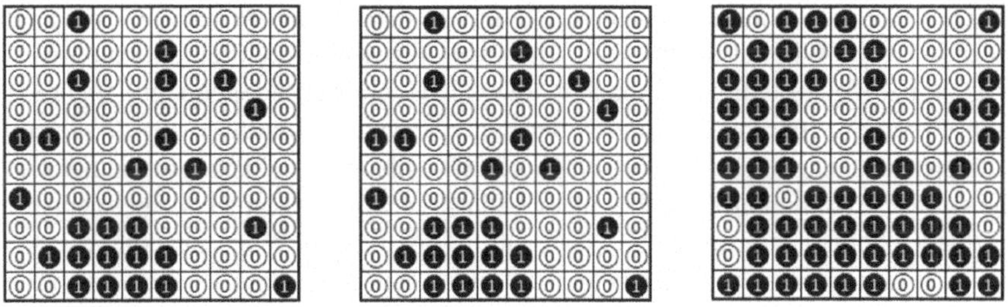

a. Subsidy, $\zeta = 0$

b. Subsidy, $\zeta = 0.5$

c. Subsidy, $\zeta = 0.91$

Fig. 10 Spatial patterns of neighborhood segregation between skilled ("1" in black circle) and unskilled workers ("0" in white circle), various levels of subsidies

level of education, which is bolstered even more by social interactions ($\theta = 0.5$). That is, local influences and lower borrowing rates complement each other in raising human capital. Figure 8b shows, however, the ambiguous relationship between borrowing rates and inequality. Lower borrowing rates appear to reduce inequality in the absence of local interactions ($\theta = 0$). Moderately strong neighborhood effects ($\theta = 0.5$), by contrast, render opposite results. Starting from $i = 5\%$, a reduction in the cost of borrowing appears to *exacerbate* initial inequality. More simulations at finer intervals for both borrowing rates and interaction strength are needed so that a more complete picture can be drawn.

As expected, in scenario B where education costs are reduced, the level of education when moderate interactions are allowed ($\theta = 0.5$) is significantly higher than without interactions ($\theta = 0$).[9] At the same time, the higher number of skilled workers

[9]Results not reported here due to space considerations.

in the interactions case brings about a dramatic increase in inequality. Thus, reduction in either education costs or borrowing rates (scenarios A or B) brings about results that are very similar in terms of aggregate output and inequality.

We next examine scenario C, in which subsidies target agents who otherwise would have to borrow in order to pursue education. Figure 9a shows that human capital investments begin to respond only when education is close to being fully subsidized. The increase, however, is dramatic (more than double) when the subsidy is at least 90% of costs, suggesting a *phase transition* in education. Phase transitions refer to the qualitative change in relationships as certain parameters acquire certain values. In this case, when subsidy is not high enough ($\zeta \leq 0$), education is impervious to transfer payments. Once the 90% breakpoint is surpassed, education becomes very responsive to additional subsidy. All the investigated cases exhibit phase transitions but, as in the other scenarios, the increase in human capital worsens the distribution of income. Notice, however, that inequality rises even *before* education starts its upward trajectory (Fig. 9b). It turns out that subsidy increases the lifetime wealth of those that are fortunate to be the recipients, and this brings about higher inequality despite the stable overall level of education.

Thus, the simulations suggest that human capital responds only to very high subsidy levels. When it does respond, however, the results are dramatic in terms of both output (because there are many more skilled workers) and inequality. Figure 10 shows the long-run spatial patterns that emerge under different levels of subsidy. Complete absence of subsidies ($\zeta = 0$) and 50% subsidy ($\zeta = 0.5$) result in not only the same level of human capital (with one-fourth of workers being educated) but also virtually identical patterns of class segregation (Fig. 10a–b). When subsidies are very high, covering over 90% of education costs ($\zeta = 0.91$), the share of skilled workers jumps to over half of the labor force. Moreover, the segregation pattern indicates the clustering of skilled workers spatially.

How does inequality rise when more workers become educated, despite the constant population? This seems to be counter-intuitive, at least at first glance. Figure 11 shows the widening wealth gap between skilled and unskilled workers as subsidies are increased. Under the no-interactions case (Fig. 11a), the original scenario ($\zeta = 0$) leads to a mean wealth (29.08) that roughly approximates the minimum and maximum values. A subsidy that effectively reduces education costs by 70% ($\zeta = 0.7$), on the other hand, results in a maximum wealth increase of 26 percentage points, while keeping the minimum at about the same level—creating a chasm between skilled and unskilled workers. The size of the skilled workers' group, however, remains the same (about one-fifth of the total labor force) as in the original scenario. When subsidy is increased further to cover 90% of costs, the number of skilled workers suddenly jumps to over 85% of the total (see Fig. 9). At the same time, the dramatic expansion in output pushes inequality higher due to the increasing isolation of the low-wage, unskilled workers.

Allowing direct interactions ($\zeta = 0.5$) brings about a larger gap at every nonzero level of subsidies (Fig. 11b under Setup 1 with $\theta = 0.5$ and TD $= 4$). It appears that a combination of neighbors' influences and high subsidies not only raises the top

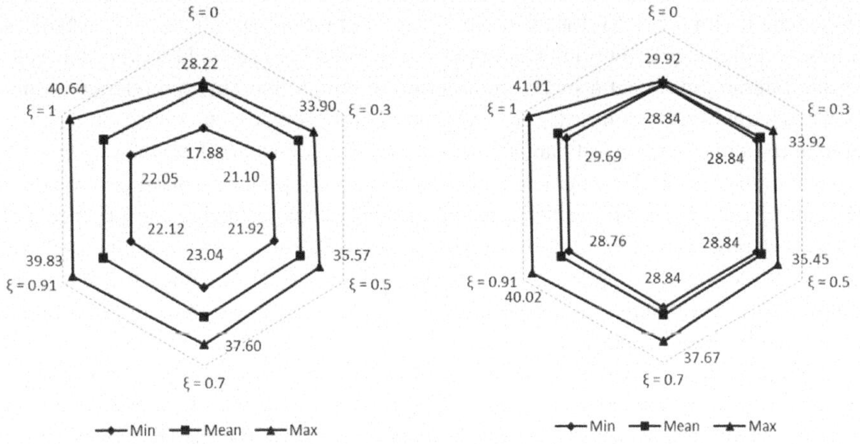

a. No direct interactions, $\theta = 0$ b. With interactions, $\theta = 0.5$

Fig. 11 Scenario A: targeted subsidies and the distribution of wealth, Setup 1

wealth as in the absence of interactions, but also *lower* those already at the bottom of the distribution.

6 Discussion and Conclusion

We have developed a model of wealth inequality in a spatial economy characterized by local interactions and initial heterogeneity. What insights can be drawn from the agent-based simulations? We found that human capital investments (and therefore output) are maximal when local interactions are moderate (i.e., $\theta = 0.5$), i.e., neither too weak nor too strong. Thus, equal considerations to own-economic calculations and to information about the neighbors' prospects produce the highest levels of education and therefore output. Stronger interactions ($\theta > 0.5$) lead to prospects in the poor (low-wealth) neighborhoods to opt out of school (though non-prospects in the wealthy neighborhoods remain in education), and this reduces overall human capital. Weaker interactions ($\theta < 0.5$), on the other hand, lead to non-prospects in wealthy neighborhoods opting out of school (though skilled-worker prospects in the poor neighborhoods remain in education), and this also reduces overall human capital.

The distributional impact of interactions exhibits a similar pattern, but of course with very different ramifications assuming that society values equity at all. Specifically, the strength of interaction ($\theta = 0.5$) that maximizes output also generates the most unequal distribution. Thus, the agent-based simulations suggest diametrically opposed directions for output and equity, should the strength of local interactions change either way. An exogenous shock leading to a more cohesive community (associated with $\theta > 0.5$), for example, lowers output but simultaneously produces a more equitable distribution. The upper tail of Fig. 5 suggests, however, that very strong interactions beget a severe contraction in output, but only a moderate reduction in inequality. By contrast, very weak interactions precipitate a mild contraction and, at the same time, significantly lower inequality. It appears that when agents are strongly interacting, the influence causing prospects in poor neighborhood to adapt (by opting out of school) overwhelms the influence that persuades non-prospects in wealthy neighborhood to adapt (by staying in school). Thus, very strong interactions produce "too much" adaptation among would-be prospects but not enough among non-prospects, which in turn is associated with a severe contraction of output coupled by relatively high inequality.

Agents in our model become susceptible to conversion (e.g., an initial non-prospect becomes educated, or vice versa) if a critical mass of *others* is present nearby. Adaptation occurs if, in addition, agents are allowed to locally interact ($\theta > 0$) and the resistance to neighborhood effects ($\pi_{\min}$) is not too high. Under a favorable set of conditions, the presence of prosperous neighbors converts a non-prospect into skilled workers because of the perceived advantages associated with being educated, and vice versa.

The size of the educated neighborhood determines the probability for non-prospects to convert. Conversion occurs if the number of educated neighbors reaches a critical level, and this makes offspring even more likely to be educated in subsequent periods. Thus, in a different context, when confronted with choices that are comparable in every aspect, social interactions might nevertheless compel people to consistently choose one over the other (Becker 1991).

The agent-based simulations reported in Sect. 5 suggest that the economic impact of educational policies depends upon the type of policy, history, and social interactions. A prime example is targeted subsidies that produce results in terms of higher human capital investments and output, but only if current subsidies are over 90% of the cost of education. Furthermore, the impact magnitude also depends upon the strength of local interactions. It is important to note, however, that in the real world not all credit-constrained persons (with few assets) are offered subsidies. This suggests one direction for future research.

The present agent-based model ignores other crucial aspects of reality, such as on-the-job training and temporary leave of absence. Training provides workers the opportunity to accumulate human capital while working on the job. A leave of absence from school, on the other hand, allows the credit-constrained Young to work on a temporary basis in order to accumulate savings, which can then be used to finance education upon their return to school. Removing these aspects of reality allows us to formulate the simplest possible model in the hope of providing the clearest insights into the interplay between inequality, human capital, and neighborhood effects.

In order to explore which policy is most effective in promoting education, the simulations must be constructed in such a way that the efficacy of different policy instruments may be compared. Comparability can be accomplished by assuming a fixed government budget used to fund a single policy implementation. This enables us to compare a policy such as reduction of education costs with that of targeted subsidy, as long as the total expenditures for both are *ex ante* identical.[10] Preliminary results (per dollar of public expenditure) suggest that cost reduction results in higher human capital investment *and* lower inequality than subsidies. Thus, reducing educational costs across the board appears effective not only in motivating people to pursue education, but also in creating a more egalitarian distribution of wealth. This preliminary result, as well as other policy implications garnered from this study, is recommended for further investigation in future studies.

References

Andergassen, R., & Nardini, F. (2007). Educational choice, endogenous inequality and economic development. *Journal of Macroeconomics, 29*, 940–958.

Becker, G. S. (1964). *Human capital: A theoretical and empirical analysis, with special reference to education*. New York: NBER.

Becker, G. S. (1991). A note on restaurant pricing and other examples of social influence on price. *Journal of Political Economy, 99*(5), 1109–1116.

Becker, G. S., Kominers, S. D., Murphy, K. M., & Spenkuch, J. L. (2018). A theory of intergenerational mobility. *Journal of Political Economy, 126*(S1), S7–S25.

Becker, G. S., & Tomes, N. (1979). An equilibrium theory of the distribution of income and intergenerational mobility. *Journal of Political Economy, 87*(6), 1153–1189.

Bénabou, R. (1996). Equity and efficiency in human capital investment: The local connection. *Review of Economic Studies, 63*(2), 237–264.

Berube, A. (2014). *All cities are not created unequal*. The Brookings Institution. https://www.brookings.edu/research/papers/2014/02/cities-unequal-berube/. Accessed February 22, 2014.

Bowles, S., Loury, G. C., & Sethi, R. (2014). Group inequality. *Journal of the European Economic Association, 12*(1), 129–152.

Card, D. (1999). The causal effect of education on earnings. In O. Ashenfelter & D. Card (Eds.), *Handbook of labor economics* (1st ed., Vol. 3A). North Holland.

Chetty, R., Hendren, N., Kline, P., & Saez, E. (2014). Where is the land of opportunity? The geography of intergenerational mobility in the United States. *The Quarterly Journal of Economics, 129*(4), 1553–1623.

Chiang, A. C. (1984). *Fundamental methods of mathematical economics* (3rd ed.). New York: McGraw-Hill.

Clark, G. (2014). *The son also rises: Surnames and the history of social mobility*. Princeton: Princeton University Press.

Corcoran, M., Gordon, R., Laren, D., & Solon, G. (1990). Effects of family and community background on economic status. *The American Economic Review, 80*(2), 362–366.

Das, M. (2007). Persistent inequality: An explanation based on limited parental altruism. *Journal of Development Economics, 84*, 251–270.

Durlauf, S. N. (1996). Neighborhood feedbacks, endogenous stratification, and income inequality. In W. Barnett, G. Gandolfo, & C. Hillinger (Eds.), *Dynamic Disequilibrium Modelling: 9th International Symposium on Economic Theory and Econometrics*. Cambridge: Cambridge University Press.

Fan, C. S. (2006). Do the rich save more? A new view based on intergenerational transfers. *Southern Economic Journal, 73*(2), 362–373.

[10]The total expenditures would be *ex-post* identical only in the trivial case where every agent invests in human capital because education has been rendered affordable to *all* agents.

Felipe, J., & Adams, F. G. (2005). "A Theory of Production" The estimation of the Cobb-Douglas function: A retrospective view. *Eastern Economic Journal, 31*(3), 427–445.
Galor, O., & Zeira, J. (1993). Income distribution and macroeconomics. *Review of Economic Studies, 60*(1), 35–52.
Glaeser, E. L., & Mare, D. C. (2001). Cities and skills. *Journal of Labor Economics, 19*(2), 316–342.
Glaeser, E. L., & Vigdor, J. L. (2012). *The end of segregated century: Racial separation in America's neighborhood, 1890–2010*. Civic Report 66, Jan 2012. Manhattan Institute for Policy Research.
Gordon, R. J. (2013). Revisiting U.S. productivity growth over the past century with a view of the future. In D. S. P. Rao & B. V. Ark (Eds.), *World economic performance: Past, present and future* (p. 317). Cheltenham: Edward Elgar.
Heckman, J. J., & Krueger, A. B. (2003). *Inequality in America: What role for human capital policies*. Cambridge: MIT Press.
Hicks, J. R. (1932). *The theory of wages*. London: Macmillan.
Howell, D. R. (2013). *The great laissez-faire experiment: American inequality and growth from an international perspective*. Center for American Progress. https://www.americanprogress.org/. Last visited February 2, 2014.
IMF. (2019). *World economic outlook: Global manufacturing downturn, rising trade barriers*. Washington, DC, October. https://www.imf.org/. Accessed December 12, 2019.
Jargowsky, P. A. (1996). Take the money and run: Economic segregation in U.S. Metropolitan Areas. *American Sociological Review, 61*(6), 984–998.
Kiyotaki, N., & Wright, R. (1998). On money as a medium of exchange. *Journal of Political Economy, 97*(4), 927–954.
Kreps, D. M. (2012). *Microeconomic foundations I: Choice and competitive markets*. Princeton: Princeton University Press.
Krueger, D., Perri, F., Pistaferri, L., & Vilante, G. L. (2010). Cross-sectional facts for macroeconomists. *Review of Economic Dynamics, 13*(1), 1–14.
Kuznets, S. (1955). Economic growth and income inequality. *The American Economic Review, 45*(1), 1–28.
Mansury, Y., & Gulyas, L. (2007). The emergence of Zipf's Law in a system of cities: An agent-based simulation approach. *Journal of Economic Dynamics and Control, 31*, 2438–2460.
Mincer, J. (1958). Investment in human capital and personal income distribution. *Journal of Political Economy, 66*(4), 281–302.
Mookherjee, D., Napel, S., & Ray, D. (2010a). Aspiration, segregation and occupational choice. *Journal of the European Economic Association, 8*(1), 139–168.
Mookherjee, D., Napel, S., & Ray, D. (2010b). Social interactions and segregation in skill accumulation. *Journal of the European Economic Association, 8*(2–3), 388–400.
Moretti, E. (2004). Estimating the social return to higher education: Evidence from longitudinal and repeated cross-sectional data. *Journal of Econometrics, 121*, 175–212.
Nakajima, T., & Nakamura, H. (2009). The price of education and inequality. *Economic Letters, 105*, 183–185.
Nomura, T. (2007). Contribution of education and educational equality to economic growth. *Applied Economics Letters, 14*, 627–630.
North, M. J., Collier, N. T., & Vos, J. R. (2006). Experiences creating three implementations of the repast agent modeling toolkit. *ACM Transactions on Modeling and Computer Simulation, 16*(1), 1–25.
OECD. (2013). *OECD economic outlook* (Vol. 2013/2). Paris: OECD Publishing. https://www.oecd.org. Accessed February 2, 2014.
OECD. (2015). *In it together: Why less inequality benefits all*. Paris: OECD Publishing. https://doi.org/10.1787/9789264235120-en. Accessed December 15, 2019.
Parker, D., & Filatova, T. (2008). A theoretical design for a bilateral agent-based land market with heterogeneous economic agents. *Computers, Environment and Urban Systems, 32*, 454–463.
Piketty, T. (2011). On the long-run evolution of inheritance: France 1820–2050. *The Quarterly Journal of Economics, 126*(3), 1071–1131.

Reardon, S. F., & Bischoff, K. (2011). Income inequality and income segregation. *American Journal of Sociology, 116*(4), 1092–1153.
Reardon, S. F., Bischoff, K., Owens, A., & Townsend, J. B. (2018). Has income segregation really increased? Bias and bias correction in sample-based segregation estimates. *Demography, 55*(6), 2129–2160.
Road. (2004). *Repast home page*. Repast organization for architecture and design, Chicago IL. https://repast.sourceforge.net/. Accessed January 13, 2014.
Roca, J. D. L., & Puga, D. (2017). Learning by working in big cities. *Review of Economic Studies, 84*(1), 106–142.
Rosenthal, S. S., & Strange, W. C. (2001). The determinants of agglomeration. *Journal of Urban Economics, 50*, 191–229.
Saez, E., & Zucman, G. (2016). Wealth inequality in the United States since 1913: Evidence from capitalized income tax data. *The Quarterly Journal of Economics, 131*(2), 519–578.
Sampson, R. J. (2012). *Great American city: Chicago and the enduring neighborhood effect*. Chicago: Chicago University Press.
Samuelson, P. A. (1958). An exact consumption loan model of interest, with or without the social contrivance of money. *Journal of Political Economy, 66*(6), 467–482.
Sen, A. (1997). *On economic inequality*. Oxford: Oxford University Press.
Schelling, T. C. (1971). Dynamic models of segregation. *Journal of Mathematical Sociology, 1*, 143–186.
Solon, G. (2002). Cross-country differences in intergenerational earnings mobility. *Journal of Economic Perspective, 16*(3), 59–66.
Solon, G. (2018). What do we know so far about multigenerational mobility? *The Economic Journal, 128*(612), F340–F352.
Stewart, E. B., Stewart, E. A., & Simons, R. L. (2007). The effect of neighborhood context on the college aspirations of African American adolescents. *American Educational Research Journal, 44*(4), 896–919.
Tesfatsion, L. (2005). *Agent-based computational economics: A constructive approach to economic theory*. https://www2.econ.iastate.edu/tesfatsi/ace.htm. Accessed January 8, 2014.
Usher, A., & Medow, J. (2010). *Global higher education rankings 2010: Affordability and accessibility in comparative perspective*. https://higheredstrategy.com/wp-content/uploads/2011/09/GHER2010_FINAL.pdf. Accessed February 13, 2014.
Walker, K. (2018). Locating neighbourhood diversity in the American metropolis. *Urban Studies, 55*(1), 116–132.
Wilson, W. J. (1987). *The truly disadvantaged*. Chicago: Chicago University Press.

Regional Economic Restructuring, Convergence, Growth, and the Environment

Regional Temporal Disaggregation on Economic Series with Macroeconomic Balance: An Entropy Econometrics-Based Model

Blanca Moreno Cuartas, Esteban Fernández Vázquez, and Geoffrey J. D. Hewings

1 Introduction

The analysis and forecasting of regional business cycles have received increasing attention in recent years due to their importance in the design of suitable policy measures. High-frequency economics series are required for business cycle analysis, but at regional level, the information is often not available at the desired frequency. For example, regional information about gross domestic product (GDP) or gross added value (GAV) is usually available only annually. The estimation of quarterly regional economic series, therefore, would be helpful for regional econometric modeling and forecasting.

Usually, the regional statistical agencies estimate individually its quarterly GDP or GAV series consistently with the annual regional aggregate. There is a wide variety of temporal disaggregation univariate methods that can be employed for disaggregating regional annual accounts, and the choice of a specific method depends on the basic information available, the mathematical procedure and the required assumptions.[1] If

[1] There is an extensive literature covering a review of other methodologies, as the document produced by European Commission (2005) or more recently by Pavía-Miralles (2010).

B. M. Cuartas (✉) · E. F. Vázquez
REGIOlab, Laboratorio de Análisis Económico Regional, Department of Applied Economics, Faculty of Economics and Business, University of Oviedo, Avda. del Cristo, s/n, 33006 Oviedo, Spain
e-mail: morenob@uniovi.es

E. F. Vázquez
e-mail: evazquez@uniovi.es

G. J. D. Hewings
Regional Economics Applications Laboratory, University of Illinois at Urbana-Champaign, 607 S. Matthew, Urbana, IL 61801-367, USA
e-mail: hewings@illinois.edu

J. Thill (ed.), *Innovations in Urban and Regional Systems*,
https://doi.org/10.1007/978-3-030-43694-0_11

quarterly national accounts (QNA) are available, a simultaneous estimation of quarterly regional series is possible, being the estimates subjected to the contemporary constraint imposed by the quarterly national aggregate.

In this chapter, we deal with the problem of deriving quarterly regional accounts (QRA) estimates for R regions simultaneously that are consistent with annual regional accounts (ARA). The method proposed allows for including quarterly national aggregates without additional assumptions by using an estimator based on entropy econometrics (EE). This approach has its roots in information theory and builds on the entropy–information measure of Shannon (1948), the classical entropy principle of Jaynes (1957a, b), which was developed to recover information from underdetermined models, and the generalized maximum entropy theory of Golan et al. (1996), which have been applied to various fields of economics and finance (e.g., Berger 1985; Avellaneda 1998) and econometrics (e.g., Maasoumi 1993; Golan et al. 1996; Ullah 1996, 2002; Moreno and López 2013).[2]

One of the main advantages of applying EE in this context is that it allows for estimating regional quarterly series without making distributional assumptions about the residuals, oppositely to the temporal disaggregation techniques commonly applied: Temporal disaggregation univariate methods have been developed assuming well-defined ARIMA structures on the error. The estimation of the parameters involved in ARIMA models requires a sufficient length of available data, which is not always possible.

Another crucial difference is that traditional disaggregation techniques are based on regression models that are highly sensitive to the choice of covariates (indicators), their number and the way of combining them. The estimator proposed in this paper allows to recover regional quarterly series for a national economic system by taking advantage of limited information and using only a covariate (indicator) related to the regional quarterly series.

The structure of the chapter is as follows. In the next section, we provide a brief review of the temporal disaggregation techniques and an outline of the Chow-Lin-based procedure. Section 3 presents the EE approach proposed for simultaneously estimating the quarterly time series for all regions of a country. In this section, we deal with the problem to keep the consistency between the quarterly regional estimates and the quarterly national and annual regional aggregates. Sections 4 and 5 present the results of a numerical experiment based on real data. In our simulation exercise, artificial regional quarterly series are generated through linear combinations of an observed regressor series plus an error term. We use these series to compare the estimates by the EE approach with those obtained by applying the Chow-Lin method. We analyze the robustness of the estimates when the common assumption of an AR(1) structure for the error is violated and the forecasting performance of the competing methods. Finally, concluding remarks and some possible directions for future research are presented in Sect. 6.

---

[2]EE was defined by Golan (2002) as "*a sub-discipline of processing information from limited and noisy data with minimal* a priori*information on the data-generating process*".

2 A Brief Review on Temporal Disaggregation Univariate Methods

The objective of any temporal disaggregation method is to estimate the high-frequency quarterly observations from an observed low-frequency time series. In order to estimate its quarterly series, different temporal disaggregation methods can be used. One of the streams of this literature bases the analysis on a purely univariate approach when no covariates are necessary. One of the first methods proposed was by Lisman and Sandee (1964), which is a purely numerical technique for constructing synthetic quarterly data based on past trends of annual data. Zani's (1970) and Denton's (1971) methods follow this same idea and extend it to the case of using a related indicator. Stram and Wei (1986) and Wei and Stram (1990) methods rely on the ARIMA representation of the series to be disaggregated. The method of Boot et al. (1967) proposed a least-squares-based technique for constructing synthetic quarterly data based on past trends in annual data.[3]

The second kind of methods is regression-based and estimates a linear relationship between the series of interest and related variables to get the best linear unbiased estimate (BLUE). The most commonly applied regression-based method was proposed by Chow and Lin (1971). These authors derived a general formulation of the disaggregation problem in the context of a linear regression model involving the observed annual series and related quarterly data. A major drawback of the Chow-Lin procedure is that the covariance matrix of the errors is unknown, being necessary some assumptions about the error generation process. Chow and Lin (1971) suppose autocorrelated errors generated by a first-order autoregressive AR(1) process.[4]

The basics of the Chow-Lin disaggregation procedure is explained below. Suppose that we have a variable available annually over n years. Let $\boldsymbol{y}$ be a $(4n \times 1)$ vector of quarterly GAV to be estimated, and assume that $\boldsymbol{y}$ can be predicted using a multiple linear regression:

$$\boldsymbol{y} = \boldsymbol{X}\boldsymbol{\beta} + \boldsymbol{u} \tag{1}$$

where $\boldsymbol{X}$ is a $(4n \times k)$ matrix of k predictor variables which are observed quarterly and $\boldsymbol{u}$ is a $(4n \times 1)$ random vector with zero mean and variance–covariance matrix $\boldsymbol{V}$ $(4n \times 4n)$.

$\boldsymbol{C}$ is the $(n \times 4n)$ aggregation matrix that transforms the $4n$ quarterly observations into n annual observations which is defined as:

[3] Cholette (1984), Hillmer and Trabelsi (1978), Trabelsi and Hillmer (1990), Guerrero (1990) and Guerrero and Martínez (1995) followed similar strategies.

[4] Most of the methods present generalizations of Chow-Lin solutions relying on assumptions about the covariance matrix of the errors. The papers by Fernández (1981) and Litterman (1983) are extensions of this method, assuming autocorrelated errors generated by a random walk and a Markov random walk-ARIMA process, respectively.

$$\boldsymbol{C} = \begin{bmatrix} 1 & 1 & 1 & 1 & 0 & 0 & 0 & 0 & \dots\dots\dots & 0 \\ 0 & 0 & 0 & 0 & 1 & 1 & 1 & 1 & \dots\dots\dots & 0 \\ & & & & & & \ddots & & & \\ 0 & \dots\dots & \dots\dots & \dots\dots & & 1 & 1 & 1 & & 1 \end{bmatrix} \tag{2}$$

Then, the Chow-Lin quarterly relationship can be converted to a regression of annual aggregates:

$$\boldsymbol{y}_a = \boldsymbol{C}\boldsymbol{y} = \boldsymbol{C}\boldsymbol{X}\boldsymbol{\beta} + \boldsymbol{C}\boldsymbol{u} = \boldsymbol{X}_a\boldsymbol{\beta} + \boldsymbol{u}_a \tag{3}$$

a being a subscript which denotes annual figures. Estimating by GLS, Eq. (3) leads to a BLUE solution for the coefficient vector:

$$\hat{\boldsymbol{\beta}}_{GLS} = \left(\boldsymbol{X}_a^{'}\boldsymbol{V}_a^{-1}\boldsymbol{X}_a\right)^{-1}\left(\boldsymbol{X}_a^{'}\boldsymbol{V}_a^{-1}\boldsymbol{y}_a\right) \tag{4}$$

where $\boldsymbol{V}_a = \left(\boldsymbol{C}\boldsymbol{V}\boldsymbol{C}'\right)$ and $\hat{\boldsymbol{u}}_a = \boldsymbol{y}_a - \boldsymbol{X}_a\hat{\boldsymbol{\beta}}_{GLS}$. In a second stage, the quarterly residuals are calculated by disaggregating the annual residuals $\boldsymbol{u}_a$ as:

$$\boldsymbol{u} = \boldsymbol{V}\boldsymbol{C}'(\boldsymbol{V}_a)^{-1}\boldsymbol{u}_a = \boldsymbol{V}\boldsymbol{C}'(\boldsymbol{V}_a)^{-1}\left(\boldsymbol{y}_a - \boldsymbol{X}_a\hat{\boldsymbol{\beta}}_{GLS}\right) \tag{5}$$

Given (4), (5), the Chow-Lin linear unbiased predictor of y is derived as:

$$\hat{\boldsymbol{y}} = \boldsymbol{X}\hat{\boldsymbol{\beta}}_{GLS} + \boldsymbol{V}\boldsymbol{C}'\left(\boldsymbol{C}\boldsymbol{V}\boldsymbol{C}'\right)^{-1}\hat{\boldsymbol{u}}_a. \tag{6}$$

The first term on the RHS of Eq. (6) gives the predicted quarterly $\boldsymbol{y}$ based on observed quarterly $\boldsymbol{X}$ and the estimated $\widehat{\boldsymbol{\beta}}_{GLS}$ from annual totals, and the second term allocates annual residuals $\widehat{\boldsymbol{u}}_a$ to the four quarters of the year.

Since $\boldsymbol{V}$ is unknown, Chow-Lin considered that $\boldsymbol{u}$ follows a stationary AR(1):

$$u_t = \rho u_{t-1} + e_t;\ |\rho| < 1 \tag{7}$$

where e_t is a white noise process with $E(e_t) = 0$ and $E\left(e_t^2\right) = \sigma^2$. Under this assumption, $\boldsymbol{V}$ takes the form:

$$\boldsymbol{V} = \frac{\sigma^2}{1-\rho}\begin{bmatrix} 1 & \rho & \rho^2 & \dots\dots & \rho^{4n-1} \\ \rho & 1 & \rho & \dots\dots & \rho^{4n-2} \\ \dots & \dots\dots & \ddots & \dots & \dots \\ \rho^{4n-1} & \dots\dots & \dots\dots & & 1 \end{bmatrix} \tag{8}$$

and the procedure needs an estimate of ρ. If the autoregressive parameter ρ is equal to zero, then $\boldsymbol{V} = \sigma^2\boldsymbol{I}_{4n}$ and $\hat{\boldsymbol{\beta}}$ reduces to the OLS estimator $\left(\boldsymbol{X}_a^{'}\boldsymbol{X}_a\right)^{-1}\left(\boldsymbol{X}_a^{'}\boldsymbol{y}_a\right)$, and the second term on the RHS of Eq. (6) amounts to allocating one quarter of the annual residual to each quarter of the year.

Generalizations of the Chow- and Lin (1971)-based methods for disaggregating multiple series when contemporaneous aggregations are also available were developed by Rossi (1982), Di Fonzo (1990), Cabrer and Pavía (1999) or Pavía-Miralles and Cabrer-Borrás (2007). In these cases, the estimated regional quarterly series must be coherent with the temporal constraint imposed in (3) but also with the contemporaneous constraint imposed by the $(4n \times 1)$ national totals: $\boldsymbol{y}^{nat} = \boldsymbol{Y}\boldsymbol{e}$, being $\boldsymbol{e}$ a $(R \times 1)$ vector of ones.

3 An Entropy Econometrics Approach for Temporal Disaggregation

In this section, we deal with the temporal disaggregation problem basing our solution on the cross-entropy (CE) procedure. The problem of estimating quarterly regional data can be described in the same terms as the matrix-balancing problem depicted in Golan (2006, p. 105), where the goal is to fill the (unknown) cells of a matrix using the information that is contained in the aggregate data of the row and column sums. Consider a country that is divided into $i = 1, \ldots, R$ regions. Keeping the same notation as in the previous section, our target is to estimate matrix $\boldsymbol{Y}$ with dimension $(4n \times R)$, where a typical element contains the variable of interest for quarter and region.

The interior cells of this matrix are unknown but some aggregate information is observable. In particular, matrix $\boldsymbol{Y}_a$ with dimension $(n \times R)$ that contains the annual regional figures is available. Note that $\boldsymbol{Y}_a = \boldsymbol{C}\boldsymbol{Y}$, being $\boldsymbol{C}$ the same aggregation matrix $(n \times 4n)$ as in the previous section. On the other hand, matrix $\boldsymbol{X}$ with the predictors is reduced now to the case where it only contains one single indicator. This means that we have a matrix $\boldsymbol{X}$ with dimension $(4n \times R)$ containing a regional variable which is observable quarterly and is considered a predictor of matrix $\boldsymbol{Y}$.

Additionally, we assume that some extra information is also available in the form of quarterly national aggregates. This assumption is not very unrealistic, since it is relatively common for statistical agencies to provide official quarterly national accounts (QNA). The estimated regional quarterly series should be coherent with the temporal constraint imposed by the equality $\boldsymbol{Y}_a = \boldsymbol{C}\boldsymbol{Y}$ but also with the contemporaneous constraint imposed by the $(4n \times 1)$ national totals $\boldsymbol{y}^{nat}$. Consequently, the following equalities are fulfilled by the y_{ti} elements of matrix $\boldsymbol{Y}$:

$$\boldsymbol{Y}_a = \boldsymbol{C}\boldsymbol{Y} \tag{9}$$

$$\boldsymbol{y}^{nat} = \boldsymbol{Y}\boldsymbol{e} \tag{10}$$

These two sets of equations reflect all we know about the elements of matrix $\boldsymbol{Y}$. The $(n \times R)$ matrix $\mathbf{Y_a}$ and the $(4n \times 1)$ vector $\boldsymbol{y}^{nat}$ contain, respectively, the annual regional and quarterly national information on the variable of interest. In such a situation, if we have available a matrix $\boldsymbol{X}$, which reflects our initial assumptions about $\boldsymbol{Y}$, the cross-entropy (CE) principle can be applied to recover the unknown y_{ti} elements. In other words, we aim at transforming an a priori matrix $\boldsymbol{X}$ into a posterior matrix $\boldsymbol{Y}$ that is consistent with the observable data $\boldsymbol{Y}_a$ and $\boldsymbol{y}^{nat}$. The solution to this type of problems is obtained by minimizing a divergence measure with matrix $\boldsymbol{X}$ subjected to the set of constraints (9) and (10), which can be written in the following terms:

$$\underset{\boldsymbol{Y}}{\text{Min}}\, D(\boldsymbol{Y}||\boldsymbol{X}) = \sum_{t=1}^{4n}\sum_{i=1}^{R} y_{ti} \ln\left(\frac{y_{ti}}{x_{ti}}\right) \tag{11}$$

The divergence measure $D(\boldsymbol{Y}\|\boldsymbol{X})$ is the Kullback and Leibler (1951) entropy divergence between the posterior and prior distributions. This idea is here extended and generalized for the cases where we allow for the presence of an error term in each cell of the target matrix, passing from a CE adjustment to a generalized CE estimation (GCE). In such a case, we assume that the y_{ti} elements are given from two sources: a signal (p_{ti}) that keeps the resemblance with the indicators x_{ti} plus a noise term (ε_{ti}). The noise components can be either positive or negative, and they are included in order to account for potential spatial heterogeneity and our uncertainty about the relationship between our variable of interest and the predictor. The uncertainty about the realizations of the errors is included in the problem by treating each element ε_{ti} as a discrete random variable with $L \geq 2$ possible outcomes contained in a vector support $\boldsymbol{v}' = \{v_1, \ldots, v_L\}$.

This vector can be used to assume certain characteristics of symmetry and kurtosis about error distribution. Usually, the vector support is uniformly and symmetrically chosen around zero with equally spaced distance discrete points. Therefore, if $L = 3$ the vector is $v' = \{-v, 0, v\}$ and if $L = 5$ the corresponding vector is $v' = \{-v, -v/2, 0, v/2, v\}$.

Under these conditions, each ε_{ti} can be defined as:

$$\varepsilon_{ti} = \sum_{l=1}^{L} w_{til} v_l;\ i = 1, \ldots, R;\ t = 1, \ldots, 4n \tag{12}$$

where w_{til} is the unknown probability of the outcome v_l for the cell *ti*. Now, the y_{ti} elements can be written as:

$$y_{ti} = p_{ti} + \varepsilon_{ti} = p_{ti} + \sum_{l=1}^{L} w_{til} v_l;\ i = 1, \ldots, R;\ t = 1, \ldots, 4n \tag{13}$$

This generalization of the estimation problem can be posed as a minimization program like:

$$\underset{\mathbf{P},\mathbf{W}}{\text{Min}}\, D(\boldsymbol{P}, \boldsymbol{W}\|\boldsymbol{X}, \boldsymbol{W}^0) = \sum_{t=1}^{4n}\sum_{i=1}^{R} p_{ti}\ln\left(\frac{p_{ti}}{x_{ti}}\right) + \sum_{l=1}^{L}\sum_{t=1}^{4n}\sum_{i=1}^{R} w_{til}\ln\left(\frac{w_{til}}{w_{til}^0}\right) \tag{14}$$

Subjected to:

$$\boldsymbol{Y}_a = \boldsymbol{C}[\boldsymbol{P} + \boldsymbol{\epsilon}] \tag{15}$$

$$\boldsymbol{y}^{nat} = [\boldsymbol{P} + \boldsymbol{\epsilon}]\boldsymbol{e} \tag{16}$$

where $\boldsymbol{\epsilon}$ stands for the $(4n \times R)$ matrix containing the ε_{ti} elements and $\boldsymbol{W}^0$ for the $L(4n \times R)$ a priori probability distributions of the ε_{ti} disturbances.

4 Evaluation by Numerical Simulation

In this section, we try to find some empirical evidence on the comparative performance of the GCE by means of a numerical simulation. Despite the huge variety of techniques to disaggregate annual into quarterly economic series, our benchmark will be the Chow-Lin procedure detailed previously since it is the most widely applied. Even when the numerical exercise carried out is mainly experimental, we try to keep some resemblance with a realistic case. In particular, given that normally the variable of interest $\boldsymbol{Y}$ is some measure of regional output—such as GDP or GAV—the context of our simulation will be the estimation of quarterly indicators of output for the set of 17 Spanish NUTS II regions.

The point of departure is the unknown elements of the $(4n \times R)$ target matrix $\boldsymbol{Y}$, which are generated by the linear equation:

$$\boldsymbol{Y} = \alpha + \beta\boldsymbol{X} + \boldsymbol{u} \tag{17}$$

Matrix $\boldsymbol{X}$ is a $(4n \times R)$ matrix that contains one single predictor for $\boldsymbol{Y}$, α and β are two scalars, and $\boldsymbol{u}$ is a $(4n \times R)$ random noise term. Note that the data in $\boldsymbol{X}$ should be some regional indicator observable quarterly. We opt for including in $\boldsymbol{X}$ regional labor figures from the Spanish Labor Force Survey (EPA), which are published quarterly. We take information of this indicator for the period 1977–2006, which leaves us with a sample of 30 years and 120 quarters. From the original regional labor figures, we extract the adjusted trend-cycle component of the series by applying the programs TRAMO and SEATS.[5]

[5]These programs are used by the Spanish Statistical Agency (INE) in the Spanish Quarterly National Accounts. See Gomez and Maravall (1997a, b) for details. Removing seasonal variations by this

The structure of the noise term plays an important role in the disaggregation problem. In the Chow-Lin procedure, it is usually assumed that it follows an AR(1) that we include here by the equation:

$$u_{ti} = \rho u_{t-1i} + e_{ti};\, e_{ti} \sim N(0, \sigma) \tag{18}$$

Parameter ρ is the autocorrelation coefficient, and σ indicates the variance of the white noise e_{ti}. Arbitrary values $\{1, 0.2\}$ for the parameters $\{\alpha, \beta\}$ are set in the linear Eq. (17), whereas originally the parameters ρ and σ have been set to 0.1 and 5, respectively. Additionally, we test the sensitivity of the results by considering alternative scenarios that are generated by values of $\rho = 0.5$ and $\sigma = 10$.

In this context, matrix $\boldsymbol{Y}$ is generated and then aggregated into matrix $\boldsymbol{Y}_a$ and vector $\boldsymbol{y}^{nat}$, which jointly with $\boldsymbol{X}$ are assumed as the only observable information. This information is processed by a GCE program as the one depicted in Eqs. (14)–(16) and compared with a standard Chow-Lin procedure. The aim of both techniques is to estimate the matrix $\boldsymbol{Y}$ for the 120 ($4n$) quarters along the 1977–2006 period and the set of 17 (R) NUTS II regions in Spain. Comparison between the performances of the two approaches is made basing on the deviation between estimates and actual values of $\boldsymbol{Y}$ basing on the percentage absolute deviation across all time periods and regions. The implementation of the GCE technique requires specifying support vectors for the ε_{ti} terms. Even when several strategies can be adopted to set these values, here we opted for setting the same vector $\boldsymbol{v}' = \{v_1, \ldots, v_L\}$ for all the ε_{ti} and considering $L = 3$ symmetric points, being the center on zero and making the bound proportional to the mean of the dependent variable. More specifically, for all the cases the vectors are defined as $\boldsymbol{v}' = \{-0.1\bar{y},\ 0,\ 0.1\bar{y}\}$, where $\bar{y}$ denotes the arithmetic mean of the variable generated by Eq. (17) across all regions and time periods. The computation has been implemented with GAMS software (General Algebraic Modeling System; see Brooke et al. 1998). In order to evaluate the deviations between our estimates and the generated true values, the mean absolute error (MAE, in %) and the mean squared error (MSE) for the quarterly regional estimations of the different models considered are computed. The results are summarized in Table 1.

The first row in Table 1 contains the statistics for the errors corresponding to a Chow-Lin (CL) estimation procedure, estimating an equation like (3) by GLS and assuming a perfect knowledge of the autocorrelation parameter ρ. The middle row reports the results to a GCE but without using the constraint in Eq. (16). Removing this restriction allows for a direct comparison with the Chow-Lin results, since they use the same amount of information. If the mean percentage of absolute error (MAE) is taken as reference, the GCE adjustment in this context yields deviations smaller on average than the Chow-Lin methods. The GCE including the aggregate of Eq. (16) is presented in the bottom row. The comparison of the GCE with and without (16) allows for evaluating the effect of this piece of information on the estimates, suggesting the results obtained a very limited improvement of the temporal disaggregation results.

---

or a different method is the usual procedure when some indicator is used in time disaggregation problems, such as in the Chow-Lin procedure.

Table 1 Deviation figures: mean absolute error (MAE, in %) and mean squared error (MSE)

	$\rho = 0.1$				$\rho = 0.5$			
	$\sigma = 5$		$\sigma = 10$		$\sigma = 5$		$\sigma = 10$	
	MAE	MSE	MAE	MSE	MAE	MSE	MAE	MSE
CL	5.24	23.51	12.13	94.03	6.13	29.64	14.40	117.90
GCE (not including the national constraint)	4.61	18.10	10.53	72.14	4.35	16.55	10.48	65.95
GCE (including the national constraint)	4.52	17.34	10.34	69.15	4.27	15.87	10.31	63.31

AR(1) structure for *u*

The general conclusions on the comparative performance of the temporal disaggregation methods are not affected by the choice of a squared error measure like the MSE.

Note that the numerical simulation described so far perfectly fits with the required situation to apply a Chow-Lin-type procedure: The temporal structure of the errors is AR(1), and the GLS estimation manages to capture without error this structure. In order to extend the experiment and check the consequences of violating these assumptions, we have modified Eq. (18) to consider a more complex structure in the residuals. Specifically, a second simulation exercise is carried out by transforming the original AR(1) process into an AR(2) structure as:

$$u_{ti} = \rho u_{t-1i} + \frac{\rho}{2} u_{t-2i} + e_{ti};\ e_{ti} \sim N(0, \sigma) \tag{19}$$

A summary of the deviation figures for this new simulation is reported in Table 2.

Not surprisingly, the Chow-Lin method in this new scenario yields worse results than in the previous case. The GCE estimates in both cases with and without the national constraint seem relatively insensitive to this modification, and the deviations are more or less the same as in the previous scenario. The corollary of these results would be that a GCE adjustment is a more conservative approach to temporal disaggregation of regional series when the assumptions on the structure of the residuals are not realistic. The lack of strong distributional assumptions on GCE limits the consequences of dealing with disturbances with structures different from the hypotheses of more traditional (e.g., Chow-Lin type) disaggregation techniques.

Table 2 Deviation figures: mean absolute error (MAE, in %) and mean squared error (MSE)

	$\rho = 0.1$				$\rho = 0.5$			
	$\sigma = 5$		$\sigma = 10$		$\sigma = 5$		$\sigma = 10$	
	MAE	MSE	MAE	MSE	MAE	MSE	MAE	MSE
CL	5.23	23.48	12.62	93.89	8.36	45.06	18.06	179.11
GCE (not including the national constraint)	4.57	17.78	10.54	70.86	4.11	14.50	9.09	57.77
GCE (including the national constraint)	4.47	17.05	10.35	68.01	4.03	13.91	8.95	55.47

AR(2) structure for u

5 Forecasting Performance Evaluation

The previous comparison between actual values and the estimates of the disaggregation techniques provides indications about their respective performance. However, it should be kept in mind that one of the most important objectives of these techniques is to produce high-frequency data to be used as input in forecasting exercises at regional scale. This section extends the simulation experiment conducted previously in order to test the forecasting performance of the proposed GCE method.

The aim of this section will be to forecast the regional quarterly values of the dependent variable for the four quarters that immediately follows the sample period, i.e., the 4 quarters of year 2007. Given that we are simulating a situation where future values will be predicted, we assume that no information on the indicator $\boldsymbol{X}$ is observable and the forecast must be based only on a univariate (ARIMA-type) analysis of past values of $\boldsymbol{Y}$.

The estimates of $\boldsymbol{Y}$ for the 1977–2006 sample period by the competing disaggregation techniques are modeled following the usual Box–Jenkins methodology. For each case—region and estimation technique—the most suitable ARIMA model is identified and taken as the basis for predicting the future values along the 4 quarters of 2007. Then, these predictions are compared with the actual (but unobservable) values generated by Eqs. (17) and (18) in this period for different short-term time horizons ($h = 1, \ldots 4$).

Table 3 shows the weighted means of the relative absolute errors in percentage for each method when the residuals follow an AR(1) structure, whereas Table 4 reports these same measures when an AR(2) process is imposed for $\boldsymbol{u}$.

As expected, when increasing the forecasting short-term time horizon the forecasting accuracy becomes generally smaller. The results suggest basically the same message as in the previous section: (i) The GCE adjustments yield forecasting errors which in general are lower than those obtained by the Chow-Lin procedure, (ii) the

Table 3 Mean relative absolute error in percentage

	Forecast horizon			
	$h = 1$	$h = 2$	$h = 3$	$h = 4$
CL	4.42	3.74	5.48	5.06
GCE (national constraint not included)	4.51	3.31	4.83	4.40
GCE (national constraint included)	4.46	3.29	4.84	4.50

AR(1) structure for u

Table 4 Mean relative absolute error in percentage

	Forecast horizon			
	$h = 1$	$h = 2$	$h = 3$	$h = 4$
CL	4.77	4.93	7.10	7.00
GCE (national constraint not included)	4.53	3.95	5.54	5.54
GCE (national constraint included)	3.91	3.11	4.75	4.40

AR(2) structure for u

impact of the national constraint in the GCE adjustment is certainly limited, and (iii) both variants of the GCE are insensitive to changes in the AR process that characterizes the residuals, oppositely to the prediction errors when the series are disaggregated by the Chow-Lin technique.

6 Conclusions and Further Research

This chapter aimed to contribute to the literature on temporal disaggregation of economic series by proposing a method based on entropy econometrics. More specifically, we approached the estimation problem as a matrix-balancing problem and suggested using a GCE procedure to solve it. This type of techniques has been previously applied in related fields (as in the regionalization and updating of input–output tables) and has the advantage of not imposing strong distributional assumptions on the noise components. It also allows for including additional information, such as quarterly national aggregates, in a straightforward way in the form of constraints in the optimization program.

The simulation experiments conducted to evaluate its performance are based on actual data from the Spanish Regional Accounts and suggest that our proposal could improve the accuracy of the estimates obtained by Chow-Lin-based methods, which are usually employed for disaggregating annual series. In general, these methods rely on assumptions about the covariance matrix of the errors ($\boldsymbol{V}$) which are unlikely to be satisfied by economic time series in real world. When the hypotheses in which those methods are based on are violated, the accuracy of the quarterly estimated series is reduced. The proposed GCE method, in contrast, is insensitive to these violations since the assumptions required are minimal.

The research presented in the chapter, however, can be generalized and extended in several ways to find more empirical evidences than those found here. First, the simulation exercise could be modified to check the sensitivity of the results to the number of regions (R) considered or the length of the annual data available (n). Moreover, since techniques based on related data could be highly sensitive to the chosen indicators ($\boldsymbol{X}$), the number of them and the way in which these are combined, it could be interesting to test the robustness of the proposed method when different situations about the related series or indicators are considered.

From a more theoretical point of view, the disaggregation technique proposed here could be extended by exploring the possibility of introducing spatial heterogeneity and spatial dependence. Previous empirical evidence suggests that when spatial structures have been taken into account in regional and local analysis, they provide valuable knowledge about the structure of the interregional relationships. In that sense, the spatial structures have been successfully introduced in information-theory-based methods for estimating disaggregated data (Anselin and Cho 2002; Peeters and Chasco 2006; Fernández Vázquez et al. 2009; Bernardini Papalia 2010) and including a spatial dimension in the proposed model would be our objective in further research.

References

Anselin, L., & Cho, W. K. T. (2002). Spatial effect and ecological inference. *Political Analysis, 10*, 276–297.

Avellaneda, M. (1998). Minimum-relative-entropy calibration of asset pricing models. *The International Journal of Theoretical and Applied Finance, 1*, 447–472.

Berger, J. (1985). *Statistical decision theory and Bayesian analysis*. New York, NY: Springer.

Bernardini Papalia, R. (2010). Incorporating spatial structures in ecological inference: An information theory approach. *Entropy, 12*, 2171–2185.

Boot, J. C. G., Feibes, W., & Lisman, J. H. C. (1967). Further methods of derivation of quarterly figures from annual data. *Applied Statistics, 16*, 67–75.

Brooke, A., Kendrick, D., Meeraus, A., & Raman, R. (1998). *GAMS, A users guide*. Washington, DC: Gams Development Corporation. https://www.gams.com/docs/gams/GAMSUsersGuide.pdf.

Cabrer, B., & Pavía, J. M. (1999). Estimating J(>1) quarterly time series in fulfilling annual and quarterly constraints. *International Advances in Economic Research, 5*, 339–350.

Cholette, P. A. (1984). Adjusting sub-annual series to yearly benchmarks. *Survey Methodology, 10*, 35–49.

Chow, G., & Lin, A. (1971). Best linear unbiased interpolation distribution and extrapolation for time series by related series. *Review of Economics and Statistics, 53*, 372–375.

Denton, F. T. (1971). Adjustment of monthly or quarterly series to annual totals: An approach based on quadratic minimization. *Journal of the American Statistical Association, 66*, 99–102.

Di Fonzo, T. (1990). The estimation of M disaggregate time series when contemporaneous and temporal aggregates are known. *The Review of Economics and Statistics, 72*, 178–182.

European Commission. (2005). *Theory and applications of univariate and multivariate models for temporal disaggregation*. Working paper and studies. https://ec.europa.eu/eurostat/ramon/statmanuals/files/KS-DT-05-031-EN.pdf.

Fernández, R. B. (1981). A methodological note on the estimation of time series. *Review of Economic and Statistics, 63*, 471–478.
Fernández Vázquez, E., Mayor, M., & Rodriguez, J. (2009). Estimating spatial autoregressive models by GME-GCE techniques. *International Regional Science Review, 32*, 148–172.
Golan, A. (2002). Information and entropy econometrics—Editor's view. *Journal of Econometrics, 107*, 1–15.
Golan, A. (2006). Information and entropy econometrics: A review and synthesis. *Foundations and Trends in Econometrics, 2*, 1–145.
Golan A, Judge G, Miller D (1996) *Maximum entropy econometrics: Robust estimation with limited data*. London: Wiley.
Gomez, V., & Maravall, A. (1997a). *Program TRAMO and SEATS: Instructions for the user*. Banco de España: Beta Version.
Gomez, V., & Maravall, A. (1997b). *Guide for using the programs TRAMO and SEATS*. Banco de España: Beta Version.
Guerrero, V. (1990). Temporal disaggregation of time series: an ARIMA-based approach. *International Statistical Review, 58*, 29–46.
Guerrero, V., & Martínez, J. (1995). A recursive ARIMA-based procedure for disaggregating a time series variable using concurrent data. *Test, 4*, 359–376.
Hillmer, S. C., & Trabelsi, A. (1978). Benchmarking of economic time series. *Journal of the American Statistical Association, 82*, 1064–1071.
Jaynes, E. T. (1957a). Information theory and statistical mechanics I. *Physics Review, 106*, 620–630.
Jaynes, E. T. (1957b). Information theory and statistical mechanics II. *Physics Review, 108*, 171–190ss.
Kullback, S., & Leibler, R. A. (1951). On information and sufficiency. *Annals of Mathematical Statistics, 22*, 79–86.
Lisman, J. H. C., & Sandee, J. (1964). Derivation of quarterly figures from annual data. *Applied Statistics, 13*, 87–90.
Litterman, R. B. (1983). A random walk. Markov model for the distribution of time series. *Journal of Business and Economic Statistics, 1*, 169–173.
Maasoumi, E. (1993). A compendium to information theory in economics and econometrics. *Econometrics Reviews, 12*, 137–181.
Moreno, B., & López, A. J. (2013). Combining economic forecasts by using a maximum entropy econometric approach. *Journal of Forecasting, 32*, 124–136.
Pavía-Miralles, J. M. (2010). A survey of methods to interpolate, distribute and extrapolate time series. *Journal of Service Science & Management, 3*, 449–463.
Pavía-Miralles, J. M., & Cabrer-Borrás, B. (2007). On estimating contemporaneous quarterly regional GDP. *Journal of Forecasting, 26*, 155–170.
Peeters, L., & Chasco, C. (2006). Ecological inference and spatial heterogeneity: An entropy-based distributionally weighted regression approach. *Papers in Regional Science, 85*, 257–276.
Rossi, N. (1982). A note on the estimation of disaggregate time series when the aggregate in known. *The Review of Economic and Statistics, 64*, 695–696.
Shannon, C. E. (1948). A mathematical theory of communication. *Bell System Technology Journal, 27*, 379–423.
Stram, D. O., & Wei, W. W. S. (1986). Temporal aggregation in the ARIMA process. *Journal of Time Series Analysis, 7*, 279–292.
Trabelsi, A., & Hillmer, S. C. (1990). Benchmarking time series with reliable benchmarks. *Applied Statistics, 39*, 367–379.
Ullah, A. (1996). Entropy, divergence and distance measures with econometric applications. *Journal of Statistical Planning and Inference, 49*, 137–162.

Ullah, A. (2002). Uses of entropy and divergence measures for evaluating econometric approximations and inference. *Journal of Econometrics, 107*, 313–326.
Wei, W. W. S., & Stram, D. O. (1990). Disaggregation of time series models. *Journal of the Royal Statistical Society Series B, 52*, 99–102.
Zani S (1970) Sui criteri di calcolo dei valori trimestrali di tendenza degli aggregate della contabilitá nazionale. *Studi e Ricerche, Fcoltá do Economia e Commercio, Universitá degli Studi di Parma, VII*, 287–349.

Spatial Allocation of European Structural Funds: What Is the Role of Strategic Interactions?

Yves Koala and Lionel Védrine

1 Introduction

The European structural policy's main objective is to ensure the economic and social cohesion within the European Union (EU). In this context, the structural funds are allocated to regions that are lagging or with demographic and industrial decline. However, we observe that several European regions do not receive the same amount of structural funds, whereas they have similar structural characteristics. Indeed, it is not possible to predict correctly the amounts received by a region based on the "socio-economic" criteria put forward by the European Commission. Moreover, the evolution of these allocations does not seem to follow redistributive logics (Dotti 2010).

According to the "*Public Choice*" theory, the procedures for the preparation and implementation of public policies could explain the distortion in the fund allocation relative to what would have been produced by "socio-economic" criteria. Some authors have tried to explain the allocation of EU funds with the politico-economic factors advanced by this school of thought (Bodenstein and Kemmerling 2012; Bouvet and Dall'erba 2010). While these studies have greatly furthered our comprehension of the process of allocation of European structural funds, the role of spatial effects in this allocation has not yet been studied.

Y. Koala
Territoires, UMR 1273, Clermont-Ferrand, France

L. Védrine (✉)
CESAER, AgroSup Dijon, INRAE, Université de Bourgogne Franche-Comté, Dijon, France
e-mail: lionel.vedrine@inrae.fr

J. Thill (ed.), *Innovations in Urban and Regional Systems*,
https://doi.org/10.1007/978-3-030-43694-0_12

Actually, the allocation of the structural funds is a bargaining process characterized by a multi-level model (Wiberg 2009). This bargaining involves the European authorities (the European Council and the European Commission), the national governments but also regional authorities, and regional interest group activities. Robert-Nicoud and Sbergami (2004) propose a new, economic geography model when the regional policy is determined by political forces. Their approach relies on a probabilistic approach of the political game. Wiberg (2009) provides a theoretical analysis of the economic impact induced by this bargaining, when the allocation process is subject to rent-seeking by regional interest groups. These studies focus their attention on the strategic interactions between the regional authorities and the European/National authorities (vertical strategic interactions).

In the context of a lobbying game, this chapter proposes to explore the role of horizontal strategic interactions that affect the spatial allocation of European structural funds. We present a rent-seeking contest model (Dixit 1987): Two regions compete for European funds, allocated by a central government. We extend the game by introducing spatial spillover between the regions. This spatial spillover plays "as a lens" to identify the direction of the strategic interactions. Hence, this chapter seeks to test if the interactions between the decisions of neighboring governments are strategic substitutes or complements (Bulow et al. 1985).

The remainder of this chapter is divided into four sections. Section 1 presents the institutional background of the cohesion policy and the traditional determinants of the European structural funds allocation. In Sect. 2, we present a theoretical model which depicts the direction of strategic interactions in a negotiation process for public grant allocation. Then, in Sect. 3, we test the direction of strategic interaction on the allocation of structural funds for the 2000–2006 programming period using spatial econometric techniques. Finally, we conclude in Sect. 4.

2 The Determinants of the European Structural Funds Allocation

2.1 *Presentation of the European Cohesion Policy*

The main objective of the European structural policy is to ensure the economic and social cohesion within the European Union. Policy was radically reformed by The Single European Act (1986) to increase the effectiveness of the three structural funds: ERDF,[1] established in 1975, the ESF,[2] and the EAGGF[3] and to provide increased financial resources.

[1]European Regional Development Fund.

[2]European Social Fund.

[3]European Fund for Agricultural Guidance and Guarantee.

The structural funds are not intended to substitute for domestic investment, whether public or private. For this purpose, any project (Regional Program) is fully financed by the structural funds. The financing is actually in the form of co-financing between private investment and public or local structural funds. The allocation of funds is programmed over a multi-year period (4–7 years) to ensure the continuity of community intervention.

For the 2000–2006 programming period, the policy was designed with three objectives. Objectives 1 and 2 are territorialized (the allocation of these two objectives are regional, while the Objective 3 allocation is national). Accordingly, we focus our analysis on these two objectives.

Objective 1 regions are eligible on the basis of a level of GDP *p.c.*[4] that is less than 75% of the European average. Nevertheless, Fig. 1 reveals that regions with similar socio-economic characteristics are not subject to the same treatment in terms of European structural funds' allocation. Figure 1 shows the allocation of European structural funds for the Objectives 1 and 2 regions according to level of regional GDP per capita (1995 euros) for 2000–2006 period. The vertical line represents the Objective 1 eligibility threshold. This figure shows that the level of European structural funds p.c. is decreasing with the level of regional income. Nevertheless, there are important inequalities in the funds allocation. For example, in the quadrant A1, we see that the Greek regions of Attiki (GR3) and Kriti (GR42) have substantially the same level of per capita income but do not receive the same amount of structural funds *p.c.* In the same quadrant, we can observe the same phenomenon for the Italian regions of Campana (ITF3) and Molise (ITF2) that have similar levels of GDP *p.c.* for different allocations of structural funds *p.c.*

Figure 2 depicts the correlation between the regional allocation of European structural funds and the unemployment rates for 2000–2006. It shows that high unemployment is associated with a high level of structural funds. Similarly to the Fig. 1, there are important inequalities in the funds allocation for same levels of unemployment rate. In the quadrant A2, Spanish regions Galicia (ES11) and Asturias (ES12) have a similar level of unemployment rate (around 7.5%), but Asturias receives more European structural funds than Galicia.

In light of these figures, it seems clear that the only notion of fairness cannot explain the amounts of European structural funds. Some authors analyze the allocation of public subsidies, taking into account the way in which the decision to allocate funds.

Bodenstein and Kemmerling (2012) were the first to demonstrate that while poorer regions receive more regional funds, "being poor" is not a "sufficient predictor" to explain the amount of funds received by a region. When examining the allocation of structural funds in various Member States, they show that the left-leaning regions are more effective to put pressure on their central governments and on the Commission in order to secure funds. Kemmerling and Bodenstein (2006) show that the intensity of electoral national elections affects the allocation of European funds. While other articles use national political data (Carrubba 1997) or only regional ones (Bodenstein

[4]*per capita.*

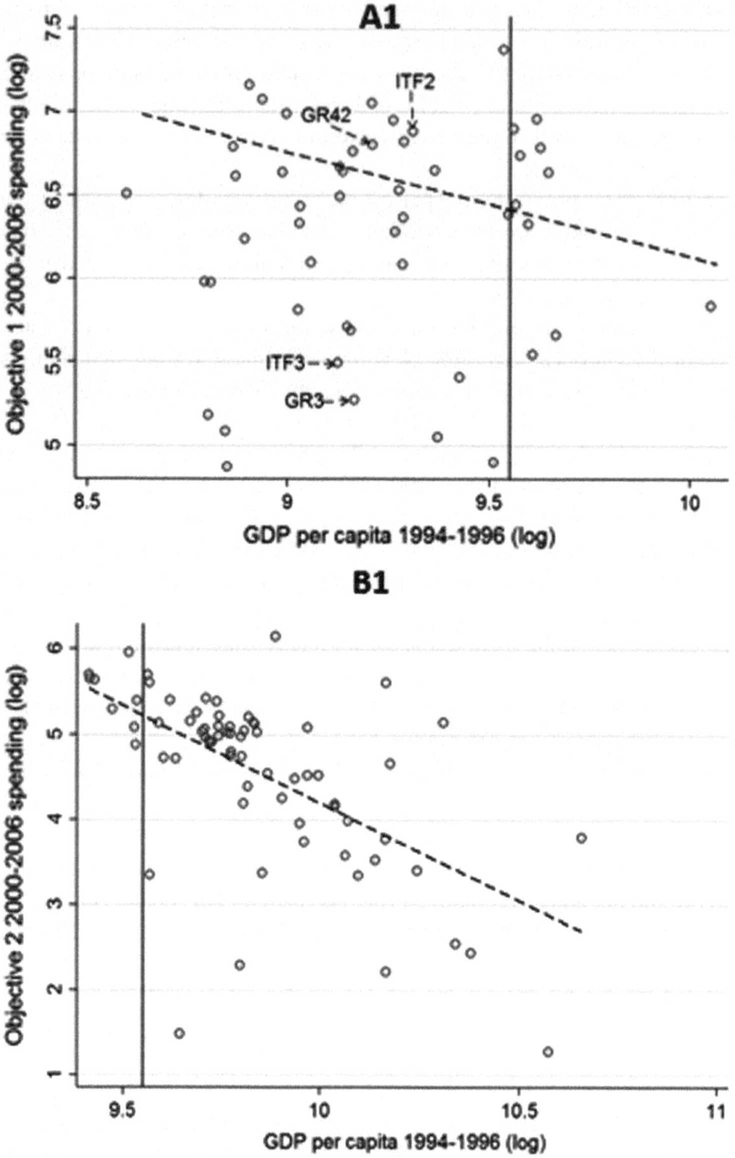

Fig. 1 Scatterplot of structural funds' allocation versus GDP p.c. (2000–2006) on logarithmic scales (1995 euros)

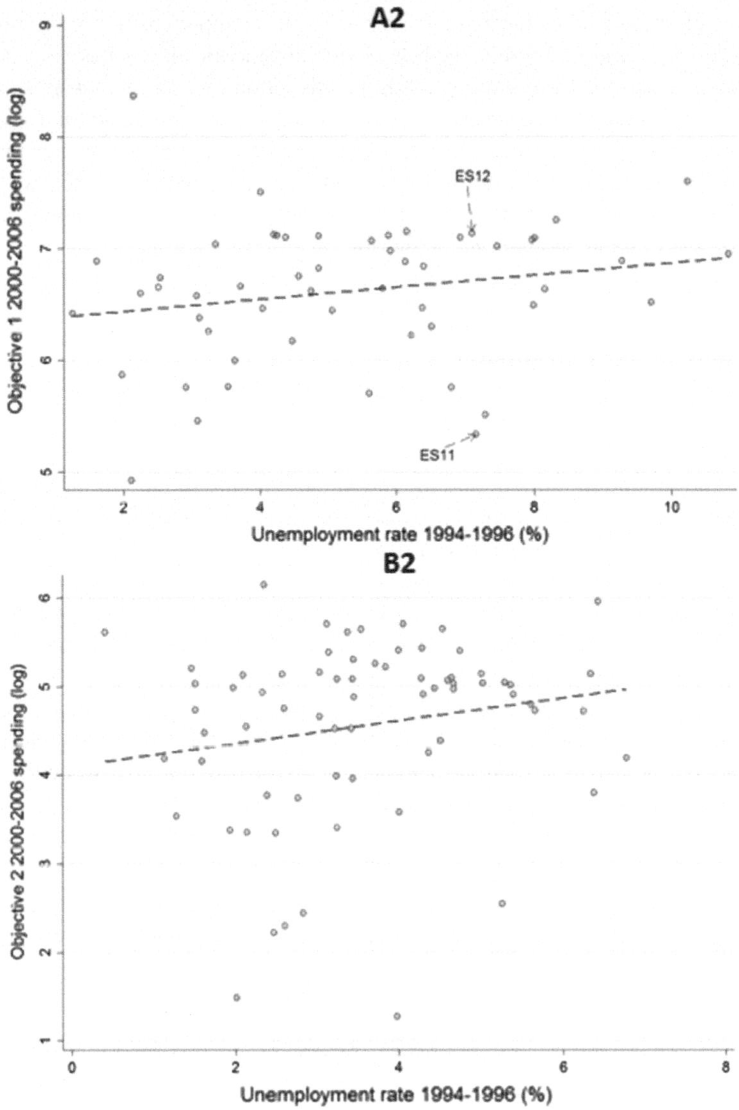

Fig. 2 Scatterplot of structural funds allocation versus unemployment rate (2000–2006)

and Kemmerling 2012), Bouvet and Dall'erba (2010) also characterized the influence of political–economic factors at the national level, in addition to the regional level specifically. Overall, their results suggest that the allocation of funds is influenced by political considerations, but that the influence of national and regional characteristics varies depending on whether the region belongs to the Objective 1 program or not.

Although these studies have greatly furthered our comprehension of the process of allocation of structural funds, the role of spatial strategic effects has not yet been studied. As described by Wiberg (2009), the allocation process of European funds is subject to negotiations between regional/national governments and the European institutions. The governance system of this policy, multi-level governance, and the obligation to co-finance EU programs explain this pattern. In this chapter, we argue the fact that such governance implies the creation of strategic interactions between the behaviors of local and national authorities.

3 Theoretical Framework

3.1 Model Description

In this section, we consider the European structural fund bargaining process as a rent-seeking contest model (Dixit 1987). We extend this game by introducing spatial spillover between the players (the regions, here).

We present a game between two regions ($i = 1, 2$) that compete for European funds, allocated by a central government (European Community). Each region exerts an effort e_i in order to attract the funds. The effort has a unique cost for each region, and the amount allocated by the European Union is a function of each region's effort. The total budget is a fixed amount (B). Thus, we can write the expected payoff (i.e., the utility provided by the grant) of each region as the difference between the share of the total budget allocated to the region i and the cost of effort:

$$\pi^1 = Bp(e_1, e_2) + \lambda B(1 - p(e_1, e_2)) - e_1 \tag{1}$$

and

$$\pi^2 = B(1 - p(e_1, e_2)) + \lambda Bp(e_1, e_2) - e_2 \tag{2}$$

where $p(e_1, e_2)$ and $(1 - p(e_1, e_2))$ are, respectively, the budget share allocated to the regions 1 and 2. The share of each region is a function of its own effort but also the effort of the other region. In order to take into account spatial spillover of public grants, we introduce an exogenous parameter λ which represents the degree of spatial spillover of public funds. Thus, the payoff of each region is directly affected by the public funds allocated in its neighborhood, weighted by the degree of spillover (λ). We assume that the spillover degree is bounded ($0 < \lambda < 1$). We capture through the introduction of this term that a region will react differently whether the funds are allocated to its neighboring regions as compared to funds allocated to regions farther away. Drawing on the literature measuring the spillover effects of public investment, we consider that the payoff of a region depends on the funds allocated in the neighboring regions, through economic spillover effects (Pereira and Roca-Sagales 2003; Bronzini and Piselli 2009). These spillover effects result from the fact that public investment provides services to producers in other regions (Ezcurra et al. 2005).

Moreover, we assume that each region's fund allocation is an increasing (decreasing) and concave (convex) function of its own (other region) effort:

$$p_1(e_1, e_2) = \frac{\partial p}{\partial e_1} > 0, \;\; p_2(e_1, e_2) = \frac{\partial p}{\partial e_2} < 0, \tag{3}$$

$$p_{11}(e_1, e_2) = \frac{\partial^2 p}{\partial e_1^2} < 0, \;\; p_{22}(e_1, e_2) = \frac{\partial^2 p}{\partial e_2^2} > 0, \tag{4}$$

We can also remark that the payoff functions are negative cross-marginal effects, due to direct competition for public funds:

$$\pi_2^1 = \frac{\partial \pi^1}{\partial e_2} = (1 - \lambda)Bp_2(e_1, e_2) < 0 \tag{5}$$

$$\pi_1^2 = \frac{\partial \pi^2}{\partial e_1} = (\lambda - 1)Bp_1(e_1, e_2) < 0 \tag{6}$$

3.2 *Nash Equilibrium of the Fund Allocation Game*

We consider that the effort choice of regions is simultaneous. We resolve the Nash equilibrium of this game, defined by the following system:

$$e_1^* = \arg\max_{e_1} \pi^1(e_1, e_2), e_2^* \text{ given } e_2^* = \arg\max_{e_2} \pi^2(e_1, e_2), e_1^* \text{ given} \tag{7}$$

The first-order conditions (FOC) for Nash equilibrium are:

$$\pi_1^1 = \frac{\partial \pi^1}{\partial e_1} = (1 - \lambda)Bp_1(e_1, e_2) - 1 = 0 \tag{8}$$

$$\pi_2^2 = \frac{\partial \pi^2}{\partial e_2} = (\lambda - 1)Bp_2(e_1, e_2) - 1 = 0 \tag{9}$$

By Eq. (3), the second-order conditions and the uniqueness of the Nash equilibrium are verified.

From these results, we define how regional governments can react to the choices of their competitors for European funds. Outside the Nash equilibrium, the partial cross-derivative of payoff functions shows that the increased effort of a region decreases the share of funds received by the other region. At the Nash equilibrium, the other government should react to this change in the effort of its competitor. So, we have to analyze how each region will react in response to a change in the effort of its competitor, at the equilibrium. Finally, we study whether the spatial spillovers affect the direction of these strategic interactions.

3.3 What is the Direction of the Strategic Interactions in This Negotiation Process?

The effort of region 1 (region 2) is a strategic substitute if its marginal payoff decreases with the effort of region 2 (region 1). They are strategic complements if the marginal payoff of one region increases with the effort of the other. Thus, we have to analyze at the equilibrium the reaction of region 1 induced by the modification of the effort produced by region 2.

We apply the envelope theorem to the FOC for Nash equilibrium in order to present the nature of interactions between regions' choices:

$$\frac{\partial e_1}{\partial e_2} = -\frac{\pi^1_{12}(e_1, e_2)}{\pi^1_{11}(e_1, e_2)} \tag{10}$$

$$\frac{\partial e_2}{\partial e_1} = -\frac{\pi^1_{21}(e_1, e_2)}{\pi^1_{22}(e_1, e_2)} \tag{11}$$

By definition of the Nash equilibrium, $\pi^1_{11}(e_1, e_2) < 0$ and $\pi^1_{12}(e_1, e_2) < 0$. Thus, the sign of the strategic interaction is the sign of cross-derivative of the marginal payoff of each region:

$$\pi^1_{12}(e_1, e_2) = \underbrace{(1-\lambda)}_{>0} Bp_{12}(e_1, e_2) \tag{12}$$

$$\pi^2_{21}(e_1, e_2) = \underbrace{(1-\lambda)}_{<0} Bp_{21}(e_1, e_2) \tag{13}$$

The sign of this relation depends on the sign of $p_{12}(e_1, e_2)$ and $p_{21}(e_1, e_2)$, which represent the curvatures of the best responses of each region induced by the modification of the effort of the other region. Theoretically, we can conclude on the direction of the interaction. The effect of the spillover degree on the reaction of region 1 is:

$$\frac{\partial \pi^1_{12}(e_1, e_2)}{\partial \lambda} = -Bp_{12}(e_1, e_2) \tag{14}$$

Similarly, the effect of the spillover degree on the reaction of region 2 is:

$$\frac{\partial \pi_{21}^{2}(e_1, e_2)}{\partial \lambda} = Bp_{21}(e_1, e_2) \tag{15}$$

We can summarize the contributions of the model by the following proposition.

Proposition

1. The direction of the strategic interactions in the allocation process depends on the curvatures of the best responses of each region induced by the modification of the effort of the other region.
2. The direction of the strategic interactions is not altered by an increase of the degree of spatial spillover between the two regions.

In order to test the direction of these interactions, we use this last result and test the direction of the spatial interactions in the European structural funds' allocation. We use the spatial spillover as a "lens" to observe strategic interactions between neighboring governments.

If the spatial interaction is negative, we can conclude that strategic interaction is also negative: Lobbying efforts of regions to attract European funds are strategic substitutes. The result is less obvious for the two other cases. The spatial interaction can be no significant for two reasons: The spatial spillover degree is zero, or there is no strategic interaction between regional in European funds' allocation. Finally, a positive spatial interaction can result from the strategic complementarities, or from similar characteristics between the region and its neighborhood. In the next section, we present several empirical specifications and test to control for the effect of spatial autocorrelation of regional characteristics.

4 Empirical Analysis

4.1 Empirical Specification

Since the focus of interest is the direction of strategic interaction between neighboring regional governments, we would like to estimate the reaction function of region i to the choice of region j. Unfortunately, we cannot observe the effort of each region itself, but we observe the equilibrium outcomes of the game, represented by the allocation of structural funds. Thus, we can estimate a linear approximation of the reaction function:

$$\text{SFpc}_i = R\left(\text{SFpc}_j | X_i\right) \tag{16}$$

According to Brueckner (2003), the consistent specification to estimate a reaction function is a spatial lag model:

$$\text{SFpc}_i = \alpha + \rho \sum_{i \neq j} w_{ij} \text{SFpc}_i + \beta_k \text{ECO}_{ki} + \gamma_k \text{POL}_{mi} + \varepsilon_i \tag{17}$$

This specification proposes to explain how the structural funds received by a region are affected by its neighboring, when we control for the socio-economic and political–economic factors of the structural funds allocation. w_{ij} is an exogenous weighting of the spatial proximity between regions. In this analysis, we define the notion of neighborliness by the concept of contiguity. The estimated slope of the reaction function is captured by the coefficient ρ. A negative slope reflects a strategic substitutability between the behaviors of neighboring regions. However, a positive effect is more difficult to interpret because there is not possible to distinguish between the effects of strategic complementarities with the effect of non unobservable similarities between neighboring regions. These interpretations remain valid under the assumption which the funds received by a region affect directly the regional the regional utility of its neighboring. Thus, the non-significance of this slope can be interpreted by the absence of strategic interaction or by the absence of spatial diffusion of funds received by a region on the regional utility of its neighborhood.

We control this test by two types of explanatory variables. The first type integrates the k socio-economic determinants (ECO_{ki}):

- the per capita income expressed in euros 1995. We use this variable to remain close to the notion of fairness. We expect a negative effect of this variable on FS p.c.
- the unemployment rate as a percentage of the working population. We expect to obtain a positive effect. In relation to the Objective 2, a region characterized by a high rate of unemployment should receive a high amount of FS p.c.
- the share of the labor force in the total population as a percentage of the total population. This variable allows us to control the demographic structure of the regional population.
- the share of agricultural employment in total employment percentage. This variable is a proxy for funds paid to rural regions by the Common Agricultural Policy. We wish to control the assumption that high agricultural regions receive less structural funds, because they receive subsidies from the CAP.

These variables explain the European structural funds of a given region by its level of per capita income, the unemployment rate, the share of the labor force in the total population, and the share of employment in agriculture total employment. These variables control for the existence of criteria of allocation based on the concept of fairness which governs the regional allocation of European structural funds.

The second type of variables introduces the m political–economic factors of the European funds allocation (POL_{mi}). We control for the intensity of electoral competition, the two mainstream political parties in the national parliament, the ability of overrepresented regions to lobby national government (A brief description of the variables with their sources is reported in Appendix 2).

4.2 The Estimation Method

We present our results by introducing step by step the socio-economic variables, the political–economic variables, and finally the spatial lag. We estimate the model without spatial autocorrelation using the OLS estimator.

The introduction of the spatial autoregressive implies a simultaneity bias. Indeed, the presence of spatial autocorrelation in a variable means that there is a functional relationship between what happens at one location in space and what happens elsewhere (Tobler 1979). Spatial autocorrelation implies that the result of each region is connected with the results of all other regions with a spatial multiplier effect. This effect is decreasing in space (first law of geography, Tobler 1979). Therefore, this spatial multiplier effect leads to biased and inefficient OLS estimators.

Thus, we use a method based on maximum likelihood (Anselin 1988) to take into account the specificity of the spatial lag model. Spatial autocorrelation can, however, be specified in other ways, for example, through a spatially autocorrelated error term (spatial error model, SEM), or with additional spatially-lagged explanatory variables (WX). To check whether the model is the best specification of spatial autocorrelation in our case, we use the estimation strategy proposed by Le Gallo (2002) and based on the LM test (Anselin et al. 1996).

4.3 Data Description

Our database is composed of 152 NUTS I/NUTS II[5] regions in 14 countries of the EU-15 for the period 2000–2006: Austria (9 regions), Belgium (11 regions), Germany (15 regions), Denmark (1 region), Spain (17 regions), Finland (5 regions), France (22 regions), Greece (13 regions), Ireland (1 area), Italy (19 regions), Netherlands (12 regions), Portugal (6 regions), Sweden (8 regions), and UK (12 regions). The list of regions and associated codes is reported in Appendix 1.

The socio-economic data come from the database "Cambridge Econometrics". Data on the regional distribution of structural funds come from the 11th Annual

[5]Nomenclature of Territorial Units for Statistics.

Report on the Structural Funds (1999). We use the data from "European Election Database"[6] to construct political and economic variables. All the variables are expressed in logarithms.

4.4 Descriptive Analysis

We start our empirical analysis with a local indicator of spatial association (LISA) in order to examine the structure of the local spatial autocorrelation in our sample. The LISA[7] informs us about the local instability of spatial autocorrelation with the detection of spatial clusters and the presence of spatial outliers.

Figure 3 suggests that there is spatial autocorrelation in the amounts of structural funds. The Moran I statistic confirms a positive spatial autocorrelation (0.30). This sign indicates that there are more regions to high SF p.c. that are located close to

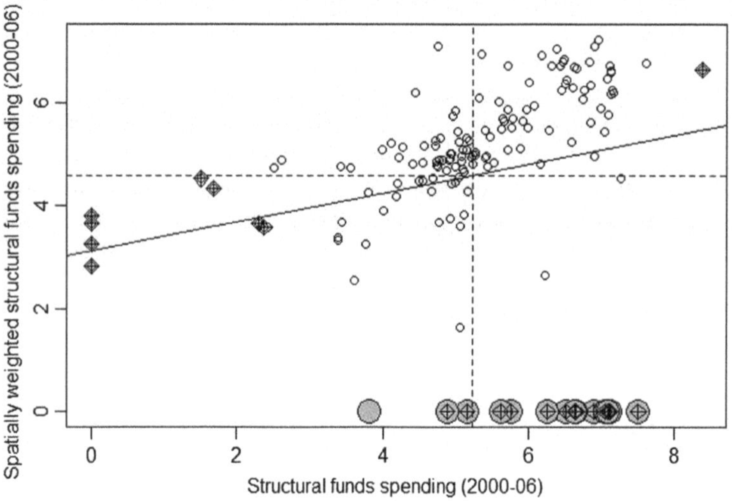

Fig. 3 Moran Scatterplot; structural funds spending in 2000–2006

[6]This database is the result of a European research project in political science. It includes the results of 330 elections in 31 European countries since the early 1990s with the Parliament European elections, the parliamentary and presidential elections of different States. The data are grouped according to the NUTS classification to facilitate comparative research on the elections for the European states.

[7]Composed of four quadrants corresponding to the four types of local spatial association between a region and its neighbors:

HH: a region associated with a high value associated with regions surrounded by high values,
BH: a region associated with a low value surrounded by regions associated with high values,
BB: a region associated with a low value is associated with regions surrounded by low values,
HB: a region associated with a high value associated with regions surrounded by low values.

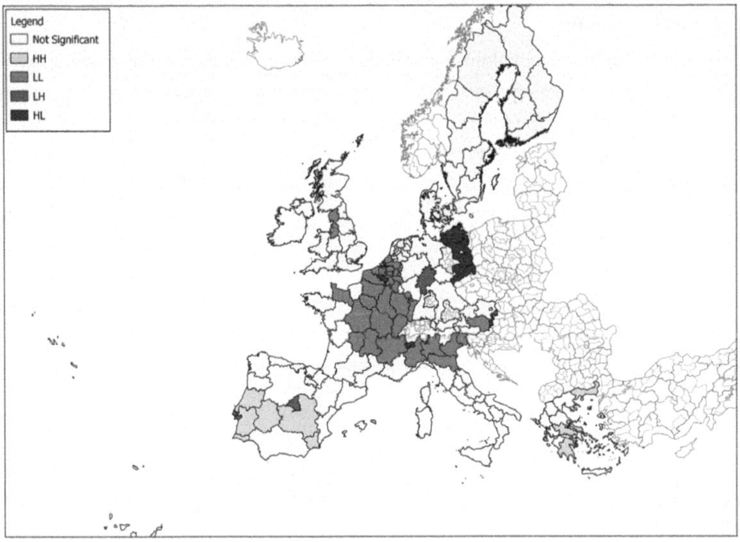

Fig. 4 Statistically significant LISA values, structural funds spending in 2000–2006

regions with similarly high SF p.c. (type HH). The situation is the same for the configuration-type BB; regions with low SF p.c. are contiguous with regions having a similar SF p.c. allocation. However, there are also numerous "atypical regions", i.e., those located near areas with different SF p.c. allocation (BH configurations and HB).

The Moran scatterplot does not inform about the significance of spatial clusters, which take into account local indicators of spatial association or *LISA*[8] in Fig. 4. This map shows the type of spatial autocorrelation regions depending on the level of funds. This map depicts significant spatial clusters. It is interpreted in the same way as the Moran scatterplot.

Figure 4 shows that there are many spatial clusters. However, we can observe that there are atypical regions ascribing to the configuration *Low–High*. In other words, according to this configuration LH, regions with low level FS p.c. are close to regions that receive a high level of SF p.c. Other regions are characterized by configurations of types, HH and LL. There were no regions of type HL.

Ultimately, our descriptive analysis confirms the need to take into account for the space in our study, namely the spatial autocorrelation of the structural funds allocation.

[8] *Local Indicator of Spatial Association,* as any statistics satisfying two criteria:

1. For each observation, *LISA* provides an indication of significant spatial clustering of similar values around each observation,
2. The sum of *LISA* associated with all observations is proportional to a global indicator of spatial association.

Table 1 Estimation results

	Economic	Political–economic	SAR	Politica–economic	SAR
Constant	0.74 (0.81)	17.13 (4.50)	19.22*** (4.37)	10.88*** (47.38)	11.70** (6.41)
GDP (log)	−0.61*** (0.17)	−2.22*** (0.44)	−2.45*** (0.44)	−1.28* (0.69)	−1.40** (0.66)
Unemployment rate (log)	1.70*** (0.23)	1.16*** (0.30)	1.14*** (0.29)	0.36 (0.44)	0.36 (0.40)
Workforce share in the total population (log)	17.65*** (3.43)	15.74*** (3.81)	17.19*** (3.68)	8.77* (4.47)	10.44** (4.62)
Agricultural employment share (log)	7.70*** (1.37)	2.16 (1.63)	1.86 (1.54)	2.81* (1.53)	2.82** (1.34)
Regional representation in national parliament (log)		0.66*** (0.17)	0.74*** (0.16)	0.43** (0.21)	0.64*** (0.20)
Electoral competition in national elections (log)		0.14 (0.13)	0.11 (0.12)	0.17 (0.12)	0.13 (0.11)
Spatial lag			−0.09** (0.04)		−0.13*** (0.04)
Objective 1 dummy				1.42*** (0.39)	1.44*** (0.33)
Member State Dummies	No	No	No	Yes	Yes
R^2	0.48	0.62		0.75	
AIG	486.65	322.70	321.5	297.73	292.73
Wald test (p.v.)	28.85 (0.00)	22.18 (0.00)	3.41 (0.06)	15.35 (0.00)	8.98 (0.00)
Spatial lag LM test (p.v.)	4.48 (0.03)	2.98 (0.08)		6.72 (0.01)	
Spatial error LM test (p.v.)	0.02 (0.88)	2.39 (0.12)		1.67 (0.20)	
Breusch–Pagan test (p.v.)	9.44 (0.00)	32.44 (0.00)	38.20 (0.00)	39.59 (0.00)	51.02 (0.00)

Standard errors in parentheses (White correction for heteroskedasticity is applied)
significant at $^{*}p < 0.10$; $^{**}p < 0.05$; $^{***}p < 0.01$

5 Results

Table 1 presents the estimation results of the determinants of the regional allocation of the structural funds. We introduce progressively into our estimates the socio-economic and political–economic factors, in order to compare our results with previous work (Table 1, the two first and fourth columns). We then introduce the spatial autoregressive term (Table 1, third and fifth columns). We check the sensitivity of our results to introduce Member States specific effects to take into account the national effects that could affect regional allocation of structural funds. Finally, we introduce a specific effect for the regions of the Objective 1 program in order to control for the specific provisions of the allocation of funds of these regions.

First, we run the estimations without the spatial effects. The Breusch–Pagan test cannot reject the presence of heteroskedasticity in our estimates. Thus, robust standard errors for heteroskedasticity are reported (White correction). Moreover, the Wald test confirms the presence of spatial autocorrelation. As expected, the regional allocation of the structural funds is negatively related with the level of GDP p.c. (Table 1, first column). This result is in favor of a redistributive allocation of the structural funds: The less developed regions tend to receive, on average, more funds. The unemployment rate positively affects the allocation of the structural funds. A region with a higher unemployment rate receives more funds than another region with a similar level of development. The workforce share in the total population and the agricultural employment positively affect the allocation of structural funds. These variables control for the productive and demographic structures of the regions.

However, these factors are not sufficient to explain precisely the regional allocation of the structural funds, because only 48% of the variability in the amounts allocated to the regions is explained by the variation of these determinants. The explanatory power of our estimate increases to 0.62 if we introduce the political–economic determinants (Table 1, second column).

In addition to the socio-economic variables, the regional representation in the national parliament also affects the allocation of European structural funds (0.66). The intensity of electoral competition in national elections does not seem related to the amount of structural funds received by a region, unlike the findings of Kemmerling and Bodenstein (2006). The introduction of these variables does not affect the estimated effect of the socio-economic factors. As expected, the GDP per capita (unemployment rate) continues to affect negatively (positively) the allocation of European structural funds. Moreover, these results are robust to the introduction of Member State dummies and Objective 1 specific effect (Table 1, fourth column). In this case, the unemployment rate ceases to significantly affect the allocation of funds. The R^2 increases to 0.75 for this last estimation; this can be explained by the fact that the Member States play a role in the process of funds allocation.

As mentioned above, these estimations suffer from spatial autocorrelation. Ignoring spatial autocorrelation leads to potentially misleading estimates and incorrect statistical inference, so we have to analyze the spatial properties of the residuals. We

Table 2 Direct and indirect impact for spatial Durbin's estimates

	Direct	Indirect	Total	Direct	Indirect	Total
GDP (log)	−3.59	0.90	−3.32	−2.04	0.78	−1.83
Unemployment rate (log)	1.67	−0.46	1.55	0.53	−0.74	0.47
Workforce share (log)	25.17	−1.89	23.28	15.32	−1.58	13.73
Agricultural employment share (log)	2.72	−0.75	2.52	4.14	−0.03	3.71
Regional representation (log)	1.08	−1.09	0.99	0.94	−1.32	0.84
Electoral competition (log)	0.17	−0.17	0.16	0.19	−0.26	0.17
Objective 1 dummy				2.11	−0.81	1.89
Member State Dummies	No	No	No	Yes	Yes	Yes

use LM test in order to determine the best specification to account for the spatial autocorrelation in our model. The results of these tests confirm that the introduction of a spatial autoregressive term is the best way to take into account spatial autocorrelation in our model (LM test, Table 1).

5.1 *The Spatial Autoregressive Model*

The third and fifth columns of the Table 1 report the results of the spatial autoregressive model. Firstly, it is important to note that the Akaike criterion, as a measure of the relative goodness of fit of our different estimates, is both improved with the introduction of the spatial autoregressive term (321.5) and the use of dummies Member States and the specific effect Objective 1 specific effect (292.73).

The GDP per capita (unemployment rate) is still negative (positive) and significant. As described by Le Sage and Pace (2008), we compute the direct (indirect) and total marginal effects of these variables (Table 2).

The total marginal effect of GDP per capita is negative. An increase of 1% of GDP per capita of a region should lead to a decrease of 1.83% (3.32% without dummies controlling by Member States and Objective 1) amounts received by this region, *ceteris paribus*. The total effect is greater than that estimated by ignoring spatial effects (respectively −1.28% and −2.22% without dummies controlling by Member States and Objective 1). The total marginal effect can be broken down by a positive direct effect and a negative marginal indirect effect: The effect of an increase in GDP p.c. in a region negatively affects the amount of EU funds received by this region, while a higher GDP p.c. positively affects other regions of amounts received by this region.

The total marginal effect of the unemployment rate is positive (Table 2). A 1% increase in the unemployment rate in a region would increase by 0.47% the amounts received by this region (1.55% without dummies controlling by Member States and Objective 1). As for GDP per capita, the total marginal effects of unemployment are higher when we introduce the spatial autoregressive term. The direct marginal effect is positive, whereas the indirect is negative.

The control variables in the sectoral and demographic structure remain significant with the introduction of the spatial autoregressive term (Table 1).

The regional representation in the national parliament also affects significantly the allocation of European structural funds, whereas the intensity of electoral competition in national elections does not (Table 1, third and fifth columns). The total marginal effect of the regional representation is still positive (0.84%). Contrary to the findings of Kemmerling and Bodenstein (2006), our results suggest that the intensity of electoral competition in each region in the national elections does not change the allocation of EU funds. However, as shown by Bouvet and Dall'erba (2010), the effect of the electoral competition is sensitive to the partisan alignment between regional and national governments. These authors find that, among all of the regions that share the same political alignment as the national government, the regions where its party's electoral margins are weaker receive more structural funds (Bouvet and Dall'erba 2010).

We note that the coefficient ρ, associated with the spatial autoregressive term, measures the direction and intensity of the interaction between the level of funds received by a region and the level achieved by its neighbors. This coefficient is negative and significant, suggesting that the spatial strategic interactions are also negative. The amount of funds received by a region would be affected by a decrease of 0.13% (0.09 without dummies controlling by Member States and Objective 1) by a 1% increase in average amounts received by its neighborhood.

This effect is significant, relative to an overall effect of competition between all European regions. In the presence of positive spatial spillover of the European funds on the utility level of the residents, it seems more profitable to a regional government to react to an increase in the lobbying effort of its neighbors, by decreasing its own lobbying effort to distort the allocation of EU funds in his favor. In this case, we can conclude that regions' lobbying efforts to attract European funds are strategic substitutes.

6 Conclusions

The aim of this chapter was to test the presence and the direction of spatial strategic interaction in the allocation of European structural funds. We constructed a rent-seeking contest model to analyze how spatial strategic interactions can affect the allocation process of European funds. We showed theoretically that spatial interactions do not alter the direction of strategic interactions. We extended the game by introducing spatial spillover between the regions. This spatial spillover plays "as a lens" to identify the direction of the strategic interactions.

Finally, we tested the direction of these strategic interactions in this allocation using spatial econometrics methods. We showed that the spatial autoregressive term is negative: An increase of European structural funds received by regions decrease the amount of funds received by its neighboring. This effect is significant, relative to an overall effect of competition between all European regions. Thus, our results suggest that the behaviors of regions are strategic substitutes in the allocation process of European funds.

The other determinants of the allocation of the European structural funds are the regional GDP p.c. and the unemployment rate. We cannot exclude a significant effect of political–economic factors: The regional representation in the national parliament seems to affect the allocation. However, we cannot confirm a significant effect of the intensity of the electoral competition unlike Kemmerling and Bodenstein (2006).

The previous studies focused their objectives on the relation between the different levels of government, especially for the political considerations (Bouvet and Dall'erba 2010). Our results improved the literature by exploring the role of the horizontal strategic interactions that affect the spatial allocation of the European structural funds, in a context of a lobbying game.

Nevertheless, this approach requires further development. Firstly, the local political considerations can also affect the allocation and the implementation of the structural funds (Dellmuth and Stoffel 2012; Bouayad-Agha et al. 2013). The role of political factors should be taken addressed at a more local level than at the regional level, and for elections other than national elections. Secondly, the study of the effects of institutional factors (such as decentralization vs. devolution of implementation of the policy) on the intensity and direction of strategic spatial interactions seems important to study (Védrine 2020).

Appendix 1: Name of Regions According Nuts1/Nuts2 Codes

Code	Region	Code	Region	Code	Region	Code	Region
Austria		ES41	Castilla y León	GR41	Voreio Aigaio	SE04	Sydsverige
AT11	Burgenland	ES42	Castilla-La Mancha	GR42	Notio Aigaio	SE06	Norra Mellansverige
AT12	Niederösterreich	ES43	Extremadura	GR43	Kriti	SE07	Mellersta Norrland
AT13	Wien	ES51	Cataluña	Ireland		SE08	Övre Norrland
AT21	Kärnten	ES52	Comunidad Valenciana	IE	Ireland	SE09	Småland med öarna
AT22	Steiermark	ES53	Illes Balears	Italy		SE0A	Västsverige
AT31	Oberösterreich	ES61	Andalucia	ITC1	Piemonte	UK	
AT32	Salzburg	Es62	Región de Murcia	ITC2	Valle d'Aosta	UKC	North East
AT33	Tirol	ES70	Canarias (ES)	ITC3	Liguria	UKD	North West
AT34	Vorarlberg	Finland		ITC4	Lombardia	UKE	Yorkshire and The Humber
Belgium		FI13	Itä-Suomi	ITD3	Veneto	UKF	East Midlands
BE10	Region of Brussels-Capital	FI18	Etelä-Suomi	ITD4	Friuli-Venezia Giulia	UKG	West Midlands
BE21	Prov. Antwerpen	FI19	Länsi-Suomi	ITD5	Emilia-Romagna	UKH	Eastern
BE22	Prov. Limburg (B)	FI1A	Pohjois-Suomi	ITE1	Toscana	UKI	London
BE23	Prov. Oost-Vlaanderen	FI20	Åland	ITE2	Umbria	UKJ	South East
BE24	Prov. Vlaams Brabant	France		ITE3	Walking	UKK	South West
BE25	Prov. West-Vlaanderen	FR10	Ile de France	ITE4	Lazio	UKL	Wales
BE31	Prov. Brabant Wallon	FR21	Champagne-Ardenne	ITF1	Abruzzo	UKM	Scotland

(continued)

(continued)

Code	Region	Code	Region	Code	Region	Code	Region
BE32	Prov. Hainaut	FR22	Picardy	ITF2	Molise	UKN	Northern Ireland
BE33	Prov. Cork	GB23	Haute-Normandie	ITF3	Campania		
BE34	Prov. Luxembourg (B)	FR24	Center	ITF4	Puglia		
BE35	Prov. Namur	FR25	Basse-Normandie	ITF5	Basilicata		
Germany		FR26	Burgundy	ITF6	Calabria		
DE1	Baden-Württemberg	FR30	Nord-Pas-de-Calais	ITG1	Sicilia		
DE2	Bayern	FR41	Lorraine	ITG2	Sardegna		
DE3	Berlin	FR42	Alsace	Netherlands			
DE4	Brandenburg	FR43	Franche-Comté	NL11	Groningen		
DE5	Bremen	FR51	Pays de la Loire	NL12	Friesland		
DE6	Hamburg	FR52	Brittany	NL13	Drenthe		
DE7	Hessen	FR53	Poitou-Charentes	NL21	Overijssel		
DE8	Mecklenburg-Vorpommern	FR61	Aquitaine	NL22	Gelderland		
DE9	Niedersachsen	FR62	Midi-Pyrenees	NL23	Flevoland		
DEA	Nordrhein-Westfalen	FR63	Limousin	NL31	Utrecht		
DEB	Rheinland-Pfalz	FR71	Rhône-Alpes	NL32	Noord-Holland		
December	Saarland	FR72	Auvergne	NL33	Zuid-Holland		
DED	Sachsen	FR81	Languedoc-Roussillon	NL34	Zeeland		
DEF	Schleswig-Holstein	FR82	Provence-Alpes-Côte d'Azur	NL41	Noord-Brabant		
DEG	Thüringen	FR83	Corsica	NL42	Limburg (NL)		
Denmark		Greece		Portugal			

(continued)

(continued)

Code	Region	Code	Region	Code	Region	Code	Region
DK00	DENMARK	GR11	Anatoliki Makedonia, Thraki	PT11	Norte		
Spain		GR12	Kentriki Makedonia	PT15	Algarve		
SS11	Galicia	GR13	Dytiki Makedonia	PT16	Centro (PT)		
ES12	Principality of Asturias	GR14	Thessalia	PT17	Lisboa		
ES13	Cantabria	GR21	Ipeiros	PT18	Alentejo		
ES21	Pais Vasco	GR22	Ionia Nisia	PT20	Região Autónoma dos Açores (PT)		
ES22	Comunidad Foral de Navarra	GR23	Dytiki Ellada	PT30	Região Autónoma da Madeira (PT)		
ES23	La Rioja	GR24	Sterea Ellada	Suede			
ES24	Aragon	GR25	Peloponnisos	SE01	Stockholm		
ES30	Comunidad de Madrid	GR30	Attiki	SE02	Östra Mellansverige		

Appendix 2: Description of Variables and Sources

Description	Source
Structural funds per capita in euro 95	11th Annual Report on the Structural Funds (1999)
GDP per capita in euro 95 p.c.	2006 Cambridge Econometrics Limited
Unemployment rate as a percentage of the labor force	2006 Cambridge Econometrics Limited
Share of the workforce in the total population, as a percentage of total population	2006 Cambridge Econometrics Limited
Share of agricultural employment in total employment, percentage	2006 Cambridge Econometrics Limited
Total number of seats of most regional p.c. in national parliament	"Eurocivic"
Electoral competition between the two major parties in the national parliamentary elections (percent)	Inspired by Borck and Owing (2003)
Distance between Brussels and the regions studied, using as a proxy for the lobbying effort. Idea: Cost of lobbying increases with the physical distance	Calculated by the authors
Dummy to control State-specific effect	Calculated by the authors

References

Anselin, L., Bera, A. K., Florax, R., & Yoon, J. M. (1996). Simple diagnostic tests for spatial dependence. *Regional Science and Urban Economics, 26*(1), 77–104.

Anselin, L. (1988). *Spatial econometrics: methods and models* (p. 284). London: Kluwer.

Bodenstein, T., & Kemmerling, A. (2012). Ripples in a rising tide: Why some EU regions receive more structural funds than others. *European Integration online Papers, 16*, article 1. https://eiop.or.at/eiop/pdf/2012-001.pdf.

Borck, R., & Owings, S. (2003). The political economy of intergovernmental grants. *Regional Science and Urban Economics, 33*(2), 139–156.

Bouayad Agha, S., Turpin, N., & Védrine, L. (2013). Fostering the development of European regions: A spatial dynamic panel data analysis of the impact of cohesion policy. *Regional Studies, 47*(9), 1573–1593.

Bouvet, F., & Dall'erba, S. (2010). European regional structural funds: How large is the influence of politics on the allocation process. *Journal of Common Market Studies, 48*, 501–528.

Bronzini, R., & Piselli, P. (2009). Determinants of long-run regional productivity with geographical spillovers: The role of R&D, human capital and public infrastructure. *Regional Science and Urban Economics, 39*(2), 187–199.

Brueckner, J. K. (2003). Strategic interaction among governments: An overview of empirical studies. *International Regional Science Review, 26*(2), 175–188.
Bulow, J., Genakoplos, J., & Klemperer, P. (1985). Multimarket oligopoly: Strategic substitutes and complements. *Journal of Political Economy, 93*, 488–511.
Carrubba, C. J. (1997). Net financial transfers in the European Union: Who gets what and why? *Journal of Politics, 59*(2), 469–496.
Dellmuth, L. S., & Stoffel, M. F. (2012). Distributive politics and intergovernmental transfers: The local allocation of European Union structural funds. *European Union Politics, 13*(3), 413–433.
Dixit, A. (1987). Strategic behavior in contests. *American Economic Review, 77*(5), 891–898.
Dixit, A., & Londregran, J. (1998). Ideology, tactics, and efficiency in redistributive politics. *Quarterly Journal of Economics, 113*(2), 497–529.
Dotti, N. (2010). *Being poor is not enough: The "non-written" factors affecting the allocation of the EU structural funds*. Paper presented at the XXXI Conferenza Italiana di Scienze Regionali. https://aisre.it/images/old_papers/Dotti1.pdf.
Ezcurra, R., Gil, C., Pascual, P., & Rapun, M. (2005). Public capital, regional productivity and spatial spillovers. *The Annals of Regional Science, 39*(3), 471–494.
Kemmerling, A., & Bodenstein, T. (2006). Partisan politics in regional redistribution: Do parties affect the distribution of EU structural funds across regions? *European Union Politics, 7*(3), 373–392.
Le Gallo, J. (2002). Econométrie spatiale: l'autocorrélation spatiale dans les modèles de régression linéaire. *Economie et Prévision, 155*(4), 139–158.
Le Sage, J., & Pace, R. K. (2008). *Introduction to spatial econometrics*. Boca Raton, FL: Chapman and Hall/CRC.
Pereira, A. M., & Roca-Sagalés, O. (2003). Spillover effects of public capital formation: Evidence from the Spanish regions. *Journal of Urban Economics, 53*(2), 238–256.
Robert-Nicoud, F., & Sbergami, F. (2004). Home-market vs vote-market effect: Location equilibrium in a probabilistic voting model. *European Economic Review, 48*(1), 155–179.
Tobler, W. R. (1979). Cellular geography. In *Philosophy in geography* (pp. 379–386). Dordrecht: Springer.
Védrine, L. (2020). Allocation of European structural funds, decentralization and strategic spatial interactions. *Regional Studies, 54*(1), 72–82.
Wiberg, M. (2009). Location equilibrium with endogenous rent seeking. *Journal of Economic Geography, 9*, 869–887.

Structural Funds, Institutional Quality and Regional Economic Convergence in EU: A Spatial Econometric Approach

Elisa Montresor, Francesco Pecci, and Nicola Pontarollo

1 Introduction

The Structural Funds, (SF), are the most important strategic tool of the European Union (EU) for promotion of regional development. Over the three last programming periods (1989–1993, 1994–1999, 2000–2006), the resources devoted to Cohesion Policy (CP) have grown to a third of the EU budget; more than 80% of which is targeted for "economic convergence."

According to Articles 158-162 of the Treaty establishing the European Community (2002), the main objective of the CP is to promote the "overall harmonious development" of the EU and to reduce the economic and social disparities between regions. To reach these objectives, European policymakers aim at fostering the endogenous development potential, focusing on regions with gross domestic product (GDP) per capita less than or equal to 75% of the EU average.

The European Commission pays close attention to the results of CP, the second main EU policy after Common Agricultural Policy, evaluating its progress in various reports. In the Third Report on Economic and Social Cohesion (European Commission 2004), it is argued that inequalities of income and employment in EU-15 have declined since the mid-1990, at both the national and regional levels. The Commission states that from 1994 to 2001, the growth of GDP in the four Cohesion countries of the EU-15—Greece, Portugal, Ireland and Spain—had an annual mean

E. Montresor · F. Pecci
Department of Economics, University of Verona, Via Cantarane 24, 37129 Verona, Italy
e-mail: elisa.montresor@univr.it

F. Pecci
e-mail: francesco.pecci@univr.it

N. Pontarollo (✉)
Department of Economics and Management, University of Brescia, Via S. Faustino 74/b, 25122 Brescia, Italy
e-mail: nicola.pontarollo@unibs.it

J. Thill (ed.), *Innovations in Urban and Regional Systems*,
https://doi.org/10.1007/978-3-030-43694-0_13

of 1% above the EU average, and that employment rates were also higher (except in the case of Greece). The GDP per capita in Objective 1 regions also increased more rapidly than in the rest of the EU. This is confirmed by Leonardi (2005, 2006) who agrees that global economic trends show signs of long-term convergence between regions. Ederveen et al. (2002) relate similar results: Convergence is taking place at the national level, but slowly, and regional convergence is occurring within countries. In a later report (European Commission 2007, p. 6), the Commission confirms that countries and regions belonging to the EU-27 are converging: "Disparities in GDP per head between regions in the EU have narrowed markedly over the past decade as growth in the least prosperous regions has outstripped that elsewhere." Bachtler and Gorzelak (2007) conclude that this statement, although generally true, masks two opposite processes: faster growth of less-developed countries and a slowdown of growth in some poorer regions within these countries.

Nevertheless, previous studies (Boldrin and Canova 2001; Fagerberg and Verspagen 1996) questioned the occurrence of regional convergence, arguing that regional disparities have remained unchanged, or even expanded (Puga 2002).

Regarding the effectiveness of CP, the empirical evidence has not been conclusive among studies. While Beugelsdijk and Eijffinger (2005), Cappelen et al. (2003), Puigcerver-Peñalver (2007), Esposti and Bussoletti (2008), Falk and Sinabell (2008), Hagen and Mohl (2008), Bähr (2008) Ramajo et al. (2008), Becker et al. (2010) and Rodríguez-Pose and Garcilazo (2015) find a positive effect of SF upon convergence, Rodríguez-Pose and Fratesi (2004), Ederveen et al. (2002, 2006), Dall'erba and Le Gallo (2007, 2008) and Mohl and Hagen (2010) conclude that SF have nonsignificant or negative effect upon regional growth of GDP per capita.

It is not possible to classify uniformly the results in function of the technique and/or types of variables used because they strongly depend on the variable choice, definition as well as the sampling.[1]

This chapter evaluates the effects of regionally targeted SF on labor productivity growth in 180 Nomenclature of Territorial Units for Statistics, level 2 (NUTS2) regions from 1989 to 2006.

The choice of a variable related to labor productivity, rather than classical income per capita, was chosen for reasons best encapsulated by the following European Commission's (2010, p. 23) statement: "Productivity growth is key to providing higher wages without losing competitiveness. It is also the main source of growth of GDP per head, and it is likely to become even more so as the share of people of working age in total population shrinks." This statement is a confirmation of Krugman's (1992) idea, according to which productivity is the main factor that ensures improvement of the standard of living for the long run.

The main contributions of this chapter to the debate on the effectiveness of CP consist of two aspects: the spatial econometric technique applied to the analysis and the inclusion of a variable comprised of a specific measure of regional institutional quality. Regarding the first contribution, the effects of SF upon regional convergence of labor productivity are assessed via a spatial filtering technique that

[1] A complete literature review can be found in Mohl and Hagen (2010).

manages both spatial dependence in residuals and the possible existence of multiple steady states, due to heterogeneous regional economic structure. Regarding the second contribution, a regional institutional quality indicator as is interacted with SF in order to assess the joint returns on growth of these public policies with respect to the associated institutions.[2]

The chapter is organized as follows. Section 2 analyzes the evolution of productivity in Europe. In Sect. 3, the empirical and spatial models are described, while in Sect. 4, the results of our analysis are illustrated. Section 5 concludes.

2 Productivity Evolution in Europe

It is well known that a slowdown in productivity growth occurred in the EU during the second half of 1990s (Ederveen et al. 2005; Van Ark 2006). Table 1 confirms these findings at the regional level. From 1989 to 1995, the mean growth was around 3% in the three samples while during the period from 1996 to 2006 productivity grew, on average, only 1.1%. In the meantime, variance decreased, showing that in first six years of the study period, growth was more heterogeneous than in the following period. Over the entire study period (1989–2006), productivity growth was 1.7% for all three samples.

Table 1 Productivity growth

	GVA/EMP growth 1989–1995	GVA/EMP growth 1996–2006	GVA/EMP growth 1989–2006
All NUTS2 regions	0.02915	0.01139	0.01691
	(0.0005)	(0.0001)	(0.0001)
Regions with a GDP per capita lower than 75% of the EU average (Objective 1 regions)	0.02731	0.01161	0.01697
	(0.0008)	(0.0001)	(0.0001)
Regions with a GDP per capita higher than or equal to 75% of the EU average (non-Objective 1 regions)	0.03000 (0.0003)	0.01129 (0.0001)	0.01689 (0.0000)

In brackets: Variance

[2]Rodríguez-Pose and Garcilazo (2015) combine regional institutional quality indicators with Cohesion Policy expenditures to assess their effects on GDP per capita growth. Studies of Beugelsdijk and Eijffinger (2005), Ederveen et al. (2006) and Bähr (2008) include institutional quality indicators and data at the national level, while Arbia et al. (2010) use measures of national institutional quality with regional data.

The remarkable point of this brief analysis is that productivity growth of regions with GDP per capita lower than 75% of the EU average (Objective 1 regions) is in line with the non-Objective 1 regions showing that a convergence process between these two groups of regions did not take place.

According to Ederveen et al. (2005), slowdown of productivity growth occurs when there is high employment growth in conjunction with investment slowdown. This process is confirmed by the results of this chapter. Correlation between Gross Value Added per employee and investment over Gross Value Added is always negative and tends to increase while correlation between Gross Value Added per employee and employment rate is positive.[3] Another explanation of productivity slowdown, according to Van Ark (2006), is in the lack of government role in setting the "rules of the game" for markets, via business regulations and policy mechanisms directly and indirectly related to innovation and technology diffusion.

Figure 1 shows the standardized Gross Value Added per employee in 1989 on the horizontal axis and the average productivity growth during the period from 1989 to 2006 on the vertical axis; the intersection corresponds to the respective mean values.

In the upper-right quadrant (HH), there are the regions where both the values of Gross Value Added per employee in 1989 and the average productivity growth from 1989 to 2006 are higher than the overall average value. Similarly, in the lower-left quadrant (LL), there are the regions in which both the values are lower than the overall average value.

In the upper-left quadrant (LH), there are the regions with the value of Gross Value Added per employee in 1989 lower than the overall average value and the average

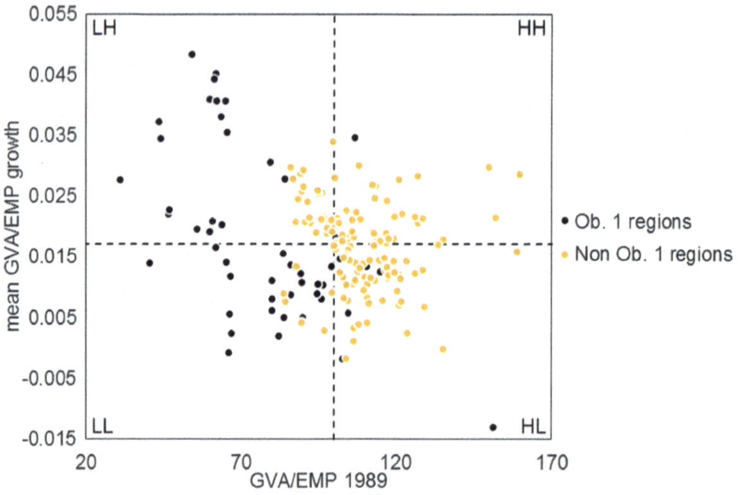

Fig. 1 Correlation between productivity growth (1989–2006) and 1989 productivity levels

[3]In 1989–2006 period, the average correlation between Gross Value Added per employee and investment over Gross Value Added is −0.43. Correlation between Gross Value Added per employee and employment rate is, on average, 0.33.

productivity growth from 1989 to 2006 higher than the overall average value. On the contrary, in the lower-right quadrant (HL), there are the regions where Gross Value Added per employee in 1989 is higher than the overall average value and the average productivity growth from 1989 to 2006 is lower than the overall average value.

In upper-right and lower-right quadrants, there are the more developed regions (non-Objective 1) while the majority of the regions of upper-left and lower-left quadrants are Objective 1.

The tendency of Objective 1 regions is to grow faster than those regions which are more developed. Their variability, however, is very high, and a case study approach could lead to understand the reasons for such varied behavior. Despite the fact that the majority of SF are devoted to Objective 1 regions, this funding alone appears to be insufficient to support regional development. It is conceivable, in fact, that a set of other factors, such as institutional quality, human capital, local entrepreneurial capacity, sectoral structure, play a crucial role in the growth process.

3 The Empirical and Spatial Model

According to Ederveen et al. (2002), the impact of CP should be assessed through case studies, simulation models or econometric estimates. The approach of this chapter is in line with the latter category, and the theoretical framework is based upon the well-known cross-sectional β-convergence model, defined as follows:

$$g_{i,T} = \frac{1}{T}\ln\left(\frac{y_{iT}}{y_{i0}}\right) = \alpha + \beta \ln y_{i0} + \mathbf{Z}\varphi + \varepsilon_i \quad (1)$$

where T is the number of years composing the time period, i the ith NUTS2 region[4] (out of 180 NUTS2 regions), y_{i0} is the initial productivity level in the year 1989 (*GVA/EMP_89*) in region i, y_{iT} is the productivity level in 2006 (*GVA/EMP_06*) in region i, $g_{i,T}$, is the average productivity growth rate and $\mathbf{Z}$ is a matrix of time variant and invariant explanatory variables. α represents a vector of constant term, β and φ are the vectors of the parameters and ε_i is the error term.

A negative relationship between the average productivity growth rate and the initial level of productivity—i.e., parameter β is significant—is the sign of a convergence process. A higher negative value of β implies that a region approaches more rapidly the steady state. A significant vector φ of coefficients of the set of control variables ($\mathbf{Z}$) means that the conditional β-convergence hypothesis holds, such as, each region converges to its own steady state. The variables are described in Table 2.

Following Mankiw et al. (1992), vector $\mathbf{Z}$ includes the physical capital investment rates (*INV/GVA*), the discount rate (*DISC*) equal to $(n_i + g_A + \delta)$—where n_i is the estimated regional employment growth, g_A is the technological progress, and δ the

[4]Regions are specified in the appendix.

Table 2 Description of variables

Variable	Description
GVA/EMP_06	GVA per employee in 2006[a]
GVA/EMP_89	GVA per employee in 1989[a]
DISC	Employment growth (mean between 1989 and 2006) + 0.03[a]
INV/GVA	Average rate of investment on GVA between 1989 and 2006 as proxy of saving rate[a]
HRSTC/POP	Mean rate of human resources in science and technology core on population between 1995 and 2006[b]
EUqogIndex	Indicator of regional institutional quality[c]
OB.1/EMP	Mean share of Structural Funds for Objective 1 (devoted to region whose GDP per capita is less than 75% of the EU average) for the whole period, over employment for the same period[d]
OB.2/EMP	Mean share of Structural Funds for Objective 2 (devoted to the regions affected by industrial crisis) for the whole period, over employment for the same period[d]
EUqogIndex × OB.1/EMP	Interaction term between regional institutional quality and Structural Funds for Objective 1 over employment
EUqogIndex × OB.2/EMP	Interaction term between regional institutional quality and Structural Funds for Objective 2 over employment

[a]Cambridge Econometrics' Database (2010)
[b]Eurostat Regio Database
[c]Charron et al. (2013)
[d]European Commission (1995, 1997, 1999, 2006)

capital depreciation rate—with $g_A + \delta = 0.05$,[5] and the human resources in science and technology core (those people who have a third-level education[6] and work in a science and technology occupation) over total population (*HRSTC/POP*).

Physical capital accumulation contributes directly to growth and, because of diminishing returns to capital, it makes that poor economies have high rates of return and therefore tend to grow faster than rich economies (Barro and Sala-i-Martin 1990).

HRSTC is a measure of human capital, accepted internationally as one of the most important indicators for assessing real and potential technological strengths of a region (OECD 2011). Furthermore, according to the European Commission (2010), it is also an indicator of innovative capacity. Intuitively, one would assume that a large share of well-trained scientists and engineers whose aim is to carry out research and development (R&D) activities, can turn research into new, commercially viable products and services that increase productivity and, as a consequence, economic growth. In this regard, one can refer to the work of Schultz (1961).

[5]See Ederveen et al. (2002), Dall'erba and Le Gallo (2008), Esposti and Bussoletti (2008).

[6]According to ISCED, tertiary programmes include academic orientation which are largely theoretical and tertiary programmes with an occupational orientation. Examples are Bachelor's degrees in many English-speaking countries, the "Diplom" in many German-speaking countries and the Licence in many French-speaking countries.

The higher are the investment rates in physical and human capital and the lower are the rates of depreciation and population growth, more a region converges rapidly to its own steady state.

In addition to these classical control variables, vector **Z** comprises an indicator of regional institutional quality (*EUqogIndex*); the SF share for Objectives 1 and 2 over employment[7] (respectively *OB.1/EMP* and *OB.2/EMP*) and a term representing the interaction between SF and regional institutional quality (*EUqogIndex* × *OB.1/EMP* and *EUqogIndex* × *OB.2/EMP*).

According to the OECD (2009), one of the main determinants of growth is the policy and institutional context, which includes aspects such as the quality of governance or the macroeconomic framework of which the regional economy is a part. Nevertheless, a convergence analysis involving regional institutional quality and CP, until now, has been performed only by Rodríguez-Pose and Garcilazo (2015), assessing their effects on GDP per capita growth. Following the cited authors, in this chapter the Quality of Government Composite Index (*EUqogIndex*) developed by Charron et al. (2013) is used. Institutions should be distinguished between formal and informal (Amin 1999). Formal institutions generally include constitutions, laws, charters and regulations, while informal institutions comprise traditions, group routines, social norms and conventions and informal networks (Amin 1999; Rodríguez-Pose and Storper 2006). A balance between formal and informal institutions determines the level of economic activity and its efficiency. Hence, efficient local institutions are believed to promote development and growth through creating the adequate conditions for investment, economic interaction and trade.

The *EUqogIndex* composite indicator refers only to formal institutions and combines data for both the national and regional level. The former category of this data is regarding control of corruption, rule of law, quality of the bureaucracy, democracy and the strength of electoral institutions; these refer to The World Bank's "World Governance Indicators" (WGI) (Kaufmann et al. 2009). The latter category is based upon a survey of approximately 34,000 EU citizens focused exclusively on three public services that are most often financed, administered or politically accounted for by sub-national authorities: education, health care and law enforcement. The authors of *EUqogIndex* aggregated the unweighted average regional Quality of Government score for each country in the survey and subtracted this from each region's individual score, which is then added to the national-level data. Therefore, each region has an adjusted score, centered around the WGI.

In this chapter, it was assumed that the *EUqogIndex* remains stable (or changes very slowly) over time (Ederveen et al. 2006).

CP, of which SF and the Cohesion Fund are the financial instruments, has the specific aim of narrowing the development disparities among regions and Member States—fostering, through SF, the goal of economic, social and territorial cohesion. The two regionally targeted SF consider are Objectives 1 and 2. The aim of Objective

[7]The share of Structural Funds over GVA is not used because, following Esposti and Bussoletti (2008) and Mohl and Hagen (2010), this can lead to simultaneity problems that might bias the results, since GVA (growth) then appears on both sides of the regression equation.

1 is promoting the development of regions with GDP per capita lower than 75% of the EU average; this effort absorbs more than 70% of the CP budget. Objective 2 has a more specific target: to support the reconversion of the regions seriously affected by industrial decline. This accounts for some 10% of the CP budget. The two Objectives are mutually exclusive.

Finally, following Ederveen et al. (2006), in order to assess the effect of institutional quality upon the effectiveness of the resources devoted to Objectives 1 and 2, an interaction term is used to explicitly capture the interaction of the SF with *EUqogIndex*, rather than investigating only the direct effect of each variable separately upon growth.

The estimated models are specified as follows:

Model 1—Base Model:

$$\begin{aligned} g_{i,T} = \frac{1}{T}\ln\left(\frac{GVA/EMP\_06_i}{GVA/EMP\_89_i}\right) &= \alpha_i + \beta_i \ln(GVA/EMP\_89_i) \\ &+ \varphi_{1i}\ln(DISC_i) + \varphi_{2i}\ln(INV/GVA_i) + \varphi_{3i}\ln(HRSTC/POP_i) \\ &+ \varphi_{4i} EUqogIndex_i + \varepsilon_i \end{aligned} \tag{2}$$

Model 2—Base model + SF:

$$\begin{aligned} g_{i,T} = \frac{1}{T}\ln\left(\frac{GVA/EMP\_06_i}{GVA/EMP\_89_i}\right) &= \alpha_i + \beta_i \ln(GVA/EMP\_89_i) \\ &+ \varphi_{1i}\ln(DISC_i) + \varphi_{2i}\ln(INV/GVA_i) + \varphi_{3i}\ln(HRSTC/POP_i) \\ &+ \varphi_{4i} EUqogIndex_i + \varphi_{5i} OB.1/EMP_i + \varphi_{6i} OB.2/EMP_i \\ &+ \varphi_{7i} EUqogIndex_i \times OB.1/EMP_i \\ &+ \varphi_{8i} EUqogIndex_i \times OB.2/EMP_i + \varepsilon_i \end{aligned} \tag{3}$$

Model 2 (*Base model + SF*, Eq. 3) takes into account the partial effect of the interaction term (holding all other variables fixed). For *OB.1/EMP*, the result is:

$$\frac{\partial g_{iT}}{\partial OB.1/EMP_i} = \varphi_{4i} + \varphi_{7i} EUqogIndex_i \tag{4}$$

implying, if $\varphi_{7i} > 0$, ceteris paribus, that regions with higher institutional quality have a greater probability that additional SF may translate into economic growth.

The aim of the interaction term is to identify when the partial effect of the dependent variable (average growth rate) with respect to an explanatory variable (SF for Objective 1 over employment) depends on the magnitude of another explanatory variable (institutional quality). In the case of positive coefficient φ_7, an increase in institutional quality yields a higher increase in the average growth rate in regions that benefit from better institutions. In order to evaluate the significance of the parameters

φ_4 and φ_7, their joint significance is tested (Wooldridge 2003). For *OB.2/EMP*, the interpretation is analogous.

The subscript i in parameters of the two models indicates that they are free to vary in each of the 180 NUTS2 regions. The spatial filtering technique adopted enables the estimation of both average and regional effects of the estimated parameters. Furthermore, according to Getis and Griffith (2002), when spatial filters are considered in combination with the ordinary least square (OLS) technique, although georeferenced data are used, residuals are not spatially autocorrelated.

When dealing with contiguous regions, it is widely demonstrated and accepted in the literature (López-Bazo et al. 2004) that residuals of growth equations are spatially autocorrelated. This finding—namely the existence of externalities across regional economies—if ignored, may lead to biased results and hence misleading conclusions.

In comparison to standard models used to address spatial autocorrelation, such as spatial autoregressive or spatial error, the advantage of spatial filtering model consists in the estimation of locally varying parameters. Furthermore, in contrast to Getis's (1995) approach that separately filters each variable, splitting its spatial from the non-spatial component, Griffith (2008) develops spatial filters exclusively on the basis of a spatial weights matrix **W**[8] and on the associated Moran Coefficient (*MC*) defined as[9]:

$$MC = \frac{n}{\sum_{i=1}^{n}\sum_{j=1}^{n} w_{i,j}} \frac{\sum_{i=1}^{n}\sum_{j=1}^{n} w_{i,j}(x_i - \overline{x}_i)(x_j - \overline{x}_j)}{\sum_{i=1}^{n}(x_i - \overline{x}_i)^2} = \frac{n}{\mathbf{1}^t\mathbf{W}\mathbf{1}} \frac{\mathbf{Y}^t\mathbf{MWMX}}{\mathbf{Y}^t\mathbf{MX}} \tag{5}$$

where i and j refer to different spatial units (i.e., NUTS2 regions centroids) of which there are n, x is the data value in each and $w_{i,j}$ is an element of **W**. The right side of Eq. 5 represents the matrix form of *MC*, where $\mathbf{M} = (\mathbf{I} - \mathbf{1}\mathbf{1}^t/n)$ is the matrix in which **I** is the identity matrix of size n-by-n, **1** is a vector of one dimension n-by-1, and the superscript t denotes the transposed matrix. The peculiarity of the **M** matrix is that it centers the vector of data value X.

Tiefelsdorf and Boots (1995) demonstrate that each eigenvalue derived from the decomposition of numerator of Eq. 5

$$\mathbf{MWM} \tag{6}$$

corresponds to a different *MC* value, once it is multiplied by the left-hand term of Eq. 5, namely $n/\mathbf{1}^t\mathbf{W}\mathbf{1}$.

With this nonparametric approach, through the use of eigenvectors, it is possible both to manage the presence of spatial autocorrelation (Getis and Griffith 2002) and, if included in a regression, to use these eigenvectors as predictors of variables not explicitly considered (Fischer and Griffith 2008). To achieve these purposes, the first

[8]The matrix that produced better results is a four-nearest neighbors row standardized matrix.

[9]To compute the *MC* statistic and spatial filters, a set of runtimes were developed in the R environment.

step is to choose a subset of eigenvectors from the entire set of eigenvectors on the basis of their eigenvalues (or MC values) ≥ 0.25 (Griffith 2003).[10] The following step, given P regressors, consists of computing the interactions terms for the P covariates multiplied by the candidate eigenvectors. From this subset, since the eigenvectors are both orthogonal and uncorrelated, a stepwise linear regression can be used to select the significant eigenvectors associated with each regressor (Tiefelsdorf and Boots 1995).

This spatial model can be considered a modified geographically weighted regression (GWR) model (Fotheringham et al. 2002) proposed by Griffith (2008). The model exploits the spatial filters through the construction of new variables created by the product between the spatial filter (i.e., the eigenvectors) and the associated regressors.

In a typical GWR model, instead of calibrating a single regression equation, a separate regression equation is generated for each observation. Each equation is then calibrated according to a distance-decay weighting scheme for each observation contained in the data set. On the contrary, in this chapter, regional coefficients are obtained calibrating a single regression through the spatial filters associated with each regressor. Each eigenvector is characterized by a MC value and hence by a geographic scale. A stepwise regression to select the significant eigenvectors for each variable enables each observation to be weighted with respect to its geographical scale, without the need of estimating separate regressions.

According to Griffith (2008), the spatial filtering technique ensures estimation of unbiased GWR-type coefficients avoiding the degrees-of-freedom problem associated with GWR. Furthermore, the more parsimonious coefficients exhibit considerably less multicollinearity (Wheeler 2007).

In a regression model where $\mathbf{Y}$ is a n-by-1 vector of the dependent variable, β_j is the jth regression coefficient and $\boldsymbol{\varepsilon}$ is a vector of the random error terms, the linear model with spatial filters incorporates a set P of regressors, $X_p = (k = 1, 2, \ldots, P)$, with a $\mathbf{k}$ set of selected eigenvectors, $\mathbf{E_k} = (k = 1, 2, \ldots, K)$. These represent different spatial models that take into account the residual spatial autocorrelation in the dependent variable and has the following form:

$$\mathbf{Y} = \beta_0 \mathbf{1} + \sum_{p=1}^{P} X_p \cdot \mathbf{1}\beta_p + \sum_{k=1}^{K} \mathbf{E_k}\beta_{E_k} + \sum_{p=1}^{P}\sum_{k=1}^{K} X_p \cdot \mathbf{E_k}\beta_{pE_k} + \boldsymbol{\varepsilon} \quad (7)$$

where $\cdot$ denotes element-wise matrix multiplication (i.e., Hadamard matrix multiplication). The regression coefficients β_0 and β_p denote global values, and the associated eigenvectors $\mathbf{E_k}$ with, respectively, coefficients β_{E_k} and β_{pE_k}, represent local modifications of these global values. The first two terms (i.e., the global attribute variable coefficients) are multiplied by the vector $\mathbf{1}$, which also is a spatial filter eigenvector. The sum of the first and third terms represents the GWR intercept, while the sum of the second and fourth ones are the regionally targeted parameters of the variables.

[10] A lower bound would cause the inclusion of eigenvectors that account for a spatial random process.

Global coefficients, needed to construct linear combinations of the eigenvectors, correspond to the impact of a variable in an average region, while eigenvectors represent local parameter estimation.

This approach, in principle independent of any economic theoretical justification, leads to a generalization of the Solow model in which each region follows this model, but their aggregate production functions and their steady states are free to vary.

4 Estimation Results

The first test performed in order to justify the use of a spatial filtering technique is the Moran's I test for the presence of spatial autocorrelation in each of the variables and in the residuals obtained from the standard OLS regressions. The results are displayed in Table 3.

The results preclude the classical assumption of spatial independence among observations and OLS residuals, validating the use of a spatial filtering technique, through which the assumption of spatial independence of observations is restored.

Table 3 demonstrates that regions with similar values are strongly clustered. In particular, the positive sign of Moran's I means that regions with high (low) values of a variable are surrounded by others with high (low) values. This is particularly evident for *EUqogIndex*, *ln(GVA/EMP_89)*, *ln(HRSTC/POP)*, *OB.1/EMP* and *EUqogIndex* $\times$ *OB.1/EMP* since each of these has a Moran's I statistic greater than 0.60 (as Moran's I varies between -1 and $+1$ for a row standardized **W** matrix). In particular, for *EUqogIndex*, the interpretation is relatively straightforward: Regional institutions are conditioned by the country to which they belong to, but this does not imply that

Table 3 Spatial autocorrelation of regression variables

Variable	Moran's I
Growth of GVA/EMP (dependent variable)	0.459***
ln(GVA/EMP_89)	0.618***
ln(DISC)	0.416***
ln(INV/GVA)	0.540***
ln(HRSTC/POP)	0.787***
EUqogIndex	0.846***
OB.1/EMP	0.772***
OB.2/EMP	0.287***
EUqogIndex × OB.1/EMP	0.772***
EUqogIndex × OB.2/EMP	0.227***
Residuals base model (Eq. 2)	0.234***
Residuals base model + SF (Eq. 3)	0.228***

Statistical significance: ***1%

they are equivalent within each country.[11] Finally, when the interaction term between *EUqogIndex* and *OB.1/EMP* is considered, the high spatial correlation shows that, jointly, these two variables are strongly related to their geographical location.

Estimation results based upon stepwise regression are presented in Table 4. They clearly show that a conditional convergence process is occurring among EU

Table 4 Global parameters of spatial filtering models

Variables	Model 1 (Base model)	Model 2 (Base model + SF)
Intercept	0.297300***	0.352600***
	(0.002296)	(0.017700)
ln(GVA/EMP_89)	−0.025780***	−0.031510***
	(0.001983)	(0.001508)
ln(DISC)	−0.005492**	−0.010900***
	(0.001895)	(0.001670)
ln(INV/GVA)	−0.006997***	−0.002179
	(0.001959)	(0.001726)
ln(HRSTC/POP)	0.021190***	0.023510***
	(0.001714)	(0.001706)
EUqogIndex	−0.000100	−0.000869
	(0.000626)	(0.000776)
OB.1/EMP		0.000006**
		(0.000002)
OB.2/EMP		−0.000035*
		(0.000002)
OB.1/EMP × EUqogIndex		0.000006***
		(0.000002)
OB.2/EMP × EUqogIndex		0.000009
		(0.000023)
Std.Breusch–Pagan test	83.575**, df = 60	42.101, df = 33
Moran's I	−0.213	−0.088
R-squared (adjusted)	0.958 (0.937)	0.920 (0.902)
Residual standard errors	0.002	0.003
AIC	−1610.436	−1546.583

In brackets: Standard errors
Statistical significance: ***1%, **5%, *10%

[11]This is evident in Italy, where the assessment of Institutions of southern regions is one of the worst in Europe, while the assessment of Institutions of the regions of northern Italy is immediately below German regions, which have very positive assessments.

regions. This is confirmed by the Wald test[12] on joint statistical significance of vector φ of coefficients of the conditioning variables. The parameter β improves from -0.0257 to -0.0315 when SF are included in the model, accelerating the convergence process. For the other parameters, in accordance with existing literature beginning with Mankiw et al. (1992), *HRSTC/POP* plays a positive role in the growth process, while the discount rate is significant and negative. Investment rate, which is negative and significant in *Model 1* (*Base model*), is not statistically significant when the additional variables are included.

Both models explain more than 90% of the variation in European regions' productivity growth, and the diagnostic tests upon residuals do not reveal spatial autocorrelation though heteroskedasticity is present in *Model 1* (*Base model*).

OB.1/EMP shows a positive impact upon growth, while resources devoted to Objective 2 have a negative influence upon the convergence process. Institutional quality is not statistically significant in either model; therefore, an independent effect upon growth of GVA per worker cannot be established. Regarding interaction between SF and institutional quality, in order to assess the impact of this interaction upon growth, as shown in Eq. (4), a Wald test upon their joint significance has to be performed. The result of the Wald test equal to 28.015 leads to reject H_0, i.e., $\varphi_4 = 0$ and $\varphi_7 = 0$, at the 1% level for *OB.1/EMP*, while for *OB.2/EMP*, and result of 4.2582 leads to reject H_0, i.e., $\varphi_4 = 0$ and $\varphi_8 = 0$, at the 5% level. Estimated coefficients are significant in both cases, showing that impact of SF must be, at least partially, attributed to the interaction with the quality of regional institutions.

While in case of Objective 1, there are no doubts regarding the positive impact of these SF on growth, even when considering the interaction term, for Objective 2, the case is different. As the two coefficients, φ_4 and φ_8, have different signs, the actual values of the variables determine whether the total effect is positive or negative. The combined effect for Objective 2, evaluated at the average value of the quality of institution index in the sample, has a value of -0.00003. Thus, the effect of the SF for a region with an average degree of institutional quality is negative.

These findings confirm that better regional institutions increase the effectiveness of SF for Objective 1 while do not modify the impact of Objective 2.

The local parameters of *Model 2* (*Base model* + *SF*) are reported in Table 5.[13] Local modifications of global, or mean, values for significant variables show that they widely vary across regions. The only variable for which this does not occur is *EUqogIndex*, the estimated value of which is constant, but not statistically significant. Coefficients associated with *OB.1/EMP* are positive in 94% of Objective 1 regions, confirming that these SF are a crucial factor for promoting convergence. The interaction term for SF for Objective 1 and regional institutional quality is positive in 83% of Objective 1 regions. This demonstrates the role of institutions for increasing the effectiveness of European regional policy in order to increase productivity growth.

---

[12]For *Model 1* (*Base model*), the Wald test is 9.949, excluding H_0 at 5% level, while for *Model 2* (*Base model* + *SF*), the Wald test is 13.598 and leading to do not accept H_0 at 1% level.

[13]In the appendix, speed of convergence of *Model 1* (*Base model*) and *Model 2* (*Base model* + *SF*) are reported.

Table 5 Local parameters spatial filtering *Model 2* (*Base model* + *SF*)

Variable	Min	First quart	Median	Mean	Third quart	Max
Intercept	0.05364	0.27021	0.32685	0.35264	0.43857	0.71667
ln(GVA/EMP_89)	−0.06824	−0.04003	−0.02887	−0.03151	−0.02340	−0.00120
ln(DISC)	−0.05056	−0.01182	−0.01088	−0.01090	−0.00946	0.02729
ln(INV/GVA)	−0.03061	−0.00377	−0.00230	−0.00218	−0.00048	0.02881
ln(HRSTC/POP)	−0.01442	0.02180	0.02356	0.02351	0.02501	0.06158
EUqogIndex	–	–	–	−0.00087	–	–
OB.1/EMP	−0.00001	4.0E−6	6.0E−6	6.0E−6	8.0E−6	1.6E−5
OB.2/EMP	−0.00063	−7.0E−6	−0.00003	−0.00003	0.00003	0.00020
EUqogIndex × OB.1/EMP	−0.00003	3.0E−6	6.0E−6	6.0E−6	0.00001	0.00005
EUqogIndex × OB.2/EMP	−0.00018	−0.00004	−4.0E−6	9.0E−6	0.00002	0.00025

The coefficients for Objective 2 and the interaction term for SF for Objective 2 and regional institutional quality are less homogeneous—they are negative in more than 50% of Objective 2 regions. This finding does not contrast with the previous finding regarding the global value because the SF for Objective 2 are able not always able to affect regional development. The resources in human capital variable have a strong positive impact on growth; the mean and the median are very close demonstrating that the coefficients do not vary substantially: The effects of human capital with respect to productivity growth, except three cases, is uniform both in sign and magnitude across European regions. Finally, the coefficients of the initial level of productivity, despite the different values of mean and median, are rather concentrated around the mean (with variance at 0.00017). There are no cases of diverging regions, and 42.5% of them have a coefficient of initial labor productivity lower than −0.03.

Figures 2 and 3 plot, respectively, the Moran scatterplot and the map of quintiles of initial level of labour productivity for *Model 2* (*Base model* + *SF*).

In Fig. 2, the first quadrant of the Moran scatterplot (HH) includes regions with high coefficient of *ln*(*GVA/EMP_89*) surrounded by others that also have high coefficient of *ln*(*GVA/EMP_89*). At the opposite, third quadrant (LL) contains regions with low coefficient of *ln*(*GVA/EMP_89*) surrounded by regions that also have low coefficient of *ln*(*GVA/EMP_89*). The second quadrant (LH) corresponds to regions with low values of the coefficient of *ln*(*GVA/EMP_89*) surrounded by others with high values. Finally, in the fourth quadrant (HL), there are regions with high coefficient of *ln*(*GVA/EMP_89*) surrounded by regions with low values.

The associations in the first and third quadrants pertain to a positive form of spatial autocorrelation between the regions and their neighbors, while the remaining two quadrants represent negative spatial autocorrelation.

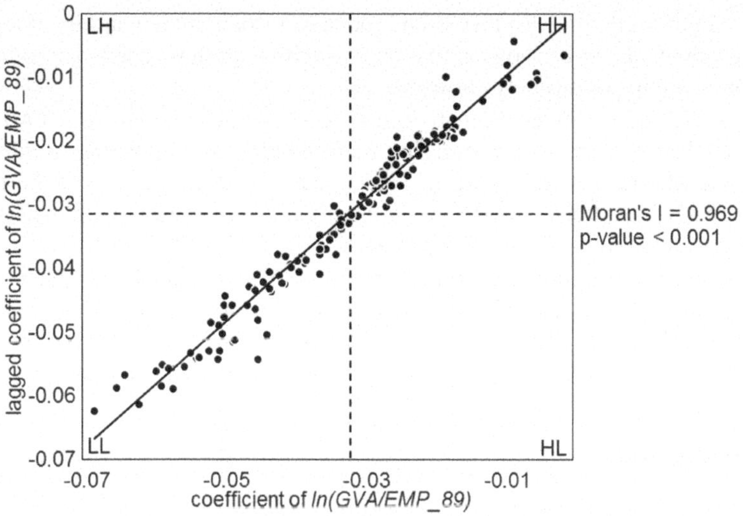

Fig. 2 Moran scatterplot of estimated coefficients of *ln(GVA/EMP_89)* in *Model 2* (*Base model + SF*)

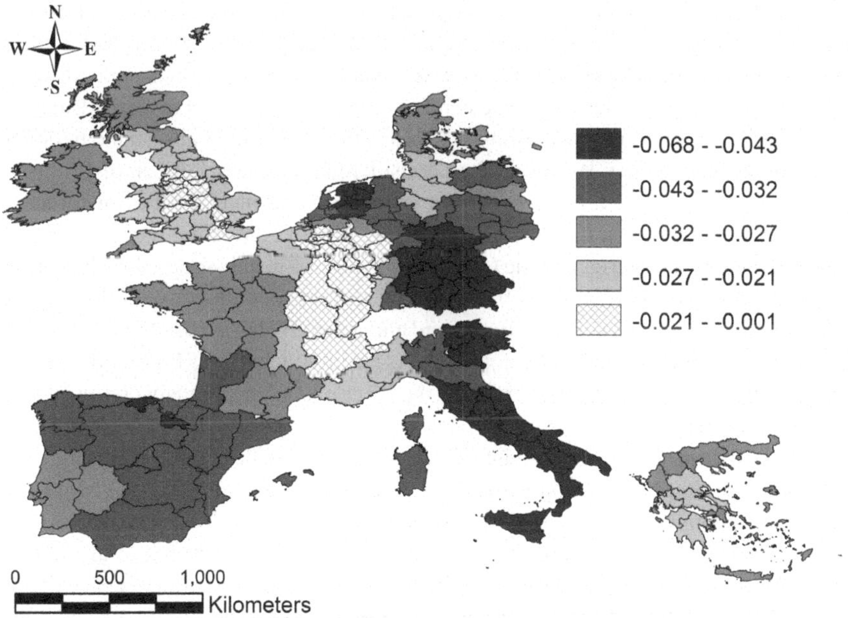

Fig. 3 Geographical distribution of EU Member States (1989) by quintile of estimated coefficient of *ln(GVA/EMP_89)* in *Model 2* (*Base model + SF*)

The Moran scatterplot displays strong spatial autocorrelation; majority of regions are in quadrants I and III showing that regions with high (low) coefficient of initial labor productivity are strongly clustered.

This is highlighted also in the map of quintiles of coefficients of *ln*(*GVA/EMP_89*) (Fig. 3). Clusters of regions converging more rapidly (corresponding to quadrant HH in the Moran scatterplot) are in Spain, Italy, Germany and the Netherlands, while Great Britain, Ireland, France, Belgium, Denmark and Greece pertain to slower convergence clusters (as in the LL quadrant).

These results, in accordance with the findings of Canova (2004) and Quah (1996, 1997), confirm that, in the EU, there are clusters of regions among which there are continuing differences in convergence rates levels.[14]

5 Conclusions

In this chapter, the impact of SF on the productivity growth of European regions has been evaluated. The spatial filtering technique adopted, besides avoiding biased results due to the presence of spatial dependence in the residuals, enables the estimation of both local parameters and convergence rates, without the necessity of imposing predetermined conditions (such as regional or country dummies variables, clubs). The results show that the effectiveness of SF upon productivity growth depends upon the allocation to particular Objectives and associated regions and is also influenced by the quality of regional institutions. SF for Objective 1, which absorb the majority of resources devoted to CP—even when considered separately from an institutional quality indicator—have a positive impact on growth and increase the convergence process. On the contrary, Funds with the precise aim of converting the regions seriously affected by industrial decline (Objective 2) show a negative sign with respect to productivity growth in roughly half of targeted regions. For Objective 2, the eligible areas within a region are very circumscribed and generally have a limited weight with respect to the overall regional development. As a consequence, this Fund shows a limited ability to invert the negative trend of those regions with seriously compromised growth potential.

The comparison with other studies adopting regional scale shows similar findings to the ones of Puigcerver-Peñalver (2007), Falk and Sinabell (2008) and Mohl and Hagen (2010) in terms of positive impact of SF for Objective 1 upon growth.

Furthermore, the results are in line with the findings of Ederveen et al. (2006), Bähr (2008) and Rodríguez-Pose and Garcilazo (2015) for the effectiveness of SF. The analogy is particularly close with Ederveen et al. (2006) and Rodríguez-Pose and Garcilazo (2015). In both, higher institutional quality increases the effectiveness of SF. Nevertheless, if corruption is concerned as the other side of the coin of institutional quality, the findings of the chapter contrast with those of Beugelsdijk and Eijffinger

[14]A case study approach would allow to understand the link between regions and the cluster to which they belong to.

(2005) that show that corruption does not undermine the effectiveness of SF with respect to growth.

According to the obtained results, SF and institutions show a kind of circular causality. Regions, having the main responsibility in the implementation of SF, need to create specific institutions in line with Community directives to reach the purpose of employing effectively SF. This process creates a *modus operandi* through which both the institutions and the regions benefit because the efficient use of SF is one of the preconditions for its assignment in the following programming period.

This is verified, especially for SF devoted to Objective 1, while it is partially true for Objective 2 because of its limited ability to influence the whole process of regional growth.

The absence of impact of institutional quality indicator upon productivity growth may be due to the construction of the variable. It refers, in fact, to the formal institutions and intentionally neglects the informal institutions, which, however, are among the factors that might contribute to regional growth.

In accordance with previous literature, the human capital indicator (HRSTC/POP) is confirmed to be a fundamental factor in fostering productivity. Its positive impact on growth, in contrast to negative or null impact of investment on physical capital, means that regional economies benefit more from investment in human, rather than in physical, capital. Looking at this phenomenon from another point of view, it implies that innovation resulting from human capital replaces more efficiently the depreciation of physical capital than what investment in physical capital does. In this regard, to gain higher productivity levels, it is more efficient to plan investment choices in human capital rather than in physical capital.

Local parameters show that SF for Objective 1 have positive signs in majority of regions, even when conditioned from institutional quality indicator. SF for Objective 2 have less uniform effects, while coefficients of resources in human capital are strongly concentrated and always positive. Moran's I and Moran scatterplot highlight clusters of regions with homogeneous values of parameter β.

This last finding, together with the significant coefficient of interaction term for Objective 1 SF and institutional quality leads to important consequences in reshaping CP for future programming periods. A progress of formal institutions leads to a more effectiveness CP implementation. It appears prudent, therefore, to accompany the support provided by CP with policies that facilitate improvement of regional institutional quality. Moreover, it should be necessary to complete the analysis upon the effectiveness of CP including the joint effect of formal and informal institutions and to understand how the latter affect the regional productivity growth.

Finally, another field of study for future research regards the comprehension of the factors that determine the success of certain regions in such a way to export, as far as possible, their best practices to other regions with lower economic performance. In this context, it is of primary importance to set up the right conditions to sustain the economic development such as the increasing of the R&D due to endogenous capabilities and territorial proximity.

Appendix: List of EU Regions Included in the Analysis and Associated Coefficient of *ln*(*GVA/EMP_89*)

Country	NUTS code	Name	Coefficient of *ln*(*GVA/EMP_89*) of *Model 2 (Base model + SF*)	Coefficient of *ln*(*GVA/EMP_89*) of *Model 1* (*Base model F*)
BE	BE10	Région de Bruxelles-Capitale	−0.00865	0.01526
BE	BE21	Prov. Antwerpen	−0.01288	0.01388
BE	BE22	Prov. Limburg	−0.01683	−0.00931
BE	BE23	Prov. Oost-Vlaanderen	−0.02198	−0.00746
BE	BE24	Prov. Vlaams-Brabant	−0.00930	0.01212
BE	BE25	Prov. West-Vlaanderen	−0.02512	−0.01372
BE	BE31	Prov. Brabant Wallon	−0.00930	0.01212
BE	BE32	Prov. Hainaut	−0.01673	0.00028
BE	BE33	Prov. Liège	−0.00987	−0.01503
BE	BE34	Prov. Luxembourg	−0.00513	−0.01270
BE	BE35	Prov. Namur	−0.00941	0.00095
DE	DE11	Stuttgart	−0.06502	−0.03573
DE	DE12	Karlsruhe	−0.04963	−0.03184
DE	DE13	Freiburg	−0.04009	−0.02895
DE	DE14	Tübingen	−0.05349	−0.03184
DE	DE21	Oberbayern	−0.05858	−0.03500
DE	DE22	Niederbayern	−0.05018	−0.03316
DE	DE23	Oberpfalz	−0.05695	−0.03535
DE	DE24	Oberfranken	−0.05868	−0.03609
DE	DE25	Mittelfranken	−0.06824	−0.03724
DE	DE26	Unterfranken	−0.06392	−0.03744
DE	DE27	Schwaben	−0.06190	−0.03468
DE	DE30	Berlin	−0.03184	−0.04930
DE	DE41	Brandenburg	−0.03147	−0.04439
DE	DE42	Brandenburg	−0.03285	−0.05236
DE	DE50	Bremen	−0.02660	−0.03265
DE	DE60	Hamburg	−0.02459	−0.03116
DE	DE71	Darmstadt	−0.04841	−0.03255
DE	DE72	Gießen	−0.04001	−0.02999
DE	DE73	Kassel	−0.04515	−0.03302

(continued)

(continued)

Country	NUTS code	Name	Coefficient of *ln*(*GVA/EMP_89*) of *Model 2* (*Base model* + *SF*)	Coefficient of *ln*(*GVA/EMP_89*) of *Model 1* (*Base model F*)
DE	DE80	Mecklenburg-Vorpommern	−0.03174	−0.03855
DE	DE91	Braunschweig	−0.03423	−0.03446
DE	DE92	Hannover	−0.02600	−0.03413
DE	DE93	Lüneburg	−0.02348	−0.03349
DE	DE94	Weser-Ems	−0.03879	−0.02609
DE	DEA1	Düsseldorf	−0.02831	−0.02092
DE	DEA2	Köln	−0.01794	−0.02204
DE	DEA3	Münster	−0.03586	−0.02425
DE	DEA4	Detmold	−0.03306	−0.03174
DE	DEA5	Arnsberg	−0.03305	−0.02725
DE	DEB1	Koblenz	−0.01651	−0.02428
DE	DEB2	Trier	−0.00507	−0.02199
DE	DEB3	Rheinhessen-Pfalz	−0.02970	−0.02814
DE	DEC0	Saarland	−0.00589	−0.02300
DE	DED1	Chemnitz	−0.03807	−0.04172
DE	DED2	Dresden	−0.03316	−0.04500
DE	DED3	Leipzig	−0.03388	−0.05044
DE	DEE0	Sachsen-Anhalt	−0.03564	−0.04078
DE	DEF0	Schleswig-Holstein	−0.02459	−0.03116
DE	DEG0	Thüringen	−0.04473	−0.03620
DK	DK0	Danmark	−0.02742	−0.02887
EL	EL11	Anatoliki Makedonia, Thraki	−0.03449	−0.03559
EL	EL12	Kentriki Makedonia	−0.03832	−0.03722
EL	EL13	Dytiki Makedonia	−0.02832	−0.02180
EL	EL14	Thessalia	−0.02674	−0.02669
EL	EL21	Ipeiros	−0.02826	−0.02708
EL	EL22	Ionia Nisia	−0.02807	−0.03972
EL	EL23	Dytiki Ellada	−0.02710	−0.04447
EL	EL24	Sterea Ellada	−0.02582	−0.04036
EL	EL25	Peloponnisos	−0.02699	−0.04804
EL	EL30	Attiki	−0.02726	−0.04189
EL	EL41	Voreio Aigaio	−0.02987	−0.02859
EL	EL42	Notio Aigaio	−0.02978	−0.03640
EL	EL43	Kriti	−0.02967	−0.03824

(continued)

(continued)

Country	NUTS code	Name	Coefficient of $ln(GVA/EMP\_89)$ of *Model 2 (Base model + SF)*	Coefficient of $ln(GVA/EMP\_89)$ of *Model 1 (Base model F)*
ES	ES11	Galicia	−0.02972	−0.00708
ES	ES12	Principado de Asturias	−0.02978	−0.00795
ES	ES13	Cantabria	−0.04389	−0.03369
ES	ES21	País Vasco	−0.04283	−0.02951
ES	ES22	Comunidad Foral de Navarra	−0.04170	−0.02842
ES	ES23	La Rioja	−0.04599	−0.03179
ES	ES24	Aragón	−0.04304	−0.03458
ES	ES30	Comunidad de Madrid	−0.04089	−0.03087
ES	ES41	Castilla y León	−0.04129	−0.03457
ES	ES42	Castilla-la Mancha	−0.03872	−0.02989
ES	ES43	Extremadura	−0.03056	−0.03322
ES	ES51	Cataluña	−0.03587	−0.03776
ES	ES52	Comunidad Valenciana	−0.03908	−0.03316
ES	ES53	Illes Balears	−0.03496	−0.03412
ES	ES61	Andalucía	−0.03587	−0.02921
ES	ES62	Región de Murcia	−0.03775	−0.02949
FR	FR10	Île de France	−0.02784	−0.01839
FR	FR21	Champagne-Ardenne	−0.01813	−0.02104
FR	FR22	Picardie	−0.02645	−0.01638
FR	FR23	Haute-Normandie	−0.02749	−0.01914
FR	FR24	Centre	−0.02800	−0.02013
FR	FR25	Basse-Normandie	−0.02724	−0.02294
FR	FR26	Bourgogne	−0.01720	−0.02707
FR	FR30	Nord	−0.02573	−0.01530
FR	FR41	Lorraine	−0.00836	−0.02221
FR	FR42	Alsace	−0.02476	−0.02621
FR	FR43	Franche-Comté	−0.01708	−0.02755
FR	FR51	Pays de la Loire	−0.02881	−0.02437
FR	FR52	Bretagne	−0.02909	−0.02755
FR	FR53	Poitou–Charentes	−0.02908	−0.02195
FR	FR61	Aquitaine	−0.03287	−0.02552
FR	FR62	Midi-Pyrénées	−0.03135	−0.03218
FR	FR63	Limousin	−0.02751	−0.02596
FR	FR71	Rhône-Alpes	−0.01573	−0.03007

(continued)

(continued)

Country	NUTS code	Name	Coefficient of $ln(GVA/EMP\_89)$ of *Model 2 (Base model + SF)*	Coefficient of $ln(GVA/EMP\_89)$ of *Model 1 (Base model F)*
FR	FR72	Auvergne	−0.02372	−0.02943
FR	FR81	Languedoc-Roussillon	−0.03018	−0.03472
FR	FR82	Provence-Alpes-Côte d'Azur	−0.02145	−0.03165
FR	FR83	Corse	−0.03360	−0.03846
IE	IE01	Border, Midland and Western	−0.02945	−0.02447
IE	IE02	Southern and Eastern	−0.02839	−0.02145
IT	ITC1	Piemonte	−0.02273	−0.03570
IT	ITC2	Valle d'Aosta	−0.01868	−0.03323
IT	ITC3	Liguria	−0.02450	−0.03897
IT	ITC4	Lombardia	−0.03287	−0.04005
IT	ITD5	Emilia-Romagna	−0.05180	−0.03658
IT	ITE1	Toscna	−0.04515	−0.04081
IT	ITE2	Umbria	−0.04630	−0.03866
IT	ITE3	Marche	−0.04347	−0.03312
IT	ITE4	Lazio	−0.04299	−0.04520
IT	ITF1	Abruzzo	−0.04827	−0.05604
IT	ITF2	Molise	−0.05309	−0.06160
IT	ITF3	Campania	−0.05309	−0.06160
IT	ITF4	Puglia	−0.05728	−0.06795
IT	ITF5	Basilicata	−0.05766	−0.06698
IT	ITF6	Calabria	−0.05936	−0.06709
IT	ITG1	Sicilia	−0.05536	−0.05978
IT	ITG2	Sardegna	−0.05045	−0.05225
IT	ITH1	Provincia Autonoma di Bolzano	−0.05446	−0.05803
IT	ITH2	Provincia Autonoma di Trento	−0.04800	−0.04856
IT	ITH3	Veneto	−0.04475	−0.04429
IT	ITH4	Friuli-Venezia Giulia	−0.03583	−0.03930
LU	LU0	Luxembourg	−0.00117	−0.01872
NL	NL11	Groningen	−0.04957	−0.02334
NL	NL12	Friesland	−0.04987	−0.02180
NL	NL13	Drenthe	−0.05030	−0.02261
NL	NL21	Overijssel	−0.05148	−0.02107

(continued)

(continued)

Country	NUTS code	Name	Coefficient of *ln*(*GVA/EMP_89*) of *Model 2 (Base model + SF*)	Coefficient of *ln*(*GVA/EMP_89*) of *Model 1* (*Base model F*)
NL	NL22	Gelderland	−0.04488	−0.01823
NL	NL23	Flevoland	−0.04948	−0.01562
NL	NL31	Utrecht	−0.04195	−0.00958
NL	NL32	Noord-Holland	−0.04076	−0.01246
NL	NL33	Zuid-Holland	−0.03385	−0.00709
NL	NL34	Zeeland	−0.02494	−0.00682
NL	NL41	Noord-Brabant	−0.02743	−0.00345
NL	NL42	Limburg	−0.02102	−0.01447
PT	PT11	Norte	−0.03289	−0.03683
PT	PT15	Algarve	−0.02936	−0.02915
PT	PT16	Centro	−0.02968	−0.03434
PT	PT17	Lisboa	−0.02910	−0.03000
PT	PT18	Alentejo	−0.02893	−0.03068
UK	UKC1	Tees Valley and Durham	−0.02386	−0.02209
UK	UKC2	Northumberland and Tyne and Wear	−0.02355	−0.02073
UK	UKD1	Cumbria	−0.02210	−0.01628
UK	UKD2	Cheshire	−0.01766	0.00291
UK	UKD3	Greater Manchester	−0.01829	0.00215
UK	UKD4	Lancashire	−0.01914	−0.00389
UK	UKD5	Merseyside	−0.01961	−0.00160
UK	UKE1	East Yorkshire and Northern Lincolnshire	−0.02391	−0.01020
UK	UKE2	North Yorkshire	−0.02187	−0.01344
UK	UKE3	South Yorkshire	−0.02110	−0.00472
UK	UKE4	West Yorkshire	−0.01989	−0.00404
UK	UKF1	Derbyshire and Nottinghamshire	−0.01904	0.00295
UK	UKF2	Leicestershire, Rutland and Northamptonshire	−0.01952	0.00173
UK	UKF3	Lincolnshire	−0.02505	−0.00954
UK	UKG1	Herefordshire, Worcestershire and Warwickshire	−0.02081	−0.01085
UK	UKG2	Shropshire and Staffordshire	−0.01910	−0.00064
UK	UKG3	West Midlands	−0.01942	−0.00163

(continued)

(continued)

Country	NUTS code	Name	Coefficient of *ln(GVA/EMP_89)* of *Model 2 (Base model + SF)*	Coefficient of *ln(GVA/EMP_89)* of *Model 1* (*Base model F*)
UK	UKH1	East Anglia	−0.02577	−0.01244
UK	UKH2	Bedfordshire and Hertfordshire	−0.01789	0.00192
UK	UKH3	Essex	−0.02151	−0.00534
UK	UKI1	Inner London	−0.01864	−0.00086
UK	UKI2	Outer London	−0.01690	0.00066
UK	UKJ1	Berkshire, Buckinghamshire and Oxfordshire	−0.02184	−0.00881
UK	UKJ2	Surrey, East and West Sussex	−0.01973	−0.00599
UK	UKJ3	Hampshire and Isle of Wight	−0.02314	−0.02022
UK	UKJ4	Kent	−0.02203	−0.00747
UK	UKK1	Gloucestershire, Wiltshire and Bristol/Bath area	−0.02410	−0.02361
UK	UKK2	Dorset and Somerset	−0.02513	−0.02655
UK	UKK3	Cornwall and Isles of Scilly	−0.02765	−0.02023
UK	UKK4	Devon	−0.02695	−0.02413
UK	UKL1	West Wales and The Valleys	−0.02684	−0.01633
UK	UKL2	East Wales	−0.02506	−0.01442
UK	UKM2	Eastern Scotland	−0.02833	−0.02089
UK	UKM3	South Western Scotland	−0.02732	−0.02014
UK	UKM5	North Eastern Scotland	−0.02695	−0.01394
UK	UKM6	Highlands and Islands	−0.02798	−0.01661
UK	UKN0	Northern Ireland	−0.02984	−0.00706

References

Amin, A. (1999). An institutionalist perspective on regional development. *International Journal of Urban and Regional Research, 23*(2), 365–378.

Arbia, G., Battisti, M., & Di Vaio, G. (2010). Institutions and geography: Empirical test of spatial growth models for European regions. *Economic Modelling, 27*(1), 12–21.

Bachtler, J., & Gorzelak, G. (2007). Reforming EU cohesion policy a reappraisal of the performance of the structural funds. *Policy Studies, 28*(4), 309–326.

Bähr, C. (2008). How does sub-national autonomy affect the effectiveness of structural funds? *Kyklos, 61*(1), 3–18.

Barro, R. J., & Sala-i-Martin, X. (1990). *Economic growth and convergence across the United States*. NBER Working Paper 3419.

Becker, S. O., Egger, P., & Von Ehrlich, M. (2010). Going NUTS: The effect of EU structural funds on regional performance. *Journal of Public Economics, 94*(9–10), 578–590.

Beugelsdijk, M., & Eijffinger, S. C. W. (2005). The effectiveness of structural policy in the European Union: An empirical analysis for the EU-15 during the period 1995–2001. *Journal of Common Market Studies, 43*(1), 37–51.

Boldrin, M., & Canova, F. (2001). Inequality and convergence in Europe's regions: Reconsidering European regional policies. *Economic Policy, 16*(32), 205–253.

Cambridge Econometrics Database 2010

Canova, F. (2004). Testing for convergence clubs: A predictive density approach. *International Economic Review, 45*(1), 49–78.

Cappelen, A., Castellacci, F., Fagerberg, J., & Verspagen, B. (2003). The impact of EU regional support on growth and convergence in the European Union. *Journal of Common Market Studies, 41*(4), 621–644.

Charron, N., Lapuente, V., & Rothstein, B. (Eds.). (2013). *Measuring the quality of government in the European Union: A comparative analysis of national and regional variation*. Cheltenham UK, Northampton USA: Edward Elgar Publishing.

Dall'erba, S., & Le Gallo, J. (2007). The impact of EU regional support on growth and employment. *Czech Journal of Economics and Finance, 57*(7–8), 325–340.

Dall'erba, S., & Le Gallo, J. (2008). Regional convergence and the impact of European structural funds 1989–1999: A spatial econometric analysis. *Papers in Regional Science, 82*(2), 219–244.

Ederveen, S., Gorter, J., De Mooij, R., & Nahuis, R. (2002). *Funds and games: The economics of European Cohesion Policy*. Occasional Paper No. 3. Brussels: European Network of European Policy Research Institutes. https://aei.pitt.edu/1839/1/ENEPRI_OP5.pdf. Accessed May 13, 2013.

Ederveen, S., Le Groot, H., & Nahuis, R. (2006). Fertile soil for structural funds? A panel data analysis of the conditional effectiveness of European Cohesion Policy. *Kyklos, 59*(1), 17–42.

Ederveen, S., Van Der Horst, A., & Tang, P. (2005). *Is the European economy a patient and the union its doctor? On jobs and growth in Europe*. ENEPRI Working Paper No. 35/April 2005. https://aei.pitt.edu/6739/1/1218_35.pdf. Accessed May 13, 2013.

Esposti, R., & Bussoletti, S. (2008). Impact of Objective 1 funds on regional growth convergence in the European Union: A panel-data approach. *Regional Studies, 42*(2), 159–173.

EuroGeographics for the administrative boundaries. https://ec.europa.eu/eurostat/web/gisco. Accessed May 13, 2013.

European Commission. (1995). *Fifth Annual Report on the implementation of the reform of structural funds 1993*, Brussels.

European Commission. (1997). *The impact of structural policies on economic and social cohesion in the Union 1989–99 a first assessment presented by country (October 1996): regional development studies*, Luxembourg.

European Commission. (1999). *The structural funds in 1998*. Tenth Annual Report, Brussels.

European Commission. (2004). *A new partnership for cohesion, convergence, competitiveness, cooperation*. Third Report on Economic and Social Cohesion. Luxembourg: Office for Official Publications of the EC.

European Commission. (2006). *18th Annual Report on implementation of the structural funds*. Luxembourg: Office for Official Publications of the EC.

European Commission. (2007). *Growing regions, growing Europe*. Fourth Report on Economic and Social Cohesion. Luxembourg: Office for Official Publications of the EC.

European Commission. (2010). *Investing in Europe's future*. Fifth Report on Economic, Social and Territorial Cohesion. Luxembourg: Office for Official Publications of the EC.

Eurostat Region Database. https://ec.europa.eu/eurostat/web/regions/data/database. Accessed May 13, 2013.

Fagerberg, J., & Verspagen, B. (1996). Heading for divergence? Regional growth in Europe reconsidered. *Journal of Common Market Studies, 34*(3), 431–448.

Falk, M., & Sinabell, F. (2008). *The effectiveness of Objective 1 structural funds in the EU15: New empirical evidence from NUTS 3 regions*. WIFO Working Papers 310. https://franz.sinabell.wifo.ac.at/papers/WP_2007_310.PDF. Accessed May 13, 2013.

Fischer, M., & Griffith, D. A. (2008). Modeling spatial autocorrelation in spatial interaction data: An application to patent citation data in the European Union. *Journal of Regional Science, 48*(5), 969–989.

Fotheringham, A. S., Brunsdon, C., & Charlton, M. (2002). *Geographically weighted regression: The analysis of spatially varying relationships*. West Sussex: Wiley.

Getis, A. (1995). Spatial filtering in a regression framework: Experiments on regional inequality, government expenditures, and urban crime. In L. Anselin & R. J. G. M. Florax (Eds.), *New directions in spatial econometrics* (pp. 172–188). Berlin: Springer.

Getis, A., & Griffith, D. A. (2002). Comparative spatial filtering in regression analysis. *Geographical Analysis, 34*(2), 130–140.

Griffith, D. A. (2003). *Spatial autocorrelation and spatial filtering: Gaining understanding through theory and scientific visualization*. Berlin: Springer.

Griffith, D. A. (2008). Spatial filtering-based contribution to a critique of geographically weighted regression (GWR). *Environment and Planning A, 40*(11), 2751–2769.

Hagen, T., & Mohl, P. (2008). *Which is the right dose of EU Cohesion Policy for economic growth?* ZEW Discussion Paper 08-104. https://ftp.zew.de/pub/zew-docs/dp/dp08104.pdf. Accessed May 13, 2013.

Kaufmann, D., Kraay, A., & Mastruzzi, M. (2009). *Governance matters VIII: Aggregate and individual governance indicators for 1996–2008*. World Bank Policy Research Working Paper No. 4978. https://www-wds.worldbank.org/servlet/WDSContentServer/WDSP/IB/2009/06/29/000158349_20090629095443/Rendered/PDF/WPS4978.pdf. Accessed May 13, 2013.

Krugman, P. (1992). *The age of diminished expectations: US economic policy in the 1980s*. Cambridge: MIT Press.

Leonardi, R. (2005). *The cohesion policy of the European Union: The building of Europe*. London: Palgrave.

Leonardi, R. (2006). The impact and added value of cohesion policy. *Regional Studies, 40*(2), 155–166.

López-Bazo, E., Vayá, E., & Artís, M. (2004). Regional externalities and growth: Evidence from European regions. *Journal of Regional Science, 44*(1), 43–73.

Mankiw, G. N., Romer, D., & Weil, D. N. (1992). A contribution to the empirics of economic growth. *Quarterly Journal of Economics, 107*(2), 407–437.

Mohl, P., & Hagen, T. (2010). Do EU structural funds promote regional growth? New evidence from various panel data approaches. *Regional Science and Urban Economics, 40*(5), 353–365.

OECD. (2009). *Investing for growth: Building innovative regions*. Background Report for the Meeting of the Territorial Development Policy Committee at Ministerial Level. Paris: OECD Publishing.

OECD. (2011). *OECD science, technology and industry scoreboard 2011*. OECD Publishing. https://www.oecd-ilibrary.org/science-and-technology/oecd-science-technology-and-industry-scoreboard-2011_sti_scoreboard-2011-en. Accessed May 13, 2013.

Puga, D. (2002). European regional policies in light of recent location theories. *Journal of Economic Geography, 2*(4), 373–406.

Puigcerver-Peñalver, M. (2007). The impact of structural funds policy on European regions' growth: A theoretical and empirical approach. *The European Journal of Comparative Economics, 4*(2), 179–208.

Quah, D. (1996). Regional convergence clusters across Europe. *European Economic Review, 40*(3–5), 951–958.

Quah, D. (1997). Empirics for growth and distribution: Stratification, polarisation and convergence clubs. *Journal of Economic Growth, 2*, 101–120.
Ramajo, J., Márquez, M., Hewings, G., & Salinas, S. (2008). Spatial heterogeneity and interregional spillovers in the European Union: Do cohesion policies encourage convergence across regions? *European Economic Review, 52*(3), 551–567.
Rodríguez-Pose, A., & Fratesi, U. (2004). Between development and social policies: The impact of European structural funds in Objective 1 regions. *Regional Studies, 38*(1), 97–113.
Rodríguez-Pose, A., & Garcilazo, E. (2015). Quality of government and the returns of investment: Examining the impact of cohesion expenditure in European regions. *Regional Studies, 49*(8), 1274–1290.
Rodríguez-Pose, A., & Storper, M. (2006). Better rules or stronger communities? On the social foundations of institutional change and its economic effects. *Economic Geography, 82*(1), 1–25.
Schultz, T. W. (1961). Investment in human capital. *The American Economic Review, 51*(11), 1–17.
Tiefelsdorf, M., & Boots, B. (1995). The exact distribution of Moran's I. *Environment and Planning A, 27*(6), 985–999.
Treaty Establishing the European Community. (2002). *Official Journal of the European Communities*. https://eur-lex.europa.eu/legal-content/EN/TXT/?uri=CELEX%3A12002E%2FTXT. Accessed May 13, 2013.
Van Ark, B. (2006). Does the European Union need to revive productivity growth? In S. Mundschenk, M. H. Stierl, U. Stierle von Schütz, & I. Traistaru (Eds.), *Competitiveness and growth in Europe. Lessons and policy implications for the Lisbon strategy* (pp. 101–126). Cheltenham: Edward Elgar Publishers.
Wheeler, D. (2007). Diagnostic tools and a remedial method for collinearity in geographically weighted regression. *Environment and Planning A, 39*(10), 2464–2481.
Wooldridge, J. M. (2003). *Introductory econometrics: A modern approach*. Mason: Thomson.

Overlapping Jurisdictions and Demand for Local Public Services: Does Spatial Heterogeneity Matter?

Marie-Estelle Binet, Alain Guengant, and Matthieu Leprince

1 Introduction

In every country with a decentralized form of government, there are multiple tiers of government services that may interact at the local level. This multi-tiered form of organization often is viewed as being economically inefficient because different levels of government may provide similar or related services. If evidence is found to support this view, it will have important implications for debates over decentralization and the optimal provision of public sector services at the local level. France is especially concerned by these debates since three main levels of local governments share the responsibilities to provide local public services. The main level in terms of public spending is the municipal level, with 56% of total local spending in 2011, the intermediate level of local government—called department—and the higher level—called regions—corresponding, respectively, to 32% and 12% of the total local spending. This overlapping structure of the French local public sector is frequently seen as responsible for the high increase of public spending in France, as suggested for example by the French Ministry of Finance (2012).

In the economics literature, inflation in total local public spending is assumed to be produced by a complementary effect, that is to say a vertical positive interaction between demands for public services provided by overlapping jurisdictions (Turnbull and Djoundourian 1993). The provision of public services then depends on the preferences of the median voter. At the lowest (municipal) level, it is assumed that the local incumbent adopts policies in the median voter's interest, taking into account public spending decisions at the next higher level (in the USA, for example, the county). A complementary relation appears as the median voter would increase

M.-E. Binet (✉) · A. Guengant
GAEL, INRAE, CNRS, Grenoble INP, University Grenoble Alpes, 38000 Grenoble, France
e-mail: marie-estelle.binet@iepg.fr

A. Guengant · M. Leprince
Univ Brest, Ifremer, CNRS, UMR 6308, AMURE, IUEM, F-29280 Plouzane, France

J. Thill (ed.), *Innovations in Urban and Regional Systems*,
https://doi.org/10.1007/978-3-030-43694-0_14

her demand for municipal public goods when public spending at the county level increases. Therefore, if expenditures at one level are a complement to expenditures at the other level, the overlapping structure increases the size of the local public sector.

This model is usually implemented empirically by estimating a demand for municipal public expenditure functions conditional upon county expenditures, in the spirit of Pollack (1971). To our knowledge, three articles have examined the existence of such a complementary effect but they do not explicitly consider the spatial dimensions of their data, thus relying on restrictions in their estimated models of public spending that might have implications on their results. We therefore try to bridge this gap in this chapter with the goal of providing a more general and flexible test of the hypothesis of vertical interactions in local spending. Studying general purpose expenditures in US municipalities and counties, Turnbull and Djoundourian (1993) and Campbell (2004) find that municipal per capita expenditures and county per capita expenditures are complementary.[1] Using panel data for Sweden over the period 1981–1986, Aronsson et al. (2000) also find a positive correlation between county and municipal expenditures. Revelli (2003) alone includes spatial autocorrelation among district expenditures in the analytical specification. He observes a spatial vertical externality among county and district expenditures and shows the importance of taking into account both vertical and horizontal externalities in the demand for public services at the lowest level.

In light of these findings, this chapter aims to expand upon the existing literature in two ways. First, we analyze for the first time the links between the two most important levels of local government (e.g., municipal and departmental) in France. The conditional demand function then is estimated on a cross-sectional sample of French municipalities and their departments, taking into account the spatial dependence among municipal public expenditures. Second, in the aforementioned studies, the authors focus on a global demand specification, with the same coefficient for all of the statistical units. The second innovative feature of the present work therefore is to introduce spatial heterogeneity in the conditional demand function that is estimated with spatial regimes.

Spatial heterogeneity is relevant when data are obtained for a cross-section of spatial units, as suggested by Anselin (1990). In practice, spatial heterogeneity can be reflected by heteroskedasticity in the error term, coefficients varying with the location, or both. Due to historical, cultural, and economic differences at the regional level, France is known to be characterized by strong spatial heterogeneity. For example, about 30% of the French gross domestic product (GDP) is produced in the Ile de France region, the area around and including the capital city of Paris. The country's territory therefore can be divided into a periphery constituted by a group of different regions and the core region 'Ile de France.'

This research is therefore an original contribution to empirical regional science literature, and uses a dataset covering all of the largest French municipalities (except Paris) and their corresponding departments for two different years: 2001 and 2005.

[1] Turnbull and Djoundourian (1993) do not confirm the complementary effect for individual service categories.

We first estimate a standard spatial error model and find a complementary effect. However, once we include spatial heterogeneity in our econometric specification (considering two different spatial divisions of France), we find evidence of independence between departmental and municipal spending decisions in most French regions. 'Ile de France' is the only French region systematically characterized by a complementary effect. Therefore, as suggested by Anselin (1990), the presence of spatial heterogeneity has implications for the generality of regional science theories and the induced policy recommendations. In our case, results suggest that the overlapping structure of local governments in France does not favor the increase in local spending once spatial heterogeneity is accounted for, contrary to what is usually stated.

The rest of this chapter proceeds as follows: In Sect. 2, we discuss the theoretical framework, and in Sect. 3, we develop the empirical methodology without including spatial heterogeneity. In Sect. 4, we present the empirical findings of the two spatial regimes specifications and in Sect. 5, we present our conclusions.

2 Conditional Demand for Municipal Public Services

Turnbull and Djoundourian (1993) and Aronsson et al. (2000) developed a model of municipal expenditures which assumes that municipal expenditures are conditioned by the expenditures at the next higher administrative level, the county. They assume that the median voter utility function depends on private consumption x and public services provided at the municipal level z_{m} and z_{c}^{0} at the county level (subscripts m and c identify municipal and county variables, respectively).

Maximizing the utility function under budget constraint, Turnbull and Djoundourian (1993) and Aronsson et al. (2000) derive the conditional demand for municipal public services:

$$z_{\mathrm{m}} = z_{\mathrm{m}}\left(y_{\mathrm{m}}^{\mathrm{m\acute{e}}*}, p_{\mathrm{m}}^{\mathrm{m\acute{e}}}, z_{\mathrm{c}}^{0}\right) \tag{1}$$

$y_{\mathrm{m}}^{\mathrm{m\acute{e}}*}$ can be measured by the income of the median municipal voter less the taxes she pays at the county level. $p_{\mathrm{m}}^{\mathrm{m\acute{e}}}$ denotes the median voter tax price:

$$p_{\mathrm{m}}^{\mathrm{m\acute{e}}} = \left(\frac{b_{\mathrm{m+c}}^{\mathrm{m\acute{e}}}}{b_{\mathrm{m}}}\right) p_{\mathrm{m}} \tag{2}$$

$b_{\mathrm{m+c}}^{\mathrm{m\acute{e}}}$ is the median voter (residential) tax base shared by the municipality and its county. b_{m} is the average municipal tax base, including the local business tax (respectively b_{c} for the county). The tax share $\frac{b_{\mathrm{m+c}}^{\mathrm{m\acute{e}}}}{b_{\mathrm{m}}}$ can be defined as the proportion of the tax burden associated with residential property. The tax price also includes unit costs of public services and the size of the population so as to deal with congestion effects.

The linkages between the two levels of local jurisdictions come from two different effects, which are called income and preferences (or taste) effects. The impact of additional county-level expenditure on the demand for the municipal public goods can be analyzed through the following first-order condition:

$$\frac{dz_{\mathrm{m}}}{dz_{\mathrm{c}}^{0}} = \underbrace{\frac{U''_{\mathrm{mc}} - p_{\mathrm{m}}^{\mathrm{m\acute{e}}} U''_{\mathrm{xc}}}{\Delta}}_{\text{preferences effect}} \underbrace{- \left(\frac{b_{\mathrm{m+c}}^{\mathrm{m\acute{e}}}}{b_{\mathrm{c}}} p_{\mathrm{c}} \right) \frac{U''_{\mathrm{xm}} - p_{\mathrm{m}}^{\mathrm{m\acute{e}}} U''_{xx}}{\Delta}}_{\text{income effect}}, \tag{3}$$

With $\Delta = 2p_{\mathrm{m}}^{\mathrm{m\acute{e}}} U''_{x\mathrm{m}} - U''_{\mathrm{mm}} - \left(p_{\mathrm{m}}^{\mathrm{m\acute{e}}}\right)^2 U''_{xx} > 0$ and U''_{ij} is the second-order partial derivatives of utility with respect to *i* and *j* (*x*: private consumption, *m*: municipal, and *c*: county). The income effect can easily be identified. If municipal public goods and services are normal, an increase in the public services at the county level financed by a tax on the median voter will reduce the median voter's income and then her demand for municipal public services. The preferences effect is indeterminate and depends on the cross partial derivatives of the utility function:

$$\left. \frac{\partial z_{\mathrm{m}}}{\partial z_{\mathrm{c}}} \right|_{\substack{p_{\mathrm{m}}^{\mathrm{m\acute{e}}} \\ y_{\mathrm{m}}^{\mathrm{m\acute{e}}}}} = \frac{U''_{\mathrm{mc}} - p_{\mathrm{m}}^{\mathrm{m\acute{e}}} U''_{x\mathrm{c}}}{\Delta}. \tag{4}$$

Services provided at the two different levels of the local public sector can be considered as complementary, substitute, or independent by the median voter. When larger allocations of county goods increase the willingness of the median voter to pay for municipal public goods, the goods are considered to be complements. This is referred to as the complementary effect. In contrast, when increases in county services decrease the marginal rate of substitution between municipal public services and private consumption, then county and municipal services are considered to be substitutes.

Formally, the conditional expenditure function estimated for municipal services can be expressed as:

$$\mathrm{EXP}_{\mathrm{m}} = \alpha_0 + \alpha_p \frac{b_{\mathrm{m+c}}^{\mathrm{m\acute{e}}}}{b_{\mathrm{m}}} + \alpha_y y_{\mathrm{m}}^{\mathrm{m\acute{e}}*} + \delta\, z_d^0 + X\beta + \varepsilon, \tag{5}$$

where ε is the error term. Price and income elasticities of demand are obtained using parameters α_p and α_y, respectively, and mean values of median income and tax share. The parameter δ accounts for the influence of public spending at a higher level (e.g., county) on municipal public expenditure. If $\hat{\delta} > 0$, the two public goods are considered to be complements whereas $\hat{\delta} < 0$ reflects a substitute relationship. We consider them to be independent or unrelated if the coefficient is not significant.

In such a specification, which does not consider spatial heterogeneity, the parameters $\alpha_0, \alpha_{\mathrm{p}}, \alpha_y, \delta$ and β are assumed to be constant across municipalities. *X* includes control variables to deal with differences in unit costs for providing local public services and the specific characteristics of a municipal population.

3 Empirical Modeling of Demand for Public Services Without Spatial Heterogeneity

We first offer a brief description of the local administrative structure of France, and of the funding sources and responsibilities of various jurisdictional levels; we then introduce the data selected for the empirical analysis. Next, we present specification tests and the estimation results of Eq. (5) are discussed.

3.1 Overview of the Organization of the Local Public Administration in France

The sub-national administrative structure of France includes three overlapping subdivisions. In order from the lower level up, there are 36,680 municipalities (with their 2599 groupings), 100 departments, and 22 metropolitan regions.[2]

Municipalities and departments, which were created in 1790 after the French Revolution, form the two lowest levels of local government in France. They finance 88% of total local spending (56% financed by municipalities and inter-municipal groups and 32% financed by departments). In 2001, the median population of a French department was 511,012 inhabitants, which is 21 times larger than the median population of a US county. Local governments finance their spending with three main revenue sources: taxes (mainly residential and business property taxes) correspond to more than 53% of fiscal revenues, while 35% are grants from the central government and 7% being debt.

Following the implementation of decentralization laws in the 1980s, municipalities have provided a wide range of major public services, including running water, garbage collection, primary education, public transport in the municipal area, and town planning. In addition, municipalities are the most important public investors in France (before the central government). Investments mainly focus on school buildings, community facilities, and municipal roads. Departments are specialized in providing decentralized welfare benefits, which account for more than 60% of their total spending (minimum benefits paid to those with no other means of support or personalized allowances for disabled people, for example).

For the period under study, the French local public sector has experienced two important reforms. First, since 2000, many municipalities have joined together to form cooperative groups, referred to in the remainder of this article as inter-communal

[2]Major administrative reforms in France in 2014 and 2015 reduced the number of metropolitan regions to 13, and the number of intercommunal groupings to 2133 in 2015, and to 1258 in 2019.

organizations, to provide services to their respective communities. In 1999, 52% of French municipalities belonged to an inter-communal organization; this figure rose to 88% in 2005 and 95% in 2011. Inter-communal organizations are able to increase the variety of services provided at the municipal level, while raising the efficiency. Second, in 2003 new decentralization laws have transferred new responsibilities to the local authorities, mainly capital spending to maintain former national roads and an income support supplement (called Revenu Minimum d'Insertion) to departments.

3.2 Data Description

Data are provided by the French Ministries of the Economy and Interior. To address the issue of heterogeneity in terms of population size, we concentrate on the largest French municipalities. Paris, Corsica, and the five overseas departments of France (Guadeloupe, Martinique, Guyane, La Réunion, and Mayotte) are excluded from the analysis due to their unique features. Sample data therefore cover 110 French municipalities with over 50,000 inhabitants and their corresponding departments (87 departments). We consider here two different years, 2001 and 2005, to check for the robustness of our results.

The dependent variable is measured by the operating expenditures plus the gross saving covering the repayment of loans coming from past investments and the self-financing of current investments. Such an approach that gives a mean value of the local budget over the electoral cycle seems desirable with cross-sectional data because investment expenditures fluctuate a lot on a year-by-year basis, with high capital expenditures voted before the election years in municipalities. We also use the same measure of aggregate expenditures per capita at the department level.

The income variable is measured by using the median income in each municipality, net of the housing tax paid to the department. The median income is computed on the basis of the distribution of the taxable income of households in each municipality. It should be noted that welfare benefits are non-taxable income. As grants from the central government are the second most important revenue stream and are allocated to departments to compensate decentralized responsibilities and to implement fiscal equalization we also consider national grant per capita as a distinct variable.

The tax share is measured by the municipal housing tax base divided by the total municipal tax base (including the local business and property taxes). This variable accounts for the direct influence of taxation on the choices of the median voter. It denotes the split of taxes between firms and households. This ratio is near 1 in residential areas, while it tends toward 0 in industrial areas.

Following the standard procedure in empirical related literature, we use control variables to deal with differences in unit costs for providing local public services and the specific characteristics of municipal populations. Initially, we used several control variables such as the number of recipients of welfare benefits or the number of secondary school pupils, and the number of social housing units. However, as non-significant values of some parameters were obtained in preliminary tests, we removed

Table 1 Data set description, French municipalities with over 50,000 inhabitants in 2001 and 2005 (Euros)

Description of variables	Mean		Min		Max	
	2001	2005	2001	2005	2001	2005
Total municipal expenditures per capita	1135	1139	664	731	2776	3137
Total expenditures per capita in departments	551	772	379	576	764	992
After-tax yearly median municipal income	10,172	9934	6707	5743	22,133	35,938
Tax share	0.20	0.25	0.11	0.12	0.35	0.41
National grant per capita	341	366	176	187	656	841
Municipal population	107,265	107,377	50,070	50,070	807,070	807,071
Number of secondary residences	1795	1795	90	90	23,560	23,560

these explanatory variables. The set of variables used in the final specification is presented in Table 1 through summary statistics, as measured in 2001 and 2005.

Table 1 shows stable values between 2001 and 2005 for most variables, except department expenditures per capita, which increased from 551 to 772 Euros, on average. This can be explained by decentralization laws enacted in 2003, which transferred new responsibilities to the departments.

3.3 *Specification Tests*

Econometric issues influence the choice of the estimator in our model (5). One issue is whether department expenditures can be considered to be exogenous in the conditional expenditures of municipalities. The second issue is whether spatial autocorrelation should be accounted for.

3.3.1 Endogeneity of Department Expenditures

In Eq. (5), department expenditures are likely to be endogenous to municipal expenditures if the department and the municipal public spending decisions occur simultaneously. We therefore have to implement a diagnostic test for endogeneity using an appropriate set of instruments.

In the spirit of the median voter model, tax share and after-tax median income, both measured at the department level, may be used as instruments for department expenditures. Since departments provide welfare benefits and support to those affected by a loss of independence, we also use the numbers of recipients of social benefits, the number of social housing units, the unemployment rate, and the percentage of people over the age of 60. The number of municipalities (urban and rural), national grants per capita, and the number of secondary school pupils also are instruments considered for department expenditures.

An instrumental variable must satisfy two requirements. It must be correlated with the endogenous variable and be orthogonal to the error term. We therefore use the Bound et al. (1995) test to select relevant instruments and perform the Sargan test to choose valid instruments.[3] The resulting set of relevant and valid instruments for department expenditures are population, unemployment rate, and the number of rural municipalities. Finally, according to the Durbin-Wu-Hausman test of endogeneity (probability greater than 0.26), we cannot reject the hypothesis that department expenditures are exogenous in the municipal expenditure equation. This means that the median municipal voter considers department spending to be as given. Therefore, we do not need to instrument the department expenditures to estimate the single Eq. (5).

3.3.2 Spatial Dependence

Our study focuses on vertical interactions between two overlapping local jurisdictions, but we also are interested in spatial horizontal interactions resulting from the behavior of incumbent municipal authorities (fiscal competition and yardstick competition). Corresponding statistical inference is based on different tests to detect spatial dependence (Moran's I, Lagrange Multipliers, Kelejian and Robinson (KR) test that does not require normality for the error terms).[4]

As the municipalities under study do not share a common border, we considered two spatial weight matrices. The first is a function of the inverse (Euclidean) distance between two municipalities. The second reflects the potential spatial interaction (in terms of competition) between two nearby and similar municipalities. It takes the value of one if the two municipalities belong to the same European NUTS1 region. We assume here that fiscal competition (through local public spending or taxes) mainly concerns municipalities located in the same European region. According to Newman and Sullivan (1988), once firms or households choose a region (the macro-localization choice), they then choose a municipality inside this region (the micro-localization choice).

Table 2 shows the diagnostics for spatial effects, considering each row-standardized weight matrix and each year.

Whatever the weight matrix or year considered, Moran's I tests are significant and we are inclined to reject the null hypothesis of absence of spatial autocorrelation in the error term (at the 10% level). SARMA tests confirm this diagnostic with the inverse distance matrix only.

---

[3]These tests are implemented with Stata10. See Baum and Schaffer (2003) for more details.

[4]Spatial data analysis is conducted with SpaceStat.

Table 2 Spatial dependence tests (probability)

Test	NUTS1 weight matrix		Inverse distance weight matrix	
Year	2001	2005	2001	2005
SARMA	0.26	0.40	0.10	0.008
Moran's I	0.07	0.03	0.06	0.0006
Lmerr	0.32	0.18	0.44	0.043
Kelejian Robinson err	0.51	0.077	0.81	0.08
LM lag	0.49	0.67	0.21	0.11
Robust LM lag	0.19	0.94	0.145	0.10
Robust LM err	0.13	0.20	0.08	0.007

If we consider the inverse distance weight matrix, as we never reject the null hypothesis for the Lmlag and RLMlag tests, we decide to dismiss the spatial lag specification, whatever the year considered. Results obtained with the Lmerr and KR err tests are more contrasted. However, we decide to retain a spatial error model (SEM) for the two years considered.

3.4 *Empirical Results*

Table 3 gives estimation results obtained for the years 2001 and 2005, using both an OLS method (with robust standard errors (se) to deal with heteroskedasticity) and a maximum likelihood estimation for the spatial error model (SEM). The best model is the SEM with the lowest value for the Akaike information criterion (AIC). The results are similar to those of other studies, with $\hat{\delta}$ positive and significant in 2001 and 2005. This may suggest that the welfare benefits provided by departments are complementary to many of the services provided by municipalities. However, the coefficient of this effect becomes smaller in 2005 (from 0.95 in 2001 to 0.35 in 2005). Therefore, we observe a decrease of the strength of the vertical interaction among municipalities and their departments between 2001 and 2005. A possible explanation could be the development of inter-communal organizations, which has increased the variety of services provided by municipalities to their jurisdiction. Therefore, it seems that the department's specificity in terms of providing the decentralized welfare repayment would become less prominent.

Concerning other results, all of the significant coefficients present the appropriate sign. Results thus reveal a negative and significant impact of population size on per capita local public expenditures. An increase of 1000 inhabitants results in a decrease of 5 Euros in per capita expenditures in 2001 (and a decrease of 4 Euros in 2005). This confirms the existence of economies of scale in consumption. The number of secondary residences is used as a proxy to capture costs of public investment in terms of tourism or leisure. Estimates of this effect show that an increase of 100 secondary

Table 3 Estimation results without spatial heterogeneity (inverse distance weight matrix)

Variables	Coefficient estimates (probability)			
Year	2001		2005	
Estimation method	OLS robust se	Spatial error model	OLS robust se	Spatial error model
Constant	38 (0.89)	95 (0.67)	174 (0.25)	205 (0.15)
National grants per capita	1.87*** (0.000)	2.05*** (0.000)	1.92*** (0.000)	2.00*** (0.000)
Tax share	−892*** (0.0001)	−1180*** (0.000)	−604** (0.04)	−766*** (0.001)
After-tax yearly median income	0.012 (0.43)	0.010 (0.33)	0.006 (0.15)	0.0059 (0.18)
Municipal population	−0.0056** (0.012)	−0.0053*** (0.003)	−0.0040*** (0.0001)	−0.0042*** (0.000)
Secondary residences	0.024** (0.019)	0.024*** (0.000)	0.033*** (0.00)	0.034*** (0.000)
Department per capita spending	1.01*** (0.000)	0.95*** (0.000)	0.36*** (0.0054)	0.35*** (0.004)
Inter-communal dummy	52 (0.32)	25 (0.61)	248*** (0.0001)	255*** (0.000)
Autoregressive coefficient		0.49** (0.03)		0.57*** (0.004)
AIC	1462	1451	1402	1398
Breusch Pagan probability	0.000***	0.0006***	0.000***	0.000***

Significance level: * for 10%, ** for 5% and *** for 1%

residences in 2005 generates an increase of around 3.4 Euros in per capita total municipal spending. Values for price and income elasticities (respectively −0.15 and 0 on average) reveal a weak elasticity of publicly provided goods and services in the largest French municipalities. Furthermore, empirical results indicate that the coefficient relating to the inter-communal organization dummy variable (which is equal to 1 when the municipality does not belong to an inter-communal organization) is positive and significant in 2005 only. Next, results show that per capita national grants received by the municipality increase public spending in a greater proportion than an equivalent rise in local income (flypaper effect).

Finally, the Breusch Pagan test reveals the presence of heteroskedasticity, and as suggested by Anselin (1992), an indication of heteroskedasticity may point to the need for a more explicit incorporation of spatial heterogeneity in the form of spatial regimes. Cross-region variations of the parameters are therefore included in the specification and are discussed in the following section.

4 Empirical Modeling of Demand for Public Services with Spatial Heterogeneity

To introduce spatial heterogeneity in the conditional demand specification (5), we use a spatial regimes model. Each regime is characterized by different values for the coefficients associated to price, income, and department expenditures. The number of regimes is defined by the spatial division of the French territory into regions. The two different spatial regimes specifications used are described and discussed in the first subsection. In the second, we present and discuss the estimation results obtained on that basis.

4.1 Spatial Division of the French Territory

French regions traditionally have been characterized by historical, cultural, and economic differences. Like other countries (see, for example, Gérard et al. (2010) in Belgium), France is not a homogeneous nation, and consequently is particularly well suited for spatial heterogeneity studies. We propose here to divide the French territory, grouping contiguous French administrative regions[5] together and by differentiating 'Ile de France' from other regions. Ile de France (IDF), which includes Paris, is one French administrative region; it usually is considered to be different from all the other regions. IDF is characterized by a high growth rate, high population density, high level of education, and a weak percentage of municipalities belonging to an inter-communal organization. Another distinctive feature is that departments and municipalities in IDF pay high levels of capital grants to the IDF region and thus co-finance large expenditures in terms of transport networks, in particular, whereas in the other French regions spending for urban and inter-urban transports are directly included in the municipal and departmental budgets.

To test the robustness of our results, we consider two different ways of dividing the territory of France (see Binet et al. (2010) for further discussion):

- IDF/North/South regimes;
- Seven French European jurisdictions.

Tests for the presence of structural instability have been implemented to justify the spatial divisions considered. The null hypothesis that the coefficients are the same in all regimes has been rejected for these two partitions of the French territory.

[5] A finer spatial division, into the 22 French administrative regions, is not possible as the numbers of municipalities and departments in each region would be too small to allow an econometric analysis.

Table 4 Average values, Ile de France, North, and South divisions in 2005

Variables	Ile de France	North	South
Total municipal expenditures per capita (Euros)	1427	953	1124
Total department expenditures per capita (Euros)	853	692	812
After-tax yearly median municipal income (Euros)	11,047	7724	8590
Tax share	0.26	0.23	0.26

4.1.1 IDF/North/South Regimes

Two of these three regimes (North, South, and IDF) include several French administrative regions and, respectively, 32 (IDF), 47 (North), and 31 (South) municipalities. The regimes also reflect the cultural, linguistic, and administrative differences that characterize regions in France, some of which date back to the Roman Empire. Furthermore, Table 4 shows that total municipal and departmental expenditures per capita are greater in jurisdictions located in IDF and in territories in the South compared with territories in the North.

Such a difference in expenditure levels can be explained by different regional traditions (higher preferences for local public services in the South and IDF compared to the North) and by the highest income level and population density in municipalities located in Ile de France. Therefore, they provide a wider range of public services than in other regions [zoo effect, Oates (1988)]. These statistics also suggest that the dependent variable under study is differentiated across space and confirms the pertinence of the IDF, North, and South spatial divisions.

4.1.2 Seven French European Jurisdictions

The French territory is divided into seven different jurisdictions for European elections. The seven jurisdictions, which are similar to NUTS1 European regions, include, respectively, 32 (IDF), 13 (East), 13 (North–West), 16 (West), 10 (South–West), 8 (Center), and 18 (South–East) municipalities. This regime provides a finer spatial division than the three-way partition discussed above. Table 5 describes public spending levels at the municipal and departmental levels in each of the seven jurisdictions.

These statistics confirm that the South, Center, and Ile de France tend to spend more than other regions. Based on these spatial divisions of the French territory, we estimate in the next subsection, spatial regimes models.

Table 5 Average values, seven jurisdictions in 2005

Variables	Total municipal expenditures per capita (Euros)	Total departmental expenditures per capita (Euros)	After-tax yearly median municipal income (Euros)	Tax share
Ile de France	1427	853	11,047	0.26
East	922	666	7785	0.24
North West	938	748	7033	0.19
West	987	663	8144	0.25
South West	1051	817	8431	0.25
Center	948	730	8036	0.24
South East	1200	817	8764	0.26

4.2 *Empirical Results*

For each regime, we estimate one specific coefficient to assess price, income, and preferences effects.

4.2.1 IDF/North/South

Table 6 gives estimation results obtained for the years 2001 and 2005, using an OLS method with robust standard errors to deal with heteroskedasticity.

The spatial regimes specification permits spatial autocorrelation to be removed, as suggested by the Moran's I and SARMA tests. Spatial autocorrelation in the error term can stem from the structural instability of coefficients of the demand model. The spatial regimes model is able to eliminate spatial autocorrelation in the error term.

Results obtained in 2001 and 2005 are robust. They show that the nature of the vertical relationships between departmental and the municipal spending decisions is prone to be different in Ile de France compared with other spatial divisions. These results reveal that complementarities between services provided by municipalities and departments concern only Ile de France (IDF), with a coefficient that becomes smaller in 2005 (from 1.02 in 2001 to 0.55 in 2005). In contrast, municipalities and their departments seem to follow independent public spending decisions in the South and North spatial divisions. As mentioned previously, these contrasting results can be explained by the high urbanization level in IDF where the range of public services is higher than in other regions. Indeed, in IDF, departments and municipalities co-finance high network and infrastructure spending (e.g., garbage collection, running water and regional transport network), which probably explains the significant complementary effect.

Whatever the year considered, non-significant coefficients obtained for the tax share reveal price elasticities equal to 0 in the South and North spatial divisions. The significant value obtained in IDF reveals a weak price elasticity equal to −0.15. In

Table 6 Estimation results with spatial (IDF/North/South) heterogeneity (probability in parentheses)

Year	2001	2005
Estimation method	OLS robust standard error	OLS robust standard error
Constant	277 (0.35)	66 (0.74)
National grants per capita	2.24*** (0.000)	2.06*** (0.000)
Municipal population	−0.0064*** (0.000)	−0.0044*** (0.0008)
Secondary residences	0.026*** (0.001)	0.021* (0.10)
Inter-communal dummy	26 (0.58)	223*** (0.006)
IDF income	0.011 (0.53)	0.004 (0.71)
North income	0.029 (0.33)	0.055*** (0.006)
South income	0.006 (0.35)	0.05** (0.012)
IDF tax share	−2398*** (0.000)	−819* (0.07)
North tax share	−337 (0.191)	−449 (0.26)
South tax share	−470 (0.135)	−162 (0.71)
IDF department per capita spending	1.02*** (0.001)	0.55*** (0.013)
North department per capita spending	−0.36 (0.40)	−0.14 (0.42)
South department per capita spending	0.28 (0.297)	−0.09 (0.66)
Adjusted R^2	0.78	0.89
AIC	1433	1390
Breusch Pagan probability	0.011**	0.000027***
Moran's I probability	0.44	0.77
SARMA probability	0.55	0.52

Significance level: *** for 1%, ** for 5% and * for 10%

contrast, income elasticities are significant only in the North and South and in 2005 (respectively, 0.44 and 0.38). Other estimated coefficients also have values similar to the one obtained before.

Overall, we find clear evidence that the vertical effects between spending decisions differ over the French territory. Results confirm the spatial heterogeneity in the preferences effect.

4.2.2 Seven French European Jurisdictions

Table 7 reports final OLS estimates (with a heteroskedastic-consistent matrix) of the spatial regimes specification resulting from the spatial divisions into seven constituencies.

Once again, we observe that both the Moran's I and SARMA tests show no evidence of spatial dependence. We also find support for spatial heterogeneity in terms of the preferences effect across French jurisdictions. We observe a highly significant complementary effect for Ile de France in 2001 and 2005 and for South–West and South–East in 2001 only. In contrast, in 2005 only, the North–West regime is characterized by a negative relationship between municipal and departmental expenditures, indicating substitutability among public spending at the two levels of local government. Other jurisdictions (Center, East, and West) show no significant vertical effect, indicating independence between local public spending decisions of the municipal and departmental tiers. This confirms the importance of spatial heterogeneity in the analysis of the preferences effect.

The coefficients associated to other variables (national grant, secondary residences) are robust and show positive effects on municipal public spending per capita. Next, we observe weak economies of scale in consumption and in production, as the coefficients associated with the population and the inter-communal dummy variable are, respectively, negative and positive.

5 Conclusions

The main purpose of this chapter was to analyze the impact of the overlapping structure of local governments on the local spending decisions. Existing studies based on data from various countries have found evidence of a complementary relation between spending at different levels of local government but they do not allow for spatial heterogeneity in the estimated models. Therefore, the lack of flexibility in spending models may lead to false evidence of complementarity and to the inappropriate policy recommendation to reduce the number of local levels of government since positive interactions between levels of government may lead to an increase in the size of the total local public sector.

In our study, we used data from 2001 and 2005 that cover the 110 largest municipalities (excluding Paris) located in 87 departments of France. First, as a benchmark,

Table 7 Estimation results with spatial heterogeneity, 7 French European jurisdictions (probability in parentheses)

Year	2001	2005
Estimation method	OLS robust standard error	OLS robust standard error
Constant	81 (0.71)	147 (0.50)
National grants per capita	1.74*** (0.000)	2.106*** (0.000)
Municipal population	−0.0042*** (0.001)	−0.0041*** (0.0026)
Secondary residences	0.021*** (0.005)	0.022 (0.118)
Inter-communal dummy	58 (0.222)	227*** (0.0056)
IDF income	0.0019 (0.88)	0.0046 (0.66)
East income	0.0028 (0.94)	0.056 (0.19)
North West income	0.079*** (0.009)	0.094*** (0.000)
West income	0.053*** (0.006)	−0.0032 (0.87)
South West income	−0.066 (0.12)	0.057*** (0.001)
Massif Central income	0.04 (0.13)	−0.0031 (0.94)
South East income	0.021 (0.16)	0.054** (0.045)
IDF tax share	−825*** (0.009)	−925** (0.048)
East tax share	−196 (0.77)	−698 (0.47)
North West tax share	−2216*** (0.000)	−787 (0.166)
West tax share	−756* (0.06)	−660 (0.47)
South West tax share	−492 (0.37)	−1188** (0.059)
Massif Central tax share	−560 (0.15)	163 (0.84)
South East tax share	−507 (0.22)	57 (0.91)

(continued)

Table 7 (continued)

Year	2001	2005
IDF department per capita spending	1.08*** (0.004)	0.45** (0.0054)
East department per capita spending	0.76 (0.37)	−0.25 (0.39)
North West department per capita spending	0.51 (0.135)	−0.56** (0.022)
West department per capita spending	0.31 (0.44)	0.53 (0.206)
South West department per capita spending	2.37*** (0.001)	0.156 (0.54)
Center department per capita spending	0.27 (0.55)	0.18 (0.55)
South East department per capita spending	0.69** (0.038)	−0.34 (0.24)
Adjusted R^2	0.81	0.88
AIC	1449	1402
Breusch Pagan probability	0.035**	0.002
Moran's I probability	0.24	0.95
SARMA probability	0.73	0.16

Significance level: *** for 1%, ** for 5% and * for 10%

we assumed the stability of regression coefficients over the observation set and we estimated a spatial error model to capture spending interactions among neighboring municipalities. In this step, we found a strong complementary effect between municipalities and departments, as suggested by previous literature.

In a second step, we estimated a demand model of local public services in a specification that controls for spatial heterogeneity. Two different spatial regimes models were estimated based on inter-regional differences in France, and the models were then compared. We thus allowed cross-region parameter variation for the preferences, price and income effects. The specification is based on two spatial divisions of the French territory, one into three groups of regions (Ile de France/South/North), and one into seven (French European jurisdictions).

We then assessed the importance of spatial heterogeneity in the analysis of the determinants of local public expenditures. The spatial regimes specification allows spatial dependence to be removed. In addition, the impact of an increase in departmental expenditures on municipal spending decisions is assessed differently if regional heterogeneity is present. Results then show that municipal and departmental spending decisions are complementary only in Ile de France (IDF). Other French regions show either independence or substitution between the services offered by the two different overlapping jurisdictions. From the perspective of policy analysis, our results show that a multi-tiered form of organization does not systematically increase the total local public spending level. This result may be seen as contributing to the ongoing debate on the multi-tiered form of administrative organization in France and in other countries.

References

Anselin, L. (1990). Spatial dependence and spatial structural instability in applied regression analysis. *Journal of Regional Science, 30*, 185–207.

Anselin, L. (1992). *SpaceStat tutorial*. Urbana-Champaign: University of Illinois.

Aronsson, T., Lundberg, J., & Wikstrom, M. (2000). The impact of regional public expenditures on the local decision to spend. *Regional Science and Urban Economics, 30*, 185–202.

Baum, C. F., & Schaffer, M. E. (2003). Instrumental variables and GMM: Estimation and testing. *The Stata Journal, 3*, 1–31.

Binet, M. E., Guengant, A., & Leprince, M. (2010). Superposition des collectivités territoriales, dépenses publiques locales et hétérogénéité spatiale: le cas des grandes villes françaises. *Revue Economique, 61*(6), 1111–1122.

Bound, J., Jaeger, D., & Baker, R. (1995). Problems with instrumental variable estimation when the correlation between the instruments and the endogeneous explanatory variable is weak. *Journal of the American Statistical Association, 90*, 443–450.

Campbell, R. J. (2004). Leviathan and fiscal illusion in local government overlapping jurisdictions. *Public Choice, 120*, 301–329.

Gérard, M., Jayet, H., & Paty, S. (2010). Tax interactions among Belgian municipalities: Do interregional differences matter? *Regional Science and Urban Economics, 40*, 336–342.

Ministry of Finance. (2012). *Rapport sur la dépense publique et son évolution, rapport annexé au projet de loi de Finances pour 2013* (p. 60).

Newman, R. J., & Sullivan, D. H. (1988). Econometric analysis of business tax impacts on industrial location: What do we know and how do we do? *Journal of Urban Economics, 23*, 215–234.

Oates, W. E. (1988). On the measurement of congestion in the provision of local public goods. *Journal of Urban Economics, 24*, 85–94.

Pollack, R. A. (1971). Conditional demand functions and the implications of separable utility. *Southern Economic Journal, 37*, 423–433.

Revelli, F. (2003). Reaction or interaction? Spatial process identification in multi-tiered government structures. *Journal of Urban Economics, 53*, 29–53.

Turnbull, G. K., & Djoundourian, S. S. (1993). Overlapping jurisdictions: Substitutes or complements? *Public Choice, 75*, 231–245.

Efficient and Super-Efficient Use of Broadband Access by the US States

Jitendra Parajuli and Kingsley E. Haynes

1 Introduction

Infrastructure investment often contributes to economic growth and development. Today's economic environment to a great extent relies on the flow and exchange of information and knowledge. This necessitates that economic agents have access to high-speed or broadband Internet. However, the mere provision of broadband infrastructure is not a sufficient condition for economic growth and development. Furthermore, as long as broadband services are not efficiently adopted and utilized, the resources for providing broadband access and services will be wasted. In an earlier study that used multiple-input/multiple-output Data Envelopment Analysis (DEA), Parajuli and Haynes (2012) found that many US states are efficient in the adoption of broadband services. In addition, it was observed that a number of states had become more efficient over the study period. While that earlier study used the classical CCR model to identify efficient states and examined various issues associated with the utilization of broadband adoption and use, the focus of this chapter is on extending the CCR model for scale assessment and the ranking of efficient states.

The organization of this chapter is as follows. The next section provides a brief literature review on infrastructure with a focus on telecommunications and broadband. Thereafter, methodology and data are discussed. The fourth section provides the results and the chapter concludes by outlining policy implications.

J. Parajuli · K. E. Haynes (✉)
Schar School of Policy and Government, George Mason University, 3351 Fairfax Drive, MS 3B1, Arlington, VA 22201, USA
e-mail: khaynes@gmu.edu

J. Thill (ed.), *Innovations in Urban and Regional Systems*,
https://doi.org/10.1007/978-3-030-43694-0_15

2 Literature Review and Research Question

Infrastructure is a part of the overall capital stock of a nation or region (Nijkamp 1986; Prud'homme 2005) and can be generally divided into two categories—hard and soft. Hard infrastructure is also referred to as "economic" infrastructure and includes roads, bridges, ports, and electrical and telecommunications systems. Soft infrastructure includes entities, such as parks, recreation facilities, health, and human services as well as schools and human capital (Haynes 2006).

Infrastructure has the potential to influence an economy in various ways. In general, it lowers the costs of inputs used by enterprises and enlarges the pool for goods and services, labor, and even capital markets (see Nijkamp 1986; Porter 2000; Prud'homme 2005); network effects increase its utility (see Katz and Shapiro 1985); and spillover effects influence nearby economic activities (see Anselin et al. 1997). On the other hand, because of space-specificity and immobility (Prud'homme 2005), infrastructure can have negative consequences in the development of areas that have a lower endowment of facilities (Yilmaz et al. 2001). It is subject to decreasing returns to scale and is not a sufficient condition for development (Rietveld 1989). Infrastructure is also characterized by market failures (Prud'homme 2005) that often compel governments to adopt regulatory mechanisms to deal with such situations (see Gomez-Ibanez 2003). Overall, infrastructure is an important determinant of economic growth and development. However, because of its inherent nature, infrastructure policy and its implications vary across regions and over time (see Gramlich 1994; Button 1998; Gomez-Ibanez 2003; Prud'homme 2005; Haynes 2006; Fulmer 2009).

Public investment in highways, streets, water systems, and sewers induces an increase in the rate of return to private capital that ultimately leads to economic growth and productivity gains (Aschauer 1989). Munnell (1990) observed that public capital investment has a positive and statistically significant impact on private sector output. Holtz-Eakin (1994), on the contrary, noted that after controlling for region-specific characteristics, the productivity benefits from the provision of public infrastructure are negligible. Interestingly, Baltagi and Pinnoi (1995) found that the contribution of public infrastructure capital is insignificant at the aggregate level. However, the disaggregate-level analysis revealed that private sector productivity can experience significantly positive and negative as well as insignificant impacts from public sector infrastructure investments.

A well-developed telecommunications infrastructure positively affects the economy (Cronin et al. 1991). Roller and Waverman (2001) reached a similar conclusion but found that this positive causality associated with telecommunications and economic development can only be achieved when a critical mass of telecommunications infrastructure is present, and that critical value appeared to be at almost the universal service level. Other studies, such as Dholakia and Harlam (1994), Datta and Agarwal (2004), and Ding et al. (2008), also found a positive effect of telecommunication on economic growth and development. In the USA, the improvement of regional productivity was positively influenced by the accumulation of telecommunications infrastructure (Yilmaz et al. 2001). However, the impact of telecommunications varies significantly across sectors (Greenstein and Spiller 1995; Yilmaz et al. 2001).

The Internet is part of a seamless platform of various information and communications technologies (ICTs) characterized by various capabilities (see Leiner et al. 1997). As a platform of innovation possibilities (Press 1994; Leiner et al. 1997), the Internet became an important element for stimulating economic expansion and supporting economic changes. Productivity in the business sector increases and continues to grow with new innovations. The lower transaction costs and prices as well as the downward pressure on inflation benefit consumers directly (Willis 2004). Further, the advancement of ICTs offered high-capacity and high-speed access as well as mobility that opened other possibilities, such as online education, telemedicine, electronic government, and electronic and mobile commerce.

As broadband offers a myriad of economic, political, and social benefits, broadband access and services has become an important public policy issue in the USA and elsewhere. At the federal level, the American Recovery and Reinvestment Act of 2009 set aside US$7.2 billion to promote broadband access and develop and maintain a broadband inventory map. Accordingly, the National Telecommunications and Information Administration (NTIA) of the Department of Commerce was responsible for administrating $4.7 billion to support the Broadband Technology Opportunities Program (BTOP) with the objective "to increase broadband penetration and adoption; provide broadband training and support to schools, libraries, healthcare providers, and other organizations; improve broadband access to public safety agencies; and stimulate demand for broadband." The remaining US$2.5 billion was allotted to the Rural Utilities Service of the Department of Agriculture to expand rural broadband access (see NTIA 2010). This appropriation was basically pursuant to the provisions of the Rural Electrification Act of 1936 that support the grants and loans programs for the expansion of telecommunications services in the rural areas. In addition, in March 2010, under the provisions of the Act of 2009, the FCC delivered the National Broadband Plan to Congress.

At the state level, governments have also been pushing their own broadband agenda. For example, former state governors Bob Riley, Arnold Schwarzenegger, and Tim Kaine were instrumental in creating the Alabama Broadband Initiative, the California Broadband Task Force, and the Office of Telework Promotion and Broadband Assistance as well as the Virginia Broadband Roundtable, respectively, to promote state-level broadband initiatives. Other states, such as Hawaii, Kansas, Kentucky, Maryland, Tennessee, and Vermont, have also formulated policies to increase broadband use (see National Conference of State Legislatures 2010).

Moreover, a growing number of local communities across the country also adopted policies to promote broadband. For example, in Utah, a municipal-owned optical fiber network offered high-speed Internet access, telephony, and television programs at 30 Mbps to the subscribers in 14 cities (Cherry 2006), and in Lake County of central Florida, the municipal optical fiber network offered businesses and municipal institutions broadband access (Ford and Koutsky 2005). Similarly, wireless- and cable-based broadband services were also promoted in many municipal areas (see Carlson 1999; Dingwall 2006; Jain et al. 2007).

Studies, such as Bauer (2006), Gillett et al. (2006), Crandall et al. (2007), and Kolko (2012), have suggested that there is a positive association between broadband and economic growth and development. However, infrastructure projects are fraught with the economic and political costs (see Flyvberg et al. 2003; Flyvberg 2007) and these ICT projects are also not free from such costs. For instance, Kandilov and Renkow (2010), Hudson (2010), and Kenny and Kenny (2011) noted that investments in the ICT sector made by either the government or the private sec tor does not necessarily give the expected returns. Further, if scarce resources are misallocated, it can negatively affect the overall returns to the development process and to specific infrastructure projects. It is thus important to examine the utilization of broadband before adding new capacity.

Grubesic (2010) examined the efficiency of broadband availability at the zip code level. It can be argued that services can be efficiently provided. However, if they are not efficiently used and the usage trend cannot be determined over time, it is often difficult for the government to formulate appropriate future policies as well as for private operators to make efficient investment decisions regarding broadband access.

Parajuli and Haynes (2012) addressed the issue of efficiency associated with broadband adoption and use across 48 contiguous US states from 2005 through 2007. Some states were found to be efficient (i.e., states had a DEA efficiency score of one). That finding of many efficient decision-making units (DMU) is not uncommon in DEA (see Adler et al. 2002). However, even among efficient DMUs, some DMUs appear to be more efficient than others. There are new techniques that allow for discriminating among efficient DMUs such that they could be rank-ordered. The higher ranks of such an analysis are identified as "super-efficient." Super-efficient DMUs, however, might not remain efficient over time and might not remain on the efficiency frontier. This leads to the new research question: *Are super-efficient regions (DMUs) likely to remain efficient over time?*

The rank-ordering process not only discriminates among efficient DMUs but also allows for examining the possibilities of increasing outputs (or decreasing inputs) while maintaining full efficiency. This chapter uses one of these techniques to discriminate and rank states that are efficient in broadband adoption and use creating a super-efficient group. For a broad view of these super-efficient methods, see Soushi et al. (2010).

3 Methodology and Data

Performance of a DMU is an important operational issue and the natural measure of performance is productivity, which is the ratio of output(s) to input(s). In order to measure productivity, a production frontier is used. Any DMU on the production frontier is said to be technically efficient. DMUs that are beneath the frontier are said to be technically inefficient. Thus, efficiency is a relative term and is applicable to a set of DMUs with common input(s) and output(s) (see Coelli et al. 2005; Sherman and Zhu 2006).

Efficiency has been extensively measured in a variety of situations. Haynes et al. (1998) distinguished efficient plants from inefficient ones and measured the progress of industry toward specific pollution prevention goals; Karkazis and Thanassoulis (1998) assessed the effectiveness of investment incentives in different regions of Greece; and Haslem et al. (1999) evaluated financial practice among banks operating internationally. The relative efficiency of biomedical research infrastructure (Haynes et al. 2000), airport facilities (Fernandes and Pacheco 2003), maritime ports (Tongzon 2001), and telecommunications infrastructure (Dinc et al. 1998) has also been studied. Note that special issues of *Computers, Environment and Urban Systems*, 1990, *European Journal of Operational Research*, 2001, and *Omega*, 2013 have a large number of examples on efficiency analysis.

DEA is a non-parametric approach to relative efficiency measurement based on mathematical programming. Compared to the conventional production function-based analysis, DEA makes no a-priori assumptions of functional forms and allows for a large number of variables and constraints to measure efficiency. In addition, the DEA technique is based on frontiers rather than central tendencies, which allows for identifying extreme (i.e., maximum and minimum) scores instead of means as in regression modeling. These extreme values are useful in the benchmarking process. DEA can address various modeling issues, such as isotonicity, non-concavity, economies of scale, piecewise linearity, ordinal relationships, and categorical variables (see Seiford and Thrall 1990; Cooper et al. 2007).

The Charnes, Cooper, and Rhodes model (CCR) is the simplest DEA model and assumes constant returns to scale (CRS) of activities (Charnes et al. 1978). An output-oriented CCR model in the envelopment form is given by:

$$\max \phi + \varepsilon\left(\sum_{i=1}^{m} \gamma_i^- + \sum_{r=1}^{s} \gamma_r^+\right) \tag{1}$$

subject to

$$\sum_{j=1}^{n} \lambda_j x_{ij} + \gamma_i^- = x_{i0}, \quad \text{for } i = 1, 2, \ldots, m$$

$$\sum_{j=1}^{n} \lambda_j y_{rj} - \gamma_r^+ = \phi y_{r0}, \quad \text{for } r = 1, 2, \ldots, s$$

$$\lambda_j, \gamma_i^-, \gamma_r^+ \geq 0 \; \forall \; i, r, j$$

where x is an input and m is the number of inputs (indexed by i); y is an output and s is the number of outputs (indexed by r); n is the number of DMUs (indexed by j); γ^- is the input excess and γ^+ is the output shortfall and is referred to as the DEA slacks; ε is non-Archimedean, that is, defined to be greater than zero but less than any real positive number; and λ is the weight of the DMU being evaluated. We solve the model by maximizing Eq. (1).

The Banker, Charnes, and Cooper extension (BCC) is employed to take into account the effect of variable returns to scale (VRS)—increasing returns to scale (IRS) and decreasing returns to scale (DRS) (Banker et al. 1984). The output-oriented BCC model in the envelopment form is given by Eq. (1) with its constraints in addition to the convexity constraint introduced by the BCC extension, which is given in Eq. (2):

$$\sum_{j=1}^{n} \lambda_j = 1. \tag{2}$$

The CCR and the BCC technical efficiency are often referred to as global technical efficiency and local pure technical efficiency, respectively. This implies that, if a DMU is BCC-efficient but has a low CCR score, it is operating efficiently locally, but not globally. This happens because of the scale size of the DMU. Thus, scale efficiency (SE) is related to these two different technical efficiencies as follows (Cooper et al. 2007):

$$\text{(Global) Technical Efficiency} = \text{(Local) Pure Technical Efficiency} \times \text{Scale Efficiency}$$
$$\phi^*_{\text{CCR}} = \phi^*_{\text{BCC}} \times \text{SE} \tag{3}$$

As noted earlier, in DEA, while it is not uncommon to obtain a large number of efficient DMUs, neither the CCR nor the BCC model is capable of discriminating and ranking among efficient DMUs. However, there are various extensions that can be used for ranking purpose (see Adler et al. 2002). Among them, the super-efficiency method assigns an efficiency score greater than or equal to one to all efficient DMUs. This will allow each efficient DMU to influence its input/output variables to ensure that resources are optimally utilized.

The Andersen and Petersen super-efficiency ranking technique discriminates all efficient DMUs by excluding the DMU under consideration and helps to rank efficient DMUs (Andersen and Petersen 1993). The Andersen and Petersen model is basically an extension of Eq. (1) except that in the first and second constraints, the DMU under consideration is not taken into account to calculate the super-efficiency score. This ranking methodology is used in this study for discriminating and ranking states that are efficient in the adoption and use of broadband.

This study uses the same dataset as in Parajuli and Haynes (2012) and variables are from the Census Bureau (CB), Bureau of Economic Analysis (BEA), FCC, and Inside Politics (see West 2010) at the state level for years ranging from 2005 to 2007 (Table 1).

Note that non-farm employment is based on the North American Industry Classification System (NAICS), and the total number of switches is the sum of local

Table 1 Variable description and source

Variable	Source
Input	
Annual estimate of the population	CB
Average personal income (in dollars)	BEA
Non-farm employment	BEA
Population at 25 years and over with a bachelor's degree or higher (in percentage)	CB
Total number of switches	CB
Output	
High-speed lines that offer over 200 kbps in at least one direction	FCC
Electronic government rating	Inside politics

switches, tandems, host switches, and remote switches reported by the incumbent local exchange carriers that are available through the Automated Reporting Management Information System (ARMIS) of the FCC.

Finally, since there is no standard definition of broadband and the FCC used the transmission rate of 200 kbps as a benchmark to collect broadband data, in this study, broadband implies a transmission rate over 200 kbps in at least one direction.

4 Results

Table 2 shows the CCR and BCC efficiency scores, scale efficiency (SE), and returns to scale (RTS: constant returns to scale, CRS; increasing returns to scale, IRS; decreasing returns to scale, DRS) properties for 48 DMUs (US states) having five inputs and two outputs in 2005.

In 2005, under the CRS assumptions, for the given inputs, 14 (29.17%) states were efficient in adopting and utilizing broadband. The mean CCR efficiency was 0.882, which implies that states were actually operating on an average efficiency of 88.2%. Under the VRS assumptions, 22 (45.83%) states were efficient and the average BCC efficiency score was 0.922.

Recall from Eq. (3) that global technical efficiency is the product of local pure technical efficiency and scale efficiency. Thus, Iowa, for instance, is globally inefficient (0.690) because of the inefficient local operating environment (0.710) and has an inefficient scale effect (0.972). Further, note that Iowa is characterized by IRS. This means Iowa is not operating at its optimal scale and is overutilizing its capacities. In order to become efficient, it needs to scale up its inputs and/or scale back its outputs.

Table 2 Efficiency scores, 2005

State	CCR score	BCC score	SE	RTS
Alabama (AL)	0.679	0.720	0.943	IRS
Arizona (AZ)	0.806	0.835	0.965	IRS
Arkansas (AR)	1.000	1.000	1.000	CRS
California (CA)	1.000	1.000	1.000	CRS
Colorado (CO)	0.823	0.824	0.998	DRS
Connecticut (CT)	1.000	1.000	1.000	CRS
Delaware (DE)	1.000	1.000	1.000	CRS
Florida (FL)	1.000	1.000	1.000	CRS
Georgia (GA)	0.855	0.904	0.946	IRS
Idaho (ID)	0.851	0.996	0.854	IRS
Illinois (IL)	0.818	0.841	0.972	IRS
Indiana (IN)	0.805	0.815	0.988	IRS
Iowa (IA)	0.690	0.710	0.972	IRS
Kansas (KS)	0.857	0.937	0.915	IRS
Kentucky (KY)	0.783	0.827	0.947	IRS
Louisiana (LA)	0.772	0.807	0.956	IRS
Maine (ME)	1.000	1.000	1.000	CRS
Maryland (MD)	0.855	0.864	0.989	IRS
Massachusetts (MA)	1.000	1.000	1.000	CRS
Michigan (MI)	0.941	0.946	0.995	IRS
Minnesota (MN)	0.752	0.779	0.966	IRS
Mississippi (MS)	0.959	1.000	0.959	IRS
Missouri (MO)	0.712	0.759	0.938	IRS
Montana (MT)	0.821	1.000	0.821	IRS
Nebraska (NE)	0.877	0.901	0.973	IRS
Nevada (NV)	1.000	1.000	1.000	CRS
New Hampshire (NH)	1.000	1.000	1.000	CRS
New Jersey (NJ)	1.000	1.000	1.000	CRS
New Mexico (NM)	0.706	1.000	0.706	IRS
New York (NY)	0.972	0.989	0.982	DRS
North Carolina (NC)	1.000	1.000	1.000	CRS
North Dakota (ND)	1.000	1.000	1.000	CRS
Ohio (OH)	0.899	0.906	0.992	IRS
Oklahoma (OK)	0.747	0.840	0.889	IRS
Oregon (OR)	0.896	0.969	0.925	IRS
Pennsylvania (PA)	0.805	0.807	0.997	DRS

(continued)

Table 2 (continued)

State	CCR score	BCC score	SE	RTS
Rhode Island (RI)	1.000	1.000	1.000	CRS
South Carolina (SC)	0.686	0.785	0.904	IRS
South Dakota (SD)	0.977	1.000	0.977	IRS
Tennessee (TN)	0.944	0.950	0.994	IRS
Texas (TX)	0.989	1.000	0.989	IRS
Utah (UT)	1.000	1.000	1.000	CRS
Vermont (VT)	0.898	1.000	0.898	IRS
Virginia (VA)	0.798	0.834	0.956	IRS
Washington (WA)	0.861	0.919	0.936	IRS
West Virginia (WV)	0.956	1.000	0.956	IRS
Wisconsin (WI)	0.773	0.802	0.965	IRS
Wyoming (WY)	0.787	1.000	0.787	IRS
Mean	0.882	0.922	0.957	

Source Authors' calculations

Montana, on the other hand, is globally inefficient (0.821) but locally efficient (1.000); its global inefficiency is due to its inability to utilize scale economy effectively. As in Iowa's case, Montana is also characterized by IRS and thus needs to scale up its inputs and/or scale back its outputs to become efficient.

New York is globally inefficient (0.972) due to local (0.989) and scale (0.982) inefficiencies. Unlike, Iowa and Montana, New York is underutilizing its capacities and thus its operation is characterized by DRS. It can scale down its inputs and/or scale up its outputs to become efficient.

According to Table 3, there are 14 CCR- and 22 BCC-efficient, 16 CCR- and 22 BCC-efficient, and nine CCR- and 20 BCC-efficient states in 2005, 2006, and 2007, respectively. This suggests that efficiency can either remain the same or change over time. This change in efficiency is due to the change in the input and output conditions that shifts the efficiency frontier from one time period to the other. Some DMUs stay on the efficiency frontier and remain efficient; others fall or stay below the efficiency frontier and are inefficient.

Table 4 shows the super-efficiency scores and state rankings based on the CCR model for 2005. There were 14 efficient states—Arkansas, California, Connecticut, Delaware, Florida, Maine, Massachusetts, Nevada, New Hampshire, New Jersey, North Carolina, North Dakota, Rhode Island, and Utah. California outranked them with a super-efficiency score of 1.751. California's radial change score was 57.11%, which implies that it could have reduced its broadband adoption and use by 42.89% and still remain efficient. Massachusetts, on the other hand, ranked fourteenth with a super-efficiency score of 1.004 and a radial change score of 99.60%. Massachusetts was marginally super-efficient and it could have reduced a mere 0.40% of its outputs and remain efficient.

Table 3 CCR and BCC efficiency scores, 2005–2007

	2005		2006		2007	
State	CCR score	BCC score	CCR score	BCC score	CCR score	BCC score
AL	0.679	0.720	0.716	0.758	0.701	0.890
AZ	0.806	0.835	0.960	1.000	0.911	1.000
AR	1.000	1.000	0.855	0.865	0.718	0.824
CA	1.000	1.000	1.000	1.000	1.000	1.000
CO	0.823	0.824	0.810	0.818	0.795	0.816
CT	1.000	1.000	0.942	0.943	0.943	0.947
DE	1.000	1.000	1.000	1.000	1.000	1.000
FL	1.000	1.000	1.000	1.000	1.000	1.000
GA	0.855	0.904	0.894	0.942	0.966	0.996
ID	0.851	0.996	0.879	1.000	0.759	1.000
IL	0.818	0.841	0.873	0.933	0.808	0.816
IN	0.805	0.815	1.000	1.000	0.816	0.861
IA	0.690	0.710	0.862	0.881	0.708	0.782
KS	0.857	0.937	0.819	0.863	0.718	0.752
KY	0.783	0.827	1.000	1.000	1.000	1.000
LA	0.772	0.807	0.918	0.928	0.794	0.856
ME	1.000	1.000	0.867	0.913	1.000	1.000
MD	0.855	0.864	0.833	0.833	0.859	0.894
MA	1.000	1.000	0.892	0.892	0.884	0.929
MI	0.941	0.946	0.992	0.995	1.000	1.000
MN	0.752	0.779	0.795	0.901	0.727	0.728
MS	0.959	1.000	0.806	1.000	0.625	1.000
MO	0.712	0.759	0.881	0.897	0.741	0.756
MT	0.821	1.000	1.000	1.000	0.857	1.000
NE	0.877	0.901	0.891	0.916	0.729	0.763
NV	1.000	1.000	1.000	1.000	1.000	1.000
NH	1.000	1.000	0.959	0.961	0.910	0.932
NJ	1.000	1.000	1.000	1.000	1.000	1.000
NM	0.706	1.000	0.792	1.000	0.680	1.000
NY	0.972	0.989	0.911	0.966	0.855	0.859
NC	1.000	1.000	0.935	0.939	0.903	0.948
ND	1.000	1.000	1.000	1.000	1.000	1.000
OH	0.899	0.906	0.975	0.977	0.865	0.920
OK	0.747	0.840	0.805	0.810	0.755	0.762
OR	0.896	0.969	0.960	1.000	0.824	0.881

(continued)

Table 3 (continued)

	2005		2006		2007	
State	CCR score	BCC score	CCR score	BCC score	CCR score	BCC score
PA	0.805	0.807	0.933	0.962	0.827	0.849
RI	1.000	1.000	1.000	1.000	0.991	1.000
SC	0.686	0.785	0.984	0.997	0.820	0.947
SD	0.977	1.000	0.936	1.000	0.649	0.864
TN	0.944	0.950	1.000	1.000	0.999	1.000
TX	0.989	1.000	1.000	1.000	0.959	0.966
UT	1.000	1.000	1.000	1.000	0.938	1.000
VT	0.898	1.000	1.000	1.000	0.809	1.000
VA	0.798	0.834	0.805	0.814	0.776	0.790
WA	0.861	0.919	0.898	0.909	0.842	0.866
WV	0.956	1.000	1.000	1.000	0.698	1.000
WI	0.773	0.802	0.768	0.775	0.672	0.705
WY	0.787	1.000	1.000	1.000	0.921	1.000
Mean	0.882	0.922	0.920	0.946	0.850	0.915

Source Parajuli and Haynes (2012)

Table 4 CCR-based super-efficiency and rankings, 2005

State	CCR-based super-efficiency score	Rank	Radial change score (%)
AR	1.108	8	90.25
CA	1.751	1	57.11
CT	1.048	12	95.42
DE	1.087	11	92.00
FL	1.524	2	65.62
ME	1.229	4	81.37
MA	1.004	14	99.60
NV	1.124	7	88.97
NH	1.043	13	95.88
NJ	1.100	10	90.91
NC	1.105	9	90.50
ND	1.213	5	82.44
RI	1.125	6	88.89
UT	1.338	3	74.74

Source Authors' calculations

Figure 1 shows the ranking of states based on the CCR efficiency/super-efficiency scores from 2005 through 2007. There were 14 (in 2005), 16 (in 2006), and nine (in 2007) CCR-efficient states (Table 3). Their super-efficiency scores were used for discriminating and ranking them. For the remaining inefficient states, their CCR efficiency scores were used for the ranking purpose. Moreover, states did not have the same (super-) efficiency scores over time. Thus, rankings are based on the least efficient to the most super-efficient score for the given period of time (Table 5).

In 2005, California, Florida, and Utah could have reduced their broadband adoption and use by 42.89%, 34.38%, and 25.26%, respectively, and would still have remained efficient. In 2006, California and Florida still ranked first and second, respectively, in terms of resource utilization. It was possible for California to reduce its outputs by 45.83% without becoming inefficient. Florida and North Dakota equally

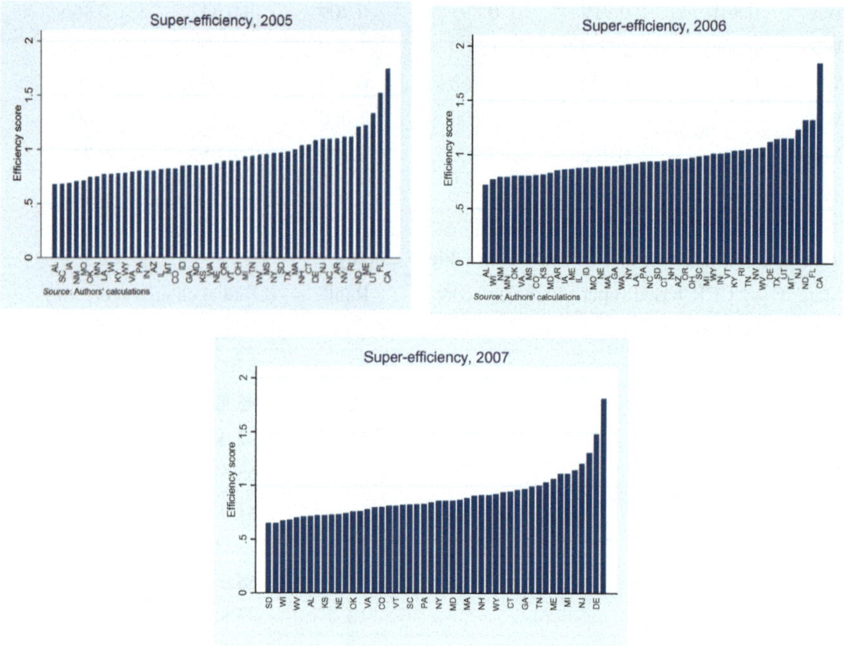

Fig. 1 Ranking of states based on the CCR efficiency scores, 2005–2007

Table 5 Top three states for efficient broadband adoption and use, 2005–2007

Ranking	2005	2006	2007
1	California	California	California
2	Florida	Florida and North Dakota	Delaware
3	Utah		Florida

Source Authors' calculations

Table 6 Super-efficiency and infeasible solutions, 2005

State	CCR super-efficiency score (Rank)	BCC super-efficiency score
AR	1.108 (8)	1.117
CA	1.751 (1)	1.961
CT	1.048 (12)	1.050
DE	1.087 (11)	1.107
FL	1.524 (2)	1.624
ME	1.229 (4)	1.254
MA	1.004 (14)	1.010
MS		1.301
MT		Big
NV	1.124 (7)	1.391
NH	1.043 (13)	1.043
NJ	1.100 (10)	1.140
NM		Big
NC	1.105 (9)	1.114
ND	1.213 (5)	Big
RI	1.125 (6)	1.282
SD		1.122
TX		1.024
UT	1.338 (3)	Big
VT		Big
WV		Big
WY		Big

Source Authors' calculations

ranked second in broadband adoption and use. Both of them could have decreased their outputs by 24.47% and would have remained efficient. In 2007, while California was still the most efficient state, Delaware pushed Florida to the third place and North Dakota was no longer in the top three places. California, Delaware, and Florida would still have been efficient even after reducing their broadband adoption and use by 44.72%, 32.43%, and 23.37%, respectively.

Table 6 shows the BCC super-efficiency scores for 2005. The BCC-based super-efficiency scores are not always feasible. Infeasibility implies that DMUs remain efficient even after reducing arbitrarily large outputs (see Dula and Hickman 1997; Xue and Harker 2002) and the infeasible solutions are indicated by "Big" in Table 6. The DMUs are not ranked because of this theoretical limitation.

5 Conclusions

The purpose of this chapter was to examine the efficiency and super-efficiency of broadband adoption and use in the 48 contiguous US states. It was found that states are not equally efficient in broadband adoption and use. In general, while some states are consistently efficient within the observation period, others are not. CCR-based super-efficiency scores allowed for discriminating efficient states and ranking them. It was found that super-efficient states can reduce their outputs and still remain efficient. However, over time, states that are super-efficient in one instant in time are not necessarily super-efficient at another instant in time.

States, such as California and Florida, which were super-efficient states throughout the study period, have adopted different strategies to increase broadband adoption and use (see California Broadband Taskforce 2008; Department of Management Services 2011). For instance, in a recent economic development agenda, the state government of California plans to develop broadband capacity for entrepreneurial activities and business growth and envisions making universal access a reality (Newsom 2011). Similarly, in 2009, the legislature of Florida passed F.S. 364.0135 that authorized its Department of Management Services to lead the broadband-related initiatives (Department of Management Services 2011), and in 2012, the Florida Public Service Commission (FPSC) addressed the issue of universal broadband access (FPSC 2012).

However, increasing broadband adoption and use is contingent upon factors, such as income, educational background/age group, and English-speaking population (see California Public Utilities Commission 2012) and some of the factors associated with relative efficiency are presented here. In addition, (broadband) Internet is a general purpose technology (GPT) that enables opportunities rather than final solutions (see Bresnahan and Trajtenberg 1995; Lipsey et al. 2005; Majumdar et al. 2010). In the absence of policies that do not take into account these issues, but merely focus on increasing broadband supply through federal or local support, the increase of actual adoption and use of broadband Internet is less likely. For instance, LaRose et al. (2011) noted that infrastructure grants do not guarantee adoption of broadband, and Jayakar and Park (2012) found that funding public computer centers in areas with low broadband availability is unlikely to generate additional usage.

Thus, broadband programs should be integrated with issues, such as income generation, education, and various development initiatives, to ensure that resources allocated for broadband are optimally utilized. Otherwise, efficient states are likely to be less efficient or even inefficient in terms of resource utilization implying that scarce resources are possibly misallocated. States need to use the appropriate mix of inputs (outputs) to maintain or improve their efficiencies before adding new broadband capacities.

There are a number of limitations in this study. First, the CCR and BCC models do not directly account for slacks (see Tone 2010) and that limitation carries forward in the Andersen-Petersen super-efficiency method. Second, as noted earlier, because of the infeasible solutions associated with super-efficiency (see Lovell and Rouse 2003; Xue and Harker 2002), it is often difficult to discriminate and rank efficient DMUs. Future research should take into account these modeling issues. Finally, depending on the availability of broadband data, the study period should be extended.

References

Adler, N., Friedman, L., & Sinuany-Stern, Z. (2002). Review of ranking methods in the data envelopment analysis context. *European Journal of Operational Research, 140*, 249–265.

Andersen, P., & Petersen, N. C. (1993). A procedure for ranking efficient units in data envelopment analysis. *Management Science, 39*, 1261–1264.

Anselin, L., Varga, A., & Acs, Z. J. (1997). Local geographic spillovers between university research and high technology innovations. *Journal of Urban Economics, 42*, 422–448.

Aschauer, D. A. (1989). Is public expenditure productive? *Journal of Monetary Economics, 23*, 177–200.

Baltagi, B. H., & Pinnoi, N. (1995). Public capital stock and state productivity growth: Further evidence from an error components model. *Empirical Economics, 20*, 351–359.

Banker, R. D., Charnes, A., & Cooper, W. W. (1984). Some models for estimating technical and scale inefficiencies in data envelopment analysis. *Management Science, 30*, 1078–1092.

Bauer, J. M. B. (2006). Broadband in the United States. In M. Fransman (Ed.), *Global broadband battles: Why the U.S. and Europe lag while Asia leads* (pp. 133–163). Stanford, CA: Stanford University Press.

Bresnahan, T. F., & Trajtenberg, M. (1995). General purpose technologies "Engines of growth"? *Journal of Econometrics, 65*, 83–108.

Button, K. J. (1998). Infrastructure investment, endogenous growth, and economic convergence. *Annals of Regional Science, 32*, 145–162.

California Broadband Taskforce. (2008). *The state of connectivity: Building innovation through broadband*. https://www.cio.ca.gov/broadband/pdf/CBTF_FINAL_Report.pdf.

California Public Utilities Commission. (2012). *California broadband report*. https://broadbandcouncil.ca.gov/WorkArea/DownloadAsset.aspx?id=100.

Carlson, S. C. (1999). A historical, economic, and legal analysis of municipal ownership of the information highway. *Rutger Computer & Technology Law Journal, 44*, 1–60.

Charnes, A., Cooper, W. W., & Rhodes, E. (1978). Measuring the efficiency of decision making units. *European Journal of Operational Research, 2*, 429–444.

Cherry, S. (2006). A Broadband Utopia. *IEEE Spectrum, 43*, 48–54.

Coelli, T. J., Rao, D. S. P., O'Donnell, C. J., & Battese, G. E. (2005). *An introduction to efficiency and productivity analysis* (2nd ed.). New York, NY: Springer.

Cooper, W. W., Seiford, L. M., & Tone, K. (2007). *Data envelopment analysis: A comprehensive text with models, applications, references and DEA-Solver software*. New York, NY: Springer.

Crandall, R. W., Lehr, W. H., & Litan, R. E. (2007). The effects of broadband deployment on output and employment: A cross-sectional analysis of U.S. data. *Issues in Economic Policy, 6*, 155–193.

Cronin, F. J., Parker, E. B., Colleran, E. K., & Gold, M. A. (1991). Telecommunications infrastructure investment and economic growth: An analysis of causality. *Telecommunications Policy, 15*, 529–535.

Datta, A., & Agarwal, S. (2004). Telecommunications and economic growth: A panel data approach. *Applied Economics, 36*, 1649–1654.

Department of Management Services. (2011). *Broadband Florida Grant Briefing Document 2011*. https://www.dms.myflorida.com/content/download/82172/471678/version/1/file/Broadband+Florida+Initiative+Brief+2011.pdf.

Dholakia, R. R., & Harlam, B. (1994). Telecommunications and economic development: Econometric analysis of the US experience. *Telecommunications Policy, 18*, 470–477.
Dinc, M., Haynes, K. E., Stough, R. R., & Yilmaz, S. (1998). Regional universal telecommunications service provisions in the US. *Telecommunications Policy, 22*, 541–553.
Ding, L., Haynes, K. E., & Liu, Y. (2008). Telecommunications infrastructure and regional income convergence in China: Panel data approaches. *Annals of Regional Science, 42*, 843–861.
Dingwall, C. (2006). Municipal broadband: Challenges and perspectives. *Federal Communications Law Journal, 59*, 67–104.
Dula, J. H., & Hickman, B. L. (1997). Effects of excluding the column being scored from the DEA envelopment LP technology matrix. *Journal of the Operational Research Society, 48*, 1001–1012.
Fernandes, E., & Pacheco, R. R. (2003). Efficient use of airport capacity. *Transportation Research Part A: Policy and Practice, 37*, 667–680.
Flyvberg, B. (2007). Policy and planning for large-infrastructure projects: Problems, causes, cures. *Environment and Planning B, 34*, 578–597.
Flyvberg, B., Skamris Holm, M. K., & Buhl, S. L. (2003). How common and how large are cost overruns in transport infrastructure projects? *Transport Reviews: A Transnational Transdisciplinary Journal, 23*, 71–88.
Ford, G. S., & Koutsky, T. M. (2005). Broadband and economic development: A Municipal Case Study from Florida. *Review of Urban and Regional Development Studies, 17*, 216–229.
FPSC. (2012). *Annual report.* https://www.psc.state.fl.us/publications/pdf/general/annualreports/2012.pdf.
Fulmer, J. E. (2009). What in the world is infrastructure? *Infrastructure Investor*, 30–32.
Gillett, S. E., Lehr, W. H., Osorio, C. A., & Sirbu, M. A. (2006). *Measuring the economic impact of broadband deployment.* https://cfp.mit.edu/publications/CFP_Papers/Measuring_bb_econ_impact-final.pdf.
Gomez-Ibanez, J. A. (2003). *Regulation infrastructure: Monopoly, contracts, and discretion.* Cambridge, MA: Harvard University Press.
Gramlich, E. M. (1994). Infrastructure investment: A review essay. *Journal of Economic Literature, 32*, 1176–1196.
Greenstein, S. M., & Spiller, P. T. (1995). Modern telecommunications infrastructure and economic activity: An empirical investigation. *Industrial and Corporate Change, 4*, 647–665.
Grubesic, T. H. (2010). Efficiency in broadband service provision: A spatial analysis. *Telecommunications Policy, 34*, 117–131.
Haslem, J. A., Scheraga, C. A., & Bedingfield, J. P. (1999). DEA efficiency profiles of U.S. banks operating internationally. *International Review of Economics & Finance, 8*, 165–182.
Haynes, K. E. (2006). *Infrastructure: The glue of megacities.* The Hague, The Netherlands: Kenniscentrum Grote Steden.
Haynes, K. E., Philogene, G. S., & Dinc, M. (2000). The relative efficiency of the general clinical research centres—An application of data envelopment analysis to medical research infrastructure. *International Journal of Healthcare Technology and Management, 2*, 403–424.
Haynes, K. E., Ratick, S., & Cummings-Saxton, J. (1998). Pollution prevention frontiers: A data envelopment simulation. In G. J. Knaap & T. J. Kim (Eds.), *Environmental program evaluation: A primer* (pp. 270–290). Urbana, IL: University of Illinois Press.
Holtz-Eakin, D. (1994). Public-sector capital and the productivity puzzle. *Review of Economics and Statistics, 76*, 12–21.
Hudson, H. E. (2010). Municipal wireless broadband: Lessons from San Francisco and Silicon Valley. *Telematics and Informatics, 27*, 1–9.
Jain, A., Mandviwalla, M., & Banker, R. D. (2007). Government as catalyst: Can it work again with wireless internet access. *Public Administration Review, 67*, 993–1005.
Jayakar, K., & Park, E. -A. (2012). Funding public computing centers: Balancing broadband availability and expected demand. *Government Information Quarterly, 29*, 50–59.
Kandilov, I. T., & Renkow, M. (2010). Infrastructure investment and rural economic development: An evaluation of USDA's broadband loan program. *Growth and Change, 41*, 165–191.

Karkazis, J., & Thanassoulis, E. (1998). Assessing the effectiveness of regional development policies in Northern Greece using data envelopment analysis. *Socio-Economic Planning Sciences, 32*, 123–137.

Katz, M. L., & Shapiro, C. (1985). Network externalities, competition, and compatibility. *American Economic Review, 75*, 424–440.

Kenny, R., & Kenny, C. (2011). Superfast broadband: Is it really worth a subsidy? *Info, 13*, 3–29.

Kolko, J. (2012). Broadband and local growth. *Journal of Urban Economics, 71*, 103–115.

LaRose, R., Strover, S., Gregg, J. L., & Straubhaar, J. (2011). The impact of rural broadband development: Lessons from a natural field experiment. *Government Information Quarterly, 28*, 91–100.

Leiner, B. M., Cerf, V. G., Clark, D. D., et al. (1997). The past and future history of the internet. *Communications of the ACM, 40*, 102–108.

Lipsey, R. G., Carlaw, K. I., & Bekar, C. T. (2005). *Economic transformations: General purpose technologies and long term economic growth*. New York: Oxford University Press.

Lovell, C. A. K., & Rouse, A. P. B. (2003). Equivalent standard DEA models to provide super-efficiency scores. *Journal of the Operational Research Society, 54*, 101–108.

Majumdar, S. K., Carare, O., & Chang, H. (2010). Broadband adoption and firm productivity: Evaluating the benefits of general purpose technologies. *Industrial and Corporate Change, 19*, 641–674.

Munnell, A. H. (1990). How does infrastructure affect regional economic performance? *New England Economic Review* (September/October), 11–32.

National Conference of State Legislatures. (2010). *State Broadband Task Forces, Commissions or Authorities and Other Broadband Resources*. https://www.ncsl.org/issues-research/telecom/state-broadband-task-forces-commissions.aspx.

Newsom, G. (2011). *An Economic Growth and Competitiveness Agenda for California*. https://www.ltg.ca.gov/docs/LGN_Econ_Agenda.pdf.

Nijkamp, P. (1986). Infrastructure and regional development: A multidimensional policy analysis. *Empirical Economics, 11*, 1–21.

NTIA. (2010). *Broadband Technology Opportunities Program (BTOP) Quarterly Program Status Report*. https://www.ntia.doc.gov/files/ntia/publications/btop_quarterlyreport_03032010.pdf.

Parajuli, J., & Haynes, K. E. (2012). Patterns of broadband efficiency in the U.S. *Growth and Change, 43*, 590–614.

Porter, M. E. (2000). Location, competition, and economic development: Local clusters in a global economy. *Economic Development Quarterly, 14*, 15–34.

Press, L. (1994). Commercialization of the internet. *Communications of the ACM, 37*, 17–21.

Prud'homme, R. (2005). Infrastructure and development. In F. Bourguignon & B. Pleskovic (Eds.), *Lessons of experience* (pp. 153–180). Washington, DC: The World Bank and Oxford University Press.

Rietveld, P. (1989). Infrastructure and regional development: A survey of multiregional economic models. *Annals of Regional Science, 23*, 255–274.

Roller, L. -H., & Waverman, L. (2001). Telecommunications infrastructure and economic development: A simultaneous approach. *American Economic Review, 91*, 909–923.

Soushi, S., Nijkamp, P., Rietveld, P., & Pels, E. (2010). A distance friction minimization approach in data envelopment analysis: A comparative study on airport efficiency. *European Journal of Operations Research, 208*, 1104–1115.

Seiford, L. M., & Thrall, R. M. (1990). Recent developments in DEA: The mathematical programming approach to frontier analysis. *Journal of Econometrics, 46*, 7–38.

Sherman, H. D., & Zhu, J. (2006). *Service productivity management: Improving service performance using Data Envelopment Analysis (DEA)*. New York, NY: Springer.

Tone, K. (2010). Variations on the theme of slacks-based measure of efficiency in DEA. *European Journal of Operational Research, 200*, 901–907.

Tongzon, J. (2001). Efficiency measurement of selected Australian and other international ports using data envelopment analysis. *Transportation Research Part A: Policy and Practice, 35*, 107–122.
West, D. (2010). *Brown Policy Reports*. https://www.insidepolitics.org/policyreports.html.
Willis, J. I. (2004). What impact will E-commerce have on the U.S. economy? *Economic Review—Federal Reserve Bank of Kansas City, 89*, 53–71.
Xue, M., & Harker, P. T. (2002). Note: Ranking DMUs with infeasible super-efficiency DEA models. *Management Science, 48*, 705–710.
Yilmaz, S., Haynes, K. E., & Dinc, M. (2001). The impact of telecommunications infrastructure investment on sectoral growth. *Australasian Journal of Regional Studies, 7*, 383–397.

Jitendra Parajuli holds his Ph.D. from the School of Policy and Government at George Mason University and is now in Nepal.

Kingsley E. Haynes is University Professor, Dean Emeritus, and holds the Hazel Endowed Chair at the Schar School of Policy and Government at George Mason University.

Elements in U.S. Households Broadband Internet Subscription Rates

Peter L. Stenberg

1 Introduction

In advanced industrialized economies, virtual activity has become recognized as a universal fact of everyday life for all but a large minority that has remained unseduced by the charms of the Internet. In North America, Europe, Japan, Korea, or other parts of the industrialized world, after the adoption rate for households has reached slightly above two-thirds, the adoption rate has, seemingly, plateaued.

Within counties, however, the rate may vary some. In America, the shortfall in adoption and use has been more acute in rural areas. Since the early 1990s, rural residents in America have been more likely than urban residents not to use the Internet, and at first, it was clearly an issue of service availability. Often it was not available locally for rural residents, so if they wanted to connect to the Internet, they needed to dial-up through toll and long-distance phone lines, an expensive proposition. As new technologies and business models were employed to connect people, Internet service became cheaper and more readily available to more households. Despite the advances over the last two decades, rural areas continue to lag. The national goal of universal Internet service, thus, remains incompletely fulfilled. The matter of choice, rather than service availability, however, has begun to be recognized as a bigger factor in determining the rural–urban subscription rate differential.

Globally, and nationally, physical distance and accessibility remain vital factors behind digital connectivity (Tranos et al. 2013). The shortfall in Internet accessibility and use will have, potentially, fundamental socio-economic consequences for individuals, businesses, governments, and regions and have been recognized by many governments. Government plans have been proffered, such as the European Commissions' Europe 2020 Strategy's Digital Agenda and the U.S. National Broadband Plan, in part to address Internet issues of adoption, availability, and use.

P. L. Stenberg (✉)
Washington, DC, USA

J. Thill (ed.), *Innovations in Urban and Regional Systems*,
https://doi.org/10.1007/978-3-030-43694-0_16

In America, federal and state policies have been implemented to increase Internet access across the country including some new programs designed explicitly to increase household Internet participation. Historically, federal and state government Internet programs have mostly leveraged private funds to increase the availability of broadband Internet service. The Rural Utility Service of the U.S. Department of Agriculture (USDA) has been a lead agency for rural Internet policy implementation through three on-going programs, namely the: (1) traditional federal rural telecommunication infrastructure program requiring all facilities to be broadband capable; (2) farm bill rural broadband program (authorized by the five-year farm bills, the Agriculture Act of 2014 is the latest of these); and (3) Community Connect Broadband Grant Program. The U.S. Department of Commerce-National Telecommunications and Information Administration (NTIA) and U. S. Department of Agriculture-Rural Utility Service (RUS) also jointly administered broadband programs resulting from the American Recovery and Reinvestment Act of 2009 that has led to, approximately, a $7 billion investment in broadband infrastructure. Recently, the Federal Communications Commission (FCC) reformed the Universal Service Fund and created the Connect America Fund that provided $300 million in phase I monies for rural broadband system development. In September 2015, the FCC announced a further $9 billion over six years in phase II monies.

The research presented in this chapter explores the factors associated with rural household subscription of the more advanced broadband Internet services. The research investigates the increase experienced in rural household broadband Internet subscriptions, socio-economic demographics of broadband Internet subscription, and the rural–urban dichotomy of broadband Internet subscription, thus showing that the provision of broadband Internet services may no longer be the critical issue in reaching universality.

2 Background

The National Telecommunications and Information Administration (NTIA) studies in the mid-1990s are the first national-level research that documented the world of Internet users. They also launched the term Digital Divide into its now familiar place in the telecommunications policy lexicon (National Telecommunications and Information Administration 1998; Greenstein and Prince 2006). The American and OECD descriptive studies offered selected national demographics of computer and Internet users and have over the years delivered snapshots of the Internet's rapid evolution from its humble origins in the academic community.

Recent studies describe the current and more static situation or examine the adoption of the largely pre-broadband era of the Internet. Household studies by Choudrie and Dwivedi (2005, 2006), Stenberg (2008) and the U.S. General Accounting Office (2001) tested socio-economic factors distinguishing adopters and non-adopters of computers and the Internet. Choudrie and Dwivedi (2005) found age, gender, and social grade were important when distinguishing between adopters and non-adopters of the Internet in British households. Their 2006 British study found that characteristics such as income and education were important factors. Nearly all studies on Internet adoption have focused on the household at the national aggregate level. The early NTIA American and the Dutton and Blank (2013) and Farrington et al. (2013) British studies, for example, described differences across many demographic and geographic groupings, not only for households but also Internet activity in the workplace. In Great Britain, Farrington et al. (2013) found that a divide manifested itself mainly between what they called "deep rural" and other areas.

Parker and Hudson (1992) were the first to raise the issue of rural–urban equal access to the Internet though when they wrote their book it was still an era where nearly no one connected to the Internet from the home. The issue that they had preternaturally recognized has continued to this day (Oden and Strover 2002; Stenberg 2013). A study recognizing the importance of rural location, however, found, after controlling for income and education attainment, rural and urban households were almost as likely to use dial-up Internet services (Stenberg 2006).

The issue, however, has further evolved, and it is no longer simply whether a household has Internet service available (Stenberg 2013; Camagni and Capello 2005), whereas at the beginning, it was simply a dial-up connection to the Internet, now the issue has become fuzzier. The technology of the service available for any given household, and hence what may be done on the Internet, has become the issue. The new technology comes under the definitional umbrella of what is called broadband technologies, also known as high-speed technologies.

Like the original dial-up Internet technologies, broadband Internet technologies are not as prevalent in rural areas as in urban areas. Unlike the dial-up technology, the economic challenges of broadband deployment are greater in rural areas. Nonetheless, rural areas have been rapidly receiving the technology. Although rural areas are still catching up to more urbanized areas, another issue has come to the fore—not all households that can acquire a broadband connection actually choose to subscribe. As a consequence, in policy forum discussions, the issue of broadband Internet availability has lessened, while the issue of why some do not choose to subscribe has risen (Malecki 2008; Stenberg et al. 2009; Stenberg and Morehart 2012). We will take a beginning look at these issues in this chapter, showing some of the dimensions of regional variance and basic analysis of the underlying reasons for observing the regional variance that exists.

3 Data Used in the Analysis

In the research presented here, we use data from the Bureau of the Census' Current Population Survey (CPS). The CPS is a monthly survey of approximately 50,000 households covering various socio-economic characteristics such as family income, employment status, and age. The computer and Internet data that we use, however, are not included in every CPS and have been collected only irregularly over the years as a supplement to one of the monthly surveys.

We use the data aggregated at the household level. The household data are not provided with geospatial coordinates although they are coded with state and metropolitan-status location codes. In other words, we are able to differentiate between households located in urban and rural areas for each state (within limits of any survey and the sampling technique employed). The definitions of metropolitan, aka urban, and nonmetropolitan, aka rural, are those used by the Bureau of the Census as defined by the U.S. Office of Management and Budget (Office of Management and Budget 2013).

Our national estimates are based on statistical analysis of the raw CPS data and a number of CPS weighting protocols. The Bureau of Census over-samples some subpopulations when conducting the CPS. The Economics and Statistics Administration of the Department of Commerce constructs weights for the survey data. The data used in the analysis come from the October 2010 monthly survey.

4 General Trends

Overall Internet subscriptions show the upward trends that other new technologies have shown in the past and resemble perhaps most closely that of cable TV's subscription growth curve (Stenberg 2006), including the recent plateauing of the subscription rates that many have recognized and can be seen to some extent in Fig. 1. Specifically, rural and urban household Internet subscriptions have increased considerably since 2000 though the rate of increase has slowed down considerably in recent years. And, as is evident, many households still do not subscribe. They do not subscribe either by choice or situation.

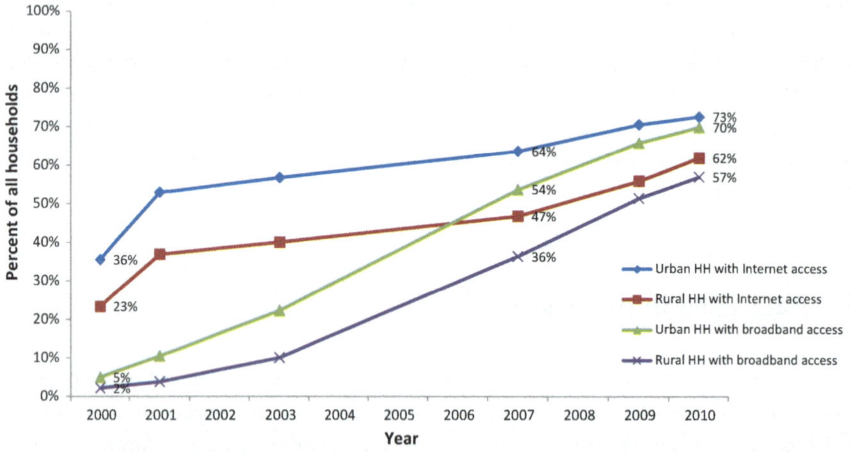

Fig. 1 Internet access in households, 2000–2010. *Note* HH means households. *Source* Stenberg (2013)

Broadband Internet service subscriptions during the same period went from nil to a point where nearly all households that subscribed to the Internet had it through high-speed technologies. Rural–urban spatial differences in overall Internet subscription rates, however, remain with roughly 73% of American urban households subscribing to home Internet connections, while only 62% of rural households do so too.

The technologies for gaining access to the Internet have been changing quite rapidly with a number of alternative broadband technologies becoming available at the same time that dial-up was becoming largely insufficient to all but the most mundane Internet activity. With the increasing sophistication of Web sites and the increasing variety of on-line products and services, accessing the Internet through broadband technologies has largely become viewed as necessary in order to fully utilize what the Web has to offer.

5 Urban–Rural Differences in Subscription Rates

While rural household Internet subscription rates remain low in comparison with urban households, the difference between urban and rural adoption rates is highly variable across the country as can be seen in the following maps, Figs. 2 and 3. The

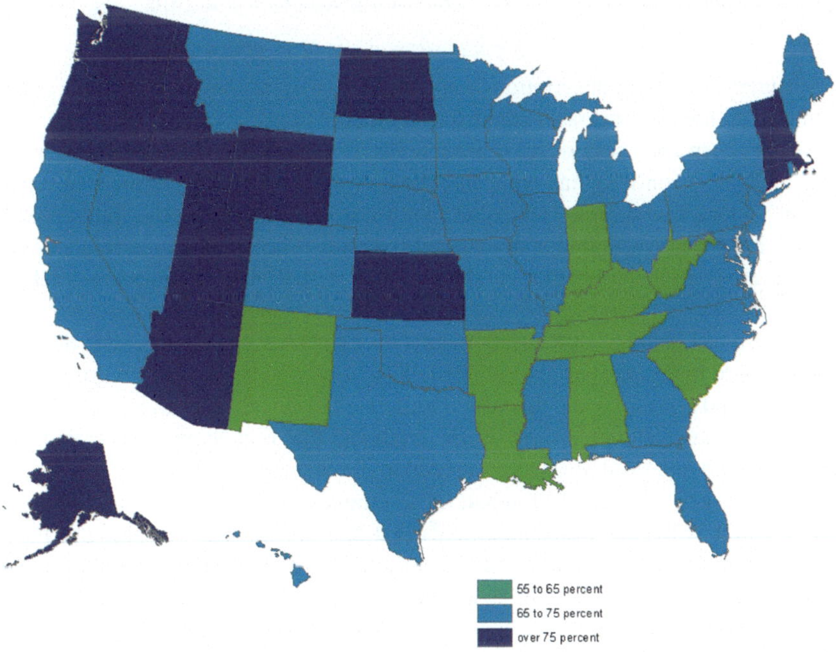

Fig. 2 Urban households with broadband internet subscriptions by state in 2010

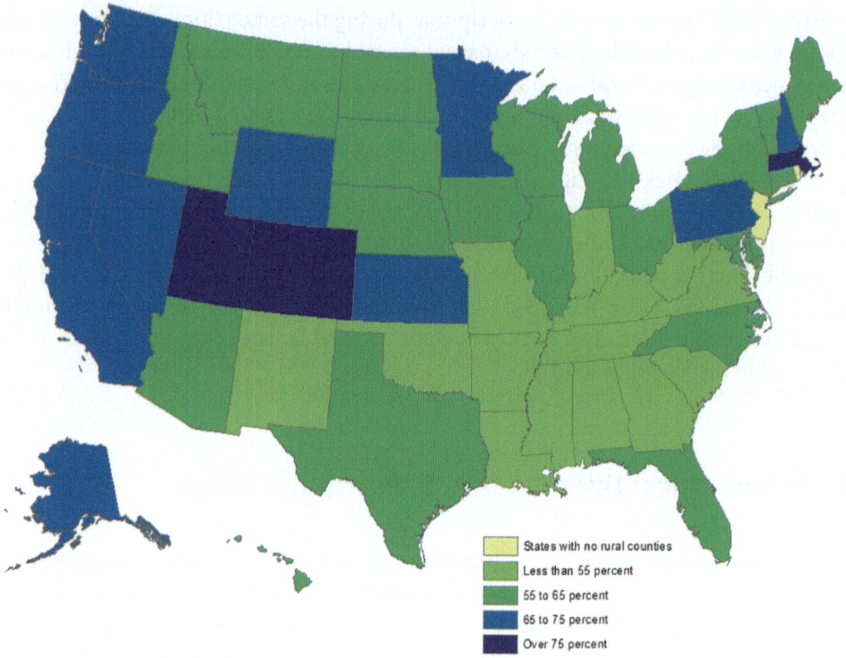

Fig. 3 Rural households with broadband internet subscriptions by state in 2010

lowest urban Internet subscriptions rates are primarily in the southern part of the country. The lowest rural rates are also in southern states.

Northeastern and western rural households, on the other hand, are, on average, more likely to go on-line than other rural households (Fig. 3). In a number of states, such as Colorado and New Hampshire, the rural household adoption rate exceeds the national urban rate significantly and substantially. The variability in rural rates of adoption suggests that more than rural isolation is at play when it comes to household subscriptions.

Once a household has purchased Internet access, however, they are most likely to have acquired high-speed access; in 2010, 96% of on-line households in urban areas had broadband service, while this penetration rate falls to 92% in rural areas (Fig. 1). This rural–urban difference supports the argument that broadband service is not as readily available in rural areas as compared to urban areas.

Analysis by Stenberg and Morehart (2012) suggests that the conversion to broadband Internet is largely from households that had preexisting, i.e., dial-up, Internet subscriptions and were not likely from households that had no existing Internet service. The analysis held no evidence to suggest that when broadband service became available households dropped their Internet subscription altogether. They either remained with their current service or, as was the most common case, moved up to broadband service. While rural northeastern and western rural households generally have higher broadband subscription rates than other parts of the country, they remain with significantly lower rates than their within region urban counterparts.

Rural broadband subscriptions have become more ubiquitous, but many challenges remain for rural service providers. By the very nature of their low population, rural areas do not exhibit the economies-to-scale that urban areas have. Provision costs, therefore, are higher than in urban areas and so tend to be subscription rates for potential customers. As much as the provider can pass on these costs, the additional costs would make broadband Internet access less affordable for businesses and consumers. Given the relative lack of competition faced by rural service providers and the growing inelasticity of demand for the services, they likely have the ability to pass on the costs (Stenberg 2006). On the other hand, the less provision costs can be passed on to potential customers, the less enticement rural service providers have to provide the service at all. Mountainous terrain and harsh weather present additional challenges in rural areas, driving up the cost of service provision in certain areas. Reliable measures of actual costs faced by rural businesses and consumers, however, are not well known and are the subject of new surveys by a number of researchers.

While broadband has increasingly become available in rural and poor areas, the issue has increasingly become a quality issue. In terms of broadband Internet service, quality means the reliability and speed of data transmission. Rural households rely more often on satellite and wireless connections instead of cable or fiber technologies than households in urban areas. Rural households, when they do subscribe to a land-based hardwired service, also use DSL, a generally slower and, arguably, less reliable technology than cable and fiber, more often than urban households (Stenberg and Morehart 2012).

Market analysis for broadband service provision (as noted by the Federal Communications Commission (FCC) and others): the lower the population density, the poorer the community, or the higher the cost of service delivery cost (due to terrain ruggedness and other challenges), the less quality of service provision. All of the characteristics of population density, relative wealth, and physical topographic challenges commonly hold, but with some notable exceptions, for rural areas. They are also consistent with the existing rural–urban dichotomy in technology provision.

As has been shown in NTIA's National Broadband Map, Native American reservations, rural poverty counties, and other counties stand out with their lower levels of service. The more densely populated areas, such as the megalopolis stretching from Washington, DC to Boston, have the highest percent of fast (4 Mbps) broadband service for households, while the low population areas, such as the Dakotas, have the lowest percentage. Wilderness areas, such as central Maine and parts of the Rockies, also show the expected low percentage.

6 Self-reported Reasons for Non-subscriptions by the Household

Historically, most of the Internet policy discussion has focused on service availability. With broadband Internet service's rapid rollout, however, it has been a sufficiently long enough period that most households do have at least one broadband service available, even where a land-based system such as fiber optics or DSL is not available. The FCC has estimated that 95% of all households have broadband service available as of December 2013. As a consequence, policy discussion has begun to shift away from the availability issue.

Policymakers are starting to recognize that not having a home Internet subscription is sometimes by choice whether it be voluntary, in one sense, such as they just do not want it, or involuntary, in another sense, such as when they cannot afford it (Fig. 4). Currently, most households that do not have an Internet subscription do not have it largely by choice.

The largest pluralities of households who do not have Internet subscriptions are those who do not want it. Rural residents, however, are slightly more likely to cite availability, or more precisely the lack of available Internet service, in their area as a reason for not subscribing. The majority of rural households say that they either do not want it or they can use it elsewhere. Only 2% stated broadband service availability as the primary reason for not subscribing.

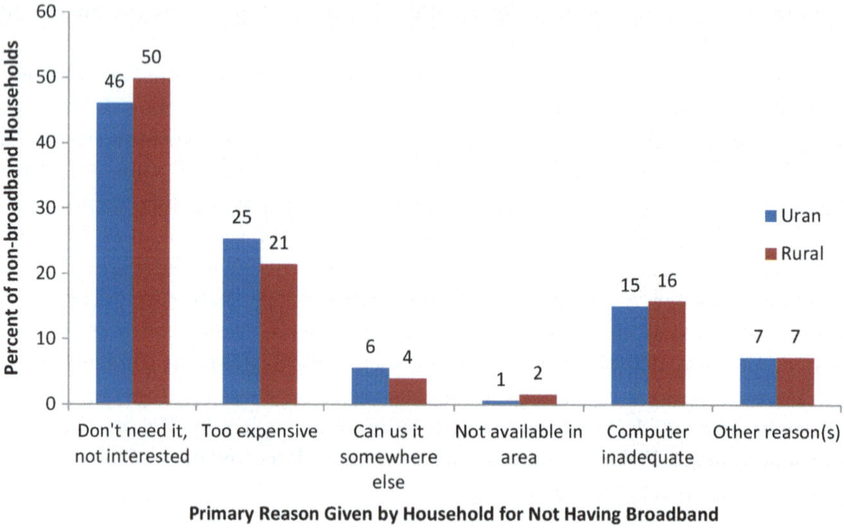

Fig. 4 Most important reason households gave for not having internet service, 2010

7 Revealed Factors in Household Internet Subscriptions

Service cost still remains a major reason cited by rural residents for not having Internet access although the decrease in the cost of broadband technologies over the last decade has had a significant impact on increasing Internet use. Federal Internet programs also have increased Internet use. Nevertheless, the pattern exhibited in Fig. 5 suggests that household income plays a significant role in household Internet subscriptions. Rural household Internet access, at any given income level, generally falls below the correspondent urban household Internet access rate. This is one indication that broadband service has not been as readily available in rural areas as in urban areas.

As stated already, once a household is purchasing Internet services, they are most likely to have broadband (Fig. 6). The gap between rural and urban households, however, remains remarkably flat, outside some data sampling noise and the off-campus college student effect at the lowest income level, when controlling for income. The result indicates that expense is not much of a factor after controlling for income, irrespective of the rurality of the household. The gap between urban and rural households would mostly, but not necessarily entirely, be a consequence of service availability. Some of the aggregate differences in adoption rate between rural and urban households would likely be as a consequence of the lower incomes found in rural households' vis-à-vis urban.

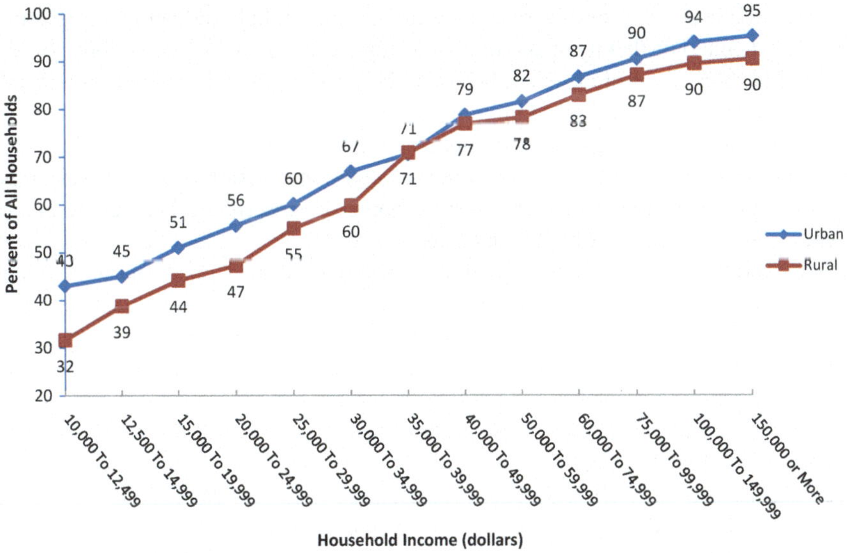

Fig. 5 Rural and urban with in-home internet access using any technology, by income, 2010

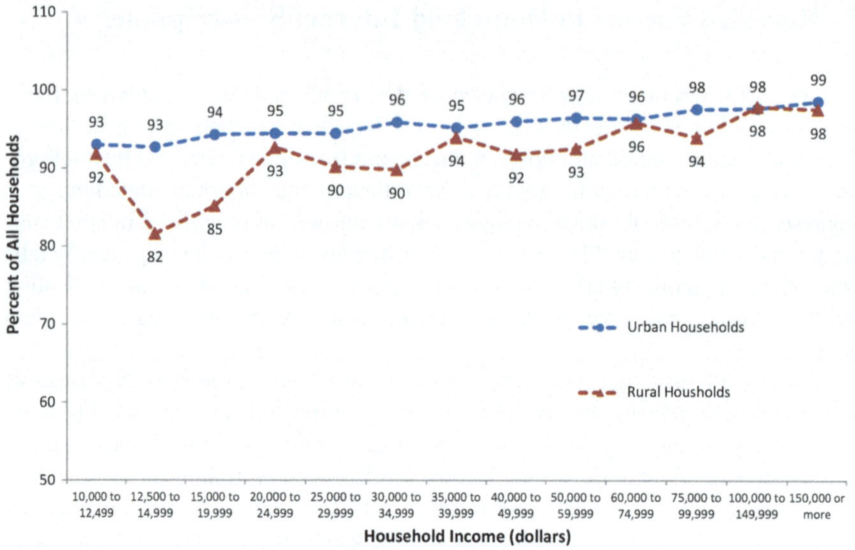

Fig. 6 Broadband as a share of in-home internet access by income, 2010

8 Logistic Regression Analysis

We explore the issue of multivariate causality in household broadband Internet subscription demand further by means of logistic regression analysis. The method allows evaluation in the light of observed behavior, in this case the selection of broadband Internet subscription. Statistical inferences are made on a model of choice behavior from US household sample data, in this case the October 2010 Bureau of the Census' Current Population Survey. The Internet adoption decision is qualitative and is postulated to be a choice between not subscribing or subscribing, i.e., having in the home either no Internet or broadband Internet service.

We hypothesize that income, age, rural–urban place of residence, and some other factors are determinants in broadband Internet subscription as some of them have already shown their influence in computer use as well as the early dial-up Internet use. Our null hypothesis is that broadband Internet use is a random event with no determinants. Furthermore, as is often the case when the dependent variable is categorical, the logit model is employed to examine factors that influence Internet adoption. The logistic specification is well suited to this type of application and has been used in similar studies. See, for example, Gloy and Akridge (2000).

It should be noted that estimates of goodness-of-fit are given in model estimations are not given here. R-squared estimates are traditionally given for OLS regressions, but they are much more controversial for categorical regressions as R-squared estimates do not exist per se. Sometimes, pseudo R-squares are estimated to proxy the R-squared value with a number of different methods often used to proxy it in categorical regression models. Nevertheless, R-squared estimates used in logistic regressions are highly controversial, with no broad acceptance of any one estimation methodology over another, and, as many statisticians argue, may be misleading and should only very carefully be used to compare models, if used at all.

9 Model Results

The results of logistic regression are shown in Table 1. The model is significant with an F-statistic greater than 500 and shows its power to predict having a household broadband Internet subscription. All of the independent variables are significant and with the expected sign and show that each is associated with broadband Internet subscription.

The model uses five income groups, each coded with whether the observation belongs to the household income group (less than $25,000, $25,000–$50,000, $50,000–$75,000, $75,000–$100,000, and over $100,000). The results show that the greater the income the more likely a broadband subscription will be obtained and that the upper income group is seven times more likely to have a broadband Internet subscription than the lowest level. Whilst prices may be perceived cheap with many able to purchase broadband service for less than 50 U.S. dollars a month, purchasing the personal computer, software, and other equipment can still be prohibitive for households of lesser means.

The greater the educational attainment, the more likely the household would purchase an Internet connection. In the model, we have educational attainment in four groups: no high school diploma, high school diploma, some college attainment, and attainment of a bachelor's degree or more. College-educated households are five times more likely to have home broadband subscription than those who have not obtained a high school diploma.

In the model, age is a continuous variable ranging from three years of age to 85. Age has a unique property that differs from income and education. As people get older, they are more likely to have an Internet subscription or at least until they reach retirement age when the likelihood starts to decrease. This is a bell-shaped curve. As a consequence, we model age with two variables: age and age-squared. These two parameters largely capture this nonlinear effect. The results for the age parameters are consistent with the bell-shaped curve and the expected positive effect of age on adoption generally. The results, however, show that this factor may not be as important as it has been noted in past research, such as Choudrie and Dwivedi (2006), with age only slightly increasing the odds if having broadband with the odds diminishing less past retirement age.

School-age children, with their exposure to the Internet in their schools as well as their increasing need to get on-line for school assignments and instruction, have been recognized, at least anecdotally, as major demand drivers for household broadband service subscriptions. The model finds that the presence of children in the household significantly increases the odds of having a broadband subscription in the household.

Table 1 Logistic regression of broadband adoption

Broadband at home	Odds ratio	Standard error	t	$P >$ $\|t\|$	95% Confidence	Interval
Family income 25 to 50 K	2.087526	0.0615735	24.95	0	1.970265	2.211765
Family income 50 to 75 K	3.496363	0.1314808	33.29	0	3.247931	3.763798
Family income 75 to 100 K	4.93101	0.2570438	30.61	0	4.452092	5.461447
Family income 100 K plus	6.961	0.3805015	35.5	0	6.253785	7.748191
Education—high school	1.901918	0.0754805	16.2	0	1.759585	2.055764
Education—some college	3.236545	0.1359027	27.97	0	2.980845	3.51418
Education—college	5.197252	0.2430401	35.24	0	4.742077	5.696118
Age in years	1.045855	0.0044221	10.6	0	1.037224	1.054559
Age-squared	0.9993266	0.0000404	−16.65	0	0.9992474	0.9994059
Household with children	1.401576	0.0453257	10.44	0	1.315496	1.49329
Houseorapt	1.620229	0.0888167	8.8	0	1.455175	1.804003
Access away from home	1.21589	0.035784	6.64	0	1.147739	1.288089
Foreigner	0.7321508	0.0367527	−6.21	0	0.6635465	0.8078481
Rural	0.8139245	0.0246702	−6.79	0	0.7669796	0.8637427
Geographic region						
Midwest	0.8770082	0.0340567	−3.38	0.001	0.8127348	0.9463645
South	0.8685122	0.0316839	−3.86	0	0.8085804	0.932886
West	1.099543	0.044084	2.37	0.018	1.016447	1.189432
_cons	0.1714402	0.0210293	−14.38	0	0.1348035	0.218034

Number of observations = 152,665
Design degrees of freedom = 152,664
$F(17, 152{,}648) = 501.12$
Prob > F = 0.0000

Stability in home environment has been postulated as a factor in adopting in-home broadband service. In the model, we use residency in a house or apartment as a proxy for this possible factor. The results indicate that living in house or apartment significantly increases the odds of having a broadband subscription.

People accessing the Internet away from home have been suggested as more likely to buy home subscriptions. Often the access is through the workplace and, to a lesser extent, libraries. The argument is as follows: the more people are familiar with **what the Internet has to offer, the more they want to have greater access**

to it. The access away from home variable shows a significant positive relationship with household subscription.

Anecdotally, it has been noted that migrant farmworkers are often not citizens and do not have broadband subscriptions. They do, however, communicate with home through smartphones. We use a non-US citizen as a proxy for this. This is not the best proxy, but the results seem to be consistent with this observation; non-citizens are much more likely not to have a broadband subscription.

We have two spatial factors in the model: (1) rural–urban location and (2) geographic region of the country. In the first, the factor is whether a household is located in a rural location or an urban location, as defined before. The model clearly shows rural households are less likely than urban households to have adopted broadband Internet. As we control for the lower income, educational attainment, and other differences in rural vis-à-vis urban households, the negative sign for rural location a sign of rural households has less broadband service availability.

Regions of the country, the Northeast, Midwest, South, and West Census Regions (U.S. Census Bureau 2012), are different, with different rural–urban spatial patterns. That pattern clearly shows in the model results. The Northeast has the most densely populated rural areas of the four regions and thus, relative to the nation, higher rates of rural household broadband service availability. The West, however, has the most urbanized rural area of the four regions and, as a consequence, has the greatest likelihood of household broadband Internet subscription. The Midwest and South have the most spread out of rural populations. Thus, if you are a Midwestern or Southern rural household, you are likely to have a greater challenge in obtaining a broadband subscription.

10 Conclusions

Obtaining broadband Internet service remains more challenging in rural areas. Rural households still are less likely to have broadband Internet service available to them than their urban peers, but nearly all rural households that have the Internet, like their urban brethren, use broadband technologies from the home.

Service availability of broadband, however, is no longer the primary reason for not having home Internet subscriptions as it has now come to pass that a much larger share of households without the Internet choose not to subscribe rather than cannot subscribe. Much of the rural–urban household variance in broadband adoption rates can be explained by the variance in household characteristics. Rural households tend to have less income and lower educational attainment, on average, than urban households. Broadband availability though greatly increased, however, remains a significant negative factor for rural residents in being able to purchase the services.

The rural underperformance in Internet use will have fundamental socio-economic consequences for individuals, businesses, governments, and regions. Federal and state policies continue to address broadband availability, but policymakers have begun to recognize this, so some new programs diverge from the original Internet infrastructure model and are designed explicitly to increase household Internet participation.

Irrespective of location, be it North America, Europe, Japan, Korea, or other parts of the industrialized world, one fact is universal—after the adoption rate has reached around two-thirds, adoption rate has plateaued. In order to understand why it has plateaued, more research is needed to better understand the dynamics of household choice with respect to broadband Internet services to better illuminate the policy discussion and program development. Ideally, this would be through longitudinal studies, but, given the currently short data time frame, other techniques will continue to offer analysis to better guide policy.

References

Camagni, R., & Capello, R. (2005). ICTs and territorial competitiveness in the era of the internet. *The Annals of Regional Science, 39*, 421–438.

Choudrie, J., & Dwivedi, Y. K. (2006). Investigating factors influencing adoption of broadband in the household. *The Journal of Computer Information Systems, 46*(4), 25–34.

Choudrie, J., & Dwivedi, Y. K. (2005). The demographics of broadband residential consumers in a British local community: The London Borough of Hillingdon. *The Journal of Computer Information Systems, 45*(4), 93–101.

Dutton, W. H., & Blank, G. (2013). *Cultures of the internet: The internet in Britain*. Oxford: University of Oxford, Oxford Internet Institute.

Farrington, J., Philip, L., Cottrill, C., Abbott, P., Blank, G., & Dutton, W. (2013). *Two-speed Britain: Rural internet use*. Oxford: University of Oxford, Oxford Internet Institute.

Gloy, B., & Akridge, J. (2000). Computer and internet adoption on large U.S. farms. *International Food and Agribusiness Management Review, 3*, 323–338.

Greenstein, S., & Prince, J. (2006). *The diffusion of the internet and the geography of the digital divide in the United States*. NBER Working Paper Series, Working Paper 123182. Cambridge, MA: National Bureau of Economic Research.

Leamer, E. E., & Storper, M. (2001). The economic geography of the internet age. Working Paper 8450. Cambridge, MA: National Bureau of Economic Research.

Malecki, E. J. (2008). *The rural digital economy: Competition and the geography of telecommunications*. Paper presented at the Economic Research Service's Broadband in the Rural Economy workshop, Washington, DC.

National Telecommunications and Information Administration & Economics and Statistics Administration. (2014). *Exploring the digital nation*. Washington, DC: U.S. Department of Commerce.

National Telecommunications and Information Administration & Economics and Statistics Administration. (2004). *A nation online: Entering the broadband age*. Washington, DC: U.S. Department of Commerce.

National Telecommunications and Information Administration & Economics and Statistics Administration. (2002). *A nation online: How Americans are expanding their use of the internet*. Washington, DC: U.S. Department of Commerce.

National Telecommunications and Information Administration. (1998). *Falling through the net: New data on the digital divide*. Washington, DC: U.S. Department of Commerce.

National Telecommunications and Information Administration. (1995). *Falling through the net*. Washington, DC: U.S. Department of Commerce.

Oden, M., & Strover, S. (2002). *Links to the future: The role of information and telecommunications technology in Appalachian Economic Development*. Washington, DC: Appalachian Regional Commission.

Office of Management and Budget. (2013). OMB Bulletin No. 13-01. Available at https://www.whitehouse.gov/sites/default/files/omb/bulletins/2013/b-13-01.pdf.

Parker, E. B. (2000). Closing the digital divide in rural America. *Telecommunications Policy, 24*(4), 281–290.

Parker, E. B., & Hudson, H. E. (1992). *Electronic byways: State policies for rural development through telecommunications*. Boulder: West View Press.

Stenberg, P. (2013). *Rural broadband at a glance* (2013 ed.). Economic Briefing Number 23. U.S. Department of Agriculture.

Stenberg, P., & Morehart, M. (2012). The existence of pent-up demand for rural broadband services: An exploration. In C. Karlsson, B. Johansson, & R. Stough (Eds.), *Entrepreneurship social capital and governance* (pp. 221–240). Northampton: Edward Elgar Publishing.

Stenberg, P., Morehart, M., Vogel, S., Breneman, V., Cromartie, J., & Brown, D. (2009). *Broadband internet's value for rural America*. Economic Research Report, ERR-78. Washington, DC: U.S. Department of Agriculture.

Stenberg, P. L. (2008). *Rural broadband internet use and the rural economy: A quasi-experimental design analysis of the broadband effect on rural economies*. Paper presented at the Economic Research Services' Broadband in the Rural Economy workshop, Washington, DC: USDA-ERS.

Stenberg, P. L. (2006). Investment and household adoption of communication and information services across the United States. In B. Johansson, C. Karlsson, R. Stough (eds.), *The emerging digital economy: Entrepreneurship, clusters and policy* (pp. 263–276). Berlin, Heidelberg, New York: Springer.

Tranos, E., Kourtit, K., & Nijkamp, P. (2013). Digital urban network connectivity: Global and Chinese internet patterns. *Papers in Regional Science, 93*(2), 409–428.

U.S. Census Bureau Census Regions and Divisions of the United States. (2012). Available at https://www2.census.gov/geo/pdfs/maps-data/maps/reference/us_regdiv.pdf.

U.S. General Accounting Office. (2001). *Characteristics and choices of internet users*. GAO-01–345. Washington, DC.

Migration Patterns in Turkey Between 1995 and 2000: Income, Employment and Distance Effects on Inter-provincial Migration

Turgay Kerem Koramaz and Vedia Dokmeci

1 Introduction

After the 1950s, a vast amount of migration from rural to urban areas was mainly caused by the availability of industrial or service jobs in the cities, and a lack of rural employment opportunities, which resulted in tremendous urban growth in Turkey (Sewell 1964; Tumertekin 1968; Gedik 1991; Tekeli 2008). This was not just an increase in urban population but was also a transformation of the economy, social fabric, public facilities and lifestyle, as well as in the relationship with the international community. This new settlement system simultaneously generated new patterns of growth and interaction. The rapid increase in the urbanization rate between 1950 and 1975 is almost equal to the increases experienced in the United States and England in the nineteenth century (Bairoch 1991). Yet, despite the large amount of research about internal migration at the international level, as already claimed by Halfacree and Boyle (1993), quantitative analysis of migration streams that caused this rapid urbanization in Turkey is scarce, with only a few exceptions from previous decades (Gedik 1977, 1978, 1991, 1992). Although there were several studies on internal migration in Turkey during the 1960s, their number decreased in the 1970s and has continued to decline ever since (Celik 2007). Thus, the present chapter investigates the factors that affected inter-provincial migration between 1995 and 2000 in Turkey, which is important as a transition period from the twentieth to the twenty-first century.

In general, it is believed that structural variables such as inequitable land distribution, lack of rural employment opportunities, and rural–urban employment wage and amenity gaps influence population movements in developing countries (Bilsborrow et al. 1987; Fukurai et al. 1987) as well as in developed ones (Kontuly and Schon 1994). Although China has carried out rural industrialization development

T. K. Koramaz (✉) · V. Dokmeci
Department of Urban and Regional Planning, Faculty of Architecture, Istanbul Technical University, Mete C. Taskisla, Taksim, 34437 Istanbul, Turkey
e-mail: koramaz@itu.edu.tr

J. Thill (ed.), *Innovations in Urban and Regional Systems*,
https://doi.org/10.1007/978-3-030-43694-0_17

policies in order to reduce intra-provincial migration since the 1970s, there has been no significant positive change in migration levels reported in the relevant literature and such policies are criticized for only having absorbed potential migrants who moved within the same province instead of between the provinces (Liang et al. 2002). Another study which investigates the causes of variation in the amount of migration to Chinese cities over the period of 1995–1999 (Chen and Coulson 2002) indicates that cities having higher industrial production capacity and a stronger service sector have more capability to attract migrants than others. The attributes of the quality of urban life, such as housing-market conditions and transport, have little explanatory power overall.

Studies of both developed and developing countries have emphasized the role of migration in population redistribution. Workers respond to changing economic opportunities and thereby contribute to the reallocation of labour towards a more optimal pattern to the benefit of regional development. In advanced industrialized and post-industrial economies, the natural increase of population is low, and thus, migration is an especially important explanation of regional variation of population growth. For example, in Japan, the increased and renewed concentration of population in the Tokyo metropolitan region has played an important role in its spatial economy (Portnov 1999). In the USA, research has identified the snow belt–sun belt shifts of jobs and population as key processes that have accelerated the economic development of Southern and Western states since the 1960s (Casetti 1984). In Italy, Bonifazi and Heins (2000) reviewed trends in inter-provincial migration over the last 40 years, with reference to the mobility transition model. They explained the overall changes in the national migration system, population and economic growth in the Northern part of the country. Even in the developing countries that have relatively high natural growth rates, internal migration contributes to regional development. For example, the continued growth of core regions in Ecuador, the Philippines, Egypt (McKellar and Vining 1995), and in a recent study by Agorestakis and Sidiropoulos (2010) in Albania, is the result of net migration from their peripheral regions. Fan (2005) addressed the interactive effects of the national migration system, population and economic growth, and therefore regional development as a whole in China; regional discrepancy in economic development during the 1990s has been accompanied by a marked increase in inter-provincial migration. These findings indicate the dominant migration flows, especially from relatively poor provinces to the rapidly growing regions, which depict China's migration pattern as consistent with the migration theories drawn from the experiences of Western countries.

When considering migration patterns in other countries, declining economic opportunities are significantly reported to expectably increase the circular migration where the majority of low-income migrants are short-term recurring contributors to the labour force of a new destination. Since it was confirmed in the literature on migration systems of African countries (Beguy et al. 2010; Clark et al. 2007; Collinson et al. 2009), circular migration may even reduce the contribution of permanent in-migration to urban growth in these countries (Potts 2009). Additionally, the variance of migration has been explained in a study conducted in an industrial city of Venezuela by the combined effect of distance, population distribution, income and industrial employment size (Ravuri 2004), as utilized in this study. Labour market characteristics and income inequality are reported to be significant factors in determining the interstate migration in Mexico, in a similar way (Fukurai et al. 1987; Greenwood et al. 1981).

The comparison of international studies with Turkish migration research reveals some contradictory results. According to Gedik (1977, 1979), push factors are emphasized in international rural to urban migration studies. Push factors are usually taken to include low rural incomes, inadequate infrastructure, facilities and services. However, this author discovered that other factors such as the educational and informational level of potential rural migrants, transportation and communication facilities and the existence of relatives, friends and people from their home town are at least as significant as the above-mentioned push factors. In addition, despite the expectations based on international studies that highest out-migration would be from the least developed regions, empirical findings for the period of 1965–1970 showed that the highest rural to urban migration occurred in fact within the developed regions (Gedik 1979). In a further study, Gedik (1992) pointed out that although the effects of migration on urban growth had been contrary to those found in the developed countries, as Turkish development increased between 1965 and 1985; the effects of migration increasingly came to resemble those of developed countries.

Given the state of knowledge on migration in developing countries, the purpose of this study is to investigate the recent migration trends among the provinces of Turkey with respect to the characteristics of both origins and destinations of migrants between 1995 and 2000. To this end, we study the spatial patterns and factors of net migration of Turkish provinces, their in-migration and out-migration. The provincial level, which refers to the NUTS 3 small region level in the European Union classification of economic territories, is chosen for the study because of the ready availability of national population census and migration, employment, income data published by the Turkish Statistical Institute (Url-1). The study deals with the period between 1995 and 2000 because it is aimed at investigating the patterns at the end of the twentieth century as an important time period for globalization shift influencing national policies. Additionally, there are data limitations in that the most recent census of 2011 adopted a different data collection approach that lacks official employment data at the province level.

The organization of the chapter is as follows. The population growth and urbanization rates of the provinces, as well as their in- and out-migration trends, are given in the second section. This section ends with a global regression model, predicting the effect of income per capita and employment sizes (based on economic sectors) of destination provinces in net migration size during the aforementioned period. In the third section, local variations of provincial in-migration and out-migration are investigated by regression analyses that take into consideration income and employment characteristics of both origin and destination provinces and distances among the provinces as independent variables. Through the use of these local regression analyses, variation of each province's in-migrant and out-migrant sizes can be related to individual income, employment size and distance effects. The following section introduces an additional spatial analysis method, the geographically weighted regression (hereafter, GWR) of net provincial migration, which encapsulates the local heterogeneity of the dependence migration flows on these predictors across provinces. The final section is devoted to our conclusions and to suggestions for further research.

2 Global Net Migration Patterns in Turkey

Turkey experienced a rapid urbanization process after World War II. The great contrasts between the eastern and western parts of the country in terms of high population growth rate, shortage of land in rural areas, partial mechanization of agriculture, socio-economic attractiveness of cities and especially construction of highways have encouraged considerable rural to urban migration (Potter 1993). Between 1950 and 2000, the population of the nation increased from 20,947,000 to 67,803,927 and the urbanization rate rose from 25 to 64.9% (Url-2). Rural to urban migration played a significant role in the increase of the urbanization rate and consequently in unbalanced development in favour of Turkey's more urbanized regions. When the rural population rate is considered over time, we find that this rate decreased from 75.78% in 1927 (10.3 million rural population on a 13.7 million total population) to 35.10% in 2000 (23.8 million rural population on a 67.8 million total population) (Fig. 1; Url-2). Consecutively in 2008 and 2013, two major legislative change in the status of metropolitan municipalities caused both the number of rural settlements and their population to become urban administrative quarters in metropolitan municipalities.

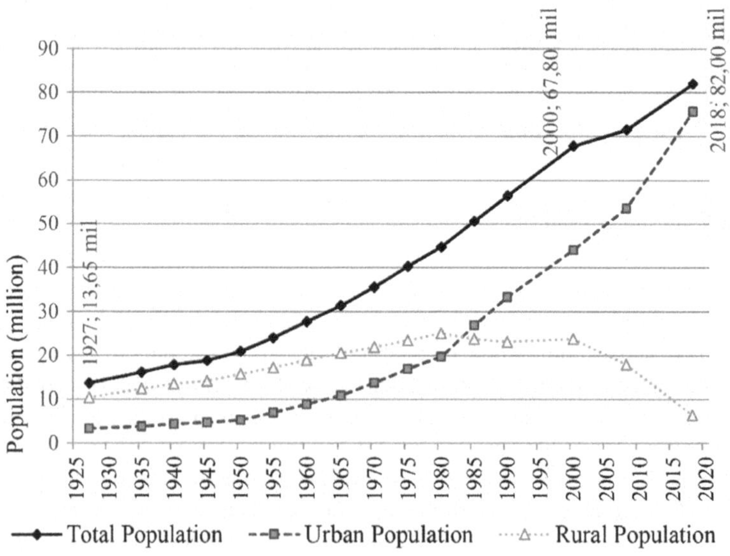

Fig. 1 Population change in Turkey

These legislative changes increased the urban population rate (as a statistical figure) artificially to 92.27% in 2018 (75.67 million urban population on a 82.00 million total population), not the actual urbanization rate. Although significant economic development was experienced at the time of the founding of the Turkish republic, the country was still facing an unbalanced development in favour of a significant increase in urban population. At that time, the periphery, referred to as the eastern parts of the country, was significantly less developed than the western parts, as evidenced by the overall migration flow directionality (Gedik 1992; Gezici and Hewings 2007).

Despite this rapid transformation of the Turkish society from rural to urban, more than half the population of 26 of the country's 81 provinces is still predominantly rural, especially in the Black Sea, Eastern and Southeastern Anatolia Regions in 2000. Although at the beginning, rural to urban migration was much higher than urban to urban migration, after the 1980s, rural to urban migration slowed and urban to urban migration almost tripled against the rural to urban migration, which mirrors the increasing urbanization rate (Tekeli 2008; Akgun et al. 2010). Between 1985 and 1990, the urban–urban ratio of migration was 77% and the rural–urban ratio was 23% (Url-1).

In this chapter, inter-provincial net migration is investigated in more detail in order to reveal migration pattern between 1995 and 2000, last period just before the aforementioned legislative changes, which deteriorated urban and rural population registries (Alkan 2015). The ten provinces with the highest positive and highest negative net migration sizes, which are calculated by the difference between in- and out-migration sizes, are given in Table 1. The net migration rate is also reported in this table by the net migration size per thousand inhabitants (Url-1). In this period, there were a total of 23 of 81 provinces with positive net migration. The spatial distribution of net migration sizes in this period is illustrated in Fig. 2. Provinces of the Marmara Region experienced high absolute migration and among the highest rates (Istanbul accounts for 407,434 net migrants with a rate of 46.09‰; Tekirdag accounts for 51,335 net migrants with the highest rate in the country at 96.81‰) (see Table 1 and Fig. 2). Other regions with relatively higher migration rates are, respectively, the Aegean Region (Izmir accounts for 120,375 net migrants with a rate of 39.88‰), the Mediterranean Region (Antalya accounts for 90,457 net migrants with the rate of 64.31‰) and the Central Anatolia Region (Ankara accounts for 90,884 net migrants with the rate of 25.59‰). The reason for this result was the higher ratio of industrial and service sector and higher job opportunities in the aforementioned provinces. All the Eastern, Southeastern and Black Sea provinces had negative net migration with the exception of Gaziantep in the Southeastern Anatolia Region (0.03‰; Url-1). Of the ten provinces with the highest negative net migration sizes, except for Adana, all have more than 40% of their population living in rural settlements.

Migration patterns in this chapter are first investigated with a global regression model, predicting the effects of variable in net migration sizes. The dependent variables used in these analyses are total in- and out-migration sizes for the period between 1995 and 2000. Independent variables for the regression analysis are income per capita and agricultural, industrial and service employment sizes. The income variable is the gross domestic product per capita in dollars ($), and employment

Table 1 Net migration between 1995 and 2000 (Url-1): top and bottom ten provinces with highest positive and negative figures

Provinces	Net migration 1995–2000	Rate (‰)
Istanbul	407,448	46.09
Izmir	120,375	39.88
Ankara	90,884	25.59
Antalya	90,457	64.31
Bursa	85,325	45.12
Tekirdag	51,335	96.81
Mugla	42,921	70.20
Aydin	21,553	25.45
Icel	18,429	12.38
Denizli	15,205	19.86
…		
Tokat	−37,172	−48.45
Hatay	−38,241	−33.87
Adana	−40,497	−23.97
Adiyamöan	−40,745	−70.23
Mardin	−42,082	−67.58
Zonguldak	−44,009	−73.82
Erzurum	−46,491	−54.78
Diyarbakir	−48,064	−40.04
Sanliurfa	−49,312	−38.90
Samsun	−51,644	−45.54

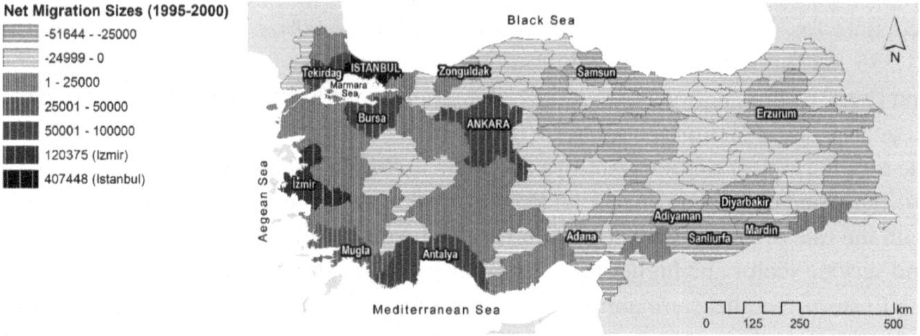

Fig. 2 Spatial distribution of net migration between 1995 and 2000

size variables are total number of employed inhabitants in the relevant sectors. All these variables are obtained from the Population Census, 2000, by Turkish Statistical Institute (Url-1).

The analysis of spatial distribution of data underscores the large disparities between the east and west of the country. For instance, the spatial distribution of the industrial sector ratio shows that only nine provinces had over 20% of their total employment in the industrial sectors. While some of the Eastern provinces have only a 2–3% industrial employment ratio, Istanbul had the largest ratio (34%) in the country. Only 29 provinces have over 30% employment in the service sector. While some of the Eastern provinces have a service employment ratio as low as 13–15%, Istanbul had the highest ratio (57%) in the country. This is due to having the largest population as well as to its advanced service functions at the national and international levels (Dokmeci and Berkoz 1994). On the other hand, 59 provinces have an agricultural employment exceeding 50% of their total, which increases to 80–85% in some of the Eastern provinces. Despite the fact that only 10% of the land is agricultural in Eastern Anatolia, 30% of the land is arable in the west. Distribution of income per capita also increases from east to west. While the lowest income was $725 per capita in Mus, on the eastern border of the country, the highest was $7556, in Kocaeli, an industrial province on the western edge of Istanbul (see Table 2). Basic descriptive statistics suggest the existence of regional co-variance in employment, income and migration.

In order to analyse global migration patterns in Turkey, the relationships between net migration, on the one hand, and income per capita and employment sizes (based on economic sectors) of destination provinces, on the other hand, are investigated by regression analysis (ordinary least squares) at the national level. The magnitude of net migration flows is taken as dependent variable; industrial, service and agricultural employment and income of provinces are taken as independent variables. The results

Table 2 Basic descriptive statistics

Variables	Minimum	Median	Maximum	Mean	Std. Dev.
In migration size (inhabitants)	7430	44,171	1,159,703	82,620.53	143,304.68
	Bayburt	Giresun	Istanbul		
Out migration size (inhabitants)	12,627	55,176	752,269	82,620.53	96,337.88
	Kilis	Kars	Istanbul		
Net migration size	−51,645	−5848	407,434	0.00	54,915.41
	Samsun	Giresun	Istanbul		
Net migration rate (‰)	−106.72	−22.55	96.81	−18.43	37.84
	Ardahan	Afyon	Tekirdağ		
Income per capita ($)	725	2107	7556	2373.33	1196.82
	Mus	Isparta	Kocaeli		
Agricultural employment size (inhabitants)	18,872	131,290	512,451	155,269.47	103,253.74
	Tunceli	Kastamonu	Konya		
Industrial employment size (inhabitants)	1586	22,860	1,332,051	57,612.42	154,662.19
	Ardahan	Bolu	Istanbul		
Service employment size (inhabitants)	8356	51,779	1,851,030	107,650.53	230,100.94
	Bayburt	Kırklareli	Istanbul		

of regression analysis indicate a strong multicollinearity problem, between industry and service employment size (Table 3). Then, two separate regression analyses, in which these two variables were included, were conducted in order to find the best model to estimate the net migration size at the provincial level. The results of these regression analyses are given in Tables 4 and 5.

According to the findings in the first OLS regression, the variables which are taken into consideration explain 85.7% of migration variance (adjusted R-squared value = 0.864) and industrial employment is the strongest characteristic (beta = 0.596) in the dependence of net migration sizes (Table 3). The industrial employment variable is again gathered as one of the strongest variable, in the following OLS regression

Table 3 Ordinary least squares (OLS) regression analysis of net inter-provincial migration-1

	Coefficient	Std. error	Std. beta coefficient	t stats.	p	Collinearity stat.	
						Tolerance	VIF
Constant	−16,982.242	6076.878		−2.795	0.007		
Income per capita	3.421	2.087	0.075	1.640	0.105	0.863	1.158
Agricultural employment size	−0.080	0.025	−0.151	−3.230	0.002	0.815	1.227
Industrial employment size	0.212	0.056	0.596	3.798	0.000	0.072	13.795
Service employment size	0.085	0.038	0.356	2.218	0.030	0.069	14.456

Independent variable = Net migration size; R-squared = 0.864; Adj. R-squared = 0.857; F = 121.003; sig. = 0.000

Table 4 Ordinary least squares (OLS) regression analysis of net inter-provincial migration-2

	Coefficient	Std. error	Std. beta coefficient	t stats.	p	Collinearity stat.	
						Tolerance	VIF
Constant	−17,359.369	6227.223		−2.788	0.007		
Income per capita	3.560	2.138	0.078	1.665	0.100	0.864	1.157
Agricultural employment size	−0.065	0.024	−0.122	−2.649	0.01	0.885	1.130
Industrial employment size	0.329	0.017	0.928	19.076	0.000	0.793	1.261

Independent variable = Net migration size; R-squared = 0.856; Adj. R-squared = 0.850; F = 151.960; sig. = 0.000

Table 5 Ordinary least squares (OLS) regression analysis of net inter-provincial migration-3

	Coefficient	Std. error	Std. beta coefficient	*t* stats.	*p*	Collinearity stat.	
						Tolerance	VIF
Constant	−18,366.4	6573.473		−2.794	0.007		
Income per capita	3.982	2.256	0.087	1.765	0.082	0.868	1.153
Agricultural employment size	−0.098	0.027	−0.184	−3.687	0.000	0.843	1.186
Service employment size	0.224	0.013	0.938	17.82	0.000	0.757	1.321

Independent variable = Net migration size; R-squared = 0.839; Adj. R-squared = 0.832; F = 133.291; sig. = 0.000

model-2, that explains 85.0% of migration variance (Table 4). This reflects similar results obtained in research on Russian migration after the 1990s (Andreanko and Guriev 2004) and in China (Chen and Coulson 2002). Service employment size is the strongest predictor in the third model of net migration, explaining 83.2% of the migration variance (Table 5). The third strongest factor (Beta = −0.151 in OLS-1; −0.122 in OLS-2; and −0.184 in OLS-3) is agricultural employment. These findings are in line with the comparative economic opportunity thesis (Fukurai et al. 1987). This thesis argues that factors such as employment opportunities are the major consideration in people's decision to move. On the other hand, in contrast to the findings of previous studies (Fukurai et al. 1987), income is not found to be significant with a 0.95 confidence interval ($p = 0.105$ in OLS-1; $p = 0.100$ in OLS-2; and $p = 0.082$ in OLS-3) as previously illustrated in China (Chen and Coulson 2002).

Violation of the homoscedasticity assumption in each model is checked with the scatter plots of standardized residuals and standardized predicted values. Fit line in the scatter plots of each OLS indicated that scores are randomly scattered through a horizontal line (R-squared = 0.000 in OLS-1; R-squared = 2.220E−16 in OLS-2 and in OLS-3).

3 Province Specific In- and Out-Migration Patterns

This section investigates origin-specific and destination-specific variations of migration sizes in relation to income, employment and distance effects by the use of local regression analysis. The method used in this section is a version of the family of spatial interaction models serving to analyse and forecast patterns of spatial interaction flows. In many uses of spatial interaction models, a distance function is included to reproduce interaction behaviour of individuals in the system (Wilson 1971; Haynes and Fotheringham 1984; Roy and Thill 2004). Dependent variables are, respectively, the amount of out-migrants and in-migrants, while independent variables include income, sectoral employment sizes and distance, as used in the global model presented in the previous section. Data on origin- or destination-specific variations in migration flows among provinces are established in a 81 × 81 non-symmetrical matrix in order to investigate each province's origin-specific and destination-specific migration patterns, by means of separate local regressions. The results are reported in the form of standardized coefficients, *t* and significance values of each variable. Additionally, multicollinearity tests are conducted for each local regression analysis and, as with the global OLS models, in 27 origin-specific regressions of out-migration and 22 destination-specific regressions of in-migration, industry and service employment variables have a collinear relationship, with variance inflation factor above 10. Then, local regressions were re-estimated with the exclusion of the less powerful employment size variable, having lower standardized coefficient and *t* values. Violation of the homoscedasticity assumption in local regression models is also checked with the scatter plots of standardized residuals and standardized predicted values.

With these local regressions, we aim to determine the dependence of each province's in- and out-migrant size on income, sectoral employment size and distance effects. As indicated in Tables 6 and 7, each province has its own regression results, which collectively represent the variation of migration flows across provinces. Migration distance between pair of provinces is used as a predictor in each local regression analysis. Distance data are obtained from the General Directorate of Highways and measured with the shortest distance (km) among the province centroids on the network of state and provincial roads. As for the global model of net migration, the 2000 population census (Url-1) is the source of data in- and out-migration, agricultural, industrial and service employment size and income per capita.

We analyse out-migration by province of origin first. For each province, the relationship between the out-migration flow, and the socio-economic characteristics of destination provinces and their distance from the origin are investigated by OLS regression analysis. Thus, 81 regressions are estimated, but only 64 local regression analyses explain a statistically significant dependence of migration flows on the predictors (adj. R-squared values > 0.50). The results of this analysis are summarized in Table 6 and Fig. 3a, where only significant coefficients of each variable are reported. Adjusted R-squared values of local regression analyses range from 0.169 to 0.971. The provinces of the Black Sea, Eastern and Southeastern Anatolia regions have high adjusted R-squared, and thus, the predictors of out-migration from these provinces can be explained to a great extent by the variables which are taken into consideration for the destination provinces. At the same time, the effect of each variable is shown on separate figures by means of the standardized coefficients. For instance, the coefficient of the industrial sector of destination provinces is relatively speaking the most important factor to affect out-migration from the Black Sea, Eastern and Southeastern Anatolia provinces and is illustrated in Fig. 3d. The reason for the stronger out-migration of population to more industrialized provinces is the desire on the part of migrants for industrial employment opportunities, which are less abundant in the province of origin than in the attractive employment centres. This was discussed in the previous section as the latent regional discrepancy that exists in Turkey. Service

Table 6 Origin-specific regression analysis of out-migration: standardized coefficients and goodness-of-fit measures

	Mean	Std. Dev.	Min.	F	sig.	R-squared of fit line (std. res.–pred. val.)	Median	F	sig.	R-squared of fit line (std. res.–pred. val.)	Max.	F	sig.	R-squared of fit line (std. res.–pred. val.)
(a) Adj. R-squared scores (n = 81)	0.734	0.224	**0.169** Burdur	9.162	0.000	0.000	**0.811** Gaziantep	115.415	0.000	0.000	**0.971** Ardahan	882.004	0.000	0.000
	Mean	Std. Dev.	Min. std. coef.	t	sig.	VIF	Median Std. coef.	t	sig.	VIF	Max. Std. coef.	t	sig.	VIF
(b) Income effect (n = 9)	0.094	0.197	**0.100** Kastamonu	−0.100	0.001	1.157	**0.114** Bayburt	2.672	0.009	1.709	**0.439** İstanbul	0.373	0.000	1.033
(c) Agricultural employment effect (n = 27)	0.045	0.201	**−0.169** Karabuk	−3.354	0.001	1.184	**−0.088** Samsun	−3.421	0.001	1.130	**0.601** Karaman	6.690	0.000	1.000
(d) Industrial employment effect (n = 39)	0.894	0.223	**−0.168** Bingol	−3.636	0.000	1.087	**0.968** Amasya	32.894	0.000	1.005	**1.041** Kastamonu	34.155	0.000	1.261
(e) Service employment effect (n = 39)	0.747	0.181	**0.428** Usak	4.704	0.000	1.198	**0.813** Antalya	14.487	0.000	1.035	**1.015** Kars	40.338	0.000	1.184
(f) Distance effect (n = 55)	−0.186	0.082	**−0.461** Osmaniye	−5.200	0.000	1.014	**−0.184** Eskisehir	−3.525	0.001	1.052	**−0.053** Ardahan	−2.600	0.011	1.129

Results are reported only when significant at the 0.05 level

Table 7 Destination-specific regression analysis of in-migration: standardized coefficients and goodness-of-fit measures

	Mean	Std. Dev.	Min.	*F*	sig.	*R*-squared of fit line (std. res. – pred. val.)	Median	*F*	sig.	*R*-squared of fit line (std. res. – pred. val.)	Max.	*F*	sig.	*R*-squared of fit line (std. res. – pred. val.)
(a) Adj. *R*-squared scores ($n = 81$)	0.654	0.203	**0.175**	9.458	0.000	2.220E−16	**0.663**	53.484	0.000	−2.220E−16	**0.952**	793.267	0.000	−2.220E−16
			Kilis				Mardin				Kars			
	Mean	Std. Dev.	Min. Std. coef.	*t*	sig.	VIF	Median Std. coef.	*t*	sig.	VIF	Max. Std. coef.	*t*	sig.	VIF
(b) Income effect ($n = 15$)	0.127	0.143	**−0.146**	−3.885	0.000	1.303	**0.180**	1.995	0.050	1.940	**0.281**	4.217	0.000	1.694
			Kastamonu				Mardin				Artvin			
(c) Agricultural employment effect ($n = 32$)	0.172	0.192	**−0.151**	−2.454	0.016	1.185	**0.192**	2.195	0.031	1.244	**0.569**	6.148	0.000	1.000
			Zonguldak				Niğde				Karaman			
(d) Industrial employment effect ($n = 25$)	0.835	0.206	**0.251**	2.510	0.014	1.145	**0.929**	22.307	0.000	1.000	**1.021**	27.457	0.000	1.262
			Usak				Balikesir				Kastamonu			
(e) Service employment effect ($n = 52$)	0.714	0.178	**0.285**	2.912	0.005	1.192	**0.774**	10.721	0.000	1.151	**1.013**	26.657	0.000	1.216
			Denizli				Bitlis				Ardahan			
(f) Distance effect ($n = 64$)	−0.265	0.102	**−0.505**	−5.585	0.000	1.031	**−0.245**	−3.478	0.001	1.995	**−0.096**	−3.797	0.000	1.071
			Gaziantep				Mus				Kars			

Results are reported only when significant at the 0.05 level

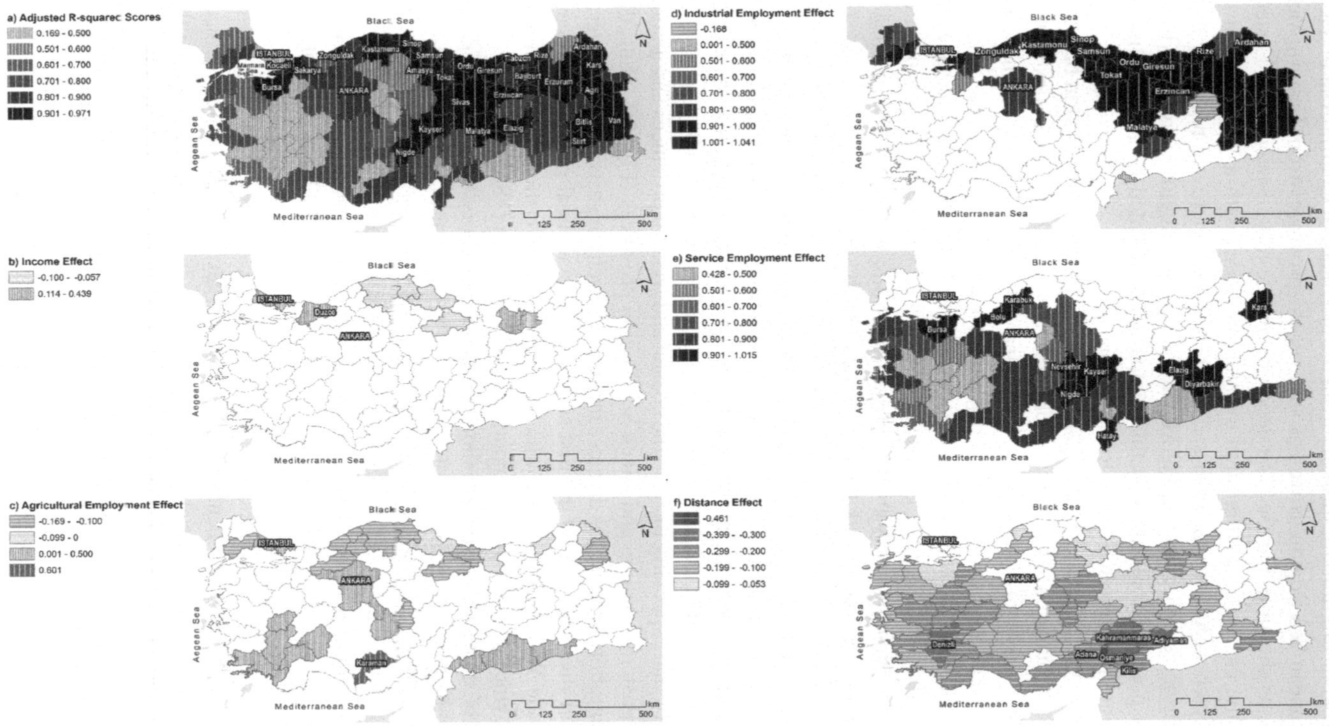

Fig. 3 Mapping of standardized coefficients and goodness-of-fit measures of origin-specific regressions (*Coefficients that are statistically significant at a 5% level are represented*)

employment characteristics of destination provinces are the second most important factor to affect out-migration, especially from the provinces of Central Anatolia and the Mediterranean Sea Region as well as some of the provinces from Eastern Anatolia regions, as shown in Fig. 3e. The deficit in service jobs, especially in Eastern Anatolia, is the cause of out-migration towards provinces with a large amount of service jobs such as Ankara, Istanbul and Izmir. Although there is a relatively high amount of service employment in the Mediterranean Sea Region in relation to its high tourism potential, provinces in this region are dominated by a "province to province" migration pattern internal to the region.

For a few provinces, there is significant positive or negative correlation between the amount of out-migration and agricultural employment at the destination provinces. They are spread throughout the country (Fig. 3c). Among them, some provinces (Gaziantep, Sanliurfa, Tekirdag Ankara and Istanbul) have a strong employment base, but the rest of the provinces with significant coefficient have relatively small urban centres only. This may be associated with return migration after individuals have accumulated some savings earned while working in the secondary or tertiary sector, and return back their province of birth.

Distance is another factor that negatively affects out-migration as observed in other countries (Fukurai et al. 1987; Pandit 1994; Lucas 2001). A number of studies have established that there is a negative relationship between migration size and the distance between the origin and destination of migration (Koramaz and Dökmeci 2016). In this study, the distance factor is significant at the 0.05 level for 55 provinces only, and especially those in the coastal and western provinces and among some of the Eastern Anatolia and Black Sea Region provinces (Fig. 3f). Thus, the amount of migration decreases as distance increases, which suggests that most migration over short distances can be the first step of the hierarchical migration process and leads to the Western provinces, which ultimately receive more migration than any other province due to their job opportunities. Ravenstein (1885) postulated that the majority of migrants move only short distances; migratory moves are made in a series of steps from close to distant locations. This pattern of population movement has been observed in some other developed and developing countries (Skeldon 1990). It is reasonable to expect that closely clustered spatial units will exhibit closer interactions than distant units. These proximate clusters or subsystems of migration can be meaningfully delineated as migration regions (Pandit 1994) as illustrated in Fig. 3f. Figure 3f depicts the common distance effect on migration with the coloured provinces having a negative significant effect. This significant effect is found in extensive coastal areas as well as in central provinces of Turkey. The rest of the provinces (white coloured provinces), which are clumped in two large spatial clusters in the north-western and far-eastern parts of Turkey, can be delineated as a different migration regime. Out-migrants in these provinces tend to choose their destination to furthest provinces (especially most populated metropolitan provinces, like Istanbul and Ankara).

While the relationship between income and net migration is not significant at the national level, origin-specific analyses reveal that income of destination provinces is a significant predictor in nine cases only (at the 0.05 level). It has mostly positive signs as in the previous studies by Chen and Coulson (2002) in China; by Kennan and Walker (2003) in the USA; and by Giannetti (2003) in Italy, who explains that skilled labour migrates to high-income provinces for possible employment. On the other hand, it sometimes has a negative sign because some people desire to migrate back to their home province, where income may be lower.

We now proceed with the analysis of in-migration for each destination province in turn. The relationships between in-migration and the socio-economic characteristics of origin provinces and distances between provinces are analysed by OLS regression, as for out-migration. The results of the analysis are summarized in Table 7 and Fig. 4; adjusted R-squared values vary between 0.175 and 0.952. According to these results, the Marmara Region, Black Sea Region and North-eastern provinces have high adjusted R-squared values, which means that their in-migration can be explained to a great extent by the variables associated with origin provinces. Also, the destination-specific impact of each variable on in-migration is shown on separate maps. We find that there is a strong correlation between in-migration to provinces of the Marmara Region and Black Sea Region and the industrial employment of origin provinces (Fig. 4d); when statistically significant, this relationship is always positive (Table 7). It is proven that industrial employment centres in Marmara Region (neighbouring provinces of Istanbul and Bursa) are also the attractive centres for migrants (dark grey coloured provinces) from the provinces having high industrial employment capacity. This result highlights the industrial employment effect on the "province to province" migration pattern. Specifically, as the most industrialized region of Turkey, the Marmara Region is relatively more dependent on industrial employment mobility among its provinces. The figures indicate that 43.5% of the entire in-migrants to these provinces are from the top ten provinces with highest industrial employment and 16.75% from Istanbul and Bursa, as the neighbouring industrial employment centres. For the other provinces, the regression analyses indicate that provinces in the Black Sea Region attract migrants, from provinces having high industrial employment opportunities. This finding may correspond to the age-group variances in migration pattern and, to say more precisely, return migration of retired population. Age-group-dependent migration figures support this statement as the Black Sea Region has dramatically more in-migrants over 45 years of age (who can be identified as retired industrial workers migrating back their home provinces), than the rest of the regions (Var et al. 2014).

Service employment of migration sources has a positive effect on the magnitude of migration flows (Table 7). The relationship between in-migration and service employment of the origin provinces is illustrated in Fig. 4e. The provinces that are the most affected by this variable are mostly clustered around Ankara, which is the capital of the country and thus has a high rate of service employment, and is an important supplier of service jobs to the surrounding provinces. This impact continues to the Eastern and Southeastern Anatolia provinces in increasing degrees. Smaller clusters of service employment impacts on the amount of in-migrants can be seen in the Aegean and Mediterranean regions due to tourism activities in nearby provinces. Service employment effect should be considered in association with industrial employment, with which there is strong collinearity at the provincial level. Provinces with

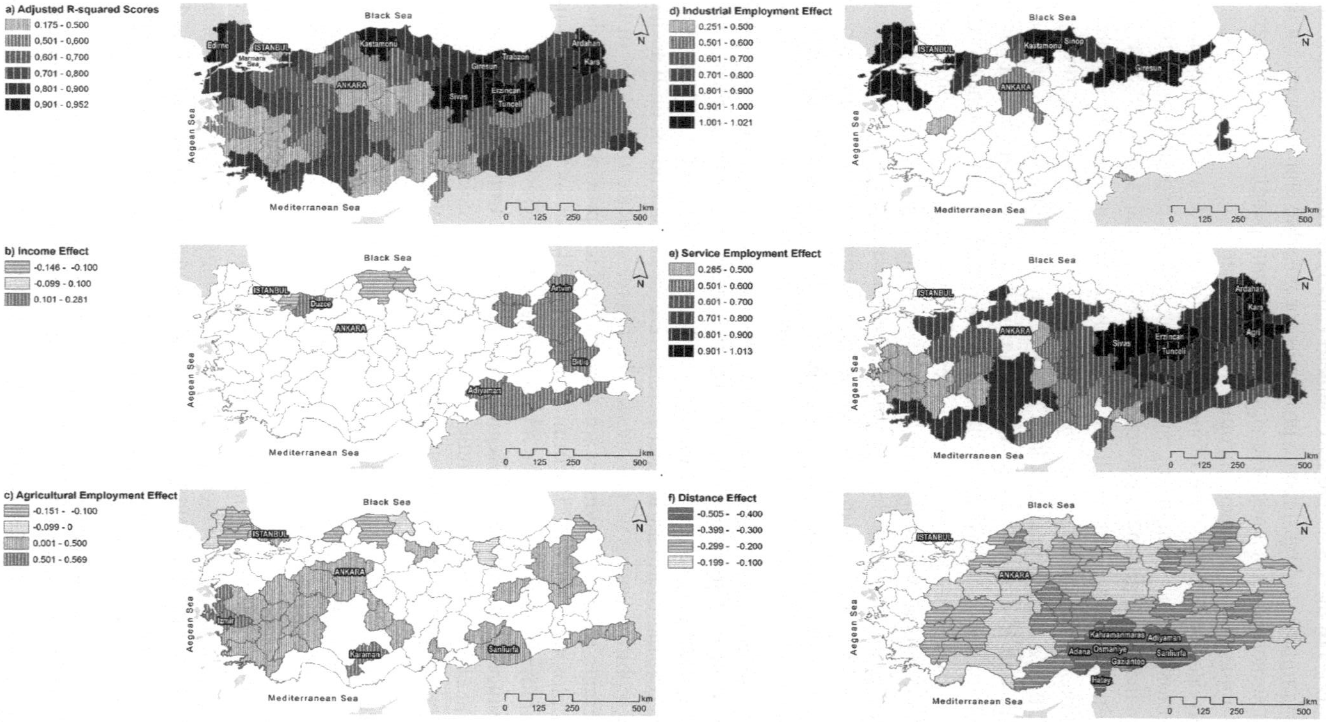

Fig. 4 Mapping of standardized coefficients and goodness-of-fit measures of destination-specific regressions (*Coefficients that are statistically significant at a 5% level are represented*)

high service employment refer to provinces that are less industrialized and a less formal service-oriented employment market; then, it can be interpreted that this type of provinces may lose their population towards attractive ones, which is also validated in origin-specific regression analysis of out-migration.

Agricultural employment is usually a generator of out-migration in provinces. Two geographic clusters where this is particularly true stand out (Fig. 4c). The first is in the Eastern Anatolia provinces in an entire agglomeration and Southeastern Anatolia provinces in a relatively dispersed pattern which represent the interaction among very agricultural provinces. These provinces may encourage both seasonal and permanent agricultural employment mobility towards provinces with high gross value added in the agriculture sector. This can be interpreted with investment capacity of the provinces as well. Large infrastructure investment in agriculture will attract in-migrants and more so from provinces with a large agricultural work force; hence, the significant coefficients are associated with these destinations, as previously reported by Hugo and Bell (1998) in Australia. According to data from the Ministry of Development (Url-3 2000), Sanliurfa received the largest public investment for agricultural infrastructure development of all provinces (in the amount of $30.5 million; in second place is Diyarbakir with $13.5 million). Thus, Sanliurfa has a higher positive coefficient as it is relatively more attractive for agriculture employment than most of the provinces in Turkey (beta = 0.148). Karaman in Central Anatolia has highest positive coefficient (beta = 0.569) but in a weak explanation power (adj R-squared = 0.315). If we take into account the regressions with high explanation power (adj R-squared scores higher than 0.500), provinces in the Aegean region (Izmir, Afyonkarahisar, Isparta and Kutahya) have the highest positive coefficients in Turkey, which may represent migration of surplus agricultural workers searching for alternative jobs in provinces with high employment growth.

The impact of income on in-migrants is only significant for 15 provinces (Fig. 4b). While in the Eastern Anatolia and Southeastern Anatolia provinces the impact of income is positive, as observed in other countries (Greenwood et al. 1981) there are twelve provinces for which the income effect is positive and three provinces, negative. The three provinces (Kastamonu, Kocaeli and Sinop) which can be more attractive for the migrants, formerly living in the provinces with less income, validate the common migration assumption of having the desire of increase in income. As for the rest of the provinces, 10 provinces in Eastern and Southeastern Anatolia regions and two provinces in the west (Sakarya and Duzce) have positive income effect for in-migration pattern. If we consider the major characteristics of these provinces, they mostly have migrants from the provinces relatively more industrialized that may refer to a return migration pattern.

The impact of distance on in-migrants is negative in the vast majority of cases, as already discussed in relation to out-migration; it is illustrated in Fig. 4f. However, the impact of distance on in-migrants is higher than on out-migrants for all the provinces, except seven provinces (Kutahya, Afyonkarahisar, Antalya, Burdur, Osmaniye, Denizli and Eskisehir), which suggests the importance of distance at the beginning of the hierarchical process of human migration. The same has also been found by He and Pooler (2002) in China, where migration mainly occurs over short distances, when distance is expressed as the crossing of a provincial boundary. It is observed that the impact of distance is effective in all of the provinces, except the Aegean and most Marmara provinces.

Thus, the results of the study illustrate that industrial and service employment in the origin and destination provinces are two major factors that affect in- and out-migration. Major findings indicate that metropolitan provinces of great employment accumulation are also closely linked to the major population flows during this study period. While the five most populated provinces in Turkey (Istanbul, Ankara, Izmir, Bursa and Konya, each having more than two million residents in 2000) account only for 32.03% of national population, they hold 49.62% of total industrial employment and 43.45% of total service employment. 6.19% of all migrants in 2000 moved towards "*and*" from these five provinces, identified here as the metropolitan provinces of great employment accumulation. According to the findings from both origin-specific analysis of out-migration and destination-specific analysis of in-migration, migration towards "*or*" from most populated and, even more strikingly, provinces of great employment accumulation constitute the major migration patterns in Turkey. In other words, major population moves are obtained from and/or to the major industrial or service employment centres, where sectoral development has been encouraged. This remark also overlaps with the data on provincial distribution of public investments in Turkey (Url-3 2000). Izmir and Kocaeli have the largest amount of public investment for industrial development; then, these provinces have relatively higher industrial and service employment sizes and subsequently in migration sizes than most of the provinces.

Despite the fast transformation of the Turkish population along the rural to urban continuum, in 59 provinces, more than 50% of employment is still in agriculture. Thus, it is necessary to pursue a public strategy for rural development that provides employment opportunities for the surplus labour in agriculture, as suggested with public investments in Turkey and in the literature covering the Chinese experience (Liang et al. 2002). Such strategies may help disadvantaged provinces in terms of industry and service employment not to lose their native population. This challenge is at the core of addressing the major condition of unbalanced development in most developing countries, which also face the problems of overcrowding and urban unemployment. Although income has been found to be an important determinant of migration in many studies, the situation is more mixed in many provinces of Turkey. The results shed light on the effects of migration and draw some important policy implications pertaining to growth of provinces.

4 Spatial Heterogeneity in Predictors of Net Provincial Migration

Origin- and destination-specific migration regression models presented in Sect. 3 indicate that the effects of income, agricultural, industrial and service employment and geographic distance on inter-provincial migration patterns vary considerably in magnitude across the national territory. To understand this heterogeneity better, we resort here to the GWR modelling approach to formally test the non-stationarity of migration patterns in Turkey and of their dependence on provincial socio-economic characteristics. The GWR model is mathematically formulated as follows:

$$y_i = c_i + \beta_1(u_i, v_i)x_{1i} + \beta_2(u_i, v_i)x_{2i} + \cdots + \beta_n(u_i, v_i)x_{ni} + \varepsilon_i$$

where y_i is the net migration size in province i; c_i is the constant term; x_i is the value of each independent variable in province i; β is the value of coefficient; and ε_i is the residuals. In this model, coefficients vary depending on the geographical coordinates (u_i, v_i) of location i. GWR extends conventional OLS regression analysis by allowing for the geographical variation of dependence relationships. This model is the local version of the regression model of net migration presented in Sect. 2 of this chapter.

As an exploratory tool of spatial analysis, GWR allows the modelling of processes that vary over space and investigate spatial variations across space (Charlton et al. 2006). Common applications of GWR include real estate markets, ecology where the attention is on the spatial changes in the relationship between species richness and environmental drivers (Foody 2005), transportation (Lim and Thill 2008), epidemiology (Nakaya et al. 2005) and migration studies (e.g. Fotheringham et al. 2002a). Using the GWR tool in the GIS Spatial Statistics Module of ArcToolbox, a spatially calibrated model, with fixed kernel type (using the Akaike Information Criterion (AIC) bandwidth method), is generated in order to estimate the relationship between spatially variable effects on net migration at the province level.

The GWR generates regression coefficients and standardized residuals for each province. The results of GWR analysis of net migration are summarized in Table 8 across all 81 provinces (bandwidth = 2.994 km; effective number = 24.433). The AIC is a recommended measure for comparing a global regression model with a GWR model (Fotheringham et al. 2002b). The AIC score for the GWR (1809.672) is approximately 164% larger than the AIC of the global OLS model (707.135). Although the GWR model has higher overall R-squared (0.948) and adjusted R-squared (0.927) scores than the global OLS model (0.857), local models representing the local variations in this applied GWR model provide a worse fit to the provincial data after accounting for differences in degrees of freedom. The reason resides with the limited number of point data available for GWR analysis (81 provinces); migration data at a finer geographic resolution would be beneficial in this respect so as to conduct the analysis on a large number of nearest neighbours to generate more robust results. While the global OLS model explains 85.7% of the variance of inter-provincial net migration, the median local R-squared obtained by the GWR is 0.716, and the mean is 0.738, representing a significant decrease in explained variance.

The spatial distribution of local R-squared values from GWR analysis is illustrated in Fig. 5a. The spatial variation in these values demonstrates how the combined statistical effect of income and employment sizes on net migration varies across provinces in Turkey. The figure identifies the provinces where GWR produces an

Table 8 Geographically weighted regression of inter-provincial net migration

	Minimum	Mean	Median	Maximum	95% confidence interval for mean values of coefficients	
					Lower bound	Upper bound
Income per capita	−10.346	−0.329	−0.552	8.763	−1.204	0.546
Agricultural employment	−0.325	−0.115	−0.111	0.003	−0.134	−0.096
Industrial employment	−0.121	0.333	0.231	1.635	0.254	0.412
Service employment	−0.235	0.036	0.068	0.222	0.014	0.059
Intercept	−22,598.930	−3207.716	−3472.459	17,982.890	−5208.184	−1207.248
Standard residuals	−3.433	−0.038	0.050	4.482	−0.290	0.215
Local *R*-squared	0.276	0.738	0.716	0.977	0.697	0.778

Independent variable = Net migration size

increase or decrease in overall model explanation compared with the original OLS model.

The *R*-squared values from the local regression models are enhanced in the provinces located in the western part of Turkey. In other words, the fit of the model based on four predictors is better in this region. On the other hand, net migration of provinces located in the central and eastern regions of the country cannot as well be associated with only income and employment size effects; local *R*-squared declines below the adjusted *R*-squared value of OLS level of 0.857, and even below 50% in Eastern Mediterranean provinces. Further research may provide a better and more complete statistical explanation of net migration in this part of Turkey, for instance including the effect of public investments or the role played by the spatial configuration of migrant options.

The GWR results enable the interpretation of the local effects of socio-economic variables on net migration. The global OLS regression model estimated the effect of industrial employment size on net migration to be statistically significant and equal to 0.212 for the country as a whole (Table 3). With the local regression estimation, the parameter for this variable ranges from −0.121 to 1.635 across provinces, with a median of 0.231 and a mean of 0.333 (Table 8). Such a local variation suggests that the relationship between industrial employment and net migration size is not stationary in the entire country. Similar results supporting the heterogeneity hypothesis are achieved also for the variables of income agricultural and service employment size. Figure 5 represents not only attractive provinces for possible migrants, but also the sectoral distribution of employment capacities. For instance, Turkey's Eastern

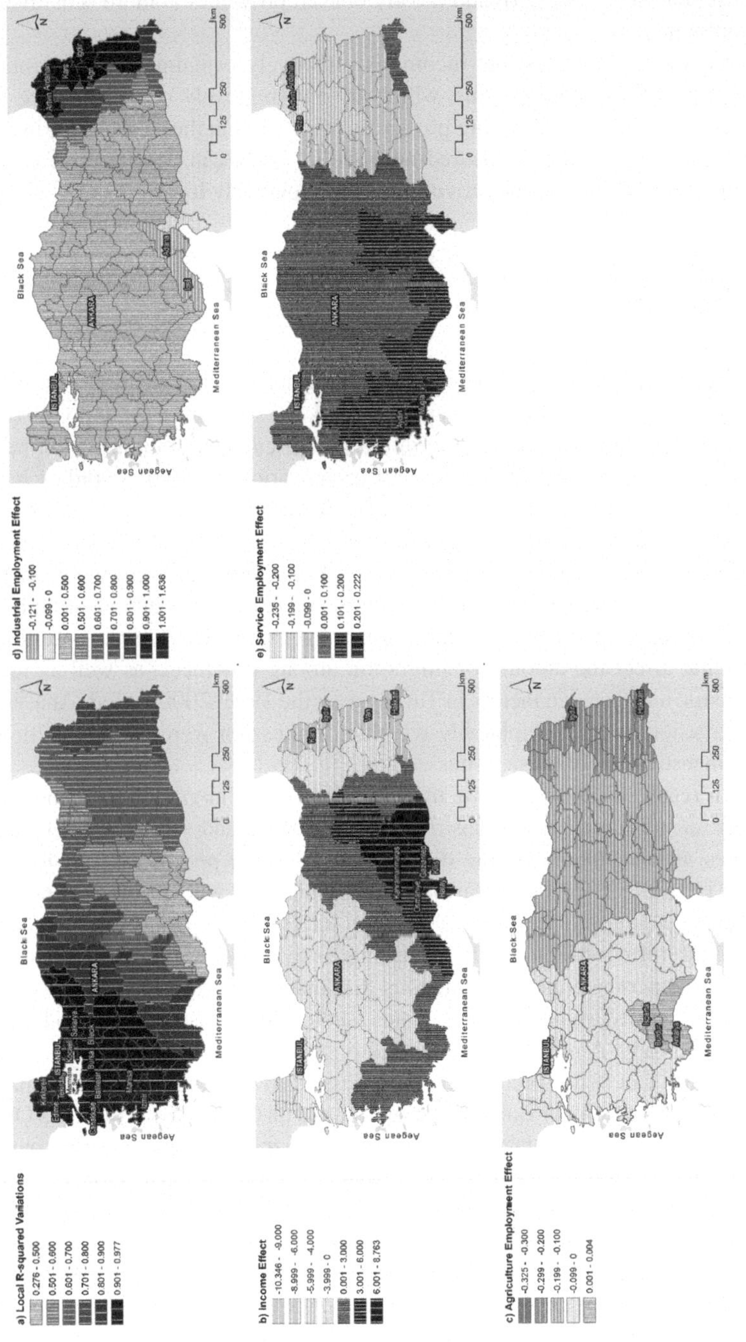

Fig. 5 Maps of local estimates for the GWR model of inter-provincial net migration

provinces are more sensitive to industrial employment when attracting migrants while the migration balance of western and mostly coastal provinces is more sensitive to service employment.

The GWR model indicates that income is effectively explaining net migration size in Turkey's near eastern regions, which are also relatively more attractive for potential migrants than neighbouring provinces (Fig. 5b). On the other hand, in far-eastern regions, this effect is measured negatively, which can be interpreted with circular migration increase to the provinces having relatively lower income.

5 Conclusions

During the past half century, Turkey's population and its urbanization rate have increased almost threefold. During this period, Turkey experienced great population mobility, including both permanent migration and circular migration, which have been related to economic development, and urbanization. Migration studies have underscored that rural to urban migration has slowed down since the 1950s and rural–urban ratios in the provinces equalized in the early 1980s. While the rural to urban migration constituted the largest portion of the migration movement earlier on, later the urban–urban ratio, in parallel with the increase in the urbanization rate, increased to almost three times the rural to urban migration. This chapter examined the factors that affect the amount of migration among provinces, as well as their socio-economic and distance factors in Turkey over the 1995–2000 period. Analyses of OLS regression and geographically weighted regression were used for a more systematic investigation of these issues.

The main result of the analysis is that inter-provincial migration decisions are influenced to a substantial extent positively by industrial and service employment opportunities and negatively by the distance between the provinces. The impact of distance is stronger for in-migrants than out-migrants with a few exceptions, which suggest that migratory moves are made in series of steps from close to distant locations which form a hierarchy. It is reasonable to expect that closely clustered provinces will experience more interactions that distant provinces. Such a remark points to the necessity to use spatial analysis, such as GIS-based GWR modelling as tested in this study. A recent publication also suggests that land use and disaster risk vulnerability can better predict spatial access of low-income groups to small-sized financial institutions in Bangladesh by the use of GWR model than the OLS regression model (Johnson et al. 2019). However, another study argues that the spatial effect of border sharing and conditional distance on international trade flows in Europe can produce more significant results with a quantile regression model which is a form of gravity model in which distance weights in quantiles are used instead of actual distances than GWR model (Lin et al. 2016). In this chapter, GWR model, which is cited as a useful spatial model, presenting a linear or a simple non-linear relationship of geographic units (Li et al. 2019) is capable of investigating spatial heterogeneity of net migration sizes in Turkey.

Based on multivariate analysis, average individual income is not found to be a strong determinant of migration flow among provinces, as found by Anjomani (2002) in the USA. While there is a positive relationship between the number of in-migrants and out-migrants, on the one hand, and income of the origin and destination provinces, on the other hand, especially in a few Eastern and Southeastern Anatolia provinces, there is a negative relationship in a few provinces of the Black Sea Region, which means people are reluctant to migrate or to return to their home province from high-income provinces. However, the GWR model indicates that income is effectively significant to explain net migration size in near eastern regions of Turkey, which are also relatively more attractive for potential migrants than neighbouring provinces. Istanbul does not take place any of the lists of regression analysis due to its Adj. R-squared being below 0.50. The variables taken into consideration here stop short of providing a robust explanation of in- and out-migrations to and from Istanbul because it stands apart from other provinces due to its national prominence, a point that will be the subject of future study.

The analytical results demonstrate how migration propensities and patterns vary between types of provinces, providing new insights into the processes through which population is redistributed throughout Turkey. Special attention is given to the role of the metropolitan provinces of Istanbul, Ankara, Izmir and Adana. These provinces serve as centres of attraction and distribution of inter-provincial migration flows and aid the understanding of the overall changes in the national migration system. After the 1980s, Turkey's transition to a market-oriented economy propelled a large percentage of migrants to large cities. Thus, inter-provincial migration can be tied to the political and economic dynamics of Turkey.

In the first decade of 2000s, economic growth with privatization programs and other neo-liberal policies have unexpectedly increased the unemployment rate at the national level (Kesici 2010) and encouraged migration to large cities more than before. While the large cities of developed countries are trying to decentralize their populations despite their high urban technology, it would not be wise for Turkey to further encourage population growth in large cities. Thus, government policy should support rural development that provides employment opportunities in order to overcome harsh regional discrepancies which may cause consecutive problems such as overcrowding, pollution, traffic congestion, high crime, social unrest, land speculation and high urban unemployment as observed in other developing countries. Therefore, the investigation of how migration patterns vary among the provinces over time is suggested for further research. This would permit the analysis of problems mentioned above as well as additional variables such as residential construction, public investments and international migration from a temporal perspective. Future work on migration in Turkey's national space will also be geared towards fully accounting for spatial dependence and spatial structure issues that may enhance the robustness and validity of statistics results as well as their policy relevance (Roy and Thill 2004; Fischer and Wang 2011). Finally, given that human migration shares many of the characteristics of complex and wicked social science problems, we will explore leveraging more fully the integration of computational procedures and visualization techniques as advocated by Fischer (2006).

References

Agorestakis, M., & Sidiropoulos, G. (2010). Population change due to geographic mobility in Albania, 1989–2001, and the repercussions of internal migration for the enlargement of Tirana. *Population, Space and Place, 13*, 471–481.

Akgun, A. A., Baycan-Levent, T., & Nijkamp, P. (2010). Business dynamics as the source of counter-urbanization: An empirical analysis of Turkey. *International Journal of Sustainable Development, 13*(1/2), 200–214.

Alkan, A. (2015). New metropolitan regime of Turkey: Authoritarian urbanization via (local) governmental restructuring. *Lex Localis—Journal of Local Self Government, 13*(3), 849–877.

Andrienko, Y., & Guriev, S. (2004). Determinants of interregional mobility in Russia. *Economics of Transition, 12*(1), 1–27.

Anjomani, A. (2002). Regional growth and interstate migration. *Socio-economic Planning Sciences, 36*(4), 239–265.

Bairoch, P. (1991). *Cities and economic development: From the dawn of History to the present*. Chicago: University of Chicago Press.

Bilsborrow, R. E., McDavitt, T. M., Kossoudji, S., & Fuller, R. (1987). The impact of origin community characteristics on rural-urban out–migration is a developing country. *Demography, 24*(2), 191–199.

Bilsborrow, R. E. (1992). The state of the art and overview of the chapters. In R. E. Bilsborrow (Ed.), *Migration, urbanization and development: New directions and issues*. New York: Springer Publishing.

Beguy, D., Bocquier, P., & Zulu, E. M. (2010). Circular migration patterns and determinants in Nairobi slum settlements. *Demographic Research, 23*, 549.

Bonifazi, C., & Heins, F. (2000). Long-term trends of internal migration in Italy. *International Journal of Population Geography, 6*, 111–131.

Casetti, E. (1984). Manufacturing productivity and snowbelt-sunbelt shifts. *Economic Geography, 60*(4), 313–324.

Celik, F. (2007). Internal migration in Turkey. *Sosyal Bilimler Enstitusu Dergisi, 22*(1), 87–109.

Charlton, M., Fotheringham, S. A., & Brunsdon, C. (2006). *NCRM methods review papers, NCRM/006. Geographically weighted regression*. Discussion Paper. Unpublished.

Chen, A., & Coulson, E. (2002). Determinants of urban migration: Evidence from Chinese cities. *Urban Studies, 39*(12), 2189–2197.

Clark, S., Collinson, M., Kahn, K., Drullinger, K., & Tollman, S. (2007). Returning home to die: Circular labour migration and mortality in South Africa. *Scandinavian Journal of Public Health, 35*(Supplement 69), 35–44.

Collinson, M., Adazu, K., White, M., & Findley, S. E. (Eds.). (2009). *The dynamics of migration. Health and livelihoods: INDEPTH network perspectives*. Aldershot: Ashgate.

Dokmeci, V., & Berkoz, L. (1994). Transformation of Istanbul from monocentric to a polycentric city. *European Planning Studies, 2*(2), 193–205.

Fan, C. C. (2005). Interprovincial migration, population redistribution, and regional development in China: 1990 and 2000 census comparisons. *The Professional Geographer, 57*(2), 295–311.

Fischer, M. M. (2006). Spatial Interaction models and the role of geographic information systems. In M. M. Fischer (Ed.), *Spatial analysis and geocomputation* (pp. 29–42). Heidelberg: Springer.

Fischer, M. M., & Wang, J. (2011). Spatial interaction models and spatial dependence. In M. M. Fischer & J. Wang (Eds.), *Spatial data analysis: Models, methods and techniques* (pp.61–70). Heidelberg: Springer.

Foody, G. M. (2005). Mapping the richness and composition of British breeding birds from coarse spatial resolution satellite sensor imagery. *International Journal of Remote Sensing, 26*(18), 3943–3956.

Fotheringham, S. A., Champion, T., Wymer, C., & Coombes, M. (2002). Measuring destination attractivity: A migration example. *International Journal of Population Geography, 6*, 391–421.

Fotheringham, S. A., Brunsdon, C., & Charlton, M. E. (2002). *Geographically weighted regression: The analysis of spatially varying relationships*. Chichester, U.K.: Wiley.

Fukurai, H., Pick, J. B., Butler, E. W., & Nag, S. (1987). An analysis of interstate migration in Mexico: Impact of origin and destination states on migration patterns. *Mexican Studies, 3*(2), 365–396.

Gedik, A. (1977). *A causal analysis of the destination choice of village-to-province-center migrants in Turkey, 1965–70*, Ph.D. dissertation. Seattle, USA: Dept. of City and Regional Planning, University of Washington.

Gedik, A. (1978). *Sizes of different migration flows in Turkey, 1965–1970, possible future directions and towards comparative analysis*. Discussion paper Series No. 78-15. Japan: Institute of Socio-economic Planning, University of Tsukuba.

Gedik, A. (1979). *Descriptive analysis of village-to-province-center migration in Turkey: 1965–70.* Discussion paper Series No. 79-22. Japan: Institute of Socio-Economic Planning, University of Tsukuba.

Gedik, A. (1991). Migration and urban growth in Turkey, 1965–70. In K. P. Strohmeir & Ch. W. Matthissen (Eds.) *Innovation and urban population dynamics*; *urban Europe ser*ies (pp. 105–122). Hampshire, UK: Avery.

Gedik, A. (1992). The effects of in- and out-migration on urban growth in Turkey (1965–85) and a comparison with the developed countriesç. *Papers in Regional Science, 71*(4), 405–419.

Gezici, F., & Hewings, G. J. D. (2007). Spatial analysis of regional inequalities in Turkey. *European Planning Studies, 15*(3), 383–403.

Giannetti, M. (2003). On the mechanics of migration decisions: Skill complementarities and endogenous price differentials. *Journal of Development Economies, 71*(2), 329–349.

Greenwood, M. J., Pick, J. R., & Siegal, B. S. (1981). Long-term trends in migration behavior in a developing country: The Case of Mexico. *Demography, 18*(3), 369–388.

Halfacree, K. H., & Boyle, P. J. (1993). The challenge facing migration research: The case for a biographical approach. *Progress in Human Geography, 17*(3), 333–348.

Haynes, K. E., & Fotheringham, A. S. (1984). *The gravity model and spatial interaction*. Beverly Hills CA: Sage Publication .

He, J., & Pooler, J. (2002). The regional concentration of China's interprovincial migration flows, 1982–90. *Population and Environment, 24*(2), 149–182.

Hugo, G., & Bell, M. (1998). The hypothesis of welfare-led migration to rural areas: The Australian case. In P. J. Boyle & K. H. Halfacree (Eds.), *Migration into rural areas: Theories and issues* (pp. 20–40). New York: Wiley.

Johnson, B. A., Scheyvens, H., Khalily, M. B., & Onishi, A. (2019). Investigating the relationships between climate hazards and spatial accessibility to microfinance using geographically-weighted regression. *International Journal of Disaster Risk Reduction, 33*, 122–130.

Kennan, J., & Walker, J. R. (2003). *The effect of expected income on individual migration decisions.* NBER Working Paper No. W9585.

Kesici, M. R. (2010). In high growth phase of Turkey economy employment and unemployment. *"IS, GUC" Industrial Relations and Human Resources Journal, 12*(3), 7–26.

Kontuly, T., & Schon, K. P. (1994). Changing Western German internal migration systems during the 2nd half of the 1980s. *Environment and Planning A, 26*(10), 1521–1543.

Koramaz, T. K., & Dökmeci, V. (2016). Impact of distance on migration in Turkey. *Migration Letters, 13*(2), 269–294.

Liang, Z., & White, M. J. (1996). Internal migration in China, 1950–1988. *Demography, 33*(3), 375–384.

Li, J., Jin, M., & Li, H. (2019). Exploring spatial influence of remotely sensed $PM_{2.5}$ concentration using a developed deep convolutional neural network model. *International Journal of Environmental Research and Public Health, 16*(3), 454.

Liang, Z., Chen, Y. P., & Gu, Y. (2002). Rural industrialization and internal migration in China. *Urban Studies, 39*, 2175–2187.

Lim, H., & Thill, J. C. (2008). Intermodal freight transportation and regional accessibility in the United States. *Environment and Planning A., 40*(8), 2006–2025.
Lin, J., Cromley, R. G., & Hanink, D. M. (2016). Mapping local regression for spatial object-pairs. *Cartography and Geographic Information Science, 43*(4), 328–337.
Lucas, R. E. B. (2001). The effects of prosperity and transportation on developing country population migration. *Journal of Economic Geography, 1*(3), 323–339.
MacKeller, F. L., & Vining, D. R. (1995). Population concentration in less developed countries: New evidence. *Papers in Regional Science, 74*(3), 259–293.
Nakaya, T., Fotheringham, S. A., Brunsdon, C., & Charlton, M. (2005). Geographically weighted Poisson regression for disease association mapping. *Statistics in Medicine, 24*, 2695–2717.
Pandit, K. (1994). Differentiating between subsystems and typologies in the analysis of migration regions: A U.S. example. *Professional Geographer, 46*(3), 334–345.
Portnov, B. A. (1999). The effect of regional inequalities on migration: A comparative analysis of Israel and Japan. *International Migration, 37*(3), 587–616.
Potter, J. J. (1993). The impact of change upon rural-urban migrants in Turkey. *Landscape and Urban Planning, 26*, 99–114.
Potts, D. (2009). The slowing of Sub-Saharan African's urbanization: Evidence and implications for urban livelihoods. *Environment and Urbanization, 21*(1), 253–259.
Ravenstein, E. G. (1885). The laws of migration. *Journal of the Royal Statistical Society, 48*, 167–227.
Ravuri, E. D. (2004). Determinants of migration to and from Bolivar State, Venezuela for 1961 and 1990: The effect of Ciudad Guayana on migration. *The Journal of Developing Areas, 37*(2), 155–167.
Roy, J. R., & Thill, J. -C. (2004). Spatial interaction modelling. *Papers in Regional Science, 83*(1), 339–361.
Sewell, G. (1964). *Squatter settlement in Turkey: Analysis of a social, political and economic problem*, Ph.D. dissertation. Cambridge, USA: Massachusetts Institute of Technology, MIT Press.
Skeldon, R. (1990). *Population mobility in developing countries*. London: Belhaven Press.
Tekeli, I. (2008). *Goc ve Otesi*. Tarih Vakfi Yayinlari: Istanbul.
Tumertekin, E. (1968). *Internal migration in Turkey*. Istanbul University, Institute of Geography Publication, No. 57.
Url-1. *Population census, 1980–2000* Turkish Statistical Institute. Website https://www.tuik.gov.tr, dated November 11, 2010.
Url-2. *Address based population registration system*, 2012 Turkish Statistical Institute. Website https://www.tuik.gov.tr, dated February 4, 2013.
Url-3. *Provincial and sectoral distribution of public investments in Turkey*, 2000. Ministry of Development. Website https://www2.dpt.gov.tr/kamuyat/ilozet.html, dated February 25, 2013.
Var, E. B., Yazgi, B., & Dokmeci, V. (2014). Age and cohort analysis of regional migration in Turkey. *Regional Science Inquiry, 6*(1), 81–94.
Wilson, A. G. (1971). A family of spatial interaction models and associated developments. *Environment and Planning, 3*(1), 1–32.

Carbon Emissions Embodied in Trade and Urban Regional Climate Policy-Making in the Shanghai Mega-Region

Zhangqi Zhong, Jean-Claude Thill, Yi Sun, and Zheng Wang

1 Introduction

Although they cover less than 1% of the earth's land surface, cities host approximately 54% of the world's population (United Nations 2014). Meanwhile, cities already contribute an estimated 67–76% of global energy consumption and 71–76% of greenhouse gas (GHG) emissions at the start of the twenty-first century (Edenhofer et al. 2014). The projected rate of growth in urbanization, particularly the rapid urbanization of developing countries, and accompanying rises in standard of living are steadily driving high growth in energy consumption and thus CO_2 emissions in urban areas (Dhakal 2010; Schulz 2010; Mi et al. 2016). Thus, proactive initiatives have been taken to promote energy conservation and emission reduction at the city level, which have become the focus of attention in countries around the globe (Kennedy et al. 2009; Zhong et al. 2017; Li et al. 2018; Mi et al. 2019b). In this context, it is essential to dispose of suitable accounts of carbon emissions that enable stakeholders to effectively implement emission reduction strategies at the urban regional scale.

According to the World Resources Institute and the World Business Council for Sustainable Development classification, emissions of a city can be classified into three scopes for the purpose of carbon emission accounting (WRI/WBCSD 2009): Scope 1 includes direct emissions such as fuel combustion, industrial processes, and

Z. Zhong
School of Economics, Zhejiang University of Finance and Economics, 310018 Hangzhou, China

J.-C. Thill (✉)
Department of Geography and Earth Sciences, University of North Carolina at Charlotte, Charlotte, NC 28223, USA
e-mail: jfthill@uncc.edu

School of Applied Economics, Renmin University of China, Beijing, China

Y. Sun · Z. Wang
Institute of Policy and Management, Chinese Academy of Sciences, 100080 Beijing, China

J. Thill (ed.), *Innovations in Urban and Regional Systems*,
https://doi.org/10.1007/978-3-030-43694-0_18

other emissions resulting from intentional and unintentional releases such as equipment leaks from joints, seals as well as coal piles, wastewater treatment, which are typically produced within the geographical boundary of the city. Scope 2 covers indirect emissions released outside the geographical boundary of the city that relate to energy consumption within the city, including electricity and district heat supplied by upstream power plants. Scope 3 emissions, also called emissions embodied in trade (EET), are associated with the extraction, production, and transportation of products or services attributable to the city through trade. In order to capture a holistic picture of a city's carbon emissions, estimation should cover all three scopes (Li et al. 2013), which is a precondition for balanced and effective mitigation actions. To date, however, GHG accounting studies have focused primarily on Scope 1 and Scope 2 sources (Satterthwaite 2008; Dhakal 2009; Kennedy et al. 2009; Hillman and Ramaswami 2010; Bi et al. 2011; Xi et al. 2011; Wang et al. 2014; Cai et al. 2018; Hung et al. 2019), whereas the comprehensive and complete analysis of emissions embodied in trade has yet to receive the attention that they deserve. In particular, more robust estimates of city carbon emissions are needed to establish the contribution of EET and to better articulate balanced climate policies including the imputation of emission reduction targets to specific causal processes operating in particular industrial sectors of particular regions. This chapter contributes to this goal by quantifying the carbon emissions embodied in trade between regional and national economies through a regional input-output (IO) analysis applied to the Shanghai metropolitan region.

While trade plays a critical role in regional economic development by providing a mechanism to efficiently allocate resources in the process of economic globalization (Amin 1999; Batabyal and Nijkamp 2003; Czarnitzki and Hottenrott 2009), a side effect is the geographic separation of consumers and the pollution emitted during the production of consumable items. Trade also provides a mechanism to shift environmental impacts to other regions (Weber and Matthews 2007; Peters and Hertwich 2008a; Scott and Barrett 2015; Chen et al. 2018; Mi et al. 2019a). Thus, part of the responsibilities for CO_2 emissions reduction that would otherwise belong to the consumer are transferred to the producer of traded goods (Batabyal and Nijkamp 2003; Schulz 2010; Liu et al. 2018). Recently, important climate policy implications stemming from using a consumption-based approach instead of a more conventional production-based approach have been discussed, while more attention has also been given to the impact of EET on the city's carbon emissions and its reduction obligation assignments (Schulz 2010; Li et al. 2013; Feng et al. 2014; Hermannsson and McIntyre 2014; Choi 2015; Mi et al. 2016; Chen and Zhu 2019).

Although the literature on the urban scale of EET (namely Scope 3) has developed considerably over the past decade, it remains largely fraught by several limitations that we seek to alleviate in this paper. First, most important studies such as Guo et al. (2012a, b, c) and Chen et al. (2013) have circumvented data availability and methodological limitations by making the assumption that commodities imported from other domestic regions have the same embodied emission intensities as local ones. Actually, Su and Ang (2013) argued that this assumption may inevitably bring biased estimations. Second, previous studies such as Larsen and Hertwich (2009),

Li et al. (2013), Zhang et al. (2014a, b), Liu et al. (2015), and Chen et al. (2018) have assumed that all products imported from various regions and countries are produced with the same technologies. However, strong evidence indicates that the embodied emission intensities vary significantly between different countries and between regions of a certain country (Chen and Chen 2011; Liu et al. 2015), which may have a strong effect on EET depending on the configuration of trade relationships. Third, the methodological limitations of single-region input-output analysis preclude us from identifying where embodied CO_2 emissions go to and come from, on an industrial sector basis. These three deficiencies may significantly handicap the fair allocation of CO_2 emissions to regions and industrial sectors and ultimately impair the formulation of climate policies and their implementation. Given the manifest risk of deficient or wrongful imputation in emission accounting, a challenging question is to determine how a vibrant national and global center of commerce should handle close trading partnerships so as to mitigate the impact of trade on imputed responsibilities in CO_2 emissions. This paper uses regionally and sectorally disaggregated data in a multi-regional input-output economic modeling framework that overcomes all these shortcomings to enable us to draw appropriate policy conclusions with regard to climate change and GHG mitigation. In particular, the IO modeling approach is effective at assessing the incidence of domestic trade on emissions on a bilateral basis and at estimating how a regional economic coalition can shield domestic trade among coalition partners from the stringent requirements of a national climate policy while preserving the economic benefits of inter-regional trade.

The city of Shanghai is selected as a case to calculate the embodiment of localized CO_2 emissions. Two accounting approaches (production accounting principle and consumption accounting principle) to CO_2 emissions are applied here and compared. The former approach focuses on CO_2 emissions occurring from economic production or activities within a region, in a way that is analogous to Gross Domestic Product (GDP) accounting; hence, emissions are imputed to the place of production. In contrast, the latter approach is concerned with emissions due to economic consumption and assigns emissions to regions based on their total consumption (including intermediate and final consumption) (Peters and Hertwich 2008a, b). The sources and flows of Shanghai's carbon emissions embodied in domestic regional imports and exports as well as international imports and exports are analyzed and we discuss the merits of a policy strategy based on regional coalitions to reduce the impact of trade on GHG emissions and climate change.

The rest of this chapter is organized as follows. The following section summarizes the literature on EET, while Sect. 3 presents important background information on Shanghai. The methodology and data are presented in Sect. 4. Section 5 discusses the results of the economic modeling analysis in the context of production- and consumption-based emission accounting approaches. Finally, the conclusions and policy implications of this research are presented in Sect. 6.

2 Literature Review

Studies on the urban scale of EET (Scope 3) are recent and remain relatively few. Schulz (2010) examined the volumes and trends of GHG emissions embodied in the products traded in Singapore from 1960 to 2000 based on life cycle analysis. He found that GHG embodied in trade exceeded the volume of direct emissions through the whole time series and that urban scale emission accounts may be of limited value to identify the most effective approaches to climate change mitigation if they focus solely on direct emissions. While the city-state of Singapore is an excellent example of open urban economy, this study failed to disaggregate the impact of trade on carbon emissions sectorally.

Sectoral decomposition is accomplished in a number of other studies that use an IO economic analysis framework. From a consumption-based perspective, Larsen and Hertwich (2009) expanded the IO table to include some sectors describing key inputs for Trondheim, Norway and then estimated the fraction of indirect GHG emissions occurring within the borders of the city. Rosado and Ferrão (2009) measured the energy embodied in household goods in Lisbon, Portugal. Using 41 sectors of the 2002 Beijing economy, Zhou et al. (2010) calculated the embodied natural resources and GHG emissions through the extension of economic IO analysis into the environmental analysis. Guo et al. (2012a, b, c) updated the embodied CO_2 emissions of Beijing for 2007, while Guo et al. (2012a, b, c) revealed the local GHG embodiment in final demand and trade, but without regard for imported emissions. Furthermore, Vause et al. (2013) calculated CO_2 emissions for Xiamen, China, and its six major economic sectors in 2007 based on both the production and consumption accounting approaches. Zhang et al. (2014a, b) explored the effects of direct trade within China on regional and national CO_2 emissions. Ji et al. (2014) studied Beijing's optimal industrial structure based on the concepts of embodied energy and embodied carbon emissions estimated on 42 sectors in 2005. Zhang et al. (2014a, b) analyzed the embodied energy consumption for the entire city of Beijing and its 30 economic sectors from 1987 to 2007. Remarkably, according to Guo et al. (2012a, b, c), Beijing was a net importer of embodied CO_2 emissions in 2007. Moreover, it is worth noting that Feng et al. (2014) explored the spatial distribution of production activities leading to CO_2 emissions across China as a consequence of final consumption in the four Chinese mega-cities of Beijing, Shanghai, Tianjin, and Chongqing, although emissions from abroad are not accounted for. Moreover, this question also can be found in Liu et al. (2015) regarding carbon emissions in trade for China's 30 provinces. Thus, the emissions embodied in international imports to a city remain to be studied.

Although the above studies have provided an important basis on which to develop sound environmental and energy policies aimed at reducing GHG emissions, shortcoming rooted in the methodology and limited data availability remain serious impediments. In particular, for both domestic and international imports, commodities have the same embodied emission intensities as local ones. In other words, it is generally assumed that imported and domestic products are all produced with the same technologies. In contrast, Chen et al. (2013) proposed a three-stage IO model of the

Beijing economy in 2007 covering 42 sectors, which incorporates databases of average carbon intensities for the world and national economies. These authors found that Beijing was a net importer of embodied CO_2 emissions. The comparison of the embodied CO_2 emissions estimated by Guo et al. (2012a, b, c) and Chen et al. (2013) reveals remarkable differences amounting to a 50% discrepancy. This large discrepancy can most likely be ascribed to the relatively low level of production technology and energy efficiency in China compared to other nations (Weber et al. 2008). Notably, however, based on version 8 of the Global Trade Analysis Project database, Feng et al. (2013) make the assumption that provincial international import matrices are sourced from countries in the same proportion as total Chinese imports; they tracked CO_2 emissions embodied in products traded among Chinese provinces and internationally.

Additionally, considering that imports and exports are divided between Mainland China and the rest of the world, Li et al. (2013) presented a systematic accounting for Macao in 2005–2009. These authors noted that total emissions are dominated by indirect emissions mainly embodied in imports, which are 3–4 times larger than direct emissions during the study period. Furthermore, it is verified that accounting under Scopes 1 and 2 cannot capture the full picture of a city's emissions and CO_2 emissions associated with trade in goods and services should be considered. However, the authors of this study assumed that all commodities imported from Mainland China have the same embodied emission intensities, which is a major shortcoming given the scale of mainland domestic trade compared to international trade, and the regional differences in production technology, industrial structure, and energy efficiency in China (Liu et al. 2012; Wang et al. 2013).

Also, under the assumption that different areas of Australia have a similar mix of factors of production, similar production structures and efficiencies, Baynes et al. (2011) used IO analysis to assess embodied energy use in Melbourne, Australia. Similar shortcomings can be found in Wu's (2011) study of Haninge, Sweden. Hermannsson and McIntyre (2014) used an extended IO model of the Glasgow region and of the rest of Scotland, and Choi (2015) analyzed GHG emissions imputable to the metropolitan areas of Atlanta, San Francisco, and Seattle. To sum up, many studies of CO_2 emissions at the scale of the urban region or city fail to account for the role of interregional and international trade, and those that do routinely make simplifying assumptions that compromise the policy formulation value of these estimates.

3 Shanghai as a Case Study

As the largest center of commerce and finance in Mainland China, Shanghai has a larger GDP than any other Chinese city and accounted for 20.10% (27.00%) of the total commodity imports (exports, respectively) of China in 2007 (CCSY 2008). In the context of the rapid growth of Shanghai's economy, energy consumption increased from 55.00 to 97.68 million tons of standard coal equivalent with an average annual growth rate of 9.70% during 2000–2007 (CESY 2008). Moreover, Shanghai's energy supply is overwhelmingly sourced from other regions (95.87 million tons of standard coal equivalent or 98.15% of the total energy demand needs) (CESY 2008). Therefore, Shanghai's energy demand has been increasing in concert with its fast economic growth and CO_2 emissions embodied in domestic trade from other regions have risen accordingly.

Additionally, Shanghai is one of the most important open coastal cities in China and has witnessed rapid changes in both international imports and exports. Specifically, commodity exports increased annually by 29.76% between 2000 and 2007, while imports increased at an annual rate of 28.49% during the same period (SSY 2008). As a result, more attention should also be paid to CO_2 emissions embodied in international trade.

Facing increasing pressure to reduce CO_2 emissions, Shanghai also set carbon-reduction targets in 2007: a 20% reduction of CO_2 emissions per unit of GDP by 2010, and an annual drop of 3–3.6% in subsequent years.[1] Whereas great strides have been made towards maintaining economic growth while also striving to meet these targets, further efforts are necessary, particularly in anticipation of tighter controls brought about by national climate change policies that continue to adjust to new political contexts. In this context, Shanghai needs to more effectively impute CO_2 emission reduction standards as part of an effective climate policy through the rigorous estimation of CO_2 emissions embodied by its urban economy, and specifically the impact of interregional and international trade.

4 Modeling Framework and Data

4.1 Modeling Framework

Input-output (IO) analysis captures direct, indirect and induced economic effects within an economy through a detailed accounting of interdependencies between inputs and outputs across economic sectors. It can thus capture indirect environmental impacts caused by upstream production and is suitable for the estimation of pollution and resource use embodiments of traded commodities (Davis and Caldeira 2010). This macroeconomic modeling technique has been recognized as an effective framework to quantify the embodied GHG and resources used in one region for the production of goods and services exported to other regions (Zhang and Folmer 1998; Wiedmann et al. 2007; Peters 2008; Wiedmann 2009; Chen et al. 2013, 2018, 2019; Su and Ang 2013, 2014; Feng et al. 2014; Liu et al. 2015; Zhong et al. 2017). In the literature, single-region input-output (SRIO) analysis and multi-region input-output (MRIO) analysis are commonly used to estimate the CO_2 emissions embodied in trade in line with the IO principles (Wiedmann et al. 2007; Peters 2008).

[1] http://www.shanghai.gov.cn/shanghai/node2314/node2319/node2404/node18120/node18121/userobject26ai11978.html.

In short, MRIO analysis is based on the monetary flows between industrial sectors and regions, considering the total economic output of each sector in each region, each sector's output produced in one region and consumed in another, and a matrix of intermediate consumption where columns reflect the input from sectors in each region required to produce one unit of output of each sector in each of the other regions (Davis and Caldeira 2010). It is superior to the SRIO analysis for analyzing the inter-regional trade effects between regions and for encapsulating the relations between different industrial sectors because it accounts explicitly for heterogeneity (Davis and Caldeira 2010; Zhong et al. 2015). Given that the paper aims to estimate CO_2 emissions embodied in trade by the Shanghai economy and to discuss the outflow and inflow of embodied CO_2 emissions between Shanghai, other areas of China and other countries, the MRIO analysis is chosen as our analytical framework.

Following the work by Peters (2008) and Peters and Hertwich (2008a), and in consideration of the contribution of international and domestic trade to the emissions of the Shanghai economy, the total CO_2 emissions occurring in Shanghai E^T_{SHA} can be explicitly decomposed into local and traded components (both regional and international) as follows:

$$E^T_{\text{SHA}} = E_{\text{SHA}} + E_{\text{SHA-}s} + E_{\text{SHA-}t} \tag{1}$$

where E_{SHA}, $E_{\text{SHA-}s}$, and $E_{\text{SHA-}t}$ denote CO_2 emissions caused by the demand of the Shanghai economy on the local production of Shanghai, emissions embodied in domestic exports (EEDE) from Shanghai to region s in China, and emissions embodied in international exports (EEIE) from Shanghai to country t, respectively. For an easier understanding, the framework of the IO table conveniently articulates these variables in relation to the three emission scopes for the purpose of carbon emission accounting, as depicted in Table 1.

The Shanghai production terms can be expressed as follows:

$$E_{\text{SHA}} = F_{\text{SHA}}(1 - A_{\text{SHA}})^{-1} Y_{\text{SHA}} \tag{2}$$

$$E_{\text{SHA-}s} = F_{\text{SHA}}(1 - A_{\text{SHA}})^{-1} e_{\text{SHA-}s} \tag{3}$$

$$E_{\text{SHA-}t} = F_{\text{SHA}}(1 - A_{\text{SHA}})^{-1} f_{\text{SHA-}t} \tag{4}$$

Table 1 Shanghai-traded components (both regional and international) and emission scopes in the IO table

	Shanghai consumption	International and other domestic consumption
Shanghai production	E_{SHA} (Scope 1)	$E_{\text{SHA-}s}$ and $E_{\text{SHA-}t}$ (Scope 3)
International and other domestic production	$E_{s\text{-SHA}}$ and $E^{\text{LB}}_{t\text{-SHA}}$ ($E^{\text{UB}}_{t\text{-SHA}}$) (Scope 2 and Scope 3)	–

where F_{SHA} is a row vector, with each element representing the emissions per unit industry output in the Shanghai economy; I is the identity matrix; A_{SHA} are the inter-industry requirements of products produced by the Shanghai economy and demanded by local industries; Y_{SHA} are the products produced and consumed by Shanghai; $e_{SHA\text{-}s}$ are the exports in domestic trade from Shanghai to region s in China; and $f_{SHA\text{-}t}$ are the international exports from Shanghai to country t.

A similar approach is used on the imports side, although some adjustments are made in the interest of tractability. First, CO_2 emissions embodied in domestic imports (EEDI) from region s to Shanghai are expressed as follows:

$$E_{s\text{-SHA}} = F_s(1 - A_s)^{-1} e_{s\text{-SHA}} \tag{5}$$

where F_s is a row vector, with each element representing the CO_2 emissions per unit industry output in region s, A_s are the inter-industry requirements of products produced in this region, and $e_{s\text{-SHA}}$ are the exports in domestic trade from region s to Shanghai. This formulation recognizes that levels of technological development vary across China's provinces, and thus so are embodied emission intensities.

Also, in line with standard IO modeling practice, the CO_2 emission factors of commodities imported from different countries should be derived from their respective IO tables (Li et al. 2013; Liu et al. 2015). However, for a city that is a major center of commerce and trade such as Shanghai, the number of trading countries can be quite large (nearly one hundred in the case of Shanghai), so that calculating embodied CO_2 intensities for each country covering various sectors can be very difficult and impractical. Furthermore, the national origin of Shanghai imports is unknown due to limitations of the MRIO tables of China. Therefore, an alternative approach based on the definition of upper and lower bounds on emissions is introduced here. Taken together, the lower and upper bounds can suitably reflect the impact of international imports on the city's emissions. Indeed, when a product is imported from countries with higher energy efficiency instead of being produced by local industries, the corresponding embodied emissions intensities are relatively low. Therefore, such traded products can significantly curtail energy consumption and even reduce CO_2 emissions. The implication is that significant environmental impacts are shifted from one region to another through trade.

The bounds on emissions embodied in international trade are determined as follows. On the one hand, Japan's CO_2 emission coefficients (F^{JAP}) are taken as the lower bound of the emission factors embodied in Shanghai's international imports. Two main reasons serve to motivate this imputation. First, given that Japan's technological level in terms of energy efficiency is highly advanced, its CO_2 emissions per unit of output are relatively low compared with Shanghai's other major trading countries such as the USA and nations of the European Union (Dhakal 2004; Qi et al. 2008). The other reason is that Japan has become one of the largest trading partners of Shanghai and the largest source of its imports since 2000 (SSY 2008).

On the other hand, because Shanghai's major foreign trading partners (USA and European Union countries) have reached a technological level in terms of energy efficiency that is generally higher than that of Shanghai, Shanghai's embodied CO_2

emission intensities (F^{SHA}) can be considered as the upper bound of the estimate of emissions embodied in Shanghai's international imports.

Accordingly, the lower bound and the upper bound of the volume of CO_2 emissions embodied in international imports (EEII-LB and EEII-UB) of Shanghai can be expressed as follows:

$$E_{t\text{-SHA}}^{LB} = F^{JAP} f_{t\text{-SHA}} \tag{6}$$

$$E_{t\text{-SHA}}^{UB} = F^{SHA} f_{t\text{-SHA}} \tag{7}$$

where $f_{t\text{-SHA}}$ are Shanghai's international imports from country t.

For Shanghai, another important parameter is the net balance of emissions embodied in domestic trade (BEEDT), which can be expressed as:

$$\mathrm{DE}_{SHA}^{BEEDT} = \sum_{s} E_{SHA\text{-}s} - \sum_{s} E_{s\text{-SHA}} \tag{8}$$

Similarly, for international trade, the lower bound and the upper bound of the balance of embodied emissions (BEEIT-LB and BEEIT-UB) are obtained as:

$$\mathrm{IE}_{SHA}^{BEEIT\text{-}UB} = \sum_{t} E_{SHA\text{-}t} - \sum_{t} E_{t\text{-SHA}}^{LB} \tag{9}$$

$$\mathrm{IE}_{SHA}^{BEEIT\text{-}LB} = \sum_{t} E_{SHA\text{-}t} - \sum_{t} E_{t\text{-SHA}}^{UB} \tag{10}$$

and the lower and the upper bounds of the balance of the emissions embodied in all trade (BEET-LB and BEET-UB) are:

$$E_{SHA}^{BEET\text{-}UB} = \mathrm{DE}_{SHA}^{BEEDT} + \mathrm{IE}_{SHA}^{BEEIT\text{-}UB} \tag{11}$$

$$E_{SHA}^{BEET\text{-}LB} = \mathrm{DE}_{SHA}^{BEEDT} + \mathrm{IE}_{SHA}^{BEEIT\text{-}LB} \tag{12}$$

As discussed earlier, it is well known in climate policy research that the imputation of responsibilities for anthropogenic emissions hinges upon the emission accounting approach, especially a consumption-based versus a production-based approach (Munksgaard and Pedersen 2001; Peters and Hertwich 2008a, b). In order to assess the impact of alternative emissions accounting approaches on specific sectors' commitments for emissions reduction at the city level, we define the production-based emissions inventory (PEI) as the total emissions occurring from economic production in Shanghai:

$$E_{SHA}^{prod} = E_{SHA}^{T} \tag{13}$$

We also define the consumption-based emissions inventory (CEI) as the total emissions occurring from economic consumption. As a result, the upper bound ($E_{\text{SHA}}^{\text{cons-UB}}$) (CEI-UB) and lower bound ($E_{\text{SHA}}^{\text{cons-LB}}$) (CEI-LB) of the total emissions stemming from economic consumption within the city of Shanghai are:

$$E_{\text{SHA}}^{\text{cons-UB}} = E_{\text{SHA}}^{T} - E_{\text{SHA}}^{\text{BEET-UB}} \tag{14}$$

$$E_{\text{SHA}}^{\text{cons-LB}} = E_{\text{SHA}}^{T} - E_{\text{SHA}}^{\text{BEET-LB}}. \tag{15}$$

With Eqs. (13)–(15), it should be noted that the balance of the CO_2 emissions embodied in all trade takes a special meaning as it captures the gap between the production-based and consumption-based emissions accounting approaches.

4.2 Data

The Chinese MRIO tables of 2007 are adopted in this paper, which constitute the most recent and most complete published data set for Shanghai and China's 29 other provinces and province-level cities.[2] They were originally assembled by Liu et al. (2012) and have since been used by many other researchers (e.g., Feng et al. 2014; Liu et al. 2015; Zhong et al. 2015). The MRIO tables provide not only data on Shanghai's imports and exports but also concerning domestic trade between Shanghai and the other 29 provinces across 30 industrial sectors. Sector classification is shown in Table 5 of Appendix.

Each sector's CO_2 emissions per unit of output is required to calculate embodied emission coefficients. Since there are no direct statistics on this matter, a proxy is used instead. First, based on the quantity of energy consumption from the energy balance table of each province (CESY 2008) and on the CO_2 emissions factors of various types of energy from the Intergovernmental Panel on Climate Change (IPCC) (2006), we calculate the CO_2 emissions (F_s) per unit of output of each sector in each province. Moreover, the Global Trade Analysis Project (GTAP)[3] provides data on CO_2 emissions for more than 87 countries and 57 industrial sectors. Based on this data set, we can obtain the Japanese CO_2 emission intensities (F^{JAP}) covering 57 sectors with the unit changed to t/Yuan based on the corresponding exchange rate; the 57 sectors are aggregated into 30 sectors for consistency with the Chinese MRIO tables. All these data pertain to 2007.

Moreover, it should be noted that there are 30 sectors in the Chinese MRIO tables of 2007, but only 6 sectors in the corresponding energy balance tables. Two data treatment schemes have been used to make these tables compatible in the recent literature. One scheme is to aggregate the input-output data to the level that matches the energy consumption data in energy balance tables. The other scheme is to disaggregate the

[2]Hong Kong, Macao, Taiwan, and Tibet are not included because of data unavailability.

[3]https://www.gtap.agecon.purdue.edu/databases/.

energy data to the level that matches the input-output data (Su and Ang 2010). However, views regarding best way to proceed are divided among scholars. For example, Machado (2000) thinks using the former may actually be better, whereas Lenzen et al. (2004) found that the latter should be better. In fact, Su and Ang (2010) pointed out that both schemes have strengths and weaknesses. The former guarantees data accuracy without incurring potential errors from imposing extra assumptions, whereas the latter retains all input-output information, which is matched with an energy data set enlarged by making certain reasonable assumptions. To preserve as much MRIO table information as possible, the latter scheme is used here.

5 Results and Discussion

5.1 Overall Emissions and Emissions Embodied in Trade

From the PEI perspective, overall emissions in Shanghai are estimated to be 163.00 million tons in 2007. By sector, values[4] (Table 2) range from 27.39% of the total PEI for Sector 14 (Smelting and Pressing of Ferrous and Nonferrous Metals) to 0.0031% of the total PEI for Sector 20, which underscores the significant emission differences among industrial sectors in Shanghai. Notably, PEI is highly concentrated on a few sectors: three sectors (Smelting and Pressing of Ferrous and Nonferrous Metals, Electric Power/Steam and Hot Water Production and Supply and Chemical Products Related Industry) alone generate 64.23% of the total PEI, which suggests that these sectors should be imposed a dominant share of emission reduction obligations during the process of economic production. Therefore, improving energy efficiency and promoting advanced emission-cutting technologies in these sectors should be effective approaches to reduce emissions in Shanghai. Second, from the perspective of the CEI, the total volumes of CEI-LB and CEI-UB for all sectors and final consumption combined are estimated to be 191.50 and 213.94 million tons, respectively (Tables 2 and 3). Interestingly, the PEI estimate is below both the lower and upper bounds based on consumption principles. Therefore, even under conservative assumptions, trade relations dictate that Shanghai's emission responsibility as a consumer greatly surpasses its responsibility on the production side. Sectorally, tremendous diversity exists between individual sectors on the CEI-LB and the CEI-UB as a result of the estimated bounds on emissions embodied in international imports, which will be discussed in detail in the next section.

The IO analysis also reveals the sheer magnitude of Shanghai's emissions that are embodied in trade. Emissions embodied in domestic trade (EED) accounts for 50.67% of Shanghai's CO_2 emissions (82.60 million tons), while the lower bound

[4]For Sectors 2, 4, 5, and 22 in the EEDE and EEIE, and for Sectors 3, 23 and 24 in the EEIE, results are zero because they are not produced nor consumed in Shanghai. Therefore, for simplicity, these results are not considered here. Some other results reported in Table 2 may also be zero for the same reasons.

Table 2 Shanghai's CO_2 emissions embodied in trade for the year 2007 by sector (Unit: 10,000t)

	PEI	CEI-LB	CEI-UB	EEDE	EEIE	EEDI	EEII-LB	EEII-UB	BEEDT	BEEIT-UB	BEEIT-LB
1	43.91	71.37	73.17	0.05	2.01	29.32	0.20	2.00	−29.28	1.82	0.01
2	0.00	0.00	0.00	0.00	0.00	0.00	0.00	0.00	0.00	0.00	0.00
3	1.97	1.62	1.55	1.17	0.00	0.72	0.10	0.03	0.45	−0.10	−0.03
4	0.00	0.00	0.00	0.00	0.00	0.00	0.00	0.00	0.00	0.00	0.00
5	0.00	0.00	0.00	0.00	0.00	0.00	0.00	0.00	0.00	0.00	0.00
6	94.81	99.22	107.66	38.81	16.08	58.35	0.95	9.39	−19.54	15.13	6.69
7	29.03	36.61	40.50	5.25	12.32	25.09	0.06	3.95	−19.84	12.26	8.36
8	8.09	29.35	30.43	4.35	1.92	27.51	0.02	1.10	−23.16	1.90	0.82
9	4.81	29.41	29.61	1.46	4.93	30.97	0.02	0.22	−29.51	4.91	4.71
10	104.71	79.55	85.53	50.67	29.03	52.35	2.19	8.17	−1.68	26.83	20.86
11	2028.90	1368.04	1764.47	526.56	204.66	61.27	9.08	405.51	465.29	195.58	−200.86
12	2520.24	2076.52	2509.97	472.94	482.86	477.34	34.76	468.20	−4.39	448.11	14.67
13	286.34	302.67	307.01	37.08	35.93	88.41	0.92	5.26	−51.33	35.00	30.66
14	4464.99	3790.37	4466.44	463.88	853.72	609.99	32.99	709.07	−146.11	820.73	144.65
15	17.24	279.53	280.32	7.24	2.83	271.45	0.89	1.68	−264.22	1.94	1.14
16	170.19	553.38	589.40	46.17	58.79	487.85	0.30	36.33	−441.68	58.49	22.46
17	93.05	321.23	333.99	32.44	32.55	293.06	0.11	12.87	−260.62	32.43	19.68
18	16.22	258.96	260.68	3.49	8.71	253.46	1.49	3.21	−249.96	7.22	5.50
19	58.19	256.83	273.37	9.17	35.00	228.97	13.83	30.37	−219.80	21.16	4.62
20	0.52	22.12	21.71	0.12	0.27	21.40	0.59	0.18	−21.28	−0.32	0.09
21	2.54	22.36	22.41	0.46	1.24	21.37	0.16	0.21	−20.90	1.08	1.03

(continued)

Table 2 (continued)

	PEI	CEI-LB	CEI-UB	EEDE	EEIE	EEDI	EEII-LB	EEII-UB	BEEDT	BEEIT-UB	BEEIT-LB
22	3484.23	3893.45	3931.43	0.00	0.00	395.58	13.65	51.62	−395.58	−13.65	−51.62
23	6.33	38.85	39.02	0.07	0.00	32.39	0.20	0.37	−32.32	−0.20	−0.37
24	138.52	669.88	680.39	28.88	0.00	558.38	1.86	12.36	−529.51	−1.86	−12.36
25	1558.48	1159.31	1290.11	327.50	436.94	332.69	32.58	163.38	−5.19	404.36	273.57
26	64.95	83.49	83.98	9.16	15.22	42.78	0.14	0.64	−33.62	15.08	14.58
27	33.49	55.19	57.21	14.33	6.10	42.13	0.00	2.02	−27.80	6.10	4.08
28	655.05	462.85	507.02	53.19	186.89	47.14	0.73	44.91	6.05	186.15	141.98
29	9.43	10.56	10.94	2.40	0.49	3.88	0.14	0.53	−1.48	0.34	−0.04
30	404.10	572.17	592.11	46.09	11.44	220.29	5.32	25.26	−174.20	6.12	−13.82

Note Sector codes are given in Table 5

Table 3 Shanghai's CO_2 emissions embodied in trade for the year 2007 by final consumption (Unit: 10,000t)

	CEI-LB	CEI-UB	SH	EEDE	EEIE	EEDI	EEII-LB	EEII-UB	BEEDT	BEEIT-UB	BEEIT-LB
31	50.12	53.08	21.79	0.00	0.00	27.89	0.44	3.40	−27.89	−0.44	−3.40
32	1348.43	1484.65	528.17	0.00	0.00	800.13	20.13	156.35	−800.13	−20.13	−156.35
33	78.32	125.04	57.40	0.00	0.00	14.02	6.90	53.63	−14.02	−6.90	−53.63
34	245.38	451.40	157.28	0.00	0.00	57.66	30.44	236.46	−57.66	−30.44	−236.46
35	882.44	889.66	418.32	0.00	0.00	463.05	1.07	8.29	−463.05	−1.07	−8.29

Note Shanghai's final consumption can be divided into five categories, namely rural consumption (code 31), urban consumption (32), government consumption (33), fixed capital formation (34), and inventories (35). It should be noted that Shanghai's final consumption consumes goods produced by Shanghai's 30 sectors, which also generates emissions. Therefore, "SH" denotes this part of emissions in Table 3 and the upper bound of Shanghai's emissions embodied in international trade (EEI) are 30.04% (48.97 million tons) and 16.27% (26.52 million tons) of Shanghai's CO_2 emissions, respectively. Overall, these results firmly indicate that domestic and international trade accounts for a large share of overall CO_2 emissions associated with the city of Shanghai and emissions embodied in trade may thus have important implications for Shanghai's emissions mitigation policy. Moreover, it should be noted that the difference between the lower bound and the upper bound of the volume of Shanghai's CO_2 EEI is very large (22.45 million tons, or 13.77% of Shanghai's CO_2 emissions in 2007), which suggests that Shanghai may still have great potential to realize substantial additional energy savings and emission reductions.

Lastly, IO results reveal contrasted trends at the interregional and international scales. First, we find that, domestically, Shanghai is a net importer of emissions owing to its heavy reliance on energy-intensive imports from other provinces as the net balance of CO_2 emissions embodied in domestic trade (BEEDT) is negative and large for Shanghai (−38.94 million tons or 23.89% of Shanghai's emissions). Conversely, the balance of CO_2 emissions embodied in international trade trends toward positive values and is smaller (BEEIT-UB and BEEIT-LB amount to 22.28 and −0.17 million tons, or 13.67% and 0.10% of Shanghai's emissions, respectively) so that, from an international perspective, Shanghai is a net exporter of emissions stemming from its strong export-led manufacturing economy. It can be concluded not only that EET should be explicitly accounted for in sound climate policy design but also that emission assessment and management should be conducted at multiple spatial scales to unravel the complexity of trade relationships.

5.2 Emissions Embodied in Trade by Sector and Final Consumption

As shown in Table 2, embodied emissions differ widely across the 30 sectors. First, from the perspective of international exports, the highest EEIE is observed for Sector 14 (Smelting and Pressing of Ferrous and Nonferrous Metals) with 853.72E+04t (34.99% of the total EEIE) while the lowest EEIE is for Sector 20 (Instruments, Artwork and Office Machinery) (0.27E+04t). From the perspective of international imports, we find that the highest EEII-UB is again in Sector 14 (709.07E+04t), accounting for 28.86% of the total EEII-UB of Shanghai, while the highest EEII-LB is for Sector 12 (Chemical Products Related Industry) (34.76E+04t), accounting for 16.37% of the total EEII-LB.

By comparing the estimated bounds on emissions embodied in international imports, we find that EEII-LB is smaller than EEII-UB for most industry sectors; also the gap between the two is large, amounting to 1845.56E+04t (or 90%) in total. Looking at individual sectors, the biggest gap is for 14 (decreased by 33.82%), followed by Sectors 12 and 11 with a difference of 433.44E+04t (21.68%) and 396.43E+04t (19.83%), respectively. However, it is worth pointing out that by comparing the global average embodied CO_2 intensities calculated by Chen et al. (2013) with embodied CO_2 emission intensities of Japan and Shanghai used to estimate the bounds on emissions, in general, the global average embodied intensities are much closer to those of Shanghai and far larger than Japan's intensities. Therefore, while the gap defined by the bounds indicates that international trade has an important impact on Shanghai's sectoral emissions, it also provides important evidence that Shanghai has ample potential to further reduce its emissions embodied in international imports in most sectors.

Second, focusing on domestic exports, results show that the biggest contributor to EEDE is Sector 11 (Petroleum Processing and Coking, Gas Production and Supply) with 526.56E+04t (24.12% of the total EEDE), while the smallest contributor is Sector 22 (Electric Power/Steam and Hot Water Production and Supply) with a mere 32.2t annually. As for domestic imports, the five biggest contributors to EEDI (Sectors 14, 24, 16, 12, and 22) account for 41.62% of the total EEDI and 15.52% of Shanghai's CO_2 emissions, respectively. Taken together, these sectors represent a disproportionate share of imports in carbon-intensive consumption. Thus, and in line with what has already been pointed out, emission reduction strategies aimed at curtailing the impact of trade on emissions should target them in priority to maximize their effectiveness.

Third, from the perspective of BEEDT, only three out of the thirty industrial sectors have a net export of EED (positive BEEDT), while twenty-four sectors have a net import of EED in 2007 (non-positive BEEDT); the latter sectors avoid CO_2 emissions in Shanghai through the consumption of goods and services that are in fact produced in other regions. This extreme imbalance points to the massive production in carbon-intensive sectors that has in effect been shifted from Shanghai to other provinces. In fact, the higher the embodied emissions exported by sector values are, the more obvious the CO_2 emissions affected by a production-based approach are. Specifically, the CO_2 emissions volume of some sectors, most notably Sector 11, will be inflated in Shanghai under a production-based accounting approach, which may cause more emissions reduction commitments to be undertaken than would be warranted if consumption is taken into account. When a sector's share of the CO_2 emissions embodied in exports is high, not only competitiveness is affected but also its participation in binding emission reductions because deep emission cuts will ultimately affect its own export and even development. The largest CO_2 exporter is Sector 11 (Petroleum Processing and Coking, Gas Production and Supply) with emissions of 465.29E+04t, followed by Sectors 28 and 3. Conversely, for most sectors with a net EED input, such as Sectors 24 (Construction Industry) and 16 (Ordinary Machinery, Equipment for Special Purpose), where trade provides a mechanism to efficiently allocate resources in the process of industrial development, environmental impacts have also been shifted outside of Shanghai, and so has a part of emissions reduction responsibilities, according to the production-based view.

In addition, from the perspective of the balance of the emissions embodied in international trade (EEI), as shown in the last two columns of Table 2, the lower bound on emissions embodied in imports estimated on the basis of Japanese CO_2 emission intensities is remarkably lower than the upper bound for almost all sectors.

Sectors 14 and 12 stand out with a shortfall of 676.07E+04t and 433.45E+04t, respectively, which identifies them as the sectors most susceptible to further reduction of embodied emissions, as we already suggested on the basis of the analysis of EEDI estimates. Also, when these statistics are compared to the total emissions stemming from economic consumption within the city of Shanghai (third and fourth columns of Table 2), it transpires that, on a sectoral basis, the volume of CO_2 emissions under a consumption-based accounting approach is several orders of magnitude smaller than under a production-based approach, which underscores the great significance of international trade to calculate the CO_2 emissions embodied by the Shanghai economy.

Finally, Table 3 displays the CO_2 emissions embodied in final consumption for the Shanghai economy. Several interesting results emerge here. First, the contribution of domestic imports to final consumption emissions (1362.75E+04t overall) dwarfs that of international imports (overall EEII-UB of 458.13E+04t and overall EEII-LB of 58.98E+04t); this happens to be the most pronounced for urban consumption (58.71% of all domestic imports) and inventories (33.98%), which are also the largest contributors to final consumption emissions in Shanghai. Second, using the gap between the lower and upper bounds on the EEII value estimates as evidence of emission inefficiencies in Shanghai, we find that the total gap between EEII-LB and EEII-UB is large (399.15E+04t). Specifically, sectors with the largest gap are fixed capital formation and urban consumption (206.02E+04t and 136.22E+04t, respectively). These estimates suggest that there is considerable residual potential for Shanghai's final consumption to reduce its CO_2 emissions.

5.3 *Sources and Flows of Emissions Embodied in Domestic Trade*

In order to reduce the impact of emissions embodied in trade and implement a targeted reduction policy in relation to the urban economy of Shanghai, it is necessary to identify the geographic sources of these emissions on the basis of the MRIO numerical results. Due to space limitation, we will focus here on analyzing the sources and flows of the emissions embodied in domestic imports and exports. Clear regional trends exist in the distribution of embodied emissions.

The total emissions embodied in domestic imports of Shanghai are 6.08E+07t. Their geographic sources are shown on the map of Fig. 1A to be concentrated in the provinces of Northern and North Central China. These provinces are indeed China's primary sources of natural resources (especially coal and oil) and many energy-intensive industries are also located there, which naturally makes them the main contributors to embodied emissions exported to Shanghai. Specifically, the top five contributors are Hebei, Zhejiang, Hubei, Jiangsu, and Henan; calculations show that the embodied emissions imported by Shanghai from these five provinces collectively account for 45.74% of the total emissions embodied in Shanghai's domestic import

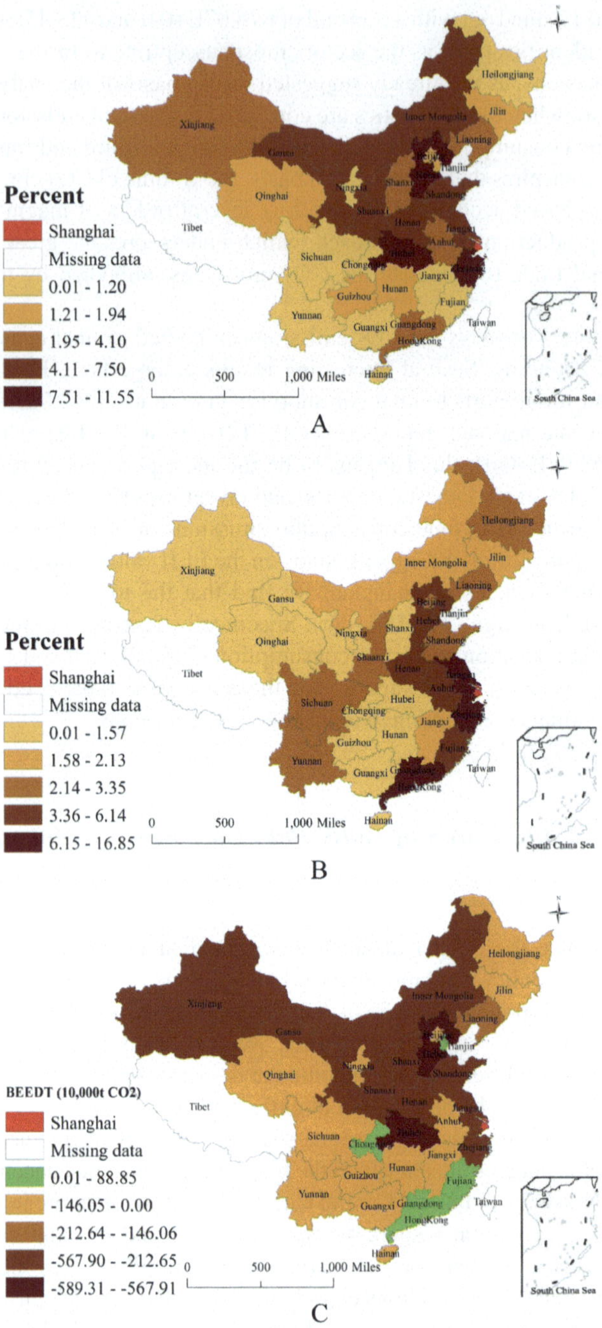
Percent
Shanghai
Missing data
0.01 - 1.20
1.21 - 1.94
1.95 - 4.10
4.11 - 7.50
7.51 - 11.55
0
500
1,000 Miles
Xinjiang
Tibet
Qinghai
Gansu
Inner Mongolia
Heilongjiang
Jilin
Liaoning
Tianjin
Shandong
Shanxi
Ningxia
Shaanxi
Henan
Anhui
Hubei
Sichuan
Hunan
Jiangxi
Guizhou
Yunnan
Guangxi
Guangdong
HongKong
Fujian
Taiwan
Hainan
South China Sea
A
Percent
Shanghai
Missing data
0.01 - 1.57
1.58 - 2.13
2.14 - 3.35
3.36 - 6.14
6.15 - 16.85
0
500
1,000 Miles
Xinjiang
Tibet
Qinghai
Gansu
Inner Mongolia
Heilongjiang
Jilin
Liaoning
Beijing
Hebei
Tianjin
Shandong
Shanxi
Ningxia
Shaanxi
Henan
Anhui
Hubei
Chongqing
Sichuan
Hunan
Jiangxi
Guizhou
Yunnan
Guangxi
Fujian
Taiwan
Hainan
South China Sea
B
BEEDT (10,000t CO2)
Shanghai
Missing data
0.01 - 88.85
-146.05 - 0.00
-212.64 - -146.06
-567.90 - -212.65
-589.31 - -567.91
0
500
1,000 Miles
Tibet
Qinghai
Gansu
Inner Mongolia
Heilongjiang
Jilin
Liaoning
Tianjin
Hebei
Shandong
Shanxi
Ningxia
Shaanxi
Henan
Anhui
Zhejiang
Sichuan
Hunan
Jiangxi
Guizhou
Yunnan
Guangxi
Guangdong
HongKong
Fujian
Taiwan
Hainan
South China Sea
C

◀**Fig. 1** Geographic sources of emissions embodied in domestic imports and exports of Shanghai. **A** Share of the volume of emissions embodied in imports from each Chinese province to the total volume of emissions embodied in Shanghai's imports from all provinces. **B** Share of the volume of emissions embodied in exports from Shanghai to each province to the total volume of emissions embodied in domestic exports from Shanghai. **C** Net balance of the CO_2 emissions embodied in domestic trade (BEEDT) in Shanghai. More detailed information can be found in Table 6 of Appendix

trade and 17.05% of Shanghai's emissions in 2007. Thus, while these provinces provide strong support for Shanghai's economic growth through thick trade relationships, under the production-based accounting principle, they have also been held accountable for a large volume of emissions embodied in products imported by Shanghai. From a consumption-based perspective, given the sharp regional concentration noted, improving energy efficiency in these regions would be very helpful to alleviate the impact of CO_2 emissions embodied in trade in the context of climate policy. Of course, beyond that, reducing the energy-intensity of Shanghai's consumption should also not be overlooked.

As far as Shanghai's domestic exports are concerned, total embodied emissions amount to 2.18E+07t. First and foremost, these embodied emissions are associated with the coastal provinces of Eastern China (Fig. 1B), which constitute the primary markets for Shanghai's products. Specifically, the top five receivers are Zhejiang, Guangdong, Jiangsu, Hebei, and Anhui, which make up 50.04% of the emissions embodied in Shanghai's domestic exports and 6.71% of its emissions. As a result of these strong trade relationships, under the consumption-based accounting principle, these provinces would be left to shoulder more responsibilities for CO_2 emission reduction than others.

Finally, Fig. 1C depicts the balance of domestic imports and exports. Tianjin, Chongqing, Guangdong, and Fujian are such that they are recipients of more emissions exported by Shanghai than Shanghai imports from them (shown in green on the map). For the other 25 provinces, the net balance is the reverse, more embodied emissions are exported to Shanghai than these provinces themselves import, which is in line with Shanghai being the premier economic and commercial center of the country, drawing on resources and on the manufacturing base of other provinces. On the whole, the majority of provinces would fare better if emissions were assessed on the basis of consumption than production. Clearly, many provinces of Northern China, such as Hebei, Gansu, Shanxi, and Inner Mongolia, have among the highest net input of embodied emissions from Shanghai. While the exports of these provinces tend to be the energy-intensive products with high emission factors, many of Shanghai's exports to these regions are deeply processed products and high-technology products with low emission. Via these trade relations, environmental impacts may be avoided in Shanghai and transferred to other provinces.

5.4 Implications for Regional Climate Policy

The identification of the regional sources of emissions embodied in the trade of an urban economy is valuable to policy makers to establish accountability. More specifically, it permits to impute responsibilities within the broader context of a comprehensive public policy on climate change that would entail the curtailment of city emissions in line with national goals, while respecting inter-regional equity considerations. However, two situations may arise from a more transparent accounting scheme rooted in consumption-based principles. First, trading regions may resort to trade protectionism that lessens the impacts of trade and protects their benefits to circumvent climate policies. The anticipated direct consequence of such action is usually the adoption of trade-based mechanisms such as border-tax adjustments by these regions (Ismer and Neuhoff 2004; Cendra 2006), which inevitably brings about trade barriers and even the risk of trade frictions in the process of regional economic development. Second, there are large social and economic costs associated with the participation in a regional climate regime (Peters and Hertwich 2008b). For regions with a large share of exports in carbon-intensive production, inter-regional trade can be viewed as an opportunity to shift production to non-participating regions to elude their responsibility to reduce emissions (Peters et al. 2007). This transfer of production may negatively impact the development of the local economy. Also, it will be much more difficult to achieve total emissions reduction target nationally. On the other hand, for regions with a large share of imports in carbon-intensive consumption, as is the case of the city of Shanghai, considering EET will compel them to face increased pressure to reduce environmental impacts, particularly emissions associated with their consumption, which may also hamper local economic development. Therefore, procedures to reduce the emission impact of trade on individual regions (particularly for Shanghai, whose economy is so intimately tied to trade) seem to be very critical at this juncture.

Regrettably, however, discussion of this issue has been inadequate so far in the regional science literature. It is our contention that effective regional solutions can be patterned after innovative frameworks advanced to reduce the impact of trade between individual nations. Victor et al. (2005) envisioned that nations intertwined by close trading relations could commit to cooperation and coordination treaties on particular emissions issues, which would make their economies more adaptive and resilient to a changing climate. Peters and Hertwich (2008a) studied groupings of countries such as the European Union and the Association of Southeast Asian Nations and concluded these nations could form coalitions to establish negotiations and implement cooperation, which would be beneficial to reduce the emission liability of individual countries due to their involvement in trade. Along the same line, according to the study conducted by Chen and Chen (2011), international cooperation of countries, such as the G7 and BRICS nations, will be conducive to drawing a globally acceptable agreement to tackle climate change.

Thus, we propose that the Yangtze River Delta (YRD) region centered on Shanghai and encompassing the four provinces of Jiangsu, Zhejiang, Anhui, and Jiangxi, in addition to Shanghai proper, would form the basis for such a coalition. The YRD exhibits strong economic complementarities and is marked by significant differences in regional economic development (NDRC 2010). More importantly, the emissions of each of the five constituting entities are significantly different and these differences are affected by trade (Guo et al. 2012a, b, c; Zhong et al. 2015). According to studies by NDRC (2010) and Li et al. (2013), promoting the free flow of services and products and strengthening regional cooperation across the YRD region in terms of emission reduction have been central elements of the climate policy of the local and central governments. However, considering the impact of EETs on each region's emissions reduction obligations within the scope of the national climate policy, a critical point is the equity of the distribution of benefits among coalition members and the scale of compensations to potential members that may be necessary to alleviate the costs and liabilities stemming from their participation. Accordingly, we continue with an analysis of the impacts of the formation of a regional coalition on Shanghai and on other prospective coalition members in terms of regional cooperation to reduce emissions.

Figure 2 provides a breakdown of the emissions embodied in imports and exports for each province before entering the coalition and for the coalition as a whole, depending on whether trade originated inside the coalition or out of it. It can be seen that Shanghai is more affected by trade individually than as a part of the coalition setup to reduce EETs on a regional basis. Specifically, in the former case, the percentages of emissions embodied in exports and emissions embodied in imports compared with total production-based emissions for Shanghai are 28.37% and 49.54%, respectively. However, when Shanghai participates in the coalition, these percentages drop greatly, reduced by 13.14% and 7.06%, respectively. The EET values are reduced if Shanghai participates in the coalition because only the EETs traded outside the coalition are still considered. Therefore, the coalition is instrumental at reducing the impacts of trade on Shanghai's assessable emissions. The EET values are generally lower for

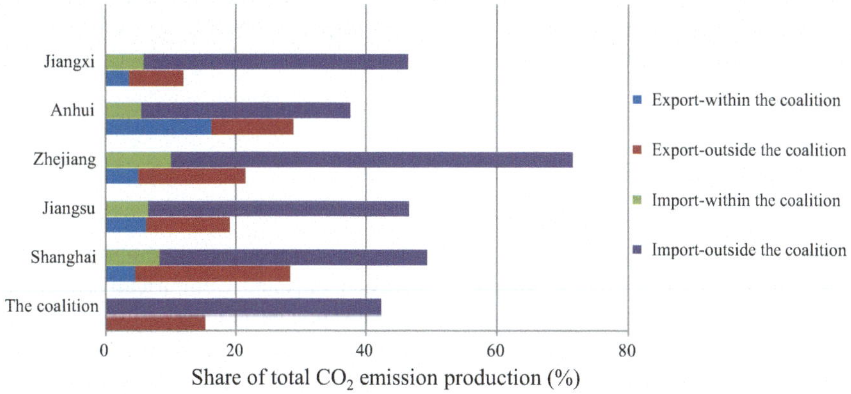

Fig. 2 Emissions embodied in exports and in imports of each province in the absence of coalition and of the coalition as a whole as a percentage of their total emission production

the coalition compared with individual members of the coalition, particularly for Shanghai. Peters and Hertwich (2008a) suggested that this situation may reduce competitiveness concerns and encourage participation.

From the perspective of each province, it is informative to look at the net balance of emissions embodied in domestic trade (BEEDT). Estimated values are reported in Table 4 on the basis of bilateral trade between pairs of provinces in the coalition. For each province, total BEEDT is also reported as the aggregation of its bilateral export and import values in tons of CO_2. The absence of a carbon trading scheme in China does not permit to directly assess the value of EETs. However, according to existing studies, CO_2 prices vary from \$10 per ton (e.g., United Nations 2010) to as much as \$1000 per ton (e.g., Anger et al. 2009). For simplicity and considering the actual situation of China, we follow Chen (2013) and assume a modest carbon price of \$12.17 per ton in this study. The total and per capita estimated market value of the net balance of emissions embodied in the trade among coalition partners are reported in Table 4. On the whole, our results indicate that trade for EETs is rather imbalanced among coalition partners. Specifically, Shanghai, Jiangxi, and Zhejiang are net EET emitters and would therefore pay their coalition partners for the right to export carbon outside of their borders under a carbon trading scheme. In sharp contrast, Jiangsu and Anhui are net receivers and would be the beneficiaries of cash payments to alleviate the costs and liabilities stemming from their participation in the coalition. Anhui is estimated to collect \$125.75 million annually or \$2.06 per capita under the assumed carbon trading scheme in the coalition. This situation results from the combination of high relative EET imports and low relative EET exports to other coalition partners. Zhejiang would spend the most overall (\$190.80 million), but Shanghai residents would shoulder the highest cost per capita (\$4.04). Given the noted imbalance in EETs among the five coalition partners, the formalization of a carbon trading scheme is envisioned as an important disposition to incentivize provinces to join the coalition. Short of this financial compensation, it would not be in the best interest of Anhui and Jiangsu to join the coalition. The compensation scheme would be instrumental in safeguarding trade between provinces and in fostering technological and industrial shifts in net emitting provinces (especially Shanghai and Zhejiang) to curtail the magnitude of their EET payouts.

6 Conclusions

The goal of this research has been to enhance our understanding of the embodiment of carbon emissions in the trade of a modern urban economy and to empirically establish the scope of differences between production-based and consumption-based emission accounting approaches. To this end, we used a robust regional economic assessment methodology and integrated it with a sectorally and regionally disaggregated analysis of emission sources. On this basis, we articulated a climate policy response that recognizes the regional dimension of EETs and preserves the economic benefits of

Table 4 Bilateral and total net balance of emissions embodied in domestic trade (BEEDT) among coalition members and total traded money for each coalition member

	Bilateral BEEDT (10,000t CO_2)					Total BEEDT (10,000t CO_2)	Total value (Unit: $10,000)	Population (10,000 persons)	Per capita ($)
	Shanghai	Jiangsu	Anhui	Jiangxi	Zhejiang				
Shanghai		−239.90	−102.60	−1.80	−271.10	−615.40	−7507.88	1858.00	−4.04
Jiangsu			−466.00	+167.50	+1021.00	+722.50	+8814.50	7624.50	+1.16
Anhui				+75.10	+489.60	+1030.70	+12,574.54	6118.00	+2.06
Jiangxi					+53.40	−189.20	−2308.24	4368.41	−0.53
Zhejiang						−1564.00	−19,080.80	5060.00	−3.77

Note In the rightmost columns, a "+" sign means a province is a net receiver of CO_2 and would receive a payment from other provinces as part of a carbon trading market. A "–" sign indicates the opposite situation inter-regional trade. The analysis was conducted in the Shanghai mega-region. Our main conclusions are as follows.

Carbon EETs were estimated with an MRIO modeling framework designed to overcome some of the most critical shortcomings of prior studies. In particular, the framework innovates by recognizing that trade flows and emission intensities embodied in domestic trade vary regionally and sectorally. As for international trade, in spite of severe data limitations on trade between Shanghai and foreign countries, we estimated lower and upper bounds on carbon emission embodied in Shanghai's foreign imports. This enhanced modeling framework has been instrumental in allowing us to extend the general discourse on EETs to underscore inter-regional and sectoral variations and articulate relevant regional climate policy issues.

Contrasting the PEI and CEI perspectives, we found that trade relations dictate that Shanghai's emission responsibility as a producer is dwarfed by its responsibility on the consumption side, even under rather conservative assumptions. The analysis of emissions embodied in trade also revealed that domestic and international trade accounts for a large share of overall carbon emissions associated with the city of Shanghai and EET is thus anticipated to have important implications for Shanghai's overall emissions mitigation policy. Domestically, Shanghai is a net importer of emissions owing to its heavy reliance on energy-intensive imports from other provinces, whereas its strong export-led manufacturing economy makes it is a net exporter of emissions at the international scale. Thus, sound climate policy design should rely on emission assessment and management at multiple spatial scales and policy concerns should more squarely focus on regional trade.

Next, the analysis showed that the relative contribution of industrial sectors to EETs is quite uneven. The biggest contributors represent a disproportionate share of imports in carbon-intensive consumption, and emission reduction strategies aimed at curtailing the impact of trade on emissions should target them in priority to maximize their effectiveness. We also found clear evidence of the massive shift in carbon-intensive production that has in effect been shifted from Shanghai to other provinces over the past two decades. The study of final consumption in the Shanghai economy pointed to its overwhelming contribution to domestic EETs; also it was found that there remains considerable residual potential for Shanghai's final consumption to reduce its carbon emissions, particularly in fixed capital formation and urban consumption.

The analysis of Shanghai's EETs has revealed clear geographic orientations in export and imports that can be traced back to the structure of trade flows in China. Shanghai's domestic imports of emissions are concentrated in the resource-rich provinces of Northern and North Central China. Under the production-based accounting principle, these provinces are held accountable for a large volume of emissions embodied in products imported by Shanghai. Conversely, emissions embodied in Shanghai's domestic exports are strongly associated with the coastal provinces of Eastern China. It is clear that, because of strong trade relationships among provinces, EETs play a strong mediating role in how the national climate policy may play out for various provinces. A production-based accounting approach is advantageous to some provinces and detrimental to others; the reverse is true of a consumption-based approach. While, depending on the prevailing emission accounting system, a region may be induced to shift carbon-intensive production to another region to circumvent production-based GHG control measures, economic development prospects may be affected as a result. We examined the properties of a regional strategy entailing the formation of an emission management and policy coalition in the Shanghai mega-region formed of the Yangtze River Delta. As part of this coalition, Shanghai would be less affected by trade than it would be as a stand-alone region. Some other coalition members would not experience similar benefits though as we noted significant imbalance in EETs among the five coalition partners. Accordingly, the formalization of a carbon trading scheme was identified as an approach to incentivize provinces to join the coalition. According to our estimates, Anhui and Jiangsu would receive large financial compensations while Shanghai, Zhejiang, and to a much lesser extent Jiangxi, would be net payees. We have argued that the compensation scheme would safeguard trade between provinces and foster technological and industrial shifts in net emitting provinces to curtail the magnitude of their EETs payouts.

We envision the present study to be extended along various promising lines of research in the future. First, replicating the study of EETs in other major urban regions, in China as well as in other nations, would allow us to establish the dependence of our conclusions on regional climate policy with respect to the dominant economic regime, the level of economic development of the local urban economy, and other macroeconomic and technology factors. Also, the analysis of regional coalitions was not disaggregated sectorally in the research reported here. Sectoral disaggregation would allow us to assess the magnitude of benefits in each industrial sector across regions, and the role played by the industrial structure and technology investment. The creation of sector-specific coalitions for the control and management of emissions is an intriguing policy question that remains to be broached.

Appendix

See Tables 5 and 6.

Table 5 Sector classification for Shanghai in the Chinese MRIO tables of 2007

Code	Sector
1	Farming, Forestry, Animal Husbandry, and Fishery
2	Coal Mining and Dressing
3	Petroleum and Natural Gas Extraction
4	Ferrous Metals Mining and Dressing
5	Nonferrous Metals Mining and Dressing
6	Food Processing, Food Production, Beverage Production, Tobacco Processing
7	Textile Industry
8	Garments and Other Fiber Products, Leather, Furs, Down and Related Products
9	Timber Processing, Bamboo, Cane, Palm and Straw Products, Furniture,
10	Papermaking and Paper Products, Printing and Record Medium Reproduction Cultural, Educational, and Sports Articles
11	Petroleum Processing and Coking, Gas Production and Supply
12	Chemical Products Related Industry
13	Nonmetal Mineral Products
14	Smelting and Pressing of Ferrous and Nonferrous Metals
15	Metal Products
16	Ordinary Machinery, Equipment for Special Purpose
17	Transportation Equipment
18	Electric Equipment and Machinery
19	Electronic and Telecommunications Equipment
20	Instruments, Artwork, and Office Machinery
21	Other Manufactured Products
22	Electric Power/Steam and Hot Water Production and Supply
23	Gas, Water Production and Supply
24	Construction
25	Transport and Storage Services
26	Wholesale, Retail Trade
27	Hotels, Catering Service
28	Leasing and Business Services
29	Research and Experimental Development
30	Other Service Industry

Table 6 The sources of the embodied CO_2 emissions in domestic imports and exports by each industry and final consumption category for the Shanghai economy in 2007

	I-SH	I-PI	I-SI	I-TI	I-31	I-32	I-33	I-34	I-35	SH-E	PI-E	SI-E	TI-E
Anhui	215.20	4.11	3.31	3.33	5.64	4.54	1.80	9.83	3.19	112.55	2.93	6.06	1.70
Beijing	95.50	1.04	1.10	3.31	2.32	1.74	31.77	8.76	0.96	73.26	12.20	2.31	7.34
Fujian	63.66	0.92	1.06	1.13	1.01	1.15	1.77	0.94	0.64	98.33	5.03	3.62	7.76
Gansu	320.43	1.86	6.15	2.99	0.64	1.30	0.79	0.34	9.25	7.75	0.91	0.30	0.57
Guangdong	173.56	2.36	2.59	4.73	1.43	2.33	1.82	8.01	2.76	262.40	8.06	13.68	5.68
Guangxi	49.36	0.68	0.78	0.53	1.71	1.06	1.50	1.04	1.01	32.59	0.97	1.27	2.34
Guizhou	111.67	0.80	2.02	0.92	1.55	1.06	4.13	2.06	2.93	24.41	0.67	0.87	2.06
Hainan	12.96	0.38	0.13	0.77	0.02	0.27	0.13	0.07	0.08	2.18	0.00	0.03	0.36
Hebei	702.00	7.23	12.74	5.67	10.97	7.27	11.05	7.75	18.21	134.11	0.84	6.39	5.20
Henan	359.34	3.32	6.87	2.49	3.53	3.19	1.24	2.56	8.32	100.55	6.24	4.91	3.44
Heilongjiang	69.14	2.25	0.61	2.42	3.94	2.87	3.40	1.96	0.40	66.09	2.56	3.35	1.80
Hubei	623.24	17.03	10.00	12.02	11.66	16.25	1.20	1.79	0.33	33.93	2.65	1.65	1.19
Hunan	117.87	1.48	1.93	1.17	3.30	2.17	3.85	2.59	2.58	34.30	1.13	1.77	0.82
Jilin	80.59	3.62	0.48	2.13	8.19	5.11	4.05	1.60	0.24	46.45	3.73	2.26	1.60
Jiangsu	455.89	8.64	7.58	7.61	7.36	9.22	2.70	6.27	3.96	216.00	10.26	10.36	8.13
Jiangxi	47.43	0.86	0.71	0.73	1.89	1.29	0.72	0.49	0.59	45.59	2.61	1.73	3.48
Liaoning	204.14	5.06	2.38	8.53	2.91	4.44	1.63	3.26	2.20	58.08	1.77	2.91	1.69
Neimeng	283.17	4.63	4.89	2.50	6.04	4.10	1.28	1.57	7.28	39.62	0.46	1.18	4.23
Ningxia	46.59	0.30	0.84	0.52	0.55	0.43	1.31	0.63	1.13	8.37	0.00	0.34	0.56
Qinghai	86.60	0.23	1.78	0.30	0.26	0.28	0.48	0.07	2.35	12.38	0.73	0.48	0.88

(continued)

Table 6 (continued)

	I-SH	I-PI	I-SI	I-TI	I-31	I-32	I-33	I-34	I-35	SH-E	PI-E	SI-E	TI-E
Shandong	304.97	6.07	5.52	3.31	3.50	3.60	0.54	1.95	6.26	57.89	16.38	2.67	2.39
Shanxi	249.13	3.42	4.32	4.90	0.32	1.73	0.67	0.67	5.93	24.55	1.39	0.89	2.01
Shananxi	290.97	2.58	5.12	4.33	1.14	1.80	8.26	9.76	7.39	53.32	0.00	2.45	2.43
Sichuan	70.52	0.77	1.21	0.40	2.03	1.19	1.17	1.72	1.74	56.58	1.31	3.19	0.32
Tianjin	83.03	0.80	1.27	1.75	0.83	0.91	3.93	5.44	1.91	109.56	5.64	1.18	19.68
Xinjiang	232.00	4.92	3.06	7.32	4.88	5.01	2.30	1.19	3.36	19.37	0.00	0.36	2.89
Yunnan	73.09	0.46	1.31	0.49	0.98	0.90	1.37	0.78	1.95	55.56	0.53	2.42	3.03
Zhejiang	639.33	14.05	10.12	13.17	11.03	14.49	1.79	12.47	2.97	368.16	9.79	20.06	4.64
Chongqing	15.49	0.16	0.15	0.55	0.37	0.27	3.32	4.45	0.10	30.69	1.20	1.30	1.80

Note "SH" denotes the abbreviated name of Shanghai. "PI, SI, and TI" denotes the abbreviated name of primary industry, secondary industry, and tertiary industry, respectively. "31–35" are Shanghai's final consumption categories. "I-SH" and "SH-E" denote the total volume of the emissions embodied in imports from other regions of China to Shanghai (Unit: 10,000t) and the total emissions embodied in exports from Shanghai to other regions of China (Unit: 10,000t). "I-PI, I-SI, I-TI, I-31, I-32, I-33, I-34, and I-35" denote the ratio between the volume of the emissions embodied in imports from each region in China to each industry and final consumption category in Shanghai and the total volume of the emissions embodied in imports from other regions of China to each industry and final consumption category in Shanghai, respectively (Unit: %). "PI-E, SI-E, and TI-E" denote the ratio between the emissions embodied in exports from each industry in Shanghai to each region in China between the total volume of the emissions embodied in exports from each industry in Shanghai to other regions of China (Unit: %)

References

Amin, A. (1999). An institutionalist perspective on regional economic development. *International Journal of Urban and Regional Research, 23*(2), 365–378.

Anger, A., Barker, T., Pollitt, H., Lindstad, H., Eyring, V., & Lee, D. S. (2009). *International shipping and market based instruments 2009*. London: International Maritime Organization.

Batabyal, A., & Nijkamp, P. (2003). The environment in regional science: An eclectic review. *Papers in Regional Science, 83*(1), 291–316.

Baynes, T., Lenzen, M., Steinberger, J. K., et al. (2011). Comparison of household consumption and regional production approaches to assess urban energy use and implications for policy. *Energy Policy, 39*(11), 7298–7309.

Bi, J., Zhang, R., Wang, H., Liu, M., & Wu, Y. (2011). The benchmarks of carbon emissions and policy implications for China's cities: Case of Nanjing. *Energy Policy, 39*(9), 4785–4794.

Cai, B., Guo, H., Cao, L., Guan, D., & Bai, H. (2018). Local strategies for China's carbon mitigation: An investigation of Chinese city-level CO_2 emissions. *Journal of Cleaner Production, 178*, 890–902.

CCSY. (2008). *China city statistical yearbook (2007)*. Beijing: China Statistical Publishing House.

Cendra, J. D. (2006). Can emissions trading schemes be coupled with border tax adjustments? An analysis vis-à-vis WTO Law. *Review of European Community and International Environmental Law, 15*(2), 131–145.

CESY. (2008). *China energy statistical yearbook (2007)*. Beijing: China Statistical Publishing House.

Chen, Z. J. (2013). *Convergence of regional carbon dioxide emissions and carbon economic policy in China performance in dynamic stochastic general equilibrium simulation*. Shanghai: East China Normal University.

Chen, Z. M., & Chen, G. Q. (2011). Embodied carbon dioxide emission at supra-national scale: A coalition analysis for G7, BRIC, and the rest of the world. *Energy Policy, 39*(5), 2899–2909.

Chen, S., & Zhu, F. (2019). Unveiling key drivers of urban embodied and controlled carbon footprints. *Applied Energy, 235*, 835–845.

Chen, G. Q., Guo, S., Shao, L., Li, J. S., & Chen, Z. M. (2013). Three-scale input–output modeling for urban economy: Carbon emission by Beijing 2007. *Communications in Nonlinear Science and Numerical Simulation, 18*(9), 2493–2506.

Chen, B., Li, J. S., Zhou, S. L., Yang, Q., & Chen, G. Q. (2018). GHG emissions embodied in Macao's internal energy consumption and external trade: Driving forces via decomposition analysis. *Renewable and Sustainable Energy Reviews, 82*, 4100–4106.

Chen, B., Wang, X. B., Li, Y. L., et al. (2019). Energy-induced mercury emissions in global supply chain networks: Structural characteristics and policy implications. *Science of the Total Environment, 670*, 87–97.

Choi, T. (2015). Understanding environmental responsibility of cites and emissions embodied in trade. *Economic Systems Research, 27*, 133–153.

Czarnitzki, D., & Hottenrott, H. (2009). Are local milieus the key to innovation performance? *Journal of Regional Science, 49*(1), 81–112.

Davis, S. J., & Caldeira, K. (2010). Consumption-based accounting of CO_2 emissions. *Proceedings of the National Academy of Sciences, 107*(12), 5687–5692.

Dhakal, S. (2004). *Urban energy use and greenhouse gas emissions in Asian mega cities-policies for a sustainable future*. Hayama, Japan: Institute for Global Environmental Strategies. Downloadable from http://enviroscope.iges.or.jp/modules/envirolib/view.php?docid=181.

Dhakal, S. (2009). Urban energy use and carbon emissions from cities in China and policy implications. *Energy Policy, 37*(11), 4208–4219.

Dhakal, S. (2010). GHG emissions from urbanization and opportunities for urban carbon mitigation. *Current Opinion in Environmental Sustainability, 2*(4), 277–283.

Edenhofer, O. R., Pichs-Madruga, Y., Sokona, E., Farahani, S., Kadner, K., Seyboth, A., et al. (Eds.). (2014). *Climate change 2014: Mitigation of climate change: Contribution of working group III to*

the fifth assessment report of the intergovernmental panel on climate change. London: Cambridge University Press.
Feng, K., Davis, S. J., Sun, L., Li, X., Guan, D., Liu, W., et al. (2013). Outsourcing CO_2 within China. *Proceedings of the National Academy of Sciences, 110*(28), 11654–11659.
Feng, K., Hubacek, K., Sun, L., & Liu, Z. (2014). Consumption-based CO_2 accounting of China's megacities: The case of Beijing, Tianjin, Shanghai and Chongqing. *Ecological Indicators, 47,* 26–31.
Guo, J., Zhang, Z., & Meng, L. (2012a). China's provincial CO_2 emissions embodied in international and interprovincial trade. *Energy Policy, 42*(3), 486–497.
Guo, S., Liu, J. B., Shao, L., Li, J. S., & An, Y. R. (2012b). Energy-dominated local carbon emissions in Beijing 2007: Inventory and input-output analysis. *The Scientific World Journal.*
Guo, S., Shao, L., Chen, H., Li, Z., Liu, J. B., Xu, F. X., et al. (2012c). Inventory and input–output analysis of CO_2 emissions by fossil fuel consumption in Beijing 2007. *Ecological Informatics, 12,* 93–100.
Hermannsson, K., & McIntyre, S. G. (2014). Local consumption and territorial based accounting for CO_2 emissions. *Ecological Economics, 104,* 1–11.
Hillman, T., & Ramaswami, A. (2010). Greenhouse gas emission footprints and energy use benchmarks for eight US cities. *Environmental Science and Technology, 44*(6), 1902–1910.
Hung, C. C., Hsu, S. C., & Cheng, K. L. (2019). Quantifying city-scale carbon emissions of the construction sector based on multi-regional input-output analysis. *Resources, Conservation and Recycling, 149,* 75–85.
IPCC. (2006). *The 2006 IPCC guidelines for national greenhouse gas inventories (2006 guidelines).* Also available at: http://www.ipcc-nggip.iges.or.jp/public/2006gl/index.html.
Ismer, R., & Neuhoff, K. (2004). *Border tax adjustments: A feasible way to address nonparticipation in emission trading* (Working Papers in Economics CWPE 0409). Cambridge, UK: Department of Applied Economics, University of Cambridge.
Ji, X., Chen, Z., & Li, J. (2014). Embodied energy consumption and carbon emissions evaluation for urban industrial structure optimization. *Frontiers of Earth Science, 8*(1), 32–43.
Kennedy, C., Steinberger, J., Gasson, B., Hansen, Y., Hillman, T., Havranek, M., et al. (2009). Greenhouse gas emissions from global cities. *Environmental Science and Technology, 43*(19), 7297–7302.
Larsen, H. N., & Hertwich, E. G. (2009). The case for consumption-based accounting of greenhouse gas emissions to promote local climate action. *Environmental Science & Policy, 12*(7), 791–798.
Lenzen, M., Pade, L., & Munksgaard, J. (2004). CO_2 multipliers in multi-region input–output models. *Economic Systems Research, 16*(4), 391–412.
Li, P. X., & Cao, Y. H. (2013). Spatial and temporal changes of industrial carbon emissions under regional industrial transfer: The case of Pan-Yangtze River Delta. *Advances in Earth Science, 28*(8), 939–947.
Li, J. S., Chen, G. Q., Lai, T. M., Ahmad, B., Chen, Z. M., Shao, L., et al. (2013). Embodied greenhouse gas emission by Macao. *Energy Policy, 59*(8), 819–833.
Li, J. S., Zhou, H. W., Meng, J., Yang, Q., Chen, B., & Zhang, Y. Y. (2018). Carbon emissions and their drivers for a typical urban economy from multiple perspectives: A case analysis for Beijing city. *Applied Energy, 226,* 1076–1086.
Liu, H. J., & Zhao, H. (2012). Empirical analysis of the regional differences of China's carbon dioxide emissions intensity. *Statistical Research, 6,* 46–50.
Liu, W. D., Chen, J., Tang, Z. P., & Yang, B. (2012). *Theory and practice for building multi-regional input-output table for 30 provinces in China in 2007.* Beijing: China Statistics Press.
Liu, Z., Feng, K., Hubacek, K., Liang, S., Anadon, L. D., Zhang, C., et al. (2015). Four system boundaries for carbon accounts. *Ecological Modelling.* https://doi.org/10.1016/j.ecolmodel.2015.02.001.
Liu, Y., Wang, Y., Mi, Z., & Ma, Z. (2018). Carbon implications of China's changing economic structure at the city level. *Structural Change and Economic Dynamics, 46,* 163–171.

Machado, G. V. (2000). Energy use, CO_2 emissions and foreign trade: An IO approach applied to the Brazilian case. In *Thirteenth International Conference on Input–Output Techniques*, Macerata, Italy, August 21–25, 2000.
Mi, Z., Zhang, Y., Guan, D., Shan, Y., Liu, Z., Cong, R., et al. (2016). Consumption-based emission accounting for Chinese cities. *Applied Energy, 184,* 1073–1081.
Mi, Z., Zheng, J., Meng, J., Zheng, H., Li, X., Coffman, D. M., et al. (2019a). Carbon emissions of cities from a consumption-based perspective. *Applied Energy, 235,* 509–518.
Mi, Z., Guan, D., Liu, Z., Liu, J., Viguié, V., Fromer, N., et al. (2019b). Cities: The core of climate change mitigation. *Journal of Cleaner Production, 207,* 582–589.
Munksgaard, J., & Pedersen, K. A. (2001). CO_2 accounts for open economies: Producer or consumer responsibility. *Energy Policy, 29*(4), 327–334.
NDRC (National Development and Reform Commission). (2010). *Regional planning for the Yangtze River Delta.* Also available at: http://www.gov.cn/gzdt/att/att/site1/20100622/001e3741a2cc0d8aa56801.pdf.
Peters, G. P. (2008). From production-based to consumption-based national emission inventories. *Ecological Economics, 65*(1), 13–23.
Peters, G. P., & Hertwich, E. G. (2008a). CO_2 embodied in international trade with implications for global climate policy. *Environmental Science and Technology, 42*(5), 1401–1407.
Peters, G. P., & Hertwich, E. G. (2008b). Post-Kyoto greenhouse gas inventories: Production versus consumption. *Climatic Change, 86*(1–2), 51–66.
Peters, G. P., Weber, C. L., Guan, D., & Hubacek, K. (2007). China's growing CO_2 emissions—A race between increasing consumption and efficiency gains. *Environmental Science and Technology, 41*(17), 5939–5944.
Qi, Y., Li, H. M., & Xu, M. (2008). Accounting embodied carbon in import and export in China. *China Population, Resources and Environment, 18,* 8–13.
Rosado, L., & Ferrão, P. (2009). Measuring the embodied energy in household goods: Application to the Lisbon City. In M. Havranek (Ed.), *ConAccount 2008: Urban metabolism: Measuring the ecological city* (pp. 159–181). Prague: Charles University Environment Center.
Satterthwaite, D. (2008). Cities' contribution to global warming: Notes on the allocation of greenhouse gas emissions. *Environment and Urbanization, 20*(2), 539–549.
Schulz, N. B. (2010). Delving into the carbon footprints of Singapore—Comparing direct and indirect greenhouse gas emissions of a small and open economic system. *Energy Policy, 38*(9), 1848–1855.
Scott, K., & Barrett, J. (2015). An integration of net imported emissions into climate change targets. *Environmental Science & Policy, 52*(10), 150–157.
SSY. (2008). *Shanghai statistical yearbook (2007).* Beijing: China Statistical Publishing House.
Su, B., & Ang, B. W. (2010). Input-output analysis of CO_2 emissions embodied in trade: The effects of spatial aggregation. *Ecological Economics, 70*(1), 10–18.
Su, B., & Ang, B. W. (2013). Input-output analysis of CO_2 emissions embodied in trade: Competitive versus non-competitive imports. *Energy Policy, 56,* 83–87.
Su, B., & Ang, B. W. (2014). Input-output analysis of CO_2 emissions embodied in trade: A multi-region model for China. *Applied Energy, 114,* 377–384.
United Nations. (2007). *City planning will determine pace of global warming.* Accessed December 2, 2008 at http://www.un.org/News/Press/docs/2007/gaef3190.doc.htm.
United Nations. (2010). *Report of the secretary-general's high-level advisory group on climate change financing.* New York: United Nations.
United Nations. (2014). *World urbanization prospects: The 2014 revision, highlights* (ST/ESA/SER.A/352). Department of Economic and Social Affairs, Population Division.
Vause, J., Gao, L., Shi, L., & Zhao, J. (2013). Production and consumption accounting of CO_2 emissions for Xiamen, China. *Energy Policy, 60*(9), 697–704.
Victor, D. G., House, J. C., & Joy, S. (2005). A Madisonian approach to climate policy. *Science, 309,* 1820–1821.

Wang, Z., Liu, X., Huang, R., Liu, H., & Zhai, S. (2013). Estimation of carbon emission requirements under balance growth in the Chinese provinces. *Bulletin of Chinese Academy of Sciences, 28*(1), 85–93.

Wang, H., Wang, Y., Wang, H., Liu, M., Zhang, Y., Zhang, R., et al. (2014). Mitigating greenhouse gas emissions from China's cities: Case study of Suzhou. *Energy Policy, 68*(5), 482–489.

Weber, C. L., & Matthews, H. S. (2007). Embodied environmental emissions in US international trade, 1997–2004. *Environmental Science and Technology, 41*(14), 4875–4881.

Weber, C. L., Peters, G. P., Guan, D., & Hubacek, K. (2008). The contribution of Chinese exports to climate change. *Energy Policy, 36*(9), 3572–3577.

Wiedmann, T. (2009). A review of recent multi-region input–output models used for consumption-based emission and resource accounting. *Ecological Economics, 69*(2), 211–222.

Wiedmann, T., Lenzen, M., Turner, K., & Barrett, J. (2007). Examining the global environmental impact of regional consumption activities—Part 2: Review of input–output models for the assessment of environmental impacts embodied in trade. *Ecological Economics, 61*(1), 15–26.

WRI/WBCSD. (2009). *The greenhouse gas protocol: A corporate accounting and reporting standard* (Revised ed.). http://www.ghgprotocol.org/.

Wu, W. (2011). *Carbon footprint: A case study on the municipality of Haninge* (Master of Science thesis). Technology and Health. http://kth.diva-portal.org/smash/get/diva2:442651/FULLTEXT01.pdf.

Xi, F., Geng, Y., Chen, X., Zhang, Y., Wang, X., Xue, B., et al. (2011). Contributing to local policy making on GHG emission reduction through inventorying and attribution: A case study of Shenyang, China. *Energy Policy, 39*(10), 5999–6010.

Zhang, Z. X., & Folmer, H. (1998). Economic modelling approaches to cost estimates for the control of carbon dioxide emissions. *Energy Economics, 20*(1), 101–120.

Zhang, L., Hu, Q., & Zhang, F. (2014a). Input-output modeling for urban energy consumption in Beijing: Dynamics and comparison. *PLoS ONE, 9*(3), e89850. https://doi.org/10.1371/journal.pone.0089850.

Zhang, Z., Guo, J., & Hewings, G. J. D. (2014b). The effects of direct trade within China on regional and national CO_2 emissions. *Energy Economics, 46,* 161–175.

Zhong, Z., Huang, R., Tang, Q., Cong, X., & Wang, Z. (2015). China's provincial CO_2 emissions embodied in trade with implications for regional climate policy. *Frontiers of Earth Science, 9*(1), 77–90.

Zhong, Z., He, L., & Wang, Z. (2017). Geographic sources and the structural decomposition of emissions embodied in trade by Chinese megacities: The case of Beijing, Tianjin, Shanghai, and Chongqing. *Journal of Cleaner Production, 158,* 59–72.

Zhou, S. Y., Chen, H., & Li, S. C. (2010). Resources use and greenhouse gas emissions in urban economy: ecological input–output modeling for Beijing 2002. *Communications in Nonlinear Science and Numerical Simulation, 15*(10), 3201–3231.

Analysis of Spatial Effects of Environmental Taxes on Water Pollution in China's Taihu Basin

Yoshiro Higano, Zhihong Shen, and Takeshi Mizunoya

1 Introduction

Lake Taihu (Tai Lake) is the third largest freshwater lake in China, with a catchment area of 36,500 km^2 and a water surface area of 2338 km^2. It provides drinking water for approximately 3 million people in the Taihu Basin located at the junction of Wuxi, Changzhou, Suzhou Cities, in Jiangsu Province and Huzhou city, in Zhejiang Province. Taihu Basin is one of the more developed areas of China, thus playing a significant role in the development of the national economy. Since the 1980s, the lake has experienced dramatic water quality degradation associated with rapid land development and the fast development of the economy. After 2003, water quality in most of Lake Taihu was below class IV of Environmental Quality Standards for Surface Water (GB 3838-2002). Although every year a large amount of the Yangtze River water is channeled into the lake to dilute the pollution, the water pollution indicators have remained high (Wang et al. 2006). On May 30, 2008, a blue–green algae bloom occurred which caused a sudden degradation of quality in Wuxi's potable water sources. This resulted in a water shortage crisis in Wuxi. Based on previous management experiences (The State Council 1998, 2002; Shen and Higano 2007), the main reasons were as follows. First, industrial development and low-efficiency environmental management caused heavy pollution in Taihu Basin. Second, the population in urban areas had increased nearly 50% from 1990 to 2006, resulting in large amounts of municipal wastewater discharge. Third, the control of non-point pollution was very difficult and required massive public funding. Fourth, the governance of water pollution lacked funds and the comprehensive management means required for coping with this enormous task.

Previous studies of basin management and environmental policies can be considered in order to solve the compounded complications stemming from such water

Y. Higano (✉) · Z. Shen · T. Mizunoya
Graduate School of Life and Environmental Sciences, University of Tsukuba, 1-1-1 Tennodai, Tsukuba 305-8572, Japan
e-mail: higano@jsrsai.jp

J. Thill (ed.), *Innovations in Urban and Regional Systems*,
https://doi.org/10.1007/978-3-030-43694-0_19

quality events. A conceptual and operational mathematical model that can be applied to a basin was proposed by Higano et al. (2002), Hirose and Higano (2000) and Shen et al. (2014). The modeling system advanced in these studies was designed to understand and clarify interactions between social–economic activities and the ecological environment, to assess the impacts and effectiveness of possible policy interventions and engineering measures for both socioeconomic development and preservation of the ecological environment and to propose an integrated optimal pollution control scheme to improve water quality of the lake and basin. A synthetic environmental policy to reduce water pollutants and greenhouse gases by means of the effective utilization of biomass resources from livestock production was analyzed by Mizunoya et al. (2007). As a new policy tool and management method, environmental taxes and integrated basin management practices have aroused the interest of relevant government agencies, scholars and experts. Historically, China has lacked environmental tax and basin management mechanisms to frame regional environmental responses. The economic effectiveness and water pollution reduction outcome of policies entailing the adoption of environmental taxes and various water purification technologies to reduce water pollutants were analyzed by Shen and Higano (2010). However, deficiencies have persisted in the spatial analysis of the issues and of proposed remedial interventions. In this study, we focus on the construction of an integrated basin management system, on the design of an environmental tax rate and of an optimized allocation of countermeasure fees to each geographic zone to determine a way to abate water pollution discharges, and achieve environmentally sustainable economic development in the Taihu Basin of China.

The rest of this chapter is organized as follows. In the next section, the overall modeling framework is introduced. In the third section, the conceptual framework of the modeling effort is explained. In the fourth section, the feasibility of an environmental tax is discussed, with a focus on the attainable level of the water emission reduction and on the extent of possible negative impacts on the economy of the basin. Finally, we summarize the important findings in this study.

2 Modeling Framework

Mass production and mass consumption are features of modern life. However, accelerated one-way consumption of natural resources is inconsistent with sustainable development in the sense that it results in emission of numerous wastes and wastewater that severely damage the ecological environment; furthermore, our society must face the eventual depletion of resources. A society that embraces recycling and environmental symbiosis needs to evolve to solve conflicts between life needs, environmental protection and energy depletion. In this chapter, we analyze the relationships of interdependence and interaction between socioeconomic activities and the ecological environment based on the extended input–output (I/O) model (Leontief 1941) and system dynamics (Forrester 1958). We build a modern integrated evaluation system (Fig. 1) which is useful to evaluate policies set to solve complex socioeconomic

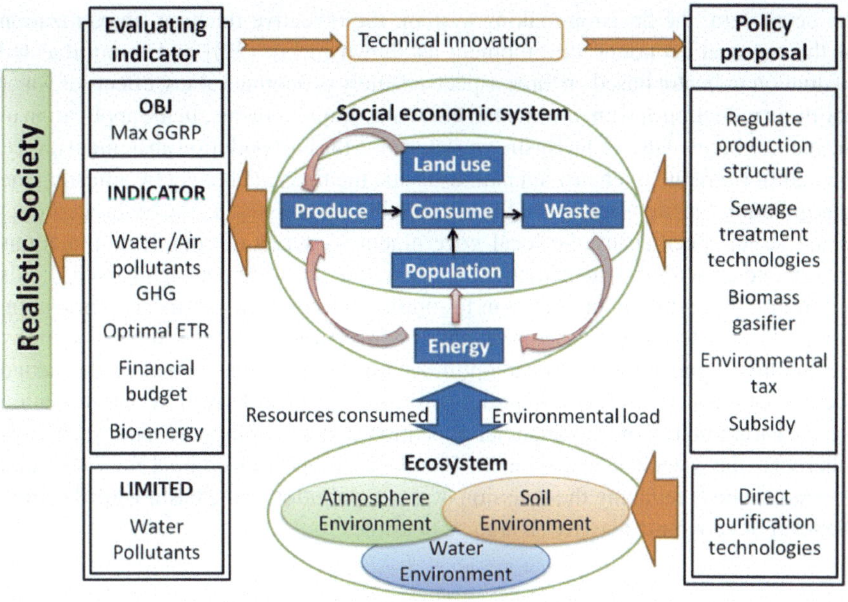

Fig. 1 Integrated evaluation system

and environmental conflicts, and we apply it to a case of basin management. We will analyze the effectiveness of policies and technologies currently available in the Taihu Basin.

In this system, we focus on how to effectively control the existing and projected water pollution problems of point sources (caused by households and industries) and non-point sources (caused by land use and others) by the analysis of different forecasted trajectories of total nitrogen (T-N), total phosphorus (T-P), and chemical oxygen demand (COD) (Table 1). It operates as a system of decisions pertaining to the selection of the optimum policy scheme, namely how much environmental tax should be charged on what objects, and according to whether the three policy options of biomass gasification, direct water treatment and land reform are adopted or not. The system is sensitive to whether there are any other factors that influence the implementation and success of adopted policies and technologies.

In order to ensure the validity and reliability of comprehensive policies across regions, the proposed approach consists in an integrated policy evaluation system. In this framework, the following elements are integrated: the socioeconomic system,

Table 1 Classification of water pollutants

Index	1	2	3
Water pollutant (*n*)	COD (chemical oxygen demand)	T-N (total nitrogen)	T-P (total phosphorus)

the ecosystem, the decision-making system, the objective function [maximization of the regional economic development as shown in Eq. (30)] and a multifaceted evaluation indicator based on three aspects, namely economic values, effects of water pollution reduction and environmental efficiency improvements. In the application to the management of the Taihu Basin, we determine the best pollution abatement target, the equilibrium rate of environmental taxes and the financial budget by adjusting the manipulation functions to control the water pollution emission in the basin according to the target specified by the local government. A series of complex simulations are conducted with the mathematical optimization software package LINGO. This programming environment enables us to simulate the complex nonlinear optimization model constructed of the environmental system and the socioeconomic system by the iterative application of linear optimization methods that represent the actual socioeconomic activities. Useful policy information can be derived by the simulation for decision making on the financial allocation and technology adoption, which is inherent to the watershed management. Finally, we propose an optimal environmental tax rate aimed at reducing the emission of water pollution in the basin while keeping the level of economic activity as high as possible.

3 Simulation Model

The dynamic simulation model consists of more than eighty mathematical equations so that only the most important ones are included in Appendix of the chapter. The structure and relationship of the various modules of the model are represented in Fig. 2.

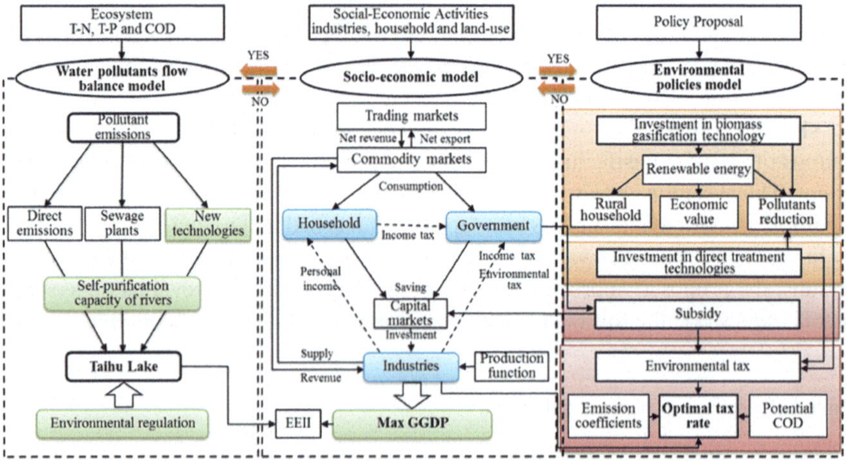

Fig. 2 Structure of the simulation models

3.1 Overview of the Modeling System

The simulation model consists of an ecosystem model, a social–economic model, an environmental policy model and an objective function. The ecosystem model is concerned with the flow balance of water pollutants; it describes how much water pollution is generated in the municipalities of each sub-basin and how the water pollutants are transported into Lake Taihu (Fig. 2). The socioeconomic model used in the simulation is based on an I/O model. The social–economic system consists of a commodity flow balance module, a value flow balance module, a capital module and a production function, which, respectively, represent the dynamic processes of production and consumption.

In this study, we assume a representative producer in each industrial sector, which behaves as if profit is to be maximized. The maximized profit (surplus profit) is zero as a usual precondition for input–output analysis with a Harrod–Domar–Leontief production function of constant return to scale and inputs, which are linearly related to output. The value added of each sector is allocated to the household sector after an environmental tax is charged according to pollution emission, and revenue tax is charged with a constant tax rate. Furthermore, we assume a representative consumer in the household sector, which has a constant propensity for saving, and her utility function is of the Cobb–Douglas type with homogeneity of the first degree. She behaves to maximize her utility, which means that the ratio of the amount of disposable revenue (it is defined as revenue after deduction of revenue tax, environmental tax if any and saving) spent for a commodity to the disposable revenue becomes a constant.

Production factors of labor and land are assumed to be abundant in the market, which is a realistic assumption in a large developing country such as China. Hence, production of each sector can be decreased or increased without considering equilibrium in the labor and land markets explicitly. The factors that confine production are capital stock and capacity of the environment. Capital stock can be accumulated via investment over time and, provided that the economy meets these constraints, it can meet the final demand of consumption, investment, etc., and can grow. The environmental policy model describes how to raise funds (such as environmental tax revenue and revenue of investment in biomass energy) and when and where to allocate funds to environmental investments and subsidies [adjusting capital stocks, production subsidies] to reduce water pollution in Lake Taihu. As a tool of economic policy, environmental tax can influence the production activities that harm the environment by levying taxes. Biomass gasification technology and water pollution purification technology are parts of the production technologies available to each region. The ripple effect on the economy caused by the new technologies may be greater than that of normal production technologies. Thus, the efficiency of production can be improved by these alternative technologies. The subsidy for adjustment of industrial structure contributes to improve the effective utilization of capital stock. Thus, the environmental tax will be used as subsidy or investment based on the environmental goal.

3.2 Model Specifications

In this study, the key factors consist of three environmental indicators as shown in Table 1, namely T-N, T-P and COD. The study area is divided into five zones (river sub-basins, Fig. 3) encompassing a total of 22 geographic areas defined on the basis of two factors (administrative entities and the hydrographic characteristic in the Taihu Basin) as shown in Table 2. The socioeconomic activities considered as sources of pollution in this model are household and industrial activities on the one hand and land uses as shown in Tables 3, 4 and 5. It is postulated that the emitted pollutants pass through the river system and eventually flow into Lake Taihu. In order to fully analyze the environmental value added and water pollution reduction ascribed to environmental policies, the simulations are organized in 33 elemental cases (Table 6) that can be grouped into three macro-cases (Cases A, B and C) characterized as follows:

- Case A: current technologies;
- Case B: adoption of biomass utilization technology, advanced wastewater treatment and subsidy for capital stock control;
- Case C: adoption of environmental tax policy integration.

Records for 2006 are used as the initial data, and the simulation running period is ten years, from 2007 to 2016.

Fig. 3 Study area

Table 2 List of targeted rivers and catchment areas

Index	Zone (i)	Index	Catchment areas (j)
1	Zhihugang	1–6	Wuxi City, and others
		7	Jiangyin City
2	Wujingang	8–12	Changzhou City, and others
		13	Wujin City
3	Nanxi	14	Yixing City
		15	Liyang City
		16	Jintan City
4	East Taihu area	17	East Taihu area
5	Tiaoxi	18–19	Huzhou City, and others
		20	Dexing County
		21	Anji County
		22	Changxing County

Table 3 Classification of land uses

Index	Land use (l, k)
1	Paddy field upland cropping
2	Mountain forest
3	City area
4	Unused land
5	Other land uses

4 Simulation Results and Discussion

4.1 Assessment of the Overall Objective

In order to assess the benefits of taxing environmental pollution, we consider three alternative sets of scenarios, each one with their own set of assumptions: no adoption of environmental tax and no new technology in Case A, adoption of some new environmental technology in Case B and adoption of an environmental tax and some new technology in Case C. Under these three scenario options, we comprehensively analyze the possibility of 0–15% T-N reduction with respect to the 2006 level. The objective according to which policy options are evaluated consists in the Green Gross Regional Product (GGRP), which deducts environmental degradation costs from the conventional Gross Regional Product (GRP).

The simulation results show an infeasible solution for Case A; also maximum pollution reduction rates for Cases B and C under sustainable economic growth in

Table 4 Classification of industries and their emission coefficients (kg/10,000 yuan)

Index	Production systems (m)	COD	T-N	T-P
1	Agriculture	0.8055	0.2165	0.0041
2	Forestry	0.0000	0.0000	0.0000
3	Livestock	4.2135	2.0085	0.0530
4	Fisheries	38.1420	15.2220	3.4762
5	Mining	0.3993	0.0283	0.0003
6	Food processing, food and beverage manufacturing and others	1.7649	0.2440	0.0029
7	Textile industry	0.8248	0.0781	0.0009
8	Textile products manufacturing	0.4001	0.0698	0.0008
9	Timber processing, furniture manufacturing and others	0.3097	0.0403	0.0005
10	Papermaking and paper products manufacturing	13.4286	0.5941	0.0071
11	Printing and record pressing manufacturing and others	19.1595	0.8665	0.0103
12	Petroleum, coking and nuclear fuel processing	0.2431	0.0810	0.0010
13	Raw chemical materials and chemical products manufacturing	1.2179	0.8039	0.0095
14	Medical and pharmaceutical products manufacturing	1.0923	0.1320	0.0016
15	Nonmetal products manufacturing	0.3068	0.0207	0.0002
16	Smelting and pressing of metals, metal products manufacturing	0.2246	0.0533	0.0006
17	Equipment and machinery manufacturing	0.0510	0.0058	0.0001
18	Water, gas, electricity production and supply and others	0.3029	0.0436	0.0005
19	Other manufacturing industry	4.3034	0.3914	0.0046
20	Other industries	0.0000	0.0000	0.0000

Source Shen and Higano (2010)

Table 5 Household wastewater disposal systems and direct water treatment technologies

Index	Household (*h*)	Direct treatment technology (tec)
1	Sewerage system	Chemical coagulation technology
2	Untreated wastewater	Purification technology of vegetation
3	Advanced processing	Biochemistry treatment technology
4	Combined treatment septic tank A	
5	Combined treatment septic tank B	

Table 6 Countermeasures against environmental degradation

Policy content	Cases A1–11	Cases B1–11	Cases C1–11
Environmental tax			×
Subsidy/investment	S1	S1–3, 6	S1–6
Reduction of the capital employed		×	×
Direct treatment technologies (DTTs, Table 7)		×	×
Biomass gasification technology (BGT, Table 8)		×	×
Reduction rate of water pollutants (%)	1–11	1–11	1–11

Notes Mark × means adoption. A, B, C of the simulation cases point to different types of policy plan as shown in Table 6
The simulation case identifier, such as Case B7 and Case A11, denotes the reduction rate of water pollutants
S1–S6 signify the subsidies or investments granted to reduce water pollutants as shown in Table 6 and Eq. (9)
The subsidy is divided into six types, as follows
S1 is the summation of the subsidy and investment for reducing water pollution emission from households
S2 is the overall subsidy in the capital control for reducing water pollution emission by high-emission industry
S3 is the overall investment in the direct treatment technology for reducing water pollution in Lake Taihu
S4 plays the same role as S2, but it is from environmental tax
S5 is the overall incentive fee for encouraging the low emission industry
S6 is the overall investment in biomass gasification technology for the effective utilization of biomass resources

Table 7 Technological parameters of direct treatment technologies (DTT)

Items	Vegetation	Biochemistry	Chemical coagulation	Unit
Construction cost	0.40	1.33	12.40	million yuan
Maintenance cost	0.08	0.07	0.32	million yuan/year
T-N reduction	1.23	1.37	3.37	kg/million yuan investment
T-P reduction	0.06	0	0.15	kg/million yuan investment
COD reduction	4.22	1.37	40.92	kg/million yuan investment
Scope of application	Lakeside and riverfront	Treat lakeside, riverfront and household wastewater	Sewage treatment plant	

Source Shen et al. (2014)

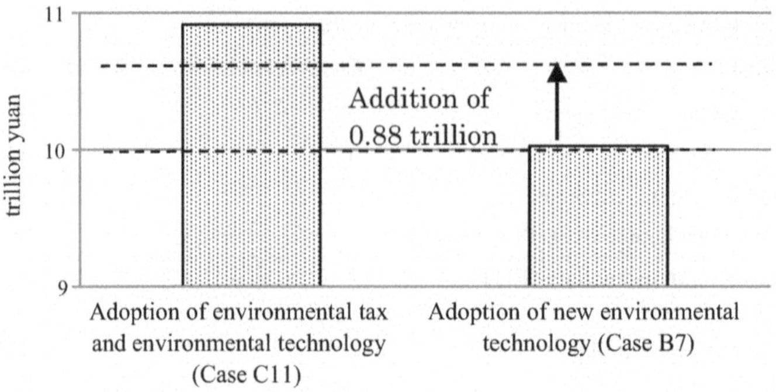

Fig. 4 Cumulative total GGRP (2007–2016)

the basin are in the range of 7 to 11% (recorded as Case B7 and Case C11, respectively) since the cumulative GGRP over ten years is 10.03 trillion and 10.91 trillion yuan, respectively (Fig. 4). Compared with a scenario that only introduces a new technology (Case B7), the introduction of an environmental tax (Case C11) not only further reduces T-N by 4% but also improves EEII by three points, increasing to 64%, while increasing the accumulated GGRP by 0.88 trillion yuan. This means that the new purification technologies that are part of the production technologies have strong ripple effects on the economy compared to general technologies. The efficiency of production can be further improved by adopting biomass gasification technology. However, the amount of biomass available for gasification in the basin is limited, and its contribution is smaller than that of the direct treatment technology. The environmental tax motivates enterprises to pay more attention to the environment and greatly promotes improvements in the efficiency of the water environment through the rational allocation of environmental tax revenues. Therefore, although the introduction of an environmental tax inhibits the development of large sewage enterprises, it promotes regional economic growth from an overall point of view. Thus, we believe that Case C11 is the best tactic of all the cases tested. This case is discussed in more detail in the rest of this chapter.

4.2 Revenue of Environmental Tax in Each Zone

According to the results pertaining to Case C11, the equilibrium rate of the environmental tax is 6.83 yuan/kg COD. As shown in Fig. 5, the environmental tax revenue is the highest in East Taihu (Zone 4) and accounts for 44–47% of the total revenue of the watershed's environmental taxes. This is because the production base of the East Taihu area is agriculture, animal husbandry and fisheries; small-scale enterprises are relatively concentrated in this zone, and hence, much larger volumes of sewage are released than in any other zones. Zhihugang (Zone 1) has the highest densities of population and industrial activities. The environmental tax revenue in this zone

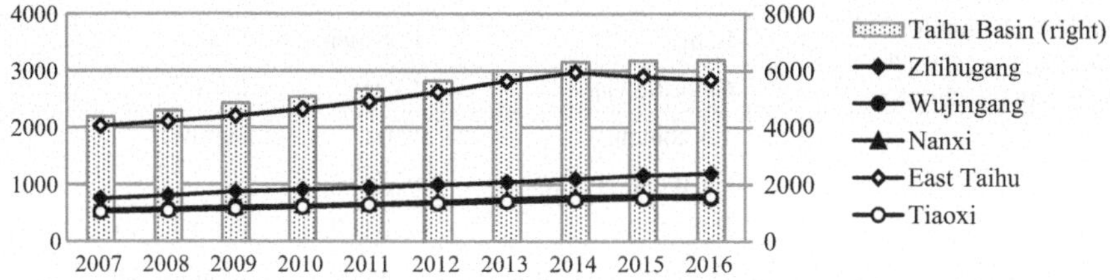

Fig. 5 Forecasted revenue from the environmental tax in each zone (million yuan) under Case C11. *Note* Revenue calculation with an optimal environmental tax rate of 6.83 yuan/kg COD

accounts for 17–18% of the watershed's total revenues. Compared to Zhihugang (Zone 1), Wujingang (Zone 2), Nanxi (Zone 3) and Tiaoxi (Zone 5) experience relatively slower paces of industrial development. However, this pace has picked up in recent years, coupled with higher concentrations of agriculture, animal husbandry and fisheries so that the environmental tax revenue of these four zones now reaches 11–12% of the watershed's total revenue.

4.3 Overall Balance of Revenue and Expenditure of the Environmental Tax Against Water Pollution

The modeling results show that expenditures on adopted countermeasures, which are to be balanced by revenues, mainly come from the environmental tax, the external subsidy from the Taihu Basin Authority, and the revenue from the regional production of biomass energy. These revenue sources amount about to 54.92, 30.87 and 3.91 billion yuan, accounting for 61.2%, 34.4% and 4.4% of total countermeasure fees revenue, respectively. The results on the expenditures show that biomass gasification technology (BGT, Table 8) is adopted in the early stages of the simulation horizon, which implies that the economic efficiency of the technology itself is higher than the other measures (Fig. 6). However, the adoption would cease in 2009, and the cumulative investment in BGT would amount to 2.56 billion yuan, accounting for 2.9% of the total revenue. The adoption reaches saturation because useable biomass resources that have not yet been tapped are limited. This implies that the adoption and introduction of BGT to utilize biomass resources of crop residues in the Taihu Basin are significantly efficient in terms of impacts on both the economy and the environment as well as the conventional biomass utilization of methane fermentation technologies currently available. Therefore, it is expected that if the utilization of livestock and forestry wastes is introduced, it will further improve the environment and promote sustainable development in the basin. From 2007 to 2016, the total amount of investment in the installation of direct treatment technologies (DTTs) and subsidies for capital to reduce water pollution emission are 35.93 billion yuan and 51.20 billion yuan, respectively. The sum of the two expenditures is more than

Table 8 Technological parameters of biomass gasification technology

Items	Parameters	Unit
Calorific value	5500	KJ/m^3
Tar content	<50	mg/m^3
Gasification efficiency	>75	%
H_2S (O_2)	<20 (<0.7)	mg/m^3 (%)
Construction costs	775,600	yuan/machine
Depreciation costs (25 years)	31,024	yuan/a machine
Operating and maintenance costs	99,265	yuan/a machine
Biomass resources consumption	500	ton/a machine
Combustible gas production capacity	2008	m^3gas/ton machine
T-P reduction	0.1048	kg/ton biomass
T-N reduction	0.52	kg/ton biomass

Source Shen and Higano (2010)
Note Biomass gasification technology is a manufacturing process for renewable energy, which converts the biomass resource into a combustible gas such as H_2, CO or CH_4. The biomass gasification system includes one unit of gasification equipment (JQ-D500), 600 M^3 storage trunk, mower and fire protection equipment, which is made in China. The supply radius of the gas transport network should be less than 1.5 km to save money. The technical parameters of the technology are as shown in Table 8

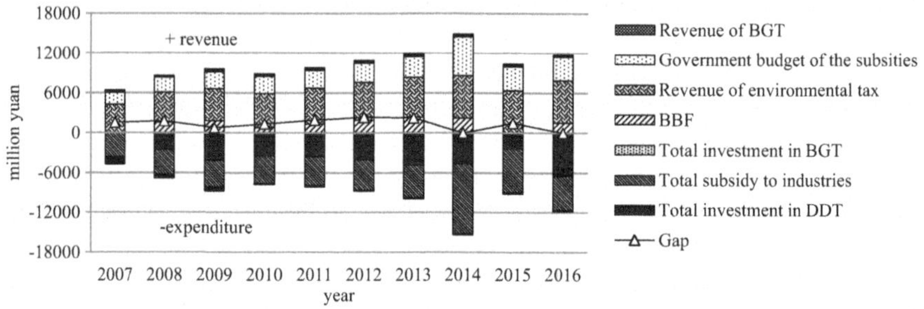

Fig. 6 Revenue and expenditure of countermeasure fees under Case C11 (million yuan)

97% of the total revenue. Accordingly, the allocation of tax revenue to enhance the adoption of countermeasures induces a decrease in emission of both non-point and point source water pollution.

Balance brought forward (BBF) is a limited surplus from environmental tax revenues realized in the previous year that is carried over and recorded at the beginning of the next year (Fig. 6). Results show that the gap between revenue and expenditure of the countermeasure fees appears to decrease only at the beginning of 2009, 2013, 2014 and 2016 and becomes zero at the end of 2016. These changes in fact describe an expenditures process of BBF in environmental investment. These indicate that BBF is useful to contribute to the longitudinal stability of environmental tax revenues and to avoid the uncertainties of environmental tax policy to meet complex and ever changing economic and environmental developments in future years. Therefore, the environmental tax policy becomes acceptable to stakeholders of the basin's economy. The integrated environmental tax policy can more effectively solve the shortages in public funds setup to support countermeasures and contribute to the smooth implementation of environmental tax policies over the long term.

4.4 Gap in the Countermeasure Fees for Each Zone

In the present study, the integrated management model of the watershed is selected to enhance the efficiency of watershed environmental improvement. We find that the zones of East Taihu, Tiaoxi and Zhihuxiang have a large gap between revenue and expenditure in countermeasure fees as shown in Fig. 7. A gap with a negative value means that the countermeasure fees in the zone are less than required to cover the expenses; the greater absolute value of the gap, the greater money shortage in water pollution countermeasure fees. Therefore, the zones of Zhihugang (Zone 1), Tiaoxi (Zone 5) and Nanxi (Zone 3) are important areas for the effective reduction in water pollution emission. Total aid to Zhihugang, Tiaoxi and Nanxi is 1847, 1541, 1488 million yuan, respectively. Conversely, a gap with a positive value means that the zone has sufficient revenue from countermeasure fees to reach their targets for water pollution reduction. The greater absolute value of the gap, the greater accumulation in water pollution countermeasure fees, such as in the East Taihu areas (18,721 million yuan in Zone 4). Thus, the complementary behaviors exhibited by among zones will improve the efficiency of the distribution of environmental tax across the Taihu Basin.

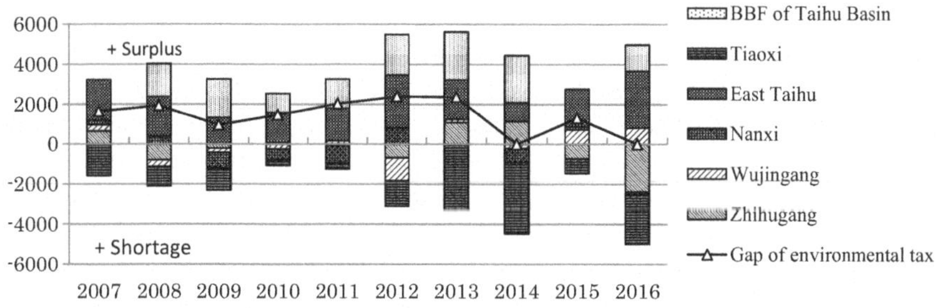

Fig. 7 Regional gap of environmental tax for water pollution (million yuan)

4.5 Countermeasure Effects on Industry and Land Reform

When compared to 2006 values, the 2016 predicted volumes of industrial pollution discharge from Zones 1 to 5 show large increases in T-N of 62.0%, 40.5%, 59.1%, 59.2% and 63.1%, and COD of 87.6%, 74.0%, 105.8%, 122.1% and 184.1%, respectively; only small increases in T-P discharges of 25.2%, 7.1%, 12.1%, 19.7% and −1.2% are forecast. This is because COD and T-N caused by manufacturing industries have more impacts on water quality than T-P. The payment of environmental tax depends on water pollution indicators. Therefore, the environmental tax has a small influence on industries that have low T-P and T-N discharges. In addition, the results show that the environmental tax can significantly inhibit water pollution discharges in livestock, fisheries, food processing, print manufacturing and other industries. However, due to population growth and demand expansion, the water pollution discharges of industries continue to increase overall.

The simulation results further show that returning farmland to a natural habitat such as forest and grassland can effectively reduce the losses by 9111 tons of T-N, 406 tons of T-P and 1925 tons of COD. No significant reduction occurred in water pollution, especially in Zones 1 and 3, which have cumulative reductions of 80 tons and 327 tons of T-P, respectively. Conversely, the study indicates an increase in T-P discharges in Zone 5, instead of a decrease. The reason for these results is that production development and population growth increase the demand for land, and residential use and production occupy most unused land or part of the forested land. All these factors weaken the effect of water pollution reduction by farmland shrinkage.

4.6 Cost-Benefit Analysis of Biomass Gasification Technology (BGT)

According to simulations, the annual accumulated gain of revenue in the Taihu Basin was 425 million yuan, with Nanxi (162 million yuan) and Tiaoxi (151 million yuan) recording the largest gains. This can be imputed to the low biomass fees and low maintenance fees even when there is a high cost of investments in equipment. Therefore, if biomass gasification technology is given priority for introduction in each zone, cumulative reductions in the basin will be 6116 tons of T-N and 1232 tons of T-P. However, due to the uneven distribution of unused biomass in farmlands of each zone, the cumulative reduction amount of T-N would show great variability. From Figs. 8 and 9, Tiaoxi (Zone 5) and Nanxi (Zone 3) are seen to have larger reductions than other zones, with 2400 tons and 2108 tons of T-N, and 484 tons and 425 tons of T-P, respectively, for the ten-year cumulative amounts. For these two zones, equipment investments are relatively higher, with 911 million yuan and 978 million yuan (Fig. 10), respectively.

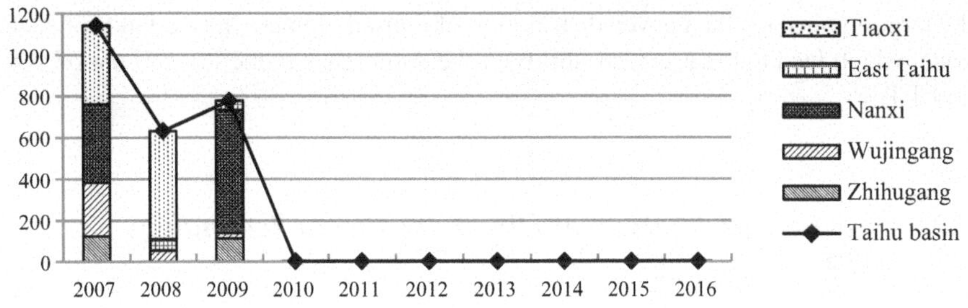

Fig. 8 Aggregate investments for BGT (million yuan)

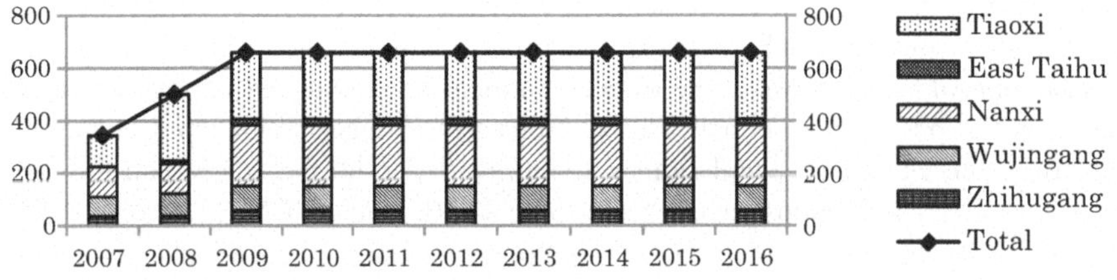

Fig. 9 Reduction in effectiveness of T-N by BGT (ton)

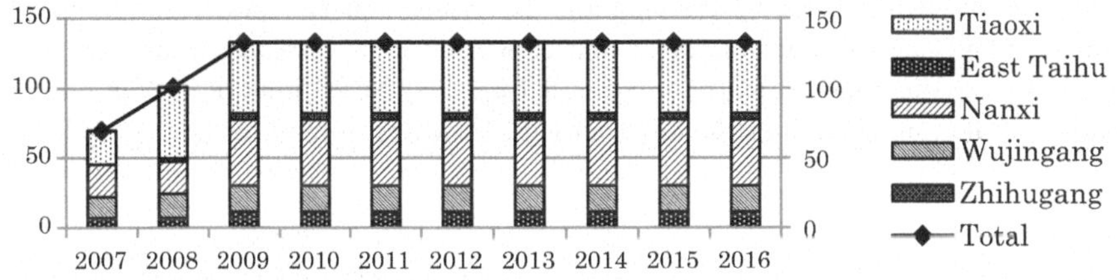

Fig. 10 Reduction in effectiveness of T-P by BGT (ton)

According to simulations, the investments in biomass gasification technology are given priority for introduction in the zones of Tiaoxi, Nanxi, Wujingang and Zhihuxiang in the first three years (Fig. 8). For these two zones, equipment investments are relatively higher, with 911 million yuan and 978 million yuan, respectively. The total of investments in Tiaoxi (Zone 5) and Nanxi (Zone 3) are about accounting to 74% of total expenditures in biomass gasification technology. The cumulative reductions in the basin will be 6116 tons of T-N and 1232 tons of T-P. The zones of Tiaoxi and Nanxi are seen to have larger reductions than other zones, with 2400 tons and 2108 tons of T-N, and 484 tons and 425 tons of T-P, respectively, for the ten-year cumulative amounts (Figs. 9 and 10). These are mainly due to (1) the low biomass fees and low maintenance fees even when there is a high cost of investments in equipment, (2) in addition, the annual accumulated gain of revenue in the Taihu Basin was 425 million yuan, with Nanxi (162 million yuan) and Tiaoxi (151 million yuan) recording the largest gains, (3) the uneven distribution of unused biomass in farmlands of each zone, which lead to the great variability in the cumulative reduction amount of T-N and T-P.

4.7 Cost-Benefit Analysis of Direct Treatment Technology (DTT)

In this study, we consider three direct treatment technologies, namely chemical coagulation, biochemistry, and vegetation. The simulations identify that the reduction in water pollutants based on DTTs tends to increase when the water environment is improved, as illustrated in Figs. 11, 12 and 13. Given to the low quality of the current purification of urban sewage, the direct treatment technology of chemical coagulation is the best way to enhance urban sewage treatment. With rapid economic development, chemical coagulation is given priority in all the zones, and the ten-year cumulative reductions are expected to be 80,500 tons of T-N, 3520 tons of T-P and 977,260 tons of COD in the basin. The total investment in chemical coagulation amount accounts for approximately 40% of the total countermeasure fees; Zhihugang (Zone 1) needs about 9179 million yuan, as the highest proportion and investments, while other zones fall between 4900 and 6100 million yuan (Fig. 14).

Compared with chemical coagulation technology, purification technology of aquatic plants has a better effect on the ecosystem of rivers and lakes (ten-year cumulative reductions estimated to be 914 tons of T-N, 45 tons of T-P and 3133 tons of COD) and beautifies the environment. However, it requires a long cycle, is

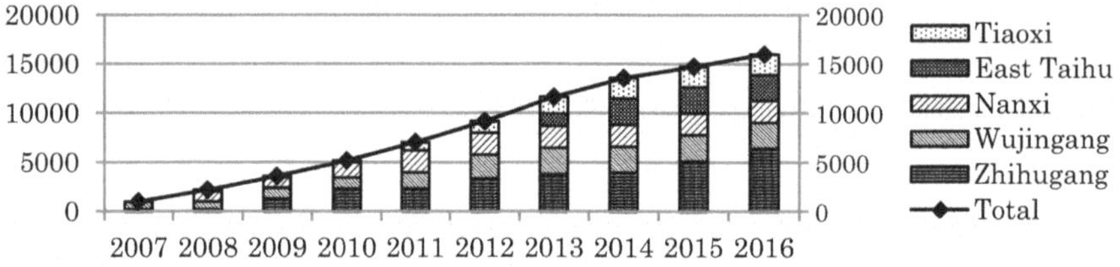

Fig. 11 Reduction in total effectiveness of T-N by DTTs (ton)

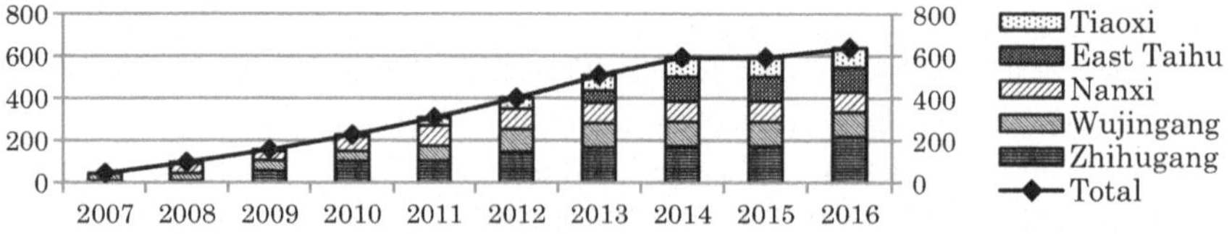

Fig. 12 Reduction in total effectiveness of T-P by DTTs (ton)

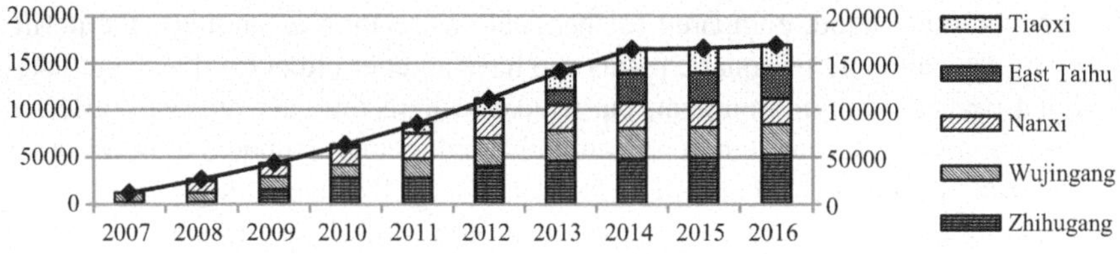

Fig. 13 Reduction in total effectiveness of COD by DTTs (ton)

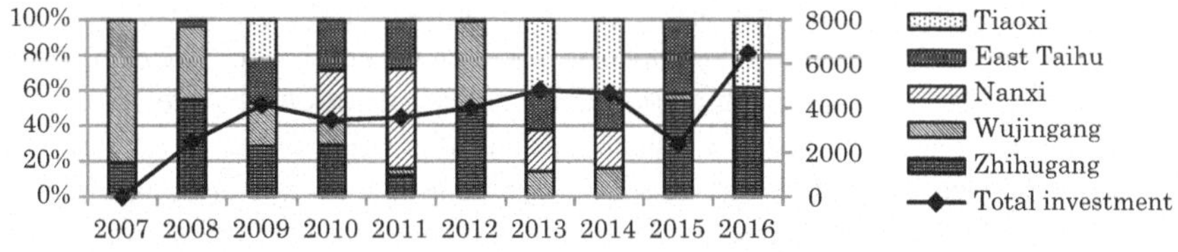

Fig. 14 Aggregate investments in DTTs (million yuan)

easily affected by natural conditions and uses 5% of the total environmental countermeasure fees (Zhihugang, 2000 million yuan and Tiaoxi, 2400 million yuan) as technological investments. Therefore, this technology will play an important role in future sewage treatment of the Taihu Basin. Finally, even if the combined biological–chemical treatment technology, which complements the other two direct treatment technologies, only affects T-N and COD, it is useful to reduce T-N and is low cost.

5 Conclusions

Based on the concept of integrated watershed management, we created a comprehensive evaluation system that recognized the existence of multiple geographic zones with their own economic and environmental characteristics. From the point of view of economic development and environmental protection, we set up an equilibrium environmental tax, optimized its allocation across zones of the study region and simulated the effects on reducing water pollution. Our main conclusions are as follows.

(1) Unused agricultural biological resources such as straw can not only indirectly reduce the discharge of 6116 tons of T-N and 1232 tons of T-P but also provide rural residents with 3.29 billion m^3 of gas per year. Therefore, the energy shortage in rural areas and air pollution caused by coal burning can be alleviated.
(2) Based on the current state of low purification of urban sewage, the direct treatment technology of chemical coagulation is the best way to enhance urban sewage treatment. With this technology, the ten-year cumulative reductions are estimated to be 80,500 tons of T-N, 3520 tons of T-P and 977,260 tons of COD.

On the other hand, compared to chemical coagulation technology, the purification technology of aquatic plants can have a better effect on rivers and lakes and beautify the environment, but it takes a long time and is easily affected by natural conditions (ten-year cumulative reductions estimated to be 914 tons of T-N, 45 tons of T-P and 3133 tons of COD). This technology will play an important role in future sewage treatment of the Taihu Lake Basin.

(3) The introduction of an environment tax system can solve the shortage of countermeasure fees used to reduce water pollution emissions in the Taihu Lake Basin. A reasonable environmental tax rate is important in the environmental tax system. Given the goal of sustainable economic growth in the basin, we found an environmental tax rate of 6.83 yuan/ton based on the comprehensive evaluation system. The cumulative countermeasure fees for disposal of water pollution increase about 55 billion yuan thanks to the imposition of the environmental tax. Through integrated watershed management, resources can be optimized, and the problem of the uneven distribution of resources in the region can also be resolved. In addition, this method can encourage enterprises to reduce sewage discharge. Compared with the non-introduction of an environmental tax, the introduction of an environmental tax can lead to exceeding the reduction targets by 4%. At the same time, it can result in an increase of 0.88 trillion yuan in revenue.

In summary, this study contributes to enhancing modeling in regional science by providing a framework of integrated evaluation system (IES, Fig. 1, the detailed modules refer to Fig. 2) which is useful to describe and estimate the spatial impacts on the ecological environment of dynamical socioeconomic activities, and the effects of adopted pollution abatement technologies and economic policies set to solve complex socioeconomic and environmental conflicts. The abundant results of the case study clearly show that the IES constitutes a very effective tool to promote a harmonious development between society, environment and economy. The framework and its implementation across a challenged basin of coastal China imply the importance of spatial analysis to realize more precise policy formulation, industry investment management and management of public finance. Moreover, the IES can also be widely applied in energy, transport, agriculture and other industries across multiple scales.

Appendix: Details of the Modeling System

The main characteristics of the models are presented in this appendix. For further information on the simulation models, please refer to Shen (2010).

1. Flow of the Market

The first equation presents the flow balance in the commodity flow balance market. In Eq. (1), the left-hand side describes supply and the right-hand side describes demand.

The total amount of production in each industry should be greater than or equal to the sum of intermediate demand and final demand, based on basic principles of the market economy. Here, consumption and investment are estimated by the sum of all private and government economic activities derived from the I–O structure expressed in the following equation:

$$X_{(t)} \geq A_1 X_{(t)} + C_{(t)} + Gc_{(t)} + I_{(t)} + Ex_{(t)} + IM_{(t)} \tag{1}$$

where $X_{(t)}$ denotes the endogenous column vectors for products of each industry (*En*); A_1 denotes the exogenous matrices of input coefficients of goods to each industry (*Ex*); $C_{(t)}$ denotes the endogenous column vectors for private consumption of goods (*En*); $Gc_{(t)}$ is the endogenous column vectors of governmental consumption of goods (*En*); $I_{(t)}$ is the endogenous column vectors for fixed capital formation by industry of each sector (*En*); $EX_{(t)}$ is the endogenous column vectors for the export of goods (*En*); and $IM_{(t)}$ is the endogenous column vectors for good imports (*En*).

2. Value Balances of Industries

The left-hand side of Eq. (2) shows the revenue and the right-hand side shows the cost. Price rates *P* of all industries are set to 1 at the benchmark year 2006:

$$P_{(t)}\tilde{X}_{(t)} = P_{(t)}A_1\tilde{X}_{(t)} + \delta\tilde{X}_{(t)} + Yh_{(t)} + \varphi\tilde{X}_{(t)} + \tau\tilde{X}_{(t)} + Te_{(t)} \tag{2}$$

where $P_{(t)}$ stands for the endogenous row vectors for prices of goods (*En*); $\tilde{X}_{(t)}$ is the diagonalized matrices from $X_{(t)}$ (*En*); δ is the exogenous row vectors for depreciation of each industrial sector (*Ex*); $Yh_{(t)}$ is the endogenous row vectors for national revenue of each industrial sector (*En*); φ is the exogenous row vectors for profit of each sector (*Ex*); τ is the exogenous row vectors for total tax of each industrial sector (*Ex*); $Te_{(t)}$ is the exogenous row vectors for the environmental tax of each industrial sector (*En*).

3. The Private Sector Consumption

The share of cost for goods or services to disposable revenue is constant. A fraction of the disposable revenue is saved, but all the rest is disbursed for consumption.

$$\ell\tilde{C}_{(t)} + S_{c(t)} + TBn_{(t)} = Yh_{(t)} \tag{3}$$

$$P_{(t)}\tilde{C}_{(t)} \leq \mathrm{ccr}\,(Yh_{(t)} + \ell\varphi\,\tilde{X}_{(t)}) \tag{4}$$

$$S_{c(t)} \leq \mathrm{scr}\,(Yh_{(t)} + \ell\varphi\,\tilde{X}_{(t)}) \tag{5}$$

$$\ell\mathrm{ccr} + \mathrm{scr} = 1 \tag{6}$$

where $\tilde{C}_{(t)}$ is the diagonalized matrices from $C_{(t)}$ (*En*); $S_{c(t)}$ is the endogenous row vectors for private saving (*En*); $TBn_{(t)}$ is the endogenous row vectors for the net investment in local bonds (*En*); ccr is the exogenous row vectors of cost share for goods in the private consumption sector (*Ex*); scr is the exogenous value for the ratio of savings to revenue (*Ex*); and ℓ is a row vector for summation.

4. The Balance of Government Revenues and Expenditures

$$T_{gi(t)} = T_{go(t)} \tag{7}$$

$$T_{gi(t)} = T_{(t)} + Te_{(t)} + X_{(t)}^{bio} + \mathrm{TS}_{(t)} + \mathrm{BBF}_{(t-1)} \tag{8}$$

$$T_{go(t)} = S_{g(t)} + S_{(t)}^{1} + S_{(t)}^{2} + S_{(t)}^{3} + S_{(t)}^{4} + S_{(t)}^{5} + S_{(t)}^{6} + \mathrm{BBF}_{(t)} \tag{9}$$

where $T_{gi(t)}$ is fiscal revenue (*En*); $T_{go(t)}$ is the fiscal expenditures (*En*); $T_{(t)}$ is the summation of government revenue from normal taxation, excluding the environmental tax revenue (*En*); $T_{e(t)}$ is the summation of environmental tax revenues $Te_{ij(t)}^{m}$, which is calculated based on the water pollution released by industry m in municipality j of zone i at time t (*En*); $X_{(t)}^{\mathrm{bio}}$ is the summation of $X_{ij(t)}^{\mathrm{bio}}$, which is the revenue from biomass gasification technology in municipality j of zone i at time t (*En*);$\mathrm{TS}_{(t)}$ is the summation of $\mathrm{TS}_{ij(t)}$, which is a subsidy for direct treatment technology in municipality j of zone i at time t (*En*); $\mathrm{BBF}_{(t)}$ is the balance brought forward (BBF) of countermeasure fees in the Taihu Basin; $S_{(t)}^{1}$ is the summation of $S_{ij(t)}^{1}$, which is the subsidy and investment for reducing water pollution emission from households in municipality j of zone i at time t (*En*); $S_{(t)}^{2}$ is the summation of $S_{ij(t)}^{2m}$, which is a subsidy in the capital control for reducing water pollution emission by high-emission industry m, in municipality j of zone i at time t (*En*); $S_{(t)}^{3}$ is the summation of $S_{ij(t)}^{3}$, which is an investment in the direct treatment technology for reducing water pollution in Lake Taihu, in municipality j of zone i at time t (*En*); $S_{(t)}^{4}$ is the summation of $S_{ij(t)}^{4m}$, which plays the same role as $S_{ij(t)}^{2m}$, but $S_{ij(t)}^{4m}$ originates from environmental taxes (*En*); $S_{(t)}^{5}$ is the summation of $S_{ij(t)}^{5m}$, which is an incentive fee for encouraging the low emission industry m in municipality j of zone i at time t, less than $S_{ij(t)}^{4m}$ (*En*); $S_{(t)}^{6}$ is the summation of $S_{ij(t)}^{6}$, which is an investment in biomass gasification technology for the effective utilization of biomass resources in municipality j of zone i at time t (*En*).

5. Models of Environmental Policy Intervention
(1) Biomass Gasification Technology

The financial budget for biomass gasification is as follows:

$$\mathrm{Tec}_{ij(t)}^{\mathrm{bio}} = \theta_1^{-1} I_{ij(t)}^{\mathrm{bio}} \tag{10}$$

$$S_{ij(t)}^{6} = \ell I_{ij(t)}^{\mathrm{bio}} + \theta_2 \ell\ell \mathrm{Tec}_{ij(t)}^{\mathrm{bio}} \tag{11}$$

$$X_{ij(t)}^{\mathrm{bio}} = P^{\mathrm{bio}} \mathrm{Gas}_{ij(t)}^{\mathrm{bio}} \tag{12}$$

$$Xn_{ij(t)}^{\mathrm{bio}} = (P_{(t)}^{\mathrm{bio}} - C_0^{\mathrm{bio}})\mathrm{Gas}_{ij(t)}^{\mathrm{bio}} \tag{13}$$

$$\mathrm{Gas}_{ij(t)}^{\mathrm{bio}} = \sigma_q^{\mathrm{bio}}\mathrm{Tec}_{ij(t)}^{\mathrm{bio}} \tag{14}$$

$$y = \ell I_{ij(t)}^{\mathrm{bio}}\left(\ell X_{ij(t)}^{\mathrm{bio}} - \theta_2 \ell \mathrm{Tec}_{ij(t)}^{\mathrm{bio}}\right)^{-1} \tag{15}$$

where $\mathrm{Tec}_{ij(t)}^{\mathrm{bio}}$ is the amount of biomass gasification in municipality j of zone i at time t (*En*); θ_1 is the setup cost for each biomass gasification unit (*Ex*); $I_{ij(t)}^{\mathrm{bio}}$ is the investment for construction of a biomass gasification unit in municipality j of zone i at time t (*En*); θ_2 is the administrative and maintenance expenses for each biomass gasification unit (*En*); $X_{ij(t)}^{\mathrm{bio}}$ is the amount of production by investment in biomass gasification in municipality j of zone i at time t (*En*); $Xn_{ij(t)}^{\mathrm{bio}}$ is the net revenue from investment in biomass gasification in municipality j of zone i at time t (*En*); $P_{(t)}^{\mathrm{bio}}$ is the market price of energy (*En*); C_0^{bio} is the biogas production cost per m^3 (*Ex*); σ_q^{bio} is the coefficient of biogas production efficiency (*Ex*); $Gas_{ij(t)}^{\mathrm{bio}}$ is the biogas capacity for biomass gasification in municipality j of zone i at time t (*En*); y is the capital recovery period (*En*).

The reduction in water pollutants and air pollutants resulting from biomass gasification is computed as follows:

$$\mathrm{BQ}_{ij(t)}^{n} = kw^n \sigma_{ij}^{\mathrm{bio}} K_{ij(t)}^{\mathrm{bio}} \tag{16}$$

$$\mathrm{Gas}_{ij(t)}^{n} = k_{\mathrm{air}}^{n} \sigma_{ij}^{\mathrm{bio}} K_{ij(t)}^{\mathrm{bio}} \tag{17}$$

where kw^n is the dissolution coefficient of Eutrophication material n in municipality j of zone i (*Ex*); k_{air}^{n} is the coefficient of exhaust reduction of air pollutant and greenhouse gas (*Ex*); $\mathrm{Gas}_{ij(t)}^{n}$ is the amount of exhaust reduction of air pollutants and GHG in municipality j of zone i at time t (*En*); $K_{ij(t)}^{\mathrm{bio}}$ is the amount of unused biomass resources in municipality j of zone i at time t (*En*).

(2) Reduction of Water Pollutants by Direct Treatment Technology

Water pollutants are usually dumped into Lake Taihu via rivers and streams. In addition, the water pollutants from materials are constantly dissolved in the lake and directly affect the quality of its water. Therefore, the introduction of direct treatment technology aimed at the direct removal of pollutants may be an effective approach. In this study, we consider three technologies, namely chemical coagulation, biochemistry and vegetation, as shown in Table 7. The direct water pollutants reduction effect depends on the level of direct treatment technologies, as well as the total amount of investment. It can be estimated as follows:

$$\mathrm{QTL}_{ij(t)}^{n} = ke_f^n \mathrm{DI}_{ij(t)}^{d} \tag{18}$$

$$S^3_{ij(t)} = \mathrm{DI}^d_{ij(t)} + \mathrm{DM}^d_{ij(t)} \tag{19}$$

where $\mathrm{QTL}^n_{ij(t)}$ is the reduction in the amount of water pollution n by direct treatment devices d in municipality j of zone i at time t (En); ke^n_f is the effect coefficient of the investment made for reducing water pollutants n by direct treatment devices (Ex); $\mathrm{DI}^d_{ij(t)}$ is the investment for setting up the direct treatment devices d in municipality j of zone i at time t (En); $\mathrm{DM}^d_{ij(t)}$ is the maintenance cost for the direct treatment devices d in municipality j of zone i at time t (En).

(3) Environmental Tax

As an available means of economic policy, an environmental tax can influence the production activities that harm the environment by levying taxes. The optimal environmental tax rate as a scale variable is simulated and analyzed in this study. Changes in tax rate vary with changes in water pollution reduction goals and changes in the water pollution emission equivalent from each industrial sector. The fundamental equations are as follows.

$$Te^m_{ij(t)} = \tau_{eco}\, Qhe^m_{ij(t)} \tag{20}$$

$$Qhe^m_{ij(t)} = \ell\left\{k^n_e X^m_{ij(t)} / Vhe^n\right\} \tag{21}$$

$$\mathrm{GFI}_{ij(t)} = Te_{ij(t)} + X^{bio}_{ij(t)} + \mathrm{TS}_{ij(t)} \tag{22}$$

$$\mathrm{GFO}_{ij(t)} = S^1_{ij(t)} + \ell S^{2m}_{ij(t)} + S^3_{ij(t)} + \ell S^{4m}_{ij(t)} + \ell S^{5m}_{ij(t)} + S^6_{ij(t)} \tag{23}$$

$$\mathrm{BBF}_{ij(t)}(\mathrm{Gap}_{ij(t)}) = \mathrm{GFI}_{ij(t)} - \mathrm{GFO}_{ij(t)} + \mathrm{BBF}_{ij(t-1)} \tag{24}$$

where $Te^m_{ij(t)}$ is the environmental tax revenue, which is calculated based on the water pollution emissions of industry m in municipality j of zone i at time t (En); τ_{eco} is the environmental tax rate, which is a scalar (En); $Qhe^m_{ij(t)}$ is the total noxious equivalents of industry m in municipality j of zone i at time t (En); Vhe^n is the equivalent conversion rate of n pollutants (COD = 1, T-N = 4 and T-P = 200 are calculated based on Shen et al. 2012); k^n_e is the emission coefficients of industry m as shown in Table 4 (Ex); $\mathrm{TS}_{(t)}$ is the summation of $\mathrm{TS}_{ij(t)}$, which is a subsidy for direct treatment technology in municipality j of zone i at time t (En); $\mathrm{BBF}_{(t)}$ is the balance brought forward (BBF) of countermeasure fees in the Taihu Basin; $\mathrm{GFI}_{ij(t)}$ is the collected funds for water pollution reduction in municipality j of zone i at time t; $\mathrm{GFO}_{ij(t)}$ is the expenditure for water pollution reduction in municipality j of zone i at time t.

(4) Production Function and Curtailment

This production function is of Harrod–Domar–Leontief type. Capital stock is the total quantity of capital used in the production of goods and services, including buildings, equipment, tools, machines and pollution abatement technologies. The capital stock, as a collective supply determinant, is especially important for the long-run expansion of the economy. Changes in environmental investment and subsidies for idling capital stock affect the effective utilization of capital stock. For example, policies to inhibit capital investment in sectors which have high pollution emission coefficients and adjust the capital stock of each industry can improve the capital efficiency and promote vigorous socioeconomic development in the basin.

$$X_{ij(t)}^{m} \leq \alpha^{m} K_{ij(t)}^{m} \tag{25}$$

$$\begin{aligned} K_{ij(t)}^{m} = {} & K_{ij(t-1)}^{m} - \delta X_{ij(t)}^{m} + II_{ij(t)}^{m} + S_{ij(t)}^{5m} - fr_{ij(t)}^{m^{-1}} (S_{ij(t)}^{2m} + S_{ij(t)}^{4m}) \\ & - Te_{ij(t)}^{m} + \mathrm{BBF}_{ij(t)}^{m} \end{aligned} \tag{26}$$

where α^{m} is the ratio of capital to production in industry m (Ex); $K_{ij(t)}^{m}$ is the production capital stock of industry m in municipality j of zone i at time t (En); fr^{m} is the subsidy rate to reduce capital investment in heavy pollution industries for water pollution reduction (Ex).

6. Environmental Efficiency Improvement Index (EEII)

The Environmental Efficiency Improvement Index is defined by comparing the driving force (DF) to the variable quantity of environmental pressure (EP) in a certain time period; it is used to evaluate how much the environment is improved in different periods. If EEII is larger or equal to zero, the increase in DF is larger than the increase in EP and the environment can be efficiently improved. If EEII is smaller than zero, the increase of DF is smaller than the increase of EP and the environment cannot be efficiently improved. The EEII is expressed as follows:

$$\mathrm{EEII}_{(t)} = \left(1 - \frac{\mathrm{EP}_{(t)}/\mathrm{DF}_{(t)}}{\mathrm{EP}_{\mathrm{ini}}/\mathrm{DF}_{\mathrm{ini}}}\right) 100\% \tag{27}$$

7. Pollutant Constraints Variable

The best limit target is assumed to be less than the capacity of the environment. Decreases in pollution from 0 to 20% are simulated to determine the best environmental policies to improve the water quality of the basin.

$$\ell Q_{ij(t)}^{n} \leq = (1 - \psi_{(t)}) Q_{2006}^{n} \tag{28}$$

$$Q_{ij(t)}^{n} = Q_{1\,ij(t)}^{n} + Q_{2\,ij(t)}^{n} + Q_{3\,ij(t)}^{n} - \mathrm{BQ}_{ij(t)}^{n} - \mathrm{QTL}_{ij(t)}^{n} \tag{29}$$

where Q_{2006}^{n} is the initial value of water pollution n, which flows into Lake Taihu (Ex); $Q_{ij(t)}^{n}$ is the amount of water pollution n that flows into Lake Taihu, in municipality j of zone i at time t (En); $\psi_{(t)}$ is the reduction target of water pollution n (Ex); $Q_{1\,ij(t)}^{n}$, $Q_{2\,ij(t)}^{n}$, $Q_{3\,ij(t)}^{n}$ are amounts of water pollution n that flow into Lake Taihu from household, industry *m* and land use, respectively, in municipality *j* of zone *i* at time *t* (*En*); $\mathrm{BQ}^{n}_{ij(t)}$ is the amount of reduced water pollution n obtained by biomass gasification technology in municipality *j* of zone *i* at time *t* (*En*); $\mathrm{QTL}^{n}_{ij(t)}$ is the amount of reduced water pollution *n* by direct treatment technology in municipality *j* of zone *i* at time *t* (*En*).

8. The Objective Function

The Green Gross Regional Product (GGRP) is an accounting system that deducts natural resources depletion costs and environmental degradation costs from the Gross Regional Product (GRP) to assess the quality of economic development (CGNASR 2004). In this study, the factor of GGDP is a GDP variant that does not include the cost of environmental protection payments for reducing water pollutant emission T-N, T-P and COD in the study area. On the basis of the maximization of the GRP and minimization of the costs of environmental management, the GGRP is selected as an economic growth indicator to assess the quality of real economic development. The best level of socioeconomic activity is to satisfy a given level of water pollution reduction obtained by maximizing GGRP under given restrictions for T-N, T-P and COD. The objective function in the model maximizes the GGRP in each zone considering capital stocks, investments for production and treatment, consumption and other related variables and constraints described above, subject to the contribution of pollutants and future changes for each sector in the zones. In other words, how the total pollution increases or decreases due to increasing GRP can be determined as one main macroeconomic indicator, with the environmental tax and the pollutants reduction targets applied in the model assuming that it will continue without any external impact on the long-term trend of economic development in the study area.

$$\max \sum_{t=1} \frac{1}{(1+\rho)^{(t-1)}} \mathrm{GGRP}_{(t)} \tag{30}$$

ρ is the social discount rate (=0.05)

$$\mathrm{GGRP}_{(t)} = \ell\ell\left\{v_{*} X_{(t)}\right\} - \mathrm{Cost}_{(t)} \tag{31}$$

$$\mathrm{Cost}_{(t)} = \ell\left\{kc_{0*} X_{(t)}\right\} + \ell\mathrm{GFO}_{ij(t)} \tag{32}$$

$\mathrm{Cost}_{(t)}$ is the investment and expenditure for environmental preservation at time *t* (*En*); kc_0 is one vector for coefficients of environmental investment and expenditure (*Ex*).

References

Forrester, J. W. (1958). Industrial dynamics: A major breakthrough for decision makers. *Harvard Business Review, 36*(4), 37–66.

Higano, Y., Mizunoya, T., Sakurai, K., & Kobayashi, S. (2002). An evaluation of water quality improvement new technology based on the social environmental system of Kasumigaura Lake Basin. In *Studies in Regional Science, ERSA Conference Paper* (p. 253). European Regional Science Association.

Hirose, F., & Higano, Y. (2000). A simulation analysis to reduce pollutants from the catchment area of Lake Kasumigaura. *Studies in Regional Science, 30*(1), 47–63.

Leontief, W. (1941). *The structure of american economy, 1919–1929: An empirical application of equilibrium analysis* (p. 181). Cambridge, MA: Harvard University Press.

Ministry of Environmental Protection of the People's Republic of China. *Environmental Quality Standards for Surface Water* (*GB 3838-2002*).

Mizunoya, T., Sakurai, K., Kobayashi, S., & Higano, Y. (2007). A simulation analysis of synthetic environment policy: Effective utilization of biomass resources and reduction of environmental burdens in Kasumigaura Basin. *Studies in Regional Science, 36*(2), 355–374.

Shen, Z. (2010). *Synthetic evaluation of environmental policies for sustainable development in the Taihu economic circle*. Ph.D. thesis (pp. 25–26, p. 72). Graduate School of Life and Environmental Sciences, University of Tsukuba.

Shen, Z., & Higano, Y. (2007). Basin management policy of water quality improvement in Taihu Valley. *Japan Association for Human and Environmental Symbiosis, 14,* 25–34.

Shen, Z., & Higano, Y. (2010). Synthetic evaluation of water pollution emission controls by environmental tax in Taihu Basin. *Japan Association for Human and Environmental Symbiosis, 17,* 99–108.

Shen, Z., Kusano, E., Chien, H., & Koyama, O. (2014). Predictive analysis of nitrogen balances resulting from the production and consumption of livestock products in the Huang-Huai-Hai Region, China. *Japan Agricultural Research Quarterly, 48*(3), 331–342.

Shen, Z., Mizunoya, T., & Higano, Y. (2012). Agriculture and sustainable development: Policies analysis of the Taihu economic circle in China. *International Journal of Foresight and Innovation Policy (IJFIP), 8*(2/3), 210–235.

The State Council. (1998). *The national ninth five-year plan and 2010 long range plan on water pollution prevention and control for the Tai Lake Basin.*

The State Council. (2002). *The national tenth five-year plan on water pollution prevention and control for the Tai Lake Basin.*

The State Environmental Protection Administration of China, SEPA and the National Bureau of Statistics of China NBS. (2006). *China green national accounting study report 2004* (*CGNASR 2004*), http://www.gov.cn/english/2006-09.

Wang, Q. G., Gu, G., & Higano, Y. (2006). Toward integrated environmental management for challenges in water environmental protection of Lake Taihu Basin in China. *Environmental Management, 37*(5), 579–588.

Climate Change and Land Use Pattern in Brazil

E. Barbosa, J. Féres, E. Haddad, and A. Paez

1 Introduction

In recent years, there has been an increasing concern among the different segments of society with the possible economic and social impacts of climate change. Evidence indicates that higher greenhouse gas concentrations in the atmosphere will have a significant effect on the planet. There is a common understanding that climate change would impact several economic sectors. Analysis of agricultural and livestock farming is important for understanding the economic impacts of climate change in the different regions of the globe, for this sector is directly affected by changes in temperature and rainfall.

According to the 2006 Agricultural Census, there were 16,567,544 people working in agricultural establishments, representing 18.9% of the employed people in Brazil. In addition, the agricultural production represented 5.8% of the Brazilian GDP in 2010 and the sector contributed to a large share of the national exports, having accounted for 37.9% of exports in 2010. Moreover, in the context of climate change, the agricultural sector plays a major role in the country because high levels of carbon dioxide (CO_2) emissions come from this sector.

Climatologists have been studying climate from a scientific perspective for many years. At first, it is important to note that the study of the impacts of climate change can be conducted according to the tools and the methodology of modern economics. And this connection between the study of climate and economics begins when climate

This chapter is based on the results of Barbosa's M.A. thesis, defended at the Department of Economics at the University of São Paulo. Haddad acted as his thesis adviser and Féres as his co-adviser; Paez also acted as his co-adviser during his visit at McMaster University, when part of the work was developed.

E. Barbosa · J. Féres · E. Haddad (✉) · A. Paez
University of São Paulo, São Paulo, Brazil
e-mail: ehaddad@usp.br

J. Thill (ed.), *Innovations in Urban and Regional Systems*,
https://doi.org/10.1007/978-3-030-43694-0_20

variables, which seemed constant over time, change and the possible effects of such changes affect people around the world in different ways.

Microeconomic tools allow for the construction of theoretical models that relate the decision-making processes in a given economy to climate variables. The microeconomic theory allows creating analytical models that relate climate variables to different options from decision-makers. Thus, by developing a model that is grounded on sound economic theory, one can use the observed data to estimate its parameters and understand the influence of climate variables on the object of the study. For the estimation of the model, modern econometric tools can be used.

The present study aims to examine how changes in climate variables—specifically temperature and precipitation—could impact the pattern of land use in Brazil, providing information on the possible effects of such changes on Brazilian regions. This paper tries to answer the following question: How would climate change impact on the land allocation decision in Brazil? In this context, an economic model is proposed, which focuses on a profit maximization problem faced by an economic agent that allocates the available resources (areas) to different uses. From the estimation of the parameters of this model and long-range forecasts of temperature and precipitation, one can create a consistent scenario for the pattern of land use in Brazil until the end of the twenty-first century.

This chapter is organized in the following way: in the next section, a brief review of the current literature on climate change effects on the agricultural sector is presented. Next, we present the economic model and a discussion on the econometric specification adopted. The subsequent section details the creation of the database used and the results of the estimated model and simulations. The last section contains the final remarks.

2 Literature Review

The study of the economic impacts caused by climate change by nature is a multidisciplinary one. The first studies that addressed the impacts of global climate change in agriculture from an economic perspective emerged in the 1980s as a result of a growing concern in several countries about the harmful effects of environmental changes. These studies sought to understand the changes in productivity that could be caused by climate change in specific regions and crops. It is interesting to note that this kind of research was closely related to studies that analyzed the impact of other factors in the agricultural sector such as the level of CO_2 or the use of irrigation or fertilizers (Moore and Negri 1992). Since these first studies, several models have been proposed, including those seeking to understand the dynamics of the land use pattern and the deforestation process. The literature can be divided in four different approaches: the production function, the Ricardian models, the fixed effects model, and the agent-based land use models (ABLUM's).

The production function approach focuses on the relationship between the production of a particular culture and climate variables such as rainfall and average temperature. Cross-section regressions make it possible to isolate the effects related to climatic variables. The results of the estimation show the sensitivity of each crop production related to the variables studied. An example of the application of the methodology can be found in Sands and Edmonds (2005), who analyzed the crops of sorghum and maize for the American case. Decker et al. (1986) and Adams (1989) used this approach to study the impacts of climate change on different crops in the USA. The main criticism on this approach is the lack of ability to capture the adaptation process of the agents.

The Ricardian approach is based on the Ricardian rent theory. This type of analysis assumes that land markets operate efficiently and, therefore, the price of the land reflects the discounted value of future farm incomes. The analysis assumes that agents have enough information to predict future earnings and calculate the discounted future income of the establishment. Another assumption is that producers maximize the allocation of the activities on their land, taking into account the economic and climate variables, so the effect of climate change would already be reflected in land prices. As farmers optimally choose input levels under the given conditions, the Ricardian approach could capture the agents' adaptation to climate change. However, this approach considers agent's adaptation as a "black box" because the model does not explicitly consider the adaptations.

Mendelsohn et al. (1994) pioneered the use of the Ricardian approach by studying the American case. One downside of this approach is that the estimation of the models could suffer from the omitted variable bias. This would occur because there would be unobserved variables with great influence on the land price, such as market-specific characteristics of each region. Thus, the effects of climate variables captured by the Ricardian approach may be incorporating the effects of omitted variables, generating biased estimation. It is important to note that this approach aims to calculate the impact of climate change on the profitability of agriculture and not on the allocation decision.

Introduced by Dêschenes and Greenstone (2007), the fixed effects model is proposed as an alternative to the omitted variable problem faced by the Ricardian approach. That is because it uses the deviations of temperature and precipitation observed in a given year relative to historical averages in order to capture the effects of climate change on profitability. When estimating this model, the authors apply the fixed effects panel methodology. This way, all the omitted variables that are constant in time will not cause bias.

The models described above seek to capture the influence of climate variables in the production of a particular crop and the profit or land price in a region. These different approaches do not take into account the decision-making process of agents. The models that attempt to capture the decision-making process are the ABLUMs models. Matthews et al. (2007) point out that the advantages of such models include taking into account social interaction, adaptation and decision-making at different levels. This type of model is not only used for the study of climate change, as is the case of Platinga (1996), where the author uses this approach to measure the impact of government subsidies on the agent's decisions. An ABLUM model could also be derived from the profit maximization problem that each farmer has to face when deciding how to allocate land with different types of crops. Following Féres et al. (2009), the optimization problem faced can be described as:

$$\text{Max} \sum_{i=1}^{m} \pi_i(p, r, n_i, X) \quad \text{s.a:} \sum_{i=1}^{m} n_i = N \tag{1}$$

where i represents the different types of crops; n_i represents the optimum areas chosen for each crop, while N is the total size of the region. There are also p, the vector of output prices; r the vector of inputs prices; and X, the climate variables vector. The main criticism of this approach is that it captures the decision process of agents today and not necessarily the way they will adapt in the future.

2.1 Previous Results for Brazil

For the Brazilian case, Sanghi et al. (1997), using a Ricardian model, concluded that the Center-West region would suffer more negative impacts from climate change then the rest of the country, mainly due to the hot and dry climate of the Brazilian *cerrado*. On the other hand, the authors argued that the southern Brazilian states could benefit due to the milder temperatures.

Evenson and Alves (1998) analyzed the pattern of land use in the face of changes in climate variables. Making a simulation for a uniform increase of 1 °C temperature and 3% in the rain, the authors concluded that this scenario would lead to a reduction of 1.84% of natural forest area and an increase of 2.76% in the areas of national pasture.

Féres et al. (2007) proposed the estimation of the fixed effects model for the Brazilian case, analyzing the impact of climate change in the profits of the agricultural sector. The authors estimated modest short-run impacts (2040–2069), a reduction of approximately 3% in agricultural income, and more significant long-run impacts (2070–2099), when the loss of profitability could reach up to 26% of farm income.

Feres et al. (2009) continued the previous work by analyzing the impacts of climate change in the land use patterns in Brazil. The results indicated an increase in pressure for deforestation in the Amazon region and the conversion of large areas of forests to pasture. The effects were strongly different in different regions of Brazil; there were increasing crop areas in the southern region, decreasing areas of pasture in the same region, and a significant decrease in crop area in the Northeast.

Although the studies did not present the same results, any comparison between them should be made with great caution due to the use of different methodologies and databases. However, the studies do show an overall negative effect in the country and different impacts in different regions of Brazil. For instance, the Center-West, North, and Northeast appear to be those which would suffer greater losses due to climate change, while the South could benefit from any increase in its average temperature.

3 Methodology

As discussed in the previous section, Féres et al. (2009) described a model for land use based on microeconomic theoretical assumptions on the profit-maximizing behavior of agents. Based on this model, this study presents an innovation that introduces specific crops as well as different types of land use. In addition, a more thorough discussion on the econometric estimation of the coefficients is proposed.

The use of models with aggregated crops in the study of impacts of climate change on land use can lead to less accurate results, since it is expected that different crops respond differently to climate changes. Also, analysis of regional impacts of climate change on the land use pattern demonstrated that the disaggregation of the different types of crops provides valuable information to be used in public policy decision-making processes (e.g., regarding infrastructure investments or mitigation policies). Thus, the following different types of land use are considered in this study: soybean, corn, sugar cane, other crops, pasture, and forest. Together, the cultivated areas of the three above-cited cultures account for approximately 60% of the crop area in Brazil. Hence, a better understanding of the impacts of climate change on land use for these three referred crops provides relevant information on the economic impacts of climate change on the Brazilian agricultural sector.

The presentation of the methodology of the land use model in the present study was divided into three parts: the economic model, the econometric specification, and the simulation, as shown in the next sub-sections.

3.1 Economic Model

The land use model adopted in the present study is the same suggested by Féres et al. (2009). This model is based on the problem of profit maximization by a representative agent. Given product prices, input costs, and the characteristic features of the climate of each region, the agent must choose the allocation of the different types of land use that maximizes profit. Or else, the agent allocates land among the different options: soybean, corn, sugar cane, other crops, pasture, and forest in the land in order to maximize profit. Then, we have the following constrained maximization problem:

$$\text{Max} \sum_{i=1}^{m} \pi_i(p, r, n_i, X) \quad \text{s.a:} \sum_{i=1}^{m} n_i = N \tag{2}$$

where the subscript i indicates the activity performed, p is the vector of prices of the products of agricultural activities, r is the vector of prices for inputs, n_i is the area allocated to activity i, X is the vector of agroclimatic variables, and N is the entire available land area. There are six types of land use in the developed model, so $m = 6$. The restriction imposed by the maximization hypothesis simply means that the sum of the areas allocated to the different activities cannot exceed the entire land area.

The following Lagrangian theorem is used to solve the problem proposed in (2):

$$L = \sum_{i=1}^{m} \pi_i(p, r, n_i, X) + \lambda\left(N - \sum_{i=1}^{m} n_i\right) \tag{3}$$

And the following first-order conditions arise:

$$\frac{\partial L}{\partial n_i} = \frac{\partial \pi_i}{\partial n_i} - \lambda = 0, i = 1, 2, 3, 4, 5 \text{ and } 6. \tag{4}$$

$$N = \sum_{i=1}^{m} n_i \tag{5}$$

From the above conditions, one can obtain optimal allocations for each type of land use (n_i^*) as a function of the price of the products, input prices, the entire land area, and the agroclimatic variables of the land area. Then, six equations determine the optimum area for each type of land use:

$n_i^*(p, r, N, X)$, for $i =$ soybean, corn, sugar cane, other crops, pasture, and forest.

When optimal allocations are substituted in the first-order conditions (5) and the expression is differentiated, the following equations are obtained:

$$\sum_{i=1}^{m} \frac{\partial n_i^*(p, r, N, X)}{\partial N} = 1 \tag{6}$$

$$\sum_{i=1}^{m} \frac{\partial n_i^*(p, r, N, X)}{\partial p} = 0 \tag{7}$$

$$\sum_{i=1}^{m} \frac{\partial n_i^*(p, r, N, X)}{\partial r} = 0 \tag{8}$$

$$\sum_{i=1}^{m} \frac{\partial n_i^*(p, r, N, X)}{\partial X} = 0 \tag{9}$$

The above equations ensure the logical consistency of the model. Equation (6) shows that the sum of the changes in allocations caused by an increase in the entire land area must be equal to the increase occurred. Equations (7)–(9) ensure that the changes in the optimal areas of the six different types of land use caused by variations in prices and agroclimatic variables have no net effect at all. In other words, if a change in temperature causes variation in an optimal crop area, e.g., sugar cane, the sum of changes in the five other types of land use must necessarily be of equal magnitude and opposite sign to the change in the sugar cane area.

3.2 Econometric Specification

The normalized quadratic functional form[1] was used to represent the profit function in order to estimate the proposed model. This type of functional form is often found in the literature and there are many reasons for using it. First, this is a flexible functional form, i.e., it does not impose any a priori restrictions on the elasticities of substitution between the various inputs. Second, the function has properties consistent with the microeconomic theory such as being homogeneous of degree one in the prices. Third, there is the fact that the optimal allocations of land use derived from this function have linear parameters, allowing the use of linear equation estimation methods. Finally, another advantage of this type of functional form compared to others is that the estimated parameters are easy to interpret.

With this function, the following equations that describe optimal allocations of different land uses to be estimated are obtained:

$$n_i^* = \beta_0^i + \sum_{f=1}^{j} \beta_{1f}^i p_f + \sum_{k=1}^{t} \beta_{2k}^i r_k + \beta_3^i N + \sum_{l=1}^{s} \beta_{4l}^i X_l + \varepsilon, i = 1, 2, 3, 4, 5 \text{ and } 6. \tag{10}$$

Subject to the following restrictions:

$$\sum_{i=1}^{6} \beta_3^i = 1 \tag{11}$$

$$\sum_{i=1}^{6} \beta_{1f}^i = 0 \tag{12}$$

$$\sum_{i=1}^{6} \beta_{2k}^i = 0 \tag{13}$$

$$\sum_{i=1}^{6} \beta_{4l}^i = 0 \tag{14}$$

$$\sum_{i=1}^{6} \beta_{1f}^i = \sum_{i=1}^{6} \beta_{1f}^f \tag{15}$$

Restrictions (11)–(14) are equivalent to those found in Eqs. (6)–(9). Restriction (15) arises from the symmetry of the normalized quadratic functional form. Therefore, there is a system of six equations to be estimated, with observation of the above described restrictions.

[1]More details on this functional form can be found in Lau (1978).

According to Matthews et al. (2007), agent-based land use models are one way to incorporate the influence of social interaction, mechanisms of adaptation, and decision making of agents to the study of land allocation for different uses. In this context, it is important to understand three different aspects of the problem: the spatial aspect, the simultaneity of the equations, and the temporal dynamics of the problem.

Kumar (2011) argues that the spatial problem must be incorporated into the study of the impacts of climate changes in agriculture for two reasons: theoretical and econometric. From a theoretical standpoint, it will be natural to think that the agents distributed across the space communicate about the activities carried out in each region and exchange experiences on production practices and techniques that generate a spatial dependence in the dependent variable (crop area). From an econometric point of view, a possible inaccuracy in the estimates must be considered due to the presence of spatial correlation in the error term of the specified equations. Therefore, an estimation method that considers the spatial factor can lead to improved efficiency of the estimation.

The simultaneous estimation of six equations of optimal allocation of land use is important because the agents' decisions occur simultaneously. The allocation of land for each type of land is made at the same time, and, thus, there is a relationship between the parameters of the equations. This relationship is described by the restrictions derived from the economic model, e.g., the parameter that captures the influence of temperature on a given land use has a relationship with the same parameter in the other equations. Moreover, the estimation of the system by a non-simultaneous approach does not consider the possible correlation among residuals.

The temporal dimension should be considered in the study of this problem in order to simulate how agents' decisions change over time. The aim of this study is to understand how agents' decisions are affected by certain variables over time. Hence, change over time within the same region is a key factor for the correct understanding of the problem. If only cross-section data over a period of time is considered, we will be using climatic variations in the different regions to explain the allocation of land use. In fact, the simulation results tend to show that if the climate pattern in a given region A becomes similar to the climate pattern in a given region B, then the use of land in region A would be similar to the use of land in region B.

Thus, in order to estimate the correct way in which the agents of region A adapt to climate change, it would be necessary to use the data from region A at different periods of time and use temporal variability of climate to explain change in land use. Fixed-effect panel models would be the most appropriate methodology for this purpose. Nevertheless, some considerations on the real relationship between climate change and land use are necessary in order to define the appropriate methodology for our purposes.

The relationship between historical average temperature and rainfall in each region and the agricultural activity is clear and direct. Some crops are more productive than others, depending on temperature and rainfall levels, and each region has a comparative advantage in producing a particular crop. The agent that makes the decision of allocation seeks to maximize profit and will choose the activities that have a comparative advantage in the region. The present study is aimed at capturing how agents would adapt to climate changes in a given region. According to the above reasoning, we try to understand how the agents in a given region A adapt to changes in temperature and rainfall in region A. This question can be answered using the variation in climate data and in land use. However, the key aspect of the decision to use the panel model would be to check whether already observed climate change would have influenced changes in the allocation of land use. Some factors indicate that the answer to this question is still non-conclusive.

The type of crop grown on a given land tends to be persistent. Several factors must be taken into consideration when a producer from a region that grows a given crop A decides to grow another crop B. For this change to occur, besides the greater expected profit, it takes time for the farmers to learn the different technology related to the new crop. Also, infrastructure consistent with the new crop is required. Climate variables could certainly influence this decision, since they would cause one crop to be more productive than others, and, consequently, would ensure higher profit to the agent. However, since climate change is a long-term phenomenon, long-term climate data of the agricultural sector is necessary to capture its impact.

Two issues arise in this type of analysis in the Brazilian case. The first issue is the reliability of older data for both land use and climate change, because the older data the less accurate they are. The second and main issue concerns the central point of our analysis: the relationship between land use and climate change. Analysis of climate data already collected shows low temporal variability in temperature and precipitation variables across the same region. The little temporal variability of climate data and the persistent decisions on the allocation of land use by agents makes it unlikely that the change in cultivated areas was related to change in temperature and rainfall. Thus, the use of the panel model carries a risk of capturing a relationship between climate changes and land use that does not really exist.

For comparison, the equations were estimated using the spatial panel data model developed by Elhorst (2003), the *spatial autoregression* (SAR) with fixed effects. The model in Eq. 16 was estimated for each of the cultures:

$$y_{it} = \delta \sum_{j=1}^{N} W_{ij} y_{jt} + X_{it}\beta + \mu_i + \varepsilon_{it} \tag{16}$$

where y_{it} is the area in the region i on the period t; δ is the autoregressive spatial coefficient; W_{ij} is an element of the spatial weights matrix W; X_{it} is the vector with the explanatory variables; μ_i is the spatial fixed effects term, and ε_{it} is the error term of the equation. For this estimation only, data from the agricultural census of 1995 and 2006 were used, and the temperature and precipitation variables were the averages observed during the last five years prior to the censuses. As seen in Table 1, the majority of the climate variables coefficients were not significant, providing further evidence that climate change over time cannot yet be used as a justification for the allocation of land use, at least over a not very long period.

It should be noted that several studies used the panel model in the assessment of climate change impacts on the profitability of the agricultural sector. In this type of

Table 1 Spatial panel regression

Equation	Soybean			Corn			Sugar Cane		
	Coef.	*t*-stat	*p*-value	Coef.	*t*-stat	*p*-value	Coef.	*t*-stat	*p*-value
Relative price sugarcane	−4762.68	−0.54	0.59	328.79	0.07	0.94	1498.84	0.56	0.57
Relative price corn	4088.69	0.82	0.41	−2326.73	−0.88	0.38	−1520.26	−1.01	0.31
Relative price soybean	−174.84	−0.04	0.97	3787.61	1.67	0.09	886.14	0.69	0.49
Relative price other crops	−0.20	−0.99	0.32	−0.07	−0.64	0.52	−0.03	−0.50	0.62
Relative price land	12.26	0.18	0.86	7.32	0.20	0.84	−3.77	−0.18	0.85
Relative price labor	−331.84	−1.60	0.11	−261.71	−2.35	0.02	−29.84	−0.47	0.64
Relative price forest	−23,005.22	−1.90	0.06	−7337.78	−1.13	0.26	1215.24	0.33	0.74
Temperature DJF	44.60	3.52	0.00	23.94	3.52	0.00	1.80	0.47	0.64
Temperature JJA	−13.15	−0.62	0.54	−16.26	−1.43	0.15	−8.29	−1.28	0.20
Temperature MAM	15.48	1.59	0.11	−2.60	−0.50	0.62	1.57	0.53	0.60
Temperature SON	6.18	0.29	0.77	18.53	1.64	0.10	−15.89	−2.48	0.01
Precipitation DJF	4767.02	1.13	0.26	2953.80	1.31	0.19	−1401.19	−1.09	0.28
Precipitation JJA	−4282.23	−0.46	0.65	−13,179.13	−2.64	0.01	13,571.89	4.80	0.00
Precipitation MAM	13,293.61	1.11	0.27	5212.22	0.81	0.42	3174.97	0.87	0.38
Precipitation SON	902.92	0.06	0.95	2477.60	0.32	0.75	−10,329.79	−2.33	0.02
Equation	Other crops			Pasture			Forest		
	Coef.	*t*-stat	*p*-value	Coef.	*t*-stat	*p*-value	Coef.	*t*-stat	*p*-value
Relative price sugarcane	7858.51	1.20	0.23	−23,952.23	−1.34	0.18	35,267.15	1.98	0.05
Relative price corn	−8884.77	−2.39	0.02	−3410.86	−0.34	0.74	49,916.16	5.01	0.00
Relative price soybean	3304.48	1.04	0.30	128.62	0.01	0.99	−37,404.85	−4.39	0.00

(continued)

Table 1 (continued)

Equation	Other crops			Pasture			Forest		
	Coef.	*t*-stat	*p*-value	Coef.	*t*-stat	*p*-value	Coef.	*t*-stat	*p*-value
Relative price other crops	0.03	0.18	0.86	0.57	1.41	0.16	−0.42	−1.04	0.30
Relative price land	45.77	0.91	0.37	34.70	0.26	0.80	−105.04	−0.78	0.44
Relative price labor	−35.34	−0.23	0.82	−584.87	−1.39	0.17	1478.32	3.53	0.00
Relative price forest	2140.65	0.23	0.81	47,353.60	1.92	0.05	−45,779.49	−1.87	0.06
Temperature DJF	−30.92	−3.24	0.00	−11.60	−0.45	0.65	−62.73	−2.46	0.01
Temperature JJA	−20.55	−1.29	0.20	77.48	1.80	0.07	16.78	0.39	0.70
Temperature MAM	18.63	2.55	0.01	−21.37	−1.08	0.28	38.24	1.94	0.05
Temperature SON	−9.77	−0.62	0.54	−72.81	−1.71	0.09	32.64	0.76	0.45
Precipitation DJF	−6620.80	−2.09	0.04	16,943.79	1.99	0.05	27,042.22	3.06	0.00
Precipitation JJA	21,139.58	3.02	0.00	889.15	0.05	0.96	55,210.17	2.94	0.00
Precipitation MAM	11,571.78	1.28	0.20	1034.43	0.04	0.97	−45,251.71	−1.86	0.06
Precipitation SON	−5980.59	−0.54	0.59	−24,150.93	−0.82	0.41	−57,879.30	−1.95	0.05

study, the temperature and precipitation data at one moment in time are considered in relation to the deviation from the historical average, and we can estimate how this deviation influenced the profitability of the sector.

In view of the question to be answered, and the above discussion, only cross-section data were estimated. By estimating parameters for cross-sectional data from the 2006 census, only with the purpose of providing a simulation in time, we assume that the current pattern of land use in Brazil represents the long-run equilibrium between agents, i.e., agents are making the best possible choices of cultivation areas given the prices and climate conditions. The simulation based on the estimation of cross-section data has a long-term character, which is exactly the type of simulation that we intend to make. The method used was the iterated seemingly unrelated regression (ISUR) model, the same used by Féres et al. (2009).

Furthermore, in an attempt to capture the spatial factor and at the same time, some effect of temporal inertia, the term W95 was added to the equation, which represents the crop area in 1995 weighted by the W spatial weight matrix. The matrix used was of the queen type. The use of a spatially and temporally delayed term is suggested by Chasco and Lopez (2004) and is aimed to capture the spatial effect throughout time. In other words, the influence of the neighbor does not occur simultaneously, but over time. Hence, besides capturing the spatial effect of the neighbor's influence, the term also captures some temporal inertia in land use in those regions. Another variable included in the model was the crop area in 1996. This term also captures the inertia in land use for different purposes.[2]

3.3 Simulation

The simulation of the areas allocated for different uses based on climate forecasts was made after the estimation of the parameters of the model, according to the following equation:

$$\hat{n}^*_{i,C} = \hat{\beta}^i_0 + \sum_{f=1}^{j} \hat{\beta}^i_{1f} p_f + \sum_{k=1}^{t} \hat{\beta}^i_{2k} r_k + \hat{\beta}^i_3 N + \sum_{l=1}^{s} \hat{\beta}^i_{4l} X_{l,C} \tag{17}$$

where $X_{l,C}$ is the vector of agroclimatic variables forecasted for the scenario C, and $\hat{n}^*_{i,C}$ is the area estimated for land use i, taking into consideration the scenario C. Then, we can analyze the changes based on the areas observed in the baseline (in this case, the year of 2006).

Furthermore, when an area allocated to a given crop was negative, the value was considered 0. In such cases, we cannot guarantee that the sum of the new areas allocated for the different land uses shall be equal to the entire land area in the

[2]Further discussion on the data used is available in the next section.

microregion. Then, adjustments are necessary in the simulation to maintain the logical consistency of the model. Based on the six new areas, the entire new land area was calculated as follows:

$$\sum_{i=1}^{6} \hat{n}^*_{i,C} = \widehat{N}^*_{j,C} \tag{18}$$

where $\widehat{N}^*_{j,C}$ represents the entire land area after simulations for region j. We can calculate the participation of each land use in the simulated region and use this result for comparison with the total area allocated for each use.

$$n^*_{i,C} = \left(\frac{\hat{n}^*_{i,C}}{\widehat{N}^*_{j,C}} \right) n^*_{i,\text{base}} \tag{19}$$

Finally, the percent change in the areas allocated for the different land uses is calculated based on the following expression:

$$\Delta n^*_i = \frac{\hat{n}^*_{i,C} - n^*_{i,\text{base}}}{n^*_{i,\text{base}}} \times 100 \tag{20}$$

4 Data

Data from 558 Brazilian microregions were used. The Brazilian microregions are a political division formed by contiguous cities within the same state. They were created in order to help the organization and planning of common public services. The area and population of the microregions vary greatly across the country, and each microregion has an average of ten municipalities. The main source of data was the Brazilian Institute of Geography and Statistics (IBGE), especially the Brazilian agricultural census of 2006. The agronomic data was supplied by Embrapa. The observed temperature and precipitation were obtained from the Climatic Center Unit (CRU/University of East Anglia) database and the CPTEC/INPE provided projections of temperature and precipitation for the period 2010–2099. The construction of the variables being used is presented below.

4.1 Land Use

The land use areas were constructed from the information of the agricultural census of 2006. The variable that defines the area for "other crops" corresponds to the sum of areas of all the permanent and temporary crops other than soybean, sugarcane, and corn. The pasture area is the sum of natural and planted pastures. Finally, the forest area is the total area of natural forests, planted forests, and unused productive land within any farmer's establishment.

4.2 *Prices of Products: Soybeans, Corn, Sugar Cane, Crop, Pasture, and Forest*

The prices of soybean, corn, and sugarcane were calculated based on average values of tons produced in each microregion. The "other crops" price was constructed as a Laspeyres index, where the national average was the base value. The index is composed of the prices and production of the major crops in terms of areas except soybean, corn, and sugarcane: rice, beans, tobacco, cassava, wheat, cotton, bananas, cocoa, coffee, and orange. The representative price of the products of the pasture activity was the average price of cattle sold in the microregion. The price of the products of forests was represented by the average per cubic (m^3) value of timber in the microregion. The hypothesis is that the price of timber is a good proxy for the opportunity cost of conserving forest areas in agricultural establishments.

Except for the price of cattle, which was observed in the agricultural census of 2006, all other prices were the average price of the five years preceding the census data (the cattle price was the not available for the period in the microregional level). The choice of using these prices instead of the prices observed in the census is related to the decision making of the agents. The future relative prices that guide the agent's decision are formed based on the recent prices that the agents observed. All prices used in the model correspond to 2006 values.

4.3 *Prices of Inputs: Land and Labor*

In this study, we considered only two inputs for agricultural production: land and labor. There is no regional data for land prices in Brazil so the land leasing price was used as a proxy. The labor price was represented by the average rural wage, given by the total monetary value of wages (in cash or goods) paid to employees or family members divided by the total number of persons employed.

4.4 *Observed Climate Variables*

Variables for temperature (°C) and precipitation (mm) were used in the model. The information was consolidated from the CRU/University of East Anglia data base and represents a 30 years average of the data from 1975 to 2005. For each season (3 months period) a different variable for temperature and precipitation was constructed. This is important because the effects of climate change in agriculture can be different depending on the season.

4.5 *Projected Climate Variables*

The temperature and precipitation projections for Brazil in the period of 2010–2099 were obtained from the CPTEC/INPE. The providing regional climates for impact studies (PRECIS) model provides projections for two emission scenarios defined by International Panel on Climate Change (IPCC), A2 and B2, and both were considered in the simulations. The A2 scenario represents a very heterogeneous world where the fertility rates converge in a slow rate between the countries. The economic development is more regional oriented and the technological change is more fragmented. The second scenario, B2, emphasizes the environmental sustainability and regional equality. In this scenario, the global population increases at a rate lower than in the A2 scenario, and we observe intermediate levels of economic development with a slower but more diverse technological change.

The PRECIS model provides monthly forecasts of climate variables; however, in the present study, 30 years averages were calculated for the following years: 2010–2039, 2040–2069, and 2070–2099. Working with averages of precipitation and temperature are adequate for two main reasons. First, it is reasonable to expect that the land allocation decisions take into account the behavior of the climate variables especially in the long run. Furthermore, when calculating the average over this relatively long period, we reduce the "uncertainty" inherent in climate projection models. The downside of working with these averages is the fact that the effects of extreme events would be diluted into the average and will not be captured by the model.

5 Results

In this section, the econometrics results of the model's estimation are presented followed by the simulations results.

5.1 *Land Use Model*

As discussed in Sect. 4, the estimation of the optimal land allocations equations was made by the ISUR methodology. The model estimated six types of land uses according to Eq. 10. Furthermore, the estimated model considered all the mentioned parametric restrictions. Table 2 presents the results of the estimation. The name of the climate variables represents the three-month period of the seasons which can be identified as December, January, February (DJF) representing the summer, March, April, May (MAM) representing the fall, June, July, August (JJA) representing the winter, and September, October, November (SON) representing the spring. The temperature and precipitation variables were significant in almost all equations. In the sugarcane and other crops equations, some variables of temperature and precipitation were not significant. This is evidence that temperature and precipitation are determinant factors in the land use allocation by farmers. It is also interesting to note the signal differences found when comparing the climatic variables for different seasons, this occurs in all the equations, justifying the use of season averages instead of the annual averages of the variables.

Both for temperature and for the precipitation in summer and winter, an increase in the magnitude of these variables favors the increase of the pasture area and the decrease in the cultivated area of all other uses. As for the spring and autumn, the effect is the opposite. The climate variables seem to affect the pasture use in an opposite way then they affect the other land uses.

One point of attention in the estimation of the model is the coefficients of the relative price variables. Most of these coefficients are not statistically significant, and the coefficient sign is not consistent with the expectations of economic theory. For example, the coefficient for the corn price shows that an increase of the corn price would reduce the area for this crop. The way the price variables were constructed (from the average in the years preceding the census) should reflect the expectation of prices in each region and, therefore, be an important factor in a land use decision. However, the model cannot adequately capture the effect of the price of crops in the land allocation decision. One possible explanation for this is the low volume of production of some crops in certain regions, which may cause distortions in relative prices. There could also be a problem of endogeneity in the price variables and subsequently an identification problem. The area designated for the production of a certain crop could influence the price of this crop product. The construction of the price variables is certainly a challenge and a point of improvement for future works.

Finally, the variable W95 was significant in all equations except for the forest and other crops equations. This demonstrates the importance of the spatial lag on the model. In this case, the neighbor's allocation decision in the previous period has an influence on land use of a region in the current period. In the equations for soybean, sugarcane, other crops, and pasture, the coefficients were found to be positive. We observe a positive influence of the neighbor's land use in the previous period in the land allocation decisions in 2006. This indicates a consolidation of this type of land use in the regions. As for corn and forests, the coefficients were negative, indicating that there was no consolidation of these crops in the regions.

Table 2 Regression results

Equation	Soybean				Corn				Sugarcane			
	Coef.	SE	*t*-stat	*p*-value	Coef.	SE	*t*-stat	*p*-value	Coef.	SE	*t*-stat	*p*-value
Relative price soybean	75,507.32	39,462.15	1.91	0.06	25,356.08	14,200.45	1.79	0.07	8283.37	3636.15	2.28	0.02
Relative price corn	−96,696.39	54,161.29	−1.79	0.07	−36,474.01	19,164.40	−1.90	0.06	−8209.29	4941.04	−1.66	0.10
Relative price sugarcane	37,809.36	49,875.59	0.76	0.45	13,757.95	17,506.12	0.79	0.43	−8415.60	5295.40	−1.59	0.11
Relative price other crops	−2.22	1.03	−2.17	0.03	−0.47	0.37	−1.28	0.20	−0.06	0.09	−0.70	0.49
Relative price forest	−22,900.91	42,741.13	−0.54	0.59	1414.88	15,800.58	0.09	0.93	−2252.55	5321.83	−0.42	0.67
Relative price land	211.67	3040.99	0.07	0.95	83.17	1040.93	0.08	0.94	−72.58	278.99	−0.26	0.80
Relative price labor	1631.84	847.77	1.92	0.05	612.37	290.44	2.11	0.04	101.21	79.63	1.27	0.20
Temperature DJF	−69,747.35	22,762.34	−3.06	0.00	−17,829.04	7826.77	−2.28	0.02	−1477.25	2091.37	−0.71	0.48
Temperature MAM	100,732.60	27,476.52	3.67	0.00	22,259.71	9426.96	2.36	0.02	3507.05	2515.97	1.39	0.16
Temperature JJA	−65,328.65	20,586.09	−3.17	0.00	−19,217.82	7067.42	−2.72	0.01	−3066.49	1891.55	−1.62	0.11

(continued)

Table 2 (continued)

Equation	Soybean				Corn				Sugarcane			
	Coef.	SE	*t*-stat	*p*-value	Coef.	SE	*t*-stat	*p*-value	Coef.	SE	*t*-stat	*p*-value
Temperature SON	39,707.11	18,692.04	2.12	0.03	16,317.79	6421.39	2.54	0.01	2118.43	1718.62	1.23	0.22
Precipitation DJF	−2101.58	324.66	−6.47	0.00	−912.46	111.59	−8.18	0.00	−45.45	29.85	−1.52	0.13
Precipitation MAM	2475.95	335.11	7.39	0.00	953.93	115.53	8.26	0.00	59.49	30.83	1.93	0.05
Precipitation JJA	−1082.50	323.23	−3.35	0.00	−491.13	110.95	−4.43	0.00	−15.08	30.12	−0.50	0.62
Precipitation SON	1760.40	346.96	5.07	0.00	645.30	119.32	5.41	0.00	26.09	31.96	0.82	0.41
AreaTot	0.05	0.01	8.17	0.00	0.02	0.00	7.54	0.00	0.00	0.00	−0.43	0.66
W95	0.10	0.03	3.32	0.00	−0.30	0.06	−5.13	0.00	0.14	0.04	3.12	0.00
_cons	−283,682.70	196,263.20	−1.45	0.15	−44,369.70	67,807.79	−0.65	0.51	−35,676.97	18,064.41	−1.97	0.05
			Rsq = 0.4422				Rsq = 0.5800				Rsq = 0.8439	
Equation	Other crops				Pasture				Forest			
	Coef.	SE	*t*-stat	*p*-value	Coef.	SE	*t*-stat	*p*-value	Coef.	SE	*t*-stat	*p*-value
Relative price soybean	9894.47	14,598.74	0.68	0.50	−146,881.60	76,638.55	−1.92	0.06	21,558.51	57,998.33	0.37	0.71

(continued)

Table 2 (continued)

Equation	Other crops				Pasture				Forest			
	Coef.	SE	*t*-stat	*p*-value	Coef.	SE	*t*-stat	*p*-value	Coef.	SE	*t*-stat	*p*-value
Relative price corn	−16,909.80	19,897.80	−0.85	0.40	140,240.40	105,925.70	1.32	0.19	22,104.52	80,118.95	0.28	0.78
Relative price sugarcane	40,212.66	18,351.04	2.19	0.03	−97,188.51	95,181.53	−1.02	0.31	3231.00	72,781.45	0.04	0.97
Relative price other crops	−0.19	0.38	−0.51	0.61	4.05	1.91	2.13	0.03	−0.96	1.44	−0.67	0.51
Relative price forest	−33,198.00	15,896.96	−2.09	0.04	103,830.60	81,428.48	1.28	0.20	−30,751.86	61,782.30	−0.50	0.62
Relative Price Land	−268.26	1106.42	−0.24	0.81	641.98	6093.80	0.11	0.92	−595.98	4604.93	−0.13	0.90
Relative price labor	−1020.26	308.66	−3.31	0.00	1733.55	1696.88	1.02	0.31	−3058.70	1282.46	−2.39	0.02
Temperature DJF	−31,752.74	8353.43	−3.80	0.00	324,819.10	45,395.69	7.16	0.00	−204,012.70	34,337.33	−5.94	0.00
Temperature MAM	35,987.36	10,030.45	3.59	0.00	−288,271.10	54,866.61	−5.25	0.00	125,784.40	41,484.32	3.03	0.00
Temperature JJA	−14,920.47	7512.97	−1.99	0.05	165,937.90	41,176.70	4.03	0.00	−63,404.51	31,126.41	−2.04	0.04

(continued)

Table 2 (continued)

Equation	Other crops				Pasture				Forest			
	Coef.	SE	*t*-stat	*p*-value	Coef.	SE	*t*-stat	*p*-value	Coef.	SE	*t*-stat	*p*-value
Temperature SON	4685.87	6823.92	0.69	0.49	−150,590.50	37,354.75	−4.03	0.00	87,761.32	28,237.92	3.11	0.00
Precipitation DJF	−133.71	119.12	−1.12	0.26	4859.89	646.72	7.51	0.00	−1666.68	488.82	−3.41	0.00
Precipitation MAM	176.81	122.68	1.44	0.15	−5798.79	666.64	−8.70	0.00	2132.62	504.11	4.23	0.00
Precipitation JJA	−81.37	118.08	−0.69	0.49	3178.13	644.68	4.93	0.00	−1508.05	487.59	−3.09	0.00
Precipitation SON	64.61	126.75	0.51	0.61	−4586.99	691.42	−6.63	0.00	2090.60	522.89	4.00	0.00
AreaTot	0.00	0.00	1.53	0.13	0.62	0.01	52.74	0.00	0.31	0.01	34.60	0.00
W95	0.05	0.03	1.70	0.09	0.02	0.00	4.24	0.00	−0.01	0.01	−1.10	0.27
_cons	171,399.00	73,113.97	2.34	0.02	−984,571.10	390,007.80	−2.52	0.01	1,150,425.00	294,856.20	3.90	0.00

5.2 *Simulations*

One should be careful in analyzing the simulation results. For instance, it is important to understand that the simulations should be interpreted as responses to the following conjecture: given the technological and productive structures for 2006 and the climate variables changes projected by the CPTEC/INPE, how would the optimal allocations of land behave once the total area occupied by each of the farms remains the same? The simulations do not capture possible changes in other variables that determine land use, such as product prices. It is natural to think that if the area of a given crop had a significant reduction, and consequently, there was a decrease in the supply of its product, its price would likely increase, providing incentive for further production of it. Thus, the simulations made here tend to overestimate the variations in the areas, since we are not capturing the second-order effects of price changes. It is also natural to think that an increase in the acreage of a particular crop would lead to an increase in the domestic production, and, therefore, to a potential decrease in prices, which in turn would induce a reduction in cultivated area.

5.2.1 Simulations—National Results

The share of each type of land use in the Brazilian microregions is described in Figs. 1, 2, 3, 4, 5 and 6.

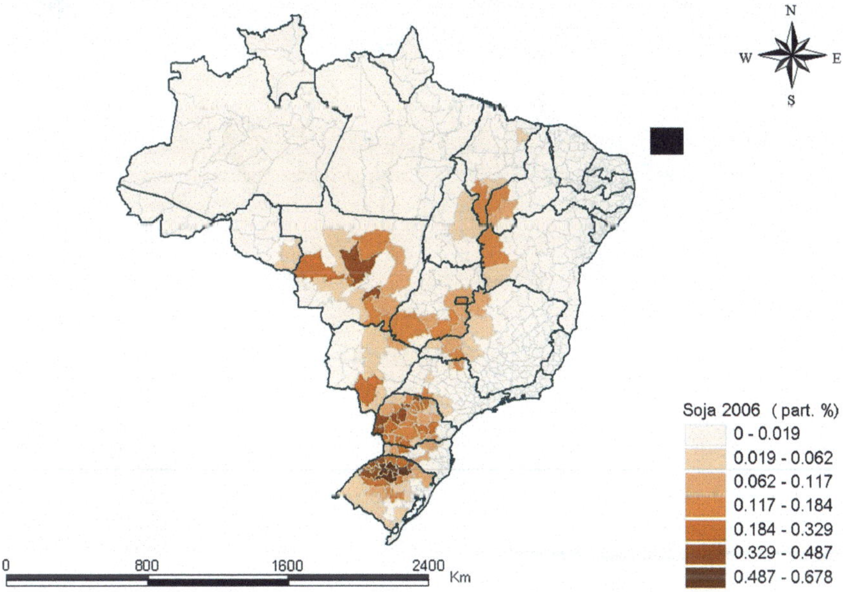

Fig. 1 Soybean share (2006)

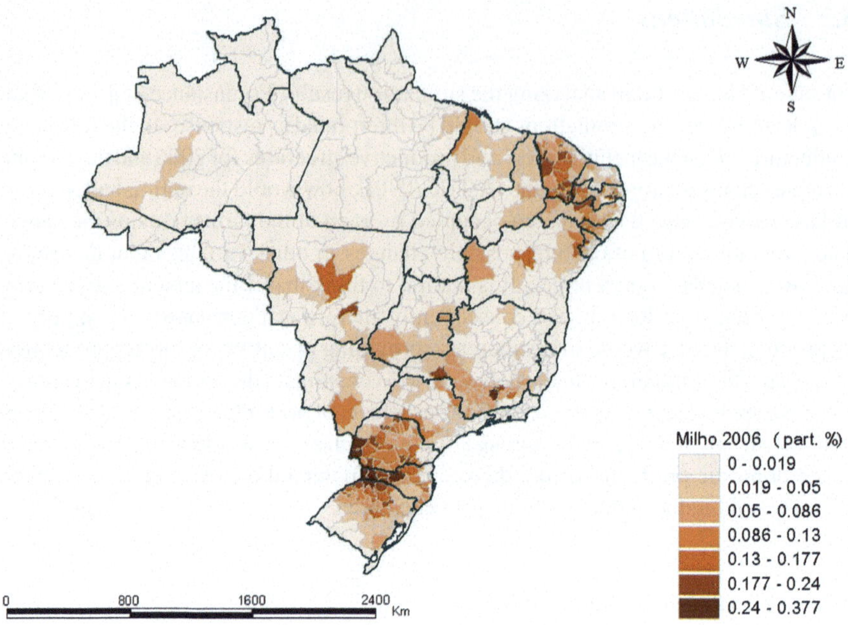

Fig. 2 Corn share (2006)

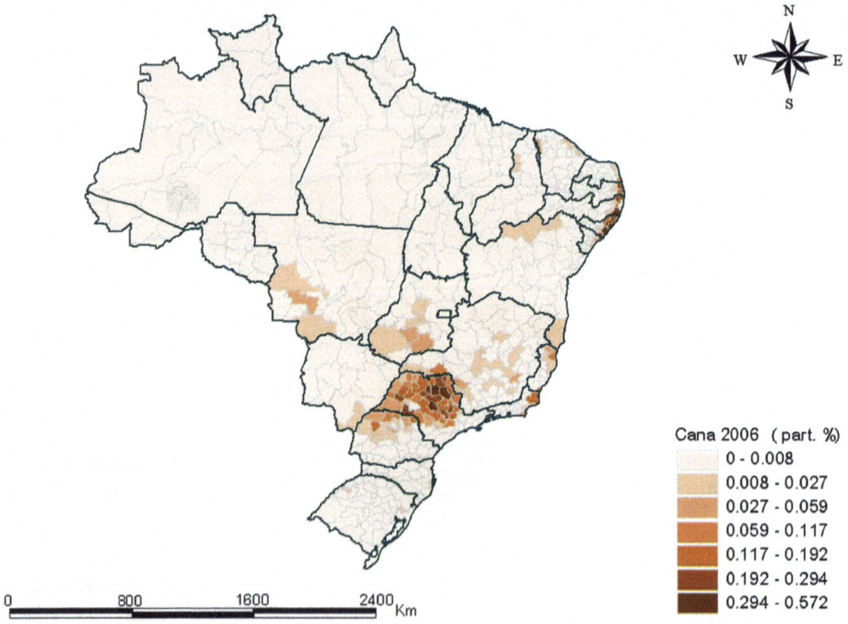

Fig. 3 Sugarcane share (2006)

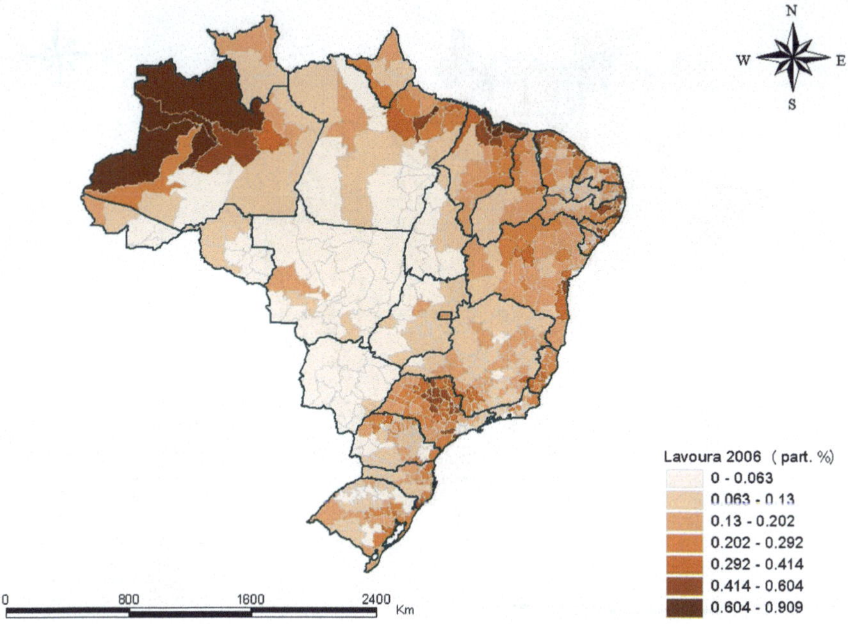

Fig. 4 Other crops share (2006)

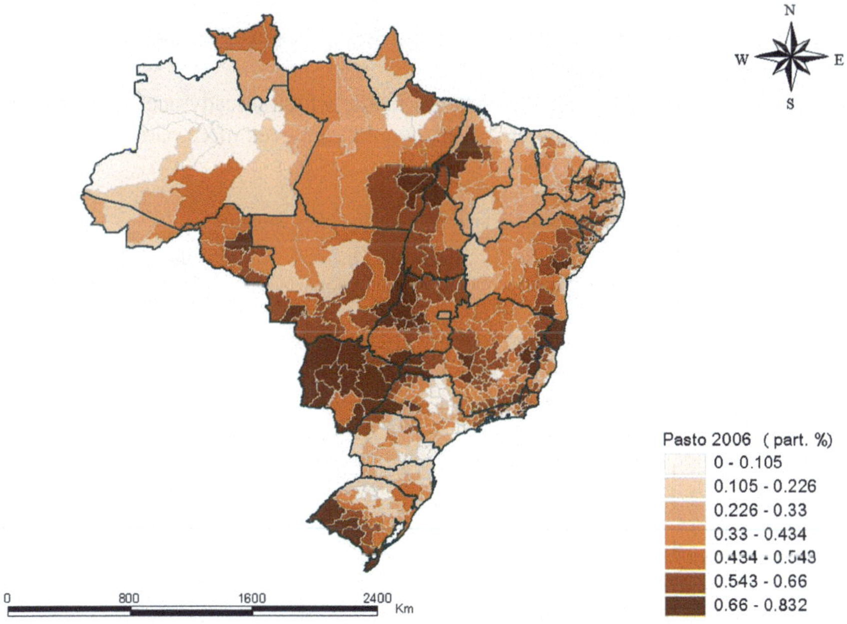

Fig. 5 Pasture share (2006)

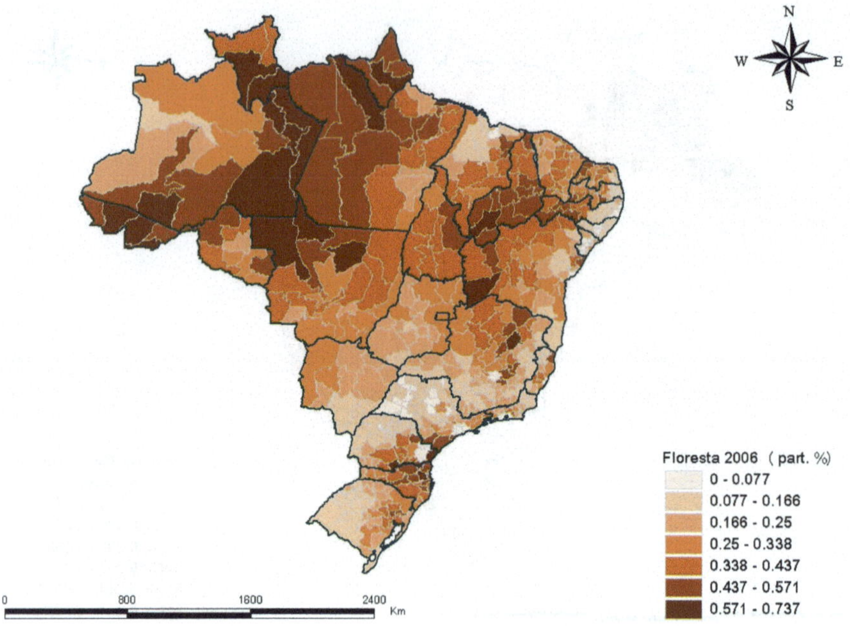

Fig. 6 Forest share (2006)

Table 3 presents the simulation results for Brazil as a whole, both in percentage and in ordinary (hectares) changes.

In both scenarios (A2 and B2), pasture lands, which already accounted for the largest share of the establishments' areas, would gain even more participation over

Table 3 Simulation results—Percentage and thousand hectares variation

	Period	Soybean	Corn	Sugarcane	O. crops	Pasture	Forest
A2	2039	83%	30%	−9%	−38%	16%	−31%
		12,983.1	3733.1	−529.9	−10,538.6	26,449.2	−32,096.9
	2069	78%	27%	−7%	−44%	24%	−42%
		12,173.9	3410.7	−424.8	−12,125.6	40,369.8	−43,403.9
	2099	81%	31%	−8%	−55%	35%	−58%
		12,631.2	3836.6	−485.9	−15,350.0	58,845.6	−59,477.5
B2	2039	81%	27%	−10%	−38%	18%	−33%
		12,618.7	3374.7	−599.3	−10,607.7	29,325.7	−34,112.2
	2069	80%	27%	−8%	−42%	20%	−36%
		12,558.6	3406.0	−466.0	−11,567.1	33,101.3	−37,032.7
	2099	84%	29%	−8%	−45%	24%	−43%
		13,189.9	3632.3	−439.6	−12,476.6	40,666.2	−44,572.2

Table 4 Land use share

	Brazil						
	Period	Soybean (%)	Corn (%)	Sugarcane (%)	O. crops (%)	Pasture (%)	Forest (%)
A2	2006	4.7	3.8	1.8	8.4	50.3	31.1
	2010–39	8.6	4.9	1.6	5.2	58.3	21.4
	2040–60	8.4	4.8	1.6	4.7	62.5	18.0
	2070–99	8.5	4.9	1.6	3.8	68.0	13.2
B2	2010–39	8.5	4.8	1.6	5.2	59.1	20.8
	2040–60	8.5	4.8	1.6	4.9	60.3	19.9
	2070–99	8.7	4.9	1.6	4.6	62.6	17.7

the years. When analyzing the absolute numbers of the pasture areas, it can be noted that the increase would be mainly due to forests areas reductions on farms. Although the reduction of forest areas and the increase of pasture areas are observed in both scenarios, it is interesting to note that the magnitude of these changes is greater in the A2 scenario. Forest areas would suffer a reduction of 58% and 43%, respectively, for the A2 and B2 scenarios. While there would be an increase of 35% and 24%, respectively, in pastures for the scenarios A2 and B2.

There would be also a sharp drop in the participation of other crops, initially, similar in both scenarios, but also of greater magnitude in the A2 scenario when the time horizon is greater. The model indicates a potential decrease of almost 40% of the area for this type of use already in the years 2010–2039 in both scenarios.

There would be a distinct pattern for sugarcane compared to soybeans and corn. The simulations show a decrease in sugarcane areas and an increased on soybean and corn areas.

Table 4 shows the change in the pattern of land use observed in the simulations, depicting how the land use structure could be affected.

5.2.2 Simulations—Regional Results

It is also important to analyze the results in a regional perspective. Tables 5, 6, 7, 8 and 9 present the results for each of the five Brazilian regions. The regional highlights show a:

- Decrease of more than 50% on forests areas in the north region and a subsequent increase in pasture areas.
- Decrease in sugar cane and forest areas in the northeast region and an increase in pasture areas.
- Decrease of more than 70% of the forests in the southeast region. Also a considerable increase in the corn and soybean areas for this region.

Table 5 North region—Simulation results—Percentage and thousand hectares variation

	Period	Soybean	Corn	Sugarcane	O. crops	Pasture	Forest
A2	2039	262%	156%	290%	−39%	36%	−42%
		589.2	461.0	61.9	−1446.4	9869.3	−9535.0
	2069	235%	129%	397%	−45%	44%	−51%
		527.3	382.2	84.8	−1647.7	12,203.1	−11,549.7
	2099	227%	160%	611%	−56%	57%	−65%
		509.8	471.6	130.4	−2036.0	15,771.2	−14,846.9
B2	2039	283%	150%	304%	−38%	36%	−43%
		635.5	444.0	64.9	−1404.6	9979.0	−9718.8
	2069	252%	139%	360%	−42%	38%	−44%
		566.0	411.1	76.9	−1533.7	10,516.7	−10,037.0
	2099	200%	114%	429%	−47%	46%	−52%
		450.2	337.3	91.6	−1734.4	12,807.8	−11,952.5

Table 6 Northeast region—Simulation results—Percentage and thousand hectares variation

	Period	Soybean	Corn	Sugarcane	O. crops	Pasture	Forest
A2	2039	13%	−63%	−59%	−55%	60%	−51%
		144.1	−2125.6	−678.0	−5475.9	21,068.5	−12,933.2
	2069	18%	61%	−57%	−59%	66%	−59%
		196.8	−2044.3	−654.1	−5801.6	23,267.2	−14,964.0
	2099	35%	−56%	−52%	−63%	75%	−71%
		388.1	−1892.3	−597.6	−6223.0	26,387.2	−18,062.4
B2	2039	−1%	−65%	−60%	−56%	62%	−53%
		−10.4	−2198.1	−679.0	−5537.2	21,943.3	−13,518.5
	2069	−10%	−66%	−59%	−59%	65%	−56%
		−107.4	−2210.3	−678.0	−5804.9	22,942.1	−14,141.4
	2039	65%	−49%	−53%	−57%	62%	−58%
		724.7	−1655.2	−609.6	−5654.0	21,810.3	−14,616.3

- Decrease of pasture lands in the south and an increase in soybean, corn, and sugarcane areas. There is also a potential increase on the forests areas in the long run.
- Decrease in the forest areas and an increase in the soybean, corn, and sugarcane areas for the center-west region.

Table 7 Southeast region—Simulation results—Percentage and thousand hectares variation

	Period	Soybean	Corn	Sugarcane	O. crops	Pasture	Forest
A2	2039	346%	61%	−9%	−36%	20%	−59%
		3244.6	1209.7	−339.2	−2465.3	5800.0	−7449.8
	2069	319%	50%	−10%	−43%	31%	−75%
		2991.8	990.5	−366.0	−2966.4	8723.1	−9373.1
	2099	278%	35%	−18%	−60%	44%	−88%
		2600.0	685.2	−652.2	−4088.0	12,461.2	−11,006.3
B2	2039	320%	50%	−12%	−38%	26%	−67%
		2996.7	983.0	−418.8 −	−2609.9	7456.6	−8407.5
	2069	306%	48%	10%	−42%	28%	−69%
		2864.7	949.0	−366.6	−2847.9	8016.8	−8616.1
	2099	292%	35%	−14%	−48%	35%	−78%
		2738.9	701.0	−490.0	−3257.5	10,082.1	−9774.4

Table 8 South region—Simulation results—Percentage and thousand hectares variation

	Period	Soybean	Corn	Sugarcane	O. crops	Pasture	Forest
A2	2039	69%	45%	73%	−10%	−58%	27%
		4667.0	1997.4	295.4	−429.9	−9272.6	2742.8
	2069	70%	48%	89%	−15%	−41%	1%
		4755.5	2114.3	363.4	−649.8	−6653.7	70.3
	2099	81%	57%	108%	−34%	−17%	−42%
		5501.4	2528.5	439.5	−1462.0	−2694.7	−4312.7
B2	2039	69%	45%	75%	−8%	−57%	24%
		4682.7	1982.3	305.8	−345.1	−9137.2	2511.4
	2069	72%	47%	86%	−12%	−52%	15%
		4910.9	2086.6	351.2	−521.2	−8390.8	1563.3
	2039	79%	51%	99%	−17%	−43%	−3%
		5408.2	2268.4	402.1	−738.6	−7005.4	−334.8

6 Final Remarks

The present study aimed to assess the expected impact of climate change on the land use patterns in Brazil. An econometric model was estimated in order to understand the climate impacts on the land allocation decision by farmers for six possible uses: soybeans, corn, sugarcane, other crops, pastures, and forests. Simulations of the optimal land allocations for the years between 2010 and 2099 were conducted based on temperature and precipitation projections.

Table 9 Center-West region—Simulation results—Percentage and thousand hectares variation

	Period	Soybean	Corn	Sugarcane	O. Crops	Pasture	Forest
A2	2039	66%	90%	21%	−23%	−2%	−15%
		4338.2	2190.6	130.0	−721.0	−1016.1	−4921.7
	2069	56%	81%	23%	−35%	5%	−24%
		3702.4	1967.9	147.1	−1060.1	2830.0	−7587.4
	2099	55%	84%	31%	−50%	12%	−35%
		3631.9	2043.7	194.0	−1541.0	6920.7	−11,249.1
B2	2039	66%	89%	20%	−23%	−2%	−15%
		4314.3	2163.6	127.8	−710.9	−915.9	−4978.9
	2069	66%	89%	24%	−28%	0%	−18%
		4324.4	2169.6	150.5	−859.4	16.5	−5801.6
	2039	59%	81%	26%	−36%	5%	−24%
		3867.9	1980.9	166.2	−1092.0	2971.2	−7894.2

The coefficients used on the simulations were estimated from the system of equations, derived from the economic model, by a simultaneous equations estimation method (ISUR). This method was chosen based on the discussion on the best way to capture the climate change impacts on land use. Although the literature shows numerous applications using panel data for estimating climate change impacts on the agriculture sector, the discussions carried out in this chapter supported the use of the cross-section estimation to solve the proposed problem. The results of the spatial panel estimation could be evidence that the panel data framework may not be appropriate to the study of the impact of climate change in land use, at least when working with a short period of time.

The simulation results suggest, at the national level, a significant expansion of the pasture area at the expense of other land uses. In particular areas of "other crops" and forests would be negatively affected. The simulations also suggested that, given the expected spatial heterogeneity of climate change, the effects can be quite different depending on the region. It is worth noting that the results suggested greater changes than those found by Féres et al. (2009), although both studies pointed to an increase in pasture areas and a decrease in forest areas. The present study shows that the possible gain in crop areas are likely to happen due to soybeans and corn rather than other types of cultivation. Regionally, the results also point to variations in the same direction as the ones found in Féres et al. (2009) with the major exceptions being the decrease of crops in favor of pasture in the North and Northeast. Both studies conclude that the South region has the smallest pressure to deforestation. The possibility to study the three major agricultural crops in Brazil (soybean, corn, and sugar cane) apart from other crops proved worthy, as the results between these three crops and the other crops varied greatly.

Despite the great magnitude of percent changes revealed in the estimates, structural changes would not be so significant. Pastures, which currently have the greater share of the land use, would remain with the largest participation in all regions.

After performing the simulations, it is important to note the limitations of the projections. The first being the low significance of the price variables, which were not statistically significant in many cases, and also presented in some cases, a different sign than expected by economic theory. The results would not be capturing the expected influence of relative prices in the land use decision. Another limiting factor in the analysis is the overestimation of the impacts. This can be due to the fact that the simulations do not take into account the second-order effects of re-allocation of land use. For example, the decrease in the cultivation area of sugarcane could probably increase the price of the product, and consequently, generate a greater incentive to the sugarcane cultivation.

Despite the limitations, the results provide important insights on the likely impacts of climate change on the Brazilian land use patterns, showing potential decreases in forest areas in favor of pasture fields.

References

Adams, R. (1989). Global climate change and agriculture: an economic perspective. *American Journal of Agricultural Economics, 71*(5), 1272–1279.

Chasco, C., & López, F. A. (2004). Modelos de regresión espacio-temporales en la estimación de la renta municipal: el caso de la Región de Murcia. In *Estudios de Economía Aplicada*, Estudios de Economía Aplicada (Vol. 22, pp. 1–24). Diciembre.

Decker, W. L., Jones, V., & Achtuni, R. (1986). *The impact of climate change from increased atmospheric carbon dioxide on American agriculture*. DOE/NBB-0077. Washington, DC: U.S. Department of Energy.

Dêschenes, O., & Greenstone, M. (2007). The economic impacts of climate change: Evidence from agricultural output and random fluctuations in weather. *American Economic Review, 97*(1), 354–385.

Elhorst, J. P. (2003). Specification and estimation of spatial panel data models. *International Regional Science Review, 26*(3), 244–268.

Evenson, R. E., & Alves, D. C. O. (1998). Technology, climate change, productivity and land use in Brazilian agriculture. *Planejamento e Políticas Públicas, 18,* 223–258.

Féres, J., Reis, E., & Speranza, J. (2007). Assessing the impact of climate change on the Brazilian agricultural sector. In *Proceedings of the 16th Annual EAERE Annual Conference*. Gothenburg: European Association of Environmental and Resource Economists.

Féres, J., Reis, E., & Speranza, J. (2009). Mudanças climática globais e seus impactos sobre os padrões de uso do solo no Brasil. In *XXXVII Encontro Nacional de Economia, 2009, Foz do Iguaçu. Anais do XXXVII Encontro Nacional de Economia*. Disponível em: http://www.anpec.org.br/encontro_2009.htm.

Kumar, K. S. (2011). Climate sensitivity of Indian agriculture: Do spatial effects matter? *Cambridge Journal of Regions, Economy and Society, 2011,* 1–15.

Lau, L. J. (1978). Applications of profit functions. In M. Fuss & D. McFadden (Eds.), *Production economics: A dual approach to theory and applications* (Vol. 1, Cap. I.3). Amsterdam: North-Holland.

Mendelsohn, R., Nordhaus, W. D., & Shaw, D. (1994). The impact of global warming on agriculture: A Ricardian analysis. *American Economic Review, 84*(4), 753–771.
Matthews, R. B., Gilbert, N. G., Roach, A., Polhill, J. G., & Gotts, N. M. (2007) Agent-based land-use models: a review of applications. *Landscape Ecology, 22*(10), 1447–1459.
Moore, M. R., & Negri, D. H. (1992). A multicrop production model of irrigated agriculture, applied to water allocation policy of the Bureau of reclamation. *Journal of Agricultural and Resource Economics, 17*, 29–43.
Platinga, A. J. (1996). The effect of agricultural policies on land use and environmental. *American Journal of Agricultural Economics, 78*(4), 1082–1091.
Sands, R. D., & Edmonds, J. A. (2005). Climate change impacts for the conterminous USA: An integrated assessment—Part 7. Economic analysis of field crops and land use with climate change. *Climate Change, 69,* 127–150.
Sanghi, A., Alves, D., Evenson, R., & Mendelsohn, R. (1997). Global warming impacts on Brazilian agriculture: Estimates of the Ricardian model. *Economia Aplicada, 1*(1), 1997.

The manufacturer's authorised representative in the EU is Springer Nature Customer Service Centre GmbH, Europaplatz 3, 69115 Heidelberg, Germany. If you have any concerns regarding our products, please contact ProductSafety@springernature.com

Printed and bound by CPI Group (UK) Ltd, Croydon, CR0 4YY
05/10/2026
02228120-0016